Adaptation Biology and Medicine

Volume 5: Health Potentials

SERIES EDITORS: **P.K. Singal and N. Takeda**

Adaptation Biology and Medicine

Volume 5: Health Potentials

EDITORS

L. Lukyanova

N. Takeda

P.K. Singal

Narosa Publishing House

New Delhi Chennai Mumbai Kolkata

Adaptation Biology and Medicine
376 pgs | 153 figs | 19 tbs

SERIES EDITORS: **P.K. Singal and N. Takeda**

EDITORS:

L. Lukyanova
Institute of General Pathology and Pathophysiology
RAMS, Moscow, Russia

N. Takeda
Department of Internal Medicine
Aoto Hospital, Jikei University School of Medicine
Tokyo, Japan

P.K. Singal
Institute of Cardiovascular Sciences
St. Boniface General Hospital Research Center
Department of Physiology Faculty of Medicine
University of Manitoba Winnipeg, Canada

NAROSA PUBLISHING HOUSE PVT. LTD.

22 Delhi Medical Association Road, Daryaganj, New Delhi 110 002
35-36 Greams Road, Thousand Lights, Chennai 600 006
306 Shiv Centre, D.B.C. Sector 17, K.U. Bazar P.O., Navi Mumbai 400 703
2F-2G Shivam Chambers, 53 Syed Amir Ali Avenue, Kolkata 700 019

www.narosa.com

ISBN 978-81-7319-767-3

Published by N.K. Mehra for Narosa Publishing House,
22 Delhi Medical Association Road, Daryaganj, New Delhi 110 002

Printed in India

This book is dedicated to Dr. Alan Hargens, Professor of Orthopaedic Surgery at the University of California, San Diego. As a President (2003-2006) and now a past President of the International Society for Adaptive Medicine and previous Editor of Adaptation Biology and Medicine, and member of the Executive Committee, Dr. Hargens played a lead role in taking the ISAM to the next level. He was Program Chair and hosted the Seventh World Congress of the ISAM in San Diego, in 2003. His research emphasizes on cardiovascular and musculoskeletal systems adaptations in microgravity and he has received many honors for his work. The Society expresses gratitude to him for his active participation and interest.

EDITORS

Preface

The International Society for Adaptive Medicine (ISAM) is an interdisciplinary group interested in adaptive changes occurring in response to different environmental factors, sustained exercise, various pathological conditions or drug administration. Its principal aim is to further the understanding of the basic mechanisms of *physiological adaptations* at molecular and cellular levels as well as at the level of complex interactions responsible for the integrated control of body functions. Another objective consists in the better understanding of *failed adaptations* responsible for disease development. Finally, studies of different cross-adaptations continue to receive unceasing attention as it was demonstrated that repeated exposures to specific stress stimuli may have applications in the treatment and prophylaxis of different diseases. These observations lead to the emergence of a new discipline of *adaptive medicine* which offers promise for therapeutic strategies in both experimental and clinical pathology.

The ISAM was created by scientists from a dozen different countries in 1990, in Freiburg, in order to provide a platform for regular confrontation of original data in the field of physiological adaptations and to promote the search for new trails in the field of adaptive medicine. At present, the Society comprises of members worldwide and its meetings bring together clinicians, scientists and engineers who constitute a unique international forum with a vibrant atmosphere. The Society also sponsors regional symposia around the world. The communications presented at some of these meetings are published in the *Adaptation Biology and Medicine* series.

The previous congresses of the International Society for Adaptive Medicine were organized in Freiburg, Moscow, Tokyo, Chandigarh, Framingham, Lyon and San Diego. The Eighth ISAM Congress, now on a three-year cycle basis, was held in Moscow, Russia, June 21-24, 2006. The present volume is based on papers from different sessions presented as oral or poster communications. Although the entire book is one continuous discussion, issues dealing with molecular and cellular aspects of adaptations have been grouped as Cardiovascular, Neurohumoral and other Systems for easy reading. An understanding of adaptive processes has tremendous potential in improving health of the population. It is hoped that the readers will find an inspiration for their reflection concerning adaptations and maladaptations in different pathologies as well as prospective developments of novel therapeutic strategies.

L. Lukyanova
N. Takeda
P.K. Singal

Acknowledgements

The Eighth World Congress of the ISAM was a resounding success in terms of the scientific exchange as well as social interactions. This scientific congress was made possible by an active participation of several international scientists including: P.K. Singal; A.R. Hargens; L.D. Lukyanova; Yu.V. Arkhipenko; D.K. Das; E.B. Manukhina. The organization of the whole congress was in the care of the Academicians: V.I. Pokrovsky, RAMS; A.I. Grigorjev, RAS and RAMS; A.A. Kubatiev, RAMS; L.D. Lukyanova, RAMS; I.B. Ushakov, RAMS and RAS.

The following institutions have supported the realization of the Eighth ISAM Congress: Russian Foundation for Basic Research; Russian Federation Pensionary Fund; Charitable Fund for Support of Achievements in the Field of Medicine "Prizvanie" (Vocation); Federal State Unitary Enterprise "State Scientific Center "NIOPIK"; Joint Stock Company Plant of Ecological Technics and Ecomeals; GREENWOOD (ZAO); AKVION Company research and development, production and promotion of Vitamin/mineral complexes, Enriched food additives Cosmetic substances; Domestic Medicines (OAO); and the Institute of Cardiovascular Sciences, Winnipeg, Canada.

An active overseeing of the organization as well as attending day-to-day needs of the delegates by Drs. L. Lukyanova and E.B. Manukhina along with their colleagues and staff made the congress an enjoyable experience. The participation as well as the cooperation of all contributors to this book, editorial help of Dr. Thomas P. Thomas, Winnipeg and secretarial help of Mrs. Pamela Vokey, Winnipeg is highly appreciated.

L. Lukyanova
N. Takeda
P.K. Singal

Contents

B. Adaptation to Neurohumoral Stresses

C. Adaptation to Other Stresses

Contributing Authors

M. Al-Khaledi
Faculty of Allied Health Sciences
Kuwait University, Kuwait

F. Al-Yatama
Faculty of Allied Health Sciences
Kuwait University, Kuwait

M. B. Anand-Srivastava
Dept. of Physiology, Faculty of Medicine
University of Montreal
C.P. 6128, Succ. Centre-Ville
Montréal, Québec, H3C 3J7 Canada

N. A. Anchishkina
Institute of General Pathology and Pathophysiology
Russian Academy of Medical Sciences
Baltiyskaya 8, Moscow 125315, Russia

S. Aou
Department of Integrative Physiology, Faculty of Medicine
Kyushu University, Fukuoka, 812-0054, Japan

Y. V. Arkhipenko
Faculty of Fundamental Medicine
M. V. Lomonosov Moscow State University
Lomonosovsky or., 31, korp. 5, Moscow 119192, Russia

Z. M. Azar
Laboratory of Cell Signaling, Montreal Diabetes Research Centre
Dept. of Medicine, Université de Montréal
Montréal, Quebec, Canada

J. E. Baker
Division of Pediatric Surgery
Medical College of Wisconsin
Milwaukee, WI 53226, USA

L. A. Bayda
Institute of Pathology and Pathophysiology
Moscow, Russian Federation

L. Belkina
Institute of General Pathology and Pathphysiology
Russian Academy of Medical Sciences
Moscow, Russia

B. Brüne
University of Frankfurt Medical School
Biochemistry I
Theodor-Stern-Kai 7, Bldg. 25B
60590 Frankfurt, Germany

B. P. Cabigas
Department of Hyperbaric Medicine
St. Luke's Medical Center
Milwaukee, WI 53215 USA

R. N. Cornelussen
Dept. of Physiology
Cardiovascular Research Institute Maastricht
Maastricht University
P.O. Box 616, 6200 MD Maastricht, The Netherlands

S. A. H. de Poot
University College Utrecht
The Netherlands

H. F. Downey
Department of Integrative Physiology
University of North Texas Health Science Center
3500 Camp Bowie Boulevard
Fort Worth, TX 76107-2699, USA

A. M. Dudchenko
Institute of General Pathology and Pathophysiology
RAMS, Moscow, Russia

K. Fujita
Laboratory of Pathophysiology
Graduate School of Pharmaceutical Sciences
Kyusku University, 3-1-1 Maidashi, Higashi-ku
Fukuoka 812-8582, Japan

E. L. Germanova
Institute of General Pathology and Pathophysiology
RAMS, Moscow, Russia

A. V. Gourine
Institute of Physiology
National Academy of Sciences of Belarus
Academicheskaya Str. 28
Minsk 220725, Belarus

V. N. Gourine
Institute of Physiology
National Academy of Sciences of Belarus
Academicheskaya Str. 28
Minsk 220725, Belarus

R. D. Grygoryan
Institute of Software Systems
National Academy of Sciences, 40, Prosp. Acad.
Glushkov, 01387, Kiev, Ukraine

A. R. Hargens
Department of Orthopaedic Surgery
University of California
San Diego, CA 92103-8894, USA

T. Hayashi
Department of Applied Mathematics and Physics
Faculty of Engineering
Tottori University, Tottori 680-8550 Japan

N. Hori
Department of Integrative Physiology, Faculty of Medicine
Kyushu University, Fukuoka, 812-0054, Japan

H. Iguchi
Shikoku Cancer Center Research Institute
Ehime, Japan

T. Ivanics
Institute of Human Physiology and Clinical Experimental Research
Semmelweis University
Budapest, Hungary

V. I. Kapelko
Russian Cardiological Research Complex, Moscow,
121552, 3-rd Cherepkovskaya Str. 15A
Moscow, Russia

T. Katafuchi
Department of Integrative Physiology,
Graduate School of Medical Sciences,
Kyushu University, Fukuoka, 812-8582, Japan

Y. Kawai
Department of Adaptation Physiology
Faculty of Medicine, Tottori University
86 Nishi-cho, Yonago 683-8503, Japan

I. P. Khomenko
Institute of Pathology and Pathophysiology
Moscow, Russia

M. Kido
Laboratory of Anatomy and Cell Biology
Graduate School of Dental Sciences
Kyushu University, Fukuoka, Japan

T. Kirillina
Institute of General Pathology and Pathphysiology
Russian Academy of Medical Sciences
Moscow, Russia

S. V. Kruglov
Institute of Pathology and Pathophysiology
Moscow, Russia

V. Lakomkin
Russian Cardiological Research-and-Production Complex
Ministry of Health, Moscow, Russia

M. Langelaar-Makkinje
Faculty of Science, Department of Biology,
Utrecht University, P.O. Box 80056
3508 TB Utrecht, The Netherlands

G. Lappas
Dept. of Physiology and Groupe de recherché sur le système nerveux autonome (GRSNA)
Faculty of Medicine, University of Montreal

M. H. Laughlin
Department of Biomedical Sciences
University of Missouri, E102 Vet Med Bldg
Columbia, MO 65211, USA

Y. Li
Dept. of Physiology and Groupe de recherché sur le système nerveux autonome (GRSNA)
Faculty of Medicine, University of Montreal

L. Ligeti
Institute of Human Physiology and Clinical Experimental Research
Semmelweis University
Budapest, Hungary

G. Limandjaja
University College Utrecht
The Netherlands

D. T. Lincoln
Entity Systems, Independent Research Foundation
53 Ashburton Street, Chapel Hill
QLD 4069, Australia

L. D. Lukyanova
Institute of General Pathology and Pathophysiology
RAMS, Moscow, Russia

M. K. Mallet
Private Practice in Psychiatry
P.O. Box 1976
Weatherford, TX 76086 USA

R. T. Mallet
Department of Integrative Physiology
University of North Texas Health Science Center
3500 Camp Bowie Boulevard
Fort Worth, TX 76107-2699, USA

I. Yu. Malyshev
Institute of Pathology and Pathophysiology
Moscow, Russia

E. B. Manukhina
Institute of Pathology and Pathophysiology
Moscow, Russia

S. Matsuo
Department of Adaptation Physiology
Faculty of Medicine, Tottori University
86 Nishi-cho, Yonago 683-8503, Japan

R. McAllister
Department of Biomedical Sciences
University of Missouri, E102 Vet Med Bldg
Columbia, MO 65211, USA

M. Z. Mehdi
Laboratory of Cell Signaling, Montreal Diabetes Research Centre
Dept. of Medicine, Université de Montréal
Montréal, Quebec, Canada

G. D. Mironova
Institute of Theoretical and Experimental Biophysics
Russian Academy of Sciences
Pushchino, Moscow Region, 142290, Russia

F. M. Mohammed
Faculty of Allied Health Sciences
Kuwait University, Kuwait

M. Y. Monakov
Institute of General Pathology and Pathophysiology
RAMS, Moscow, Russia

S. C. Newcomer
Department of Biomedical Sciences
University of Missouri, E102 Vet Med Bldg
Columbia, MO 65211, USA

M. Noda
Laboratory of Pathophysiology
Graduate School of Pharmaceutical Sciences
Kyusku University, 3-1-1 Maidashi, Higashi-ku
Fukuoka 812-8582, Japan

H. Ohnishi
Department of Adaptation Physiology
Faculty of Medicine, Tottori University
86 Nishi-cho, Yonago 683-8503, Japan

Y. Oomura
Department of Integrative Physiology, Faculty of Medicine
Kyushu University, Fukuoka, 812-0054, Japan

D. A. Pokidyshev
Institute of Pathology and Pathophysiology
Moscow, Russia

E. Popkova
Institute of General Pathology and Pathphysiology
Russian Academy of Medical Sciences
Moscow, Russia

J. A. Post
Faculty of Science, Department of Biology,
Utrecht University, P.O. Box 80056
3508 TB Utrecht, The Netherlands

A. T. Proshin
P. K. Anokhin Institute of Normal Physiology
RAMS, Moscow, Russia

M. G. Pshennikova
Institute of General Pathology and Pathophysiology
Baltijskaya 8, Moscow 125315, Russia

M.-G. Ryou
Department of Integrative Physiology
University of North Texas Health Science Center
3500 Camp Bowie Boulevard
Fort Worth, TX 76107-2699, USA

K. Sasaki
Department of Life Science, Faculty of Engineering
Toyama University, Toyama 930-0887, Japan

T. G. Sazontova
Faculty of Fundamental Medicine,
M. V. Lomonosov Moscow State University
Lomonsovsky pr., 31, korp. 5, Moscow 119192, Russia

E. Schnitzer
Institute of Biochemistry I
Faculty of Medicine
Johann Wolfgang Goethe-University
60590, Frankfurt, Germany

T. Seike
Laboratory of Pathophysiology
Graduate School of Pharmaceutical Sciences
Kyusku University, 3-1-1 Maidashi, Higashi-ku
Fukuoka 812-8582, Japan

M. V. Shimkovich
Institute of Pathology and Pathophysiology
Moscow, Russia

P. K. Singal
Institute of Cardiovascular Sciences, St. Boniface General Hospital Research Centre,
University of Manitoba, Winnipeg, Canada

E. Smirnova
Institute of General Pathology and Pathphysiology
Russian Academy of Medical Sciences
Moscow, Russia

L. H. E. H. Snoeckx
Dept. of Physiology
Cardiovascular Research Institute Maastricht
Maastricht University
P.O. Box 616, 6200 MD Maastricht, The Netherlands

A. K. Srivastava
Research Centre, CHUM-Angus Campus
2901, Rachel East
Montreal, Quebec H1W 4A4, Canada

Z. I. Storojeva
P. K. Anokhin Institute of Normal Physiology,
RAMS, Moscow, Russia

K.V. Sudakov
Anokhin Institute of Normal Physiology
Russian Academy of Medical Sciences
Moscow, Russia

V. S. Sukhorukov
Moscow Institute of Pediatrics & Children's Surgery
Taldomskaya, 2, Moscow, 125412, Russia

K. Tagawa
Department of Regional Environment
Faculty of Regional Sciences
Tottori University, Tottori 680-8550 Japan

T. Tanaka
Laboratory of Anatomy and Cell Biology
Graduate School of Dental Sciences
Kyushu University, Fukuoka, Japan

E. N. Tkachuk
Hypoxia Medical Academy
Ivankovskoe shosse 3, Moscow 123367, Russia

V. G. Tsyplenkova
Russian Cardiological Research Complex
Moscow, Russia

T.A. Tzybina
Institute of General Pathology and Pathophysiology
RAMS, Moscow, Russia

M. Usacheva
Institute of General Pathology and Pathphysiology
Russian Academy of Medical Sciences
Moscow, Russia

G. J. van der Vusse
Dept. of Physiology
Cardiovascular Research Institute Maastricht
Maastricht University
P.O. Box 616, 6200 MD Maastricht, The Netherlands

O. N. Vorontsova
Institute of Pathology and Pathophysiology
Moscow, Russian Federation

F. A. C. Wiegant
Faculty of Science, Department of Biology,
Utrecht University, P.O. Box 80056
3508 TB Utrecht, The Netherlands

G. Wikman
Swedish Herbal Institute
Göteborg, Sweden

E. I. Zakharova
Institute of General Pathology and Pathophysiology
RAMS, Moscow, Russia

O. M. Zelnina
Institute of Pathology and Pathophysiology
Moscow, Russia

T. A. Zenina
Institute of Pathology and Pathophysiology
Moscow, Russian Federation

J. Zhou
Institute of Biochemistry I
Faculty of Medicine
Johann Wolfgang Goethe-University
60590, Frankfurt, Germany

A. G. Zhukova
Institute of General Pathology and Pathophysiology
Russian Academy of Medical Sciences
Baltiyskaya 8, Moscow 125315, Russia

G. R. Zubieta-Calleja
High Altitude Pathology Institute (IPPA)
La Paz, Bolivia

L. Zubieta-Calleja
High Altitude Pathology Institute (IPPA)
La Paz, Bolivia

N. Zubieta-Calleja
High Altitude Pathology Institute (IPPA)
La Paz, Bolivia

G. Zubieta-Castillo
High Altitude Pathology Institute (IPPA)
La Paz, Bolivia

Adaptation Biology and Medicine (Volume 5: Health Potentials)
Editors: L. Lukyanova, N. Takeda and P.K. Singal

1. Tyrosine Kinases as Upstream Regulators of H_2O_2-induced Signaling

Zeina M. Azar, Mohamad Z. Mehdi and Ashok K. Srivastava*

Laboratory of Cell Signaling, Montreal Diabetes Research Centre, Research Centre, Centre hospitalier de l'Université de Montréal (CHUM)-Angus Campus and Department of Medicine, Université de Montréal, Montreal Quebec, Canada

Abstract

Reactive oxygen species (ROS) have emerged in the recent years as important players in the pathogenesis of cardiovascular disorders, such as hypertension, atherosclerosis and cardiac hypertrophy as well as in the vascular complications of diabetes and in cancer. ROS are known to modify the cell components by oxidation and to activate signaling pathways responsible for cell growth, hypertrophy, proliferation, and migration as well as survival in a variety of cell types. The most important of these pathways are the mitogen-activated protein kinases (MAPKs) and the phosphatidylinositol-3-kinase (PI3-K)/protein kinase B (PKB) pathways. The precise molecular mechanisms by which cells respond and adapt to ROS remain to be fully elucidated. However, a role for receptor and non-receptor protein tyrosine kinases (PTKs) in mediating some of the effects of ROS in these signaling events has been suggested. Therefore, the aim of this chapter is to highlight the involvement of PTKs in modulating ROS (H_2O_2)-induced signaling.

Keywords: Pathogenesis, restenosis, angioplasty, hydroxyl radical, cyclooxygenase and cytokines

Introduction

Reactive oxygen species are now believed to play a critical role in the pathogenesis of various diseases, including atherosclerosis, hypertension, and restenosis after angioplasty (Dhalla *et al.*, 2000; Fortuno *et al.*, 2005; Pollock and Pollock, 2005) as well as diabetes (Houstis *et al.*, 2006) and cancer (reviewed in (Dreher and Junod, 1996)). Therefore, there is a lot of interest to understand the mechanism of ROS generation, its elimination and identification of its cellular targets.

ROS are formed as intermediates in redox reactions in almost all cell types, leading from molecular oxygen (O_2) to water (H_2O) (Fridovich, 1997). They are classified into superoxide anion ($^{\cdot}O_2^-$), hydroxyl radical ($^{\cdot}OH$) and hydrogen peroxide (H_2O_2). Their major intracellular source is the mitochondria which converts 1-2% of consumed O_2 into $^{\cdot}O_2^-$ (Boveris and Chance, 1973). There are many other sources for $^{\cdot}O_2^-$ generation, such as xanthine oxidase, cyclooxygenase, lipoxygenase, heme oxygenases, peroxidases, as well as hemoproteins such as heme and hematin, whereas the main source for $^{\cdot}O_2^-$ in the vascular wall is believed to be the NAD(P)H oxidase (reviewed in (Touyz, 2004)). The

*Address for Correspondence: Ashok K. Srivastava, Ph.D. Research Centre, CHUM-Angus Campus 2901, Rachel East Montreal (Quebec) H1W 4A4 Canada Tel.: (514) 890 8000 ext. 23604 Fax: (514) 412 7648 E-mail: ashok.srivastava@umontreal.ca

univalent reduction of O_2 leads to $^{\cdot}O_2^-$, which is relatively unstable because of its unpaired electron (Schafer and Buettner, 2001). Under physiological conditions, the enzyme superoxide dismutase (SOD) converts $^{\cdot}O_2^-$ to hydrogen peroxide (H_2O_2) which is more stable than $^{\cdot}O_2^-$. It is scavenged by catalase and glutathione peroxidase to produce H_2O (Schafer and Buettner, 2001). In the presence of metal-containing molecules such as Fe^{2+}, H_2O_2 can also be reduced to generate the hydroxyl radical ($^{\cdot}OH$) that causes damage to the cell components (Fridovich, 1997).

Normally, the rate of ROS production is balanced by the rate of their elimination. However, in pathological conditions, a disequilibrium between ROS generation and elimination results in increased ROS bioavailability leading to oxidative stress (Zalba *et al.*, 2001).

The precise molecular mechanism by which oxidative stress contributes to the pathogenesis of diseases remains unknown. However, number of studies using mainly H_2O_2 as mimicker of ROS have shown that H_2O_2 activates several key components of growth-promoting and proliferative signaling pathways (Blanc *et al.*, 2003; Blanc *et al.*, 2004; Mehdi *et al.*, 2005; Wang *et al.*, 2000; Zhuang and Schnellmann, 2004). Among these pathways, H_2O_2-induced activation of mitogen-activated protein kinases (MAPKs) and phosphatidylinositol-3 kinase (PI3-K)/protein kinase B (PKB/Akt) pathways have been studied in some detail due to their adaptive/regulatory role in mediating growth, proliferation, migration, survival and death responses of the cell.

Furthermore, several lines of evidence have indicated that many agonists, such as vasoactive peptides (Daou and Srivastava, 2004), growth factors (Mills et al, 1998) and cytokines (Lo *et al.*, 1996) mediate their responses through generation of ROS in various cell types and this is believed to be an important mechanism by which ROS could contribute to the pathophysiology of different diseases.

The MAPK Pathway

MAPK are a family of serine/threonine protein kinases, classically associated with cell growth, proliferation, differentiation and death (Kyosseva, 2004; Pearson *et al.*, 2001). MAPK are activated in response to a variety of external stimuli such as growth factors, hormones and stress, as well as vasoactive peptides and ROS (reviewed in (Seger and Krebs, 1995; Sugden, 2001)). Of the major mammalian MAP kinases, extracellular signal-regulated kinases (ERK1/2), p38MAP kinase (p38mapk), c-Jun N-terminal kinases (JNK) and ERK5 are the best characterized. MAPK are activated by MAPK kinase (MAPKK also known as MEK) which, in turn are activated by MAPKK kinase (MAPKKK also known as MEKK or Raf). The MAPKKK/Raf are serine/threonine kinases and are activated by phosphorylation and/or by their interaction with small GTP-binding protein of Ras/Rho family through adaptor proteins Son of sevenless/Growth factor receptor binding protein 2 ((Sos)/Grb2)). Activated MAPKKK/Raf phosphorylate and activate MAPKK/MEK, which then phosphorylate MAPK on Thr and Tyr residues located in the activation loop of the kinases. ERK1/2, phosphorylated by MEK1/2 (MAP/ERK kinase), is a major growth signaling kinase, whereas p38mapk and JNK, phosphorylated by MEK3/6 and MEK4/7 respectively, influence cell survival, apoptosis, differentiation and inflammation (Pearson *et al.*, 2001). ERK5, regulated by MEK5, is involved in protein synthesis, cell cycle progression and cell proliferation (Abe *et al.*, 1996; English *et al.*, 1999; Nicol *et al.*, 2001).

H_2O_2 has been shown to activate the MAPK pathway in different cell types, including Rat-2 fibroblasts (Esposito *et al.*, 2003), rabbit renal proximal tubular cells (Zhuang and Schnellmann,

2004), Chinese hamster ovary (CHO) cells (Mehdi *et al.*, 2005), rat cardiomyocytes and heart fibroblasts (Purdom and Chen, 2005) as well as VSMC (Blanc *et al.*, 2003; Blanc *et al.*, 2004; Tabet *et al.*, 2005). Moreover, ROS generation was shown to be critical in the activation of MAPK by AngII (Touyz *et al.*, 2004; Ushio-Fukai *et al.*, 1998) and ET-1 (Daou and Srivastava, 2004; Touyz *et al.*, 2004) in VSMC.

The PI3-K/PKB Pathway

PI3-K/PKB pathway is another key cellular signaling pathway that plays an important role in cell growth, survival, proliferation and gene expression. PKB, also known as Akt, exists as three isoforms: PKBa/Akt1, PKBb/Akt2 and PKBg/Akt3; each isoform possesses an amino-terminal PH domain, a kinase domain and a carboxy-terminal regulatory domain. PKB is activated in response to insulin (Farese *et al.*, 2005; Van der Heide *et al.*, 2005), AngII (Saward and Zahradka, 1997), ET-1 (Daou and Srivastava, 2004) and many other growth factors (Dolloff *et al.*, 2005; Duan *et al.*, 2000; Markadieu *et al.*, 2005; Zaka *et al.*, 2005) in a phosphatidylinositol-3-kinase (PI3-K)-dependent manner.

PI3-K is a heterodimeric lipid kinase, composed of a 85-kDa (p85) regulatory subunit and a 110-kDa (p110) catalytic subunit. The p85 subunit contains the src homology-2 (SH-2) domain and interacts with phosphorylated tyrosine residues on receptor or other docking proteins, leading to stimulation of the 110-kDa catalytic subunit. p110 catalyzes the phosphorylation of phosphatidylinositol (PI), PI 4-phosphate and PI 4,5-biphosphate in the 3' position of the inositol ring. This reaction promotes the generation of PI 3'-P, PI 3, 4-biphosphate (PI-3,4-P_2) and PI-3, 4, 5-triphosphate (PIP_3) (Leevers *et al.*, 1999).

Binding of PIP_3 to PKB recruits it to the plasma membrane for phosphorylation by phosphoinositide-dependent kinases 1 (PDK-1) and 2 (PDK-2). PDK-1 phosphorylates PKB at Threonine 308 (Thr^{308}) in the catalytic domain while putative PDK-2 phosphorylates it at Serine 473 (Ser^{473}) in the C-terminal regulatory domain of PKB (reviewed in (Whiteman *et al.*, 2002)). Activated PKB phosphorylates several downstream substrates, such as glycogen synthase kinase-3β (GSK-3β), Forkhead transcription factor (FKHR), Bcl-2-associated death (BAD), Caspase 9, mammalian target of rapamycin (mTOR), nuclear factor-κB (NF-κB) and endothelial nitric oxide synthase (eNOS) (reviewed in (Song *et al.*, 2005; Whiteman *et al.*, 2002)). Phosphorylated form of these substrates regulates diverse cellular functions, such as glucose transport, cell growth, gene expression, cell survival and death as well as protein synthesis (Saward and Zahradka, 1997) (reviewed in (Datta *et al.*, 1999)).

H_2O_2 has been shown to enhance the phosphorylation of PKB on Ser^{473} in many cell types including NIH3T3, MCF-7 and malignant HeLa cells (Wang *et al.*, 2000), Rat-2 fibroblasts (Esposito *et al.*, 2003), CHO cells (Mehdi *et al.*, 2005), renal cells (Zhuang and Schnellmann, 2004), cardiomyocytes and heart fibroblasts (Purdom and Chen, 2005) as well as VSMC (Blanc *et al.*, 2003; Blanc *et al.*, 2004). Moreover, an intermediary role of ROS in mediating PKB activation by AngII (Ushio-Fukai *et al.*, 1999) and ET-1 (Daou and Srivastava, 2004) in VSMC has also been demonstrated.

The precise upstream mechanisms by which ROS activate MAPK and PI3-K/PKB pathways are not fully characterized. However, a potential role for both growth factor receptors (Frank *et al.*, 2003; Purdom and Chen, 2005; Tabet *et al.*, 2005; Wang *et al.*, 2000; Zhuang, S. and Schnellmann, R.G., 04), and non-receptor protein tyrosine kinases PTKs (Esposito *et al.*, 2003; Mehdi *et al.*, 2005; Tabet *et al.*, 2005; Zhuang and Schnellmann, 2004) as signal transducers of H_2O_2-induced responses has been suggested.

Involvement of the Epidermal Growth Factor Receptor (EGFR)-PTK

The EGFR is a receptor tyrosine kinase that is ubiquitously expressed in a variety of cell types, with the most abundant expression in epithelial cells and many cancer cells (Carpenter, 2000; Mendelson, 2000; Prenzel *et al.*, 2001). It contains an extracellular ligand binding domain, a single transmembrane domain and a cytoplasmic tyrosine kinase autophosphorylation and regulatory domain (reviewed in (Normanno *et al.*, 2005)). EGFR undergoes homodimerization or heterodimerization to induce autophosphorylation of tyrosine residues in the tyrosine kinase domain in response to ligand binding (Mendelson, 2000; Schlessinger, 2002). Dimerization activates the intrinsic tyrosine kinase activity of the intracellular domain at different residues, resulting in the recruitment of the SH-containing domain proteins, which trigger downstream events. The phosphorylation of EGFR on tyrosine 1068 (Tyr^{1068}) is followed by recruitment of the adaptor protein Grb2, leading to the activation of Ras/ERK1/2 pathway.

The EGFR can also undergo phosphorylation in a ligand-independent manner by a process called transactivation and trigger the response of many agonists, such as H_2O_2. Thus, H_2O_2 has been shown to enhance tyrosine phosphorylation of EGFR in HeLa cells (Wang *et al.*, 2000), endothelial cells (Chen *et al.*, 2001), renal cells (Zhuang and Schnellmann, 2004) as well as in VSMC (Frank *et al.*, 2003; Rao, 1996; Saito *et al.*, 2002). Whether H_2O_2 directly activates the intrinsic tyrosine kinase activity of EGFR or modulates signaling molecules that transactivate the receptor is still unclear. There are some reports proposing that ROS may exert their effects through targeting the cysteine regions of the active sites of tyrosine phosphatases, which in turn activates tyrosine kinases (Finkel, 1998). In fact, H_2O_2 was shown to inhibit the dephosphorylation of the EGFR by inhibiting a tyrosine phosphatase (Knebel *et al.*, 1996). Some other reports suggest that ROS may activate receptor tyrosine kinases by generating growth factors, such as heparin-binding EGF-like growth factor (HB-EGF) through metalloprotease-dependent cleavage (Frank *et al.*, 2003).

The tyrosine phosphorylation of the EGFR by ROS is accompanied by its association with Grb2 leading to activation of ERK1/2. In fact, by using the AG1478 as specific inhibitor of the EGFR (Levitzki and Gazit, 1995), several studies have shown that activation of ERK1/2 by H_2O_2 was dependent on EGFR in various cell types (Rao, 1996; Zhuang and Schnellmann, 04; Lee *et al.*, 2005; Purdom and Chen, 2005; Tabet *et al.*, 2005). However, AG1478 failed to block H_2O_2-induced p38mapk phosphorylation in VSMC derived from mesenteric arteries (Tabet *et al.*, 2005) as well as ERK1/2 phosphorylation in A10 VSMC (Azar *et al.*, 2006b) (Figure 1).

Similarly, studies on the involvement of EGFR transactivation in H_2O_2-induced PKB phosphorylation gave also divergent results: for example, EGFR transactivation was implicated in H_2O_2-induced PKB phosphorylation in HeLa cells (Wang *et al.*, 2000), but not in renal cells (Zhuang and Schnellmann, 2004), PC12 cells (Banno *et al.*, 2005) or A10 VSMC (Azar *et al.*, 2006a) (Figure 1).

Involvement of the Insulin-like Growth Factor Type 1 Receptor (IGF-1R)-PTK

The IGF-1R is a transmembrane protein tyrosine kinase that shares structural and functional homology with the insulin receptor. The mature receptor is a tetramer consisting of 2 extracellular α-chains and 2 intracellular β-chains (Hernandez-Sanchez *et al.*, 1997). The β-chains include an intracellular tyrosine kinase domain that is believed to be essential for most of the receptor's biologic effects (Delafontaine, 1995). Binding of IGF-1 or Insulin (at very high, unphysiological concentrations) induces the activation of PTK domain of IGF-1Rβ subunit which in turn activates the autophosphorylation of the receptor (reviewed in (Adams *et al.*, 2000)).

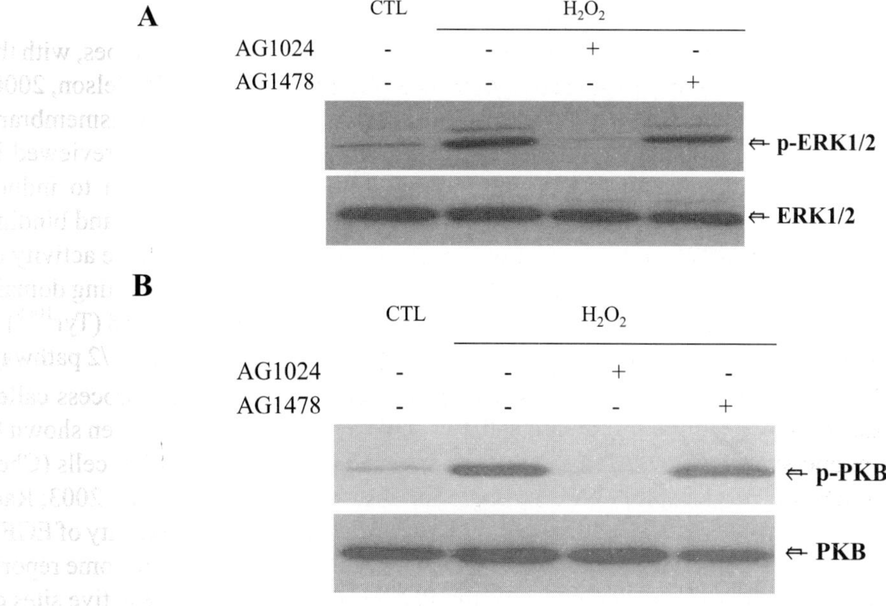

Fig. 1. H_2O_2-induced ERK1/2 and PKB phosphorylation are dependent on IGF-1R-PTK but not on EGFR-PTK in A10 VSMC.
Confluent, serum-starved A10 cells were pretreated without (–) or with (+) 1 µM AG1024 or 10 µM AG1478 for 30 minutes followed by stimulation in the absence (CTL) or presence of 500 µM H_2O_2 for 10 minutes. The cells were lysed, and equal amounts of protein were separated on 7.5% SDS-PAGE. ERK1/2 (A) and PKB (B) phosphorylation were detected by immunoblotting with phospho-specific antibodies (Thr^{202}-Tyr^{204}/Thr^{185}-Tyr^{187}) for ERK1/2 and Ser^{473} for PKB (top panel of each section). Blots were also analyzed for total ERK1/2 and PKB (bottom panels). (A) Adapted from (Azar *et al.*, 2007) (B) Adapted from (Azar *et al.*, 2006).

One of the earliest steps in signal transduction initiated by the IGF-1R is the phosphorylation of adaptor/docking proteins such as insulin receptor substrate (IRS-1 or IRS-2), Shc and Grb2 (LeRoith *et al.*, 1995; White, 1998). IRS-1, an important substrate for both the insulin and the IGF-1 receptor, contains multiple tyrosine phosphorylation sites that recognize and bind SH2-containing signaling molecules, such as Grb2, Nck, the p85 subunit of PI3-K and the SH2 domain-containing tyrosine phosphatase-2 (SHP-2) (LeRoith *et al.*, 1995). Of these, the binding of Grb2/Sos to tyrosine-phosphorylated IRS-1 activates Ras, which then stimulates the Raf-1/MAPK cascade (Rozakis-Adcock *et al.*, 1992). Shc can also interact directly with IGF-1R (Giorgetti *et al.*, 1994). After tyrosine phosphorylation of Shc, it recruits the Grb2/Sos complex and activates the Ras/Raf-1/MEK/ERK pathway (Rozakis-Adcock *et al.*, 1992). The activated IGF-1R also triggers the activation of PI3-K and its downstream targets PKB and p70s6k (Oldham and Hafen, 2003; Zheng and Clemmons, 1998).

Use of AG1024, the specific inhibitor of IGF-1R-PTK (Parrizas *et al.*, 1997) have suggested a role of this receptor-PTK in H_2O_2-induced ERK1/2 phosphorylation, because AG1024 pretreatment significantly blocked the effect of H_2O_2 on ERK1/2 phosphorylation in cultured VSMC from mesenteric arteries (Tabet *et al.*, 2005). This observation was further confirmed by Azar *et al.* who have also shown that not only ERK1/2 but also PKB phosphorylation by H_2O_2 was blocked by AG1024 in A10 VSMC (Azar *et al.*, 2006a; Azar *et al.*, 2006b) **(Figure 1)**. Moreover, these authors provided further proof for the involvement of IGF-1R in H_2O_2 signaling by showing that IGF-1R was tyrosine phosphorylated in response to H_2O_2 and that this phosphorylation was also blocked by AG1024 (Azar *et al.*, 2006a).

Involvement of the Platelet-derived Growth Factor Receptor (PDGFR)-PTK

The PDGFR is a receptor tyrosine kinase expressed in many cell types. It exists in two isoforms: PDGFRα and PDGFRβ that display affinity for the different isoforms of PDGF family PDGF-AA, PDGF-AB, PDGF-BB, PDGF-CC and PDGF-DD (reviewed in (Board and Jayson, 2005)). Tissue culture and in vivo mouse models studies have suggested that PDGFRα and PDGFRβ activate distinct signaling pathways. Downstream targets of the PDGFRβ include ERK, PKB and small G proteins including Rho and Rac-1, which ultimately mediate PDGF-induced responses such as cell cycle progression, migration, and survival (reviewed in (Rosenkranz and Kazlauskas, 1999)). Similar to EGF, binding of PDGF to its receptor on the cell surface induces its dimerization and autophosphorylation of the tyrosine kinase domain which in turn recruits and activates SH2 domain containing proteins such as Grb2, Src, p85 subunit of PI3-K and phospholipase C γ (PLCγ). The Tyr^{751} in the kinase domain of PDGFRβ is the docking site for PI3-K. Tyr^{740} is also important for the activation of PI3-K by PDGFRβ while PDGF-stimulated PLCγ signaling is dependent on the phosphorylation at the two sites Tyr^{1009} and Tyr^{1021} (reviewed in (Board and Jayson, 2005)).

Several studies have shown that PDGFR undergoes tyrosine phosphorylation by ROS in different cell types such as MCF10A (Esposito *et al.*, 2003), NIH3T3 fibroblasts (Catarzi *et al.*, 2005) and VSMC prepared from the thoracic aorta of Sprague-Dawley rats (Saito *et al.*, 2002). H_2O_2 was found to enhance the Tyr^{1021} phosphorylation of PDGFRβ in a Src- and PKCδ-dependent manner (Saito *et al.*, 2002). On the other hand, unlike the EGFR transactivation by ROS, PDGFR transactivation was shown to be independent of metalloproteases effect (Saito *et al.*, 2002). An implication of PDGFR in ROS-induced activation of MAPK pathway in VSMC derived from mesenteric arteries has been studied by using AG1295, a selective inhibitor of PDGFR kinase, and it was found that this inhibitor blocked only the phosphorylation of ERK1/2 but had no effect on p38mapk phosphorylation in response to H_2O_2 (Tabet *et al.*, 2005). Conversely, AG1295 failed to inhibit ERK1/2 and PKB phosphorylation in response to H_2O_2 in A10 VSMC, while it blocked completely the PDGF-induced responses (Azar *et al.*, 2006b) (Figure 2). It thus appears that cell-specific distinct triggering mechanisms contribute to H_2O_2-induced responses on these signaling components.

Involvement of the Src-family-PTKs (SFKs)

The SFKs, a family of non-receptor tyrosine kinases, consists of nine structurally-related peptide members: c-Src, Fyn and Yes are widely expressed in most tissues, whereas the others (Lck, Lyn, Hck, Fgr, Blk, Yrk) have a more restricted distribution (reviewed in (Yeatman, 2004)). SFK are 52–62 kDa proteins composed of six distinct functional regions: the SH4 domain, the unique region, the SH3 domain, the SH2 domain, the catalytic domain (SH1), and a short C-terminal tail containing a negative regulatory tyrosine residue (Tyr^{530} in humans) (reviewed in (Yeatman, 2004)). The SH2 and SH3 domains are important for molecular interactions that regulate Src catalytic activity, Src localization, and recruitment of substrates, whereas the SH1 kinase domain contains autophosphorylation site,

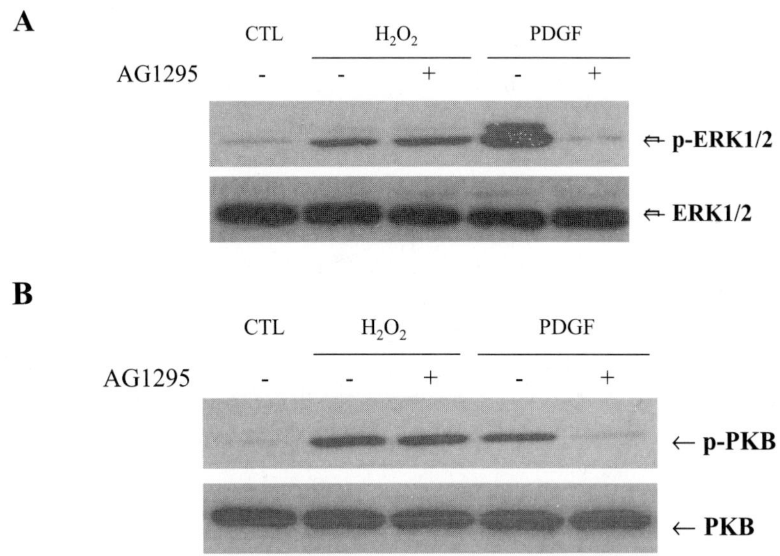

Fig. 2. H_2O_2-induced ERK1/2 and PKB phosphorylation are independent of PDGFR-PTK in A10 VSMC. Confluent, serum-starved A10 cells were pretreated without (–) or with (+) 10 μM AG1295 for 30 minutes followed by stimulation in the absence (CTL) or presence of 500 μM H_2O_2 for 10 minutes, or 1ng/mL PDGF for 5 minutes. The cells were lysed, and equal amounts of protein were separated on 7.5% SDS-PAGE. ERK1/2 (A) and PKB (B) phosphorylation was detected by immunoblotting with phospho-specific antibodies as described in figure 1 (top panel of each section). Blots were also analyzed for total ERK1/2 and PKB (bottom panels). Adapted from (Azar *et al.*, 2007).

which is important for regulation of its catalytic activity. This site of phosphorylation corresponds to Tyr^{416} in mouse c-Src and Tyr^{419} in human's, which is not phosphorylated in inactive wild type Src, but is constitutively phosphorylated in activated Src (reviewed in (Yeatman, 2004)).

SFKs can be activated by various extra cellular stimuli, including oxidative stress (Azar *et al.*, 2006; Mehdi *et al.*, 2005; Rosado *et al.*, 2004; Zhuang and Schnellmann, 2004) in various cell types. In the last few years, much attention has been given to the critical role of c-Src as a mediator of GPCR- and H_2O_2-induced responses: transactivation of EGFR by H_2O_2 was shown to be c-Src-dependent in renal cells (Zhuang and Schnellmann, 2004). Similarly, PDGFR transactivation by ROS seems to be Src-dependent in VSMC (Saito *et al.*, 2002) and in NIH3T3 fibroblasts (Catarzi *et al.*, 2005). Recently, H_2O_2-induced tyrosine phosphorylation of Src has been shown to be IGF-1R-dependent in A10 VSMC (Azar *et al.*, 2006) (Figure 4).

Moreover, Src is not only involved in H_2O_2-induced transactivation of RTK, but also in triggering the activation of MAPK and PI3-K/PKB pathways in many cell types. Pharmacological approaches have suggested a role for c-Src in mediating H_2O_2-induced ERK1/2 phosphorylation in mesenteric arteries VSMC (Tabet *et al.*, 2005), in CHO-IR cells (Mehdi *et al.*, 2005) and in renal cells (Lee *et al.*, 2005; Zhuang and Schnellmann, 2004). Moreover, c-Src-PTK activity appears to be required in mediating H_2O_2-induced PKB phosphorylation in different cell types such as Rat-2 fibroblasts (Esposito *et al.*, 2003), renal cells (Zhuang and Schnellmann, 2004), CHO-IR (Mehdi *et al.*, 2005) and in A10 VSMC (Azar *et al.*, 2006a) (Figure 3). In contrast, diethylmaleate, which is known to induce an

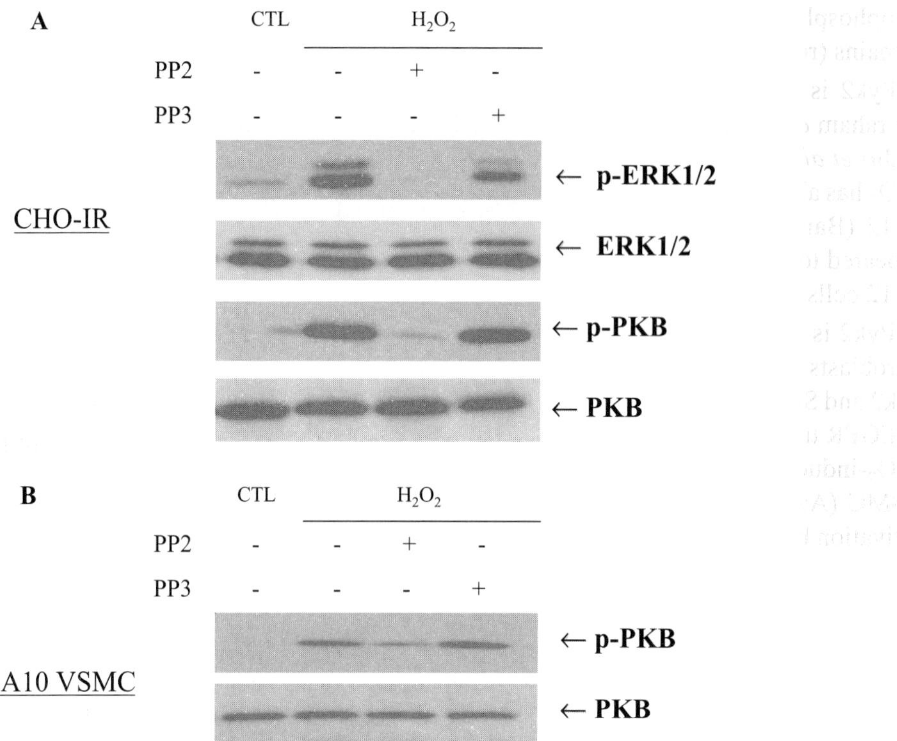

Fig. 3. Requirement of c-Src in H_2O_2-induced ERK1/2 and PKB phosphorylation in CHO cells (A) and PKB phosphorylation in A10 VSMC (B).
Confluent, serum-starved CHO (A) or A10 cells (B) were pretreated without (–) or with (+) 10 μM PP2 or its inactive analogue PP3 for 30 minutes followed by stimulation in the absence (CTL) or presence of 1 mM (A) or 500 μM H_2O_2 (B) for 10 minutes. The cells were lysed, and equal amounts of protein were separated on 7.5% SDS-PAGE. The cells were lysed, and equal amounts of protein were separated on 7.5% SDS-PAGE. ERK1/2 (A) and PKB (A)(B) phosphorylation was detected by immunoblotting with phospho-specific antibodies as described in figure 1. Blots were also analyzed for total ERK1/2 and PKB. (A) Adapted from (Mehdi *et al.*, 2005) (B) Adapted from (Azar *et al.*, 2006).

increase in intracellular ROS levels, has enhanced ERK1/2 phosphorylation in a Src-independent fashion in Rat-2 fibroblasts (Esposito *et al.*, 2003). Conversely, a recent report has shown that H_2O_2 markedly inactivates Src *in vivo*, but not *in vitro* and that a reduced phosphorylation of the tyrosine residue in the activation loop is responsible for the inactivation *in vivo* (Tang *et al.*, 2005). Therefore, it appears that H_2O_2 exerts both activation and inhibitory effect on Src and triggers the responses of several agonists in a cell-specific manner and differently *in vitro* and *in vivo*.

Involvement of the Proline-rich Tyrosine Kinase 2 (Pyk2)

Pyk2 is a non-receptor tyrosine kinase, also known as related adhesion focal tyrosine kinase (RAFTK) (Avraham *et al.*, 1995) or cell adhesion kinase beta (CAKβ) (Sasaki, H. *et al.*, 95) or calcium-dependent tyrosine kinase (Yu *et al.*, 1996). It contains a central catalytic domain and two non-catalytic domains, the N- and the C-terminal domains. The N-terminal domain contains a tyrosine autophosphorylation site at Tyr^{402}, which serves as a docking site for SFKs to bind via their SH2 domains (reviewed in (Avraham *et al.*, 2000)).

Pyk2 is activated by phosphorylation on tyrosine residues ($Tyr^{402,\ 579,\ 580}$ and ^{881}) (reviewed in (Avraham *et al.*, 2000)) in a Ca^{2+}- and PKC-dependent fashion in PC12 (Lev *et al.*, 1995) and VSMC (Sabri *et al.*, 1998) in response to stimulation by GPCR agonists. Direct stimulation with exogenous H_2O_2 has also been shown to enhance the tyrosine phosphorylation of Pyk2 in many cell types, such as PC12 (Banno *et al.*, 2005) and VSMC (Azar *et al.*, 2006a; Frank *et al.*, 2000). Moreover, Pyk2 appeared to be a critical mediator of PKB activation by H_2O_2 in both VSMC (Frank *et al.*, 2003) and PC12 cells (Banno *et al.*, 2005).

Pyk2 is known to interact physically with Src in PC12 cells (Park *et al.*, 2004), mouse embryonic fibroblasts (Andreev *et al.*, 2001), and VSMC (Sabri *et al.*, 1998). This complex formation between Pyk2 and Src is important in mediating GPCR-induced activation of many downstream signaling, such as EGFR transactivation in mouse fibroblasts (Andreev *et al.*, 2001) and VSMC (Eguchi *et al.*, 1999). H_2O_2-induced phosphorylation of Pyk2 has recently been shown to be dependent on IGF-1R in A10 VSMC (Azar *et al.*, 2006a) (Figure 4). Whether Pyk2 activation is required for ERK1/2 and PKB activation by H_2O_2 in A10 cells remains to be clarified.

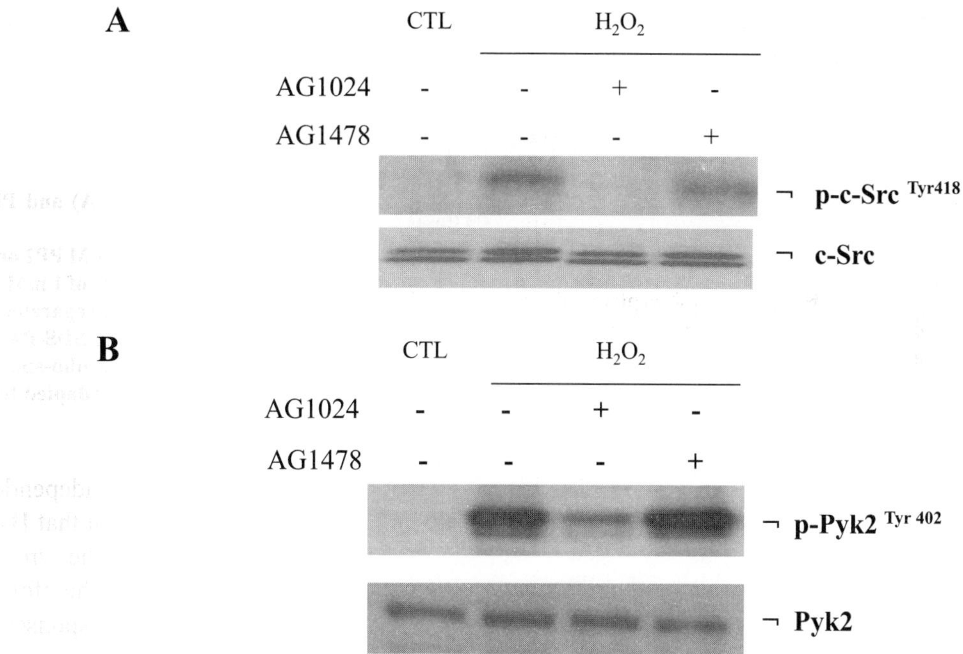

Fig. 4. **H_2O_2-induced Src and Pyk2 tyrosine phosphorylation are dependent on IGF-1R-PTK in A10 VSMC. Confluent, serum-starved A10 cells were pretreated without (–) or with (+) 1 μM AG1024 or AG1478 for 30 minutes followed by stimulation in the absence (CTL) or presence of 500 μM H_2O_2 for 10 minutes. The cells were lysed, and equal amounts of protein were separated on 7.5% SDS-PAGE. c-Src (A) and Pyk2 (B) phosphorylation was detected by immunoblotting with phospho-specific Tyr^{418} c-Src antibody (top panel of section (A)) and Tyr^{402} Pyk2 antibody (top panel of section (B)). Blots were also analyzed for total c-Src and total Pyk2 (bottom panels). Adapted from (Azar *et al.*, 2006).**

Involvement of Protein Tyrosine Phosphatases (PTPases) and Phosphatase and Tensin Homolog Deleted on Chromosome Ten (PTEN) in Triggering H_2O_2 Response

The precise mechanism by which H_2O_2 activates receptor- and non-receptor-PTKs is still unclear. However, in view of the ability of H_2O_2 to inhibit PTPases, such as PTP-1B (Lee *et al.*, 1998), and SH-2 domain-containing tyrosine phosphatase-2 (SHP-2) (Meng *et al.*, 2002), it has been postulated that H_2O_2 treatment can shift the equilibrium of the phosphorylation-dephosphorylation cycle, culminating in a net increase of tyrosine phosphorylation of PTKs and/or other proteins (Azar *et al.*, 2006a; Mehdi *et al.*, 2005). The activated PTKs promote the assembly of signaling components essential to trigger the ERK1/2 and PKB signaling pathways, which are critical to regulate various physiological responses (Fig. 5). In addition, several laboratories have reported that PTEN, which catalyzes PIP_3 dephosphorylation, is a direct target of H_2O_2 and becomes inactivated by H_2O_2 treatment (Lee *et al.*, 2002). Inactivation of PTEN, which results in an increase of PIP_3 levels, has been attributed to PKB activation in response to several agents known to generate ROS (Seo *et al.*, 2005). However, our studies showing that H_2O_2 treatment enhanced PI3-K activity in CHO-IR cells and that Wortmannin, an inhibitor of PI3-K, suppressed H_2O_2-induced PKB phosphorylation suggested that, in addition to PTEN inhibition, stimulation of PI3-K by H_2O_2 contributes to this response (Mehdi *et al.*, 2005). Moreover, multiple findings demonstrating that pharmacological inhibition of PTKs blocked H_2O_2-induced phosphorylation of PKB and ERK1/2 indicate an intermediary role of PTKs in mediating H_2O_2-induced signaling events.

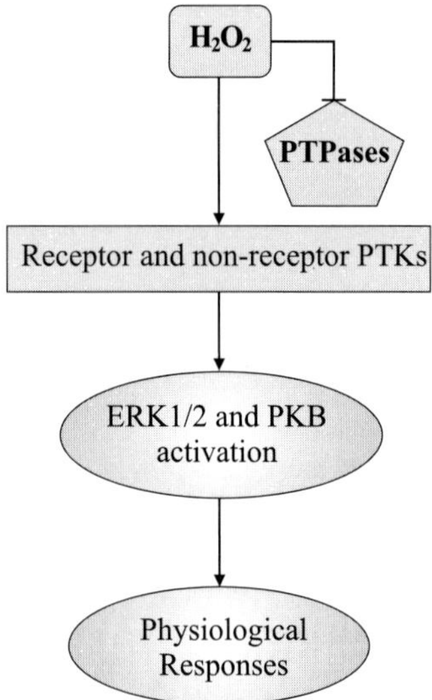

Fig. 5. Schematic model showing the possible mechanism by which H_2O_2 activates ptorein tyrosine kinases. H_2O_2 has been shown to activate ERK1/2 and PKB in many cell types. By using pharmacological inhibitors, a potential role for both receptor and non-receptor PTKs in these events has been suggested. Since H_2O_2 is a potent inhibitor of protein tyrosine phosphatases (PTPases), it is possible that H_2O_2 can shift the equilibrium of phosphorylation and dephosphorylation cycle resulting in the net increase in tyrosine phosphorylation of the catalytic activity of various PTKs. Adapted from (Azar *et al.*, 2007).

Conclusion

There is an emerging body of evidence to suggest that H_2O_2 induces the activation of several signaling pathways, including MAPK and PI3-K/PKB pathways, with an intermediary role of receptor and non-receptor PTKs. Among the receptor tyrosine kinases, the transactivation of EGF-receptor tyrosine kinases has been examined in some detail. However, more recent work have recognized an involvement of IGF-1R and PDGFR in this process. In addition, members of Src family of non-receptor-PTKs as well as Pyk2 have also been identified as upstream regulators of H_2O_2-induced adaptive/regulatory signal transduction events.

References

1. Abe J, Kusuhara M, Ulevitch RJ, Berk BC, Lee JD. Big mitogen-activated protein kinase 1 (BMK1) is a redox-sensitive kinase. *J Biol Chem.* 1996; 271:16586-16590.
2. Adams TE, Epa VC, Garrett TP, Ward CW. Structure and function of the type 1 insulin-like growth factor receptor. *Cell Mol. Life Sci.* 2000; 57:1050-1093.
3. Andreev J, Galisteo ML, Kranenburg O, Logan SK, Chiu ES, Okigaki M, Cary LA, Moolenaar WH, Schlessinger J. Src and Pyk2 mediate G-protein-coupled receptor activation of epidermal growth factor receptor (EGFR) but are not required for coupling to the mitogen-activated protein (MAP) kinase signaling cascade J Biol Chem., 2001; 276: 20130-20135.
4. Avraham H, Park SY, Schinkmann K, Avraham S. RAFTK/Pyk2-mediated cellular signalling Cell Signal., 2000; 12:123-133.
5. Avraham S, London R, Fu Y, Ota S, Hiregowdara D, Li J, Jiang S, Pasztor LM, White RA, Groopman JE. Identification and characterization of a novel related adhesion focal tyrosine kinase (RAFTK) from megakaryocytes and brain.1 *J Biol Chem.* 1995; 270:27742-27751.
6. Azar ZM, Mehdi MZ, Srivastava AK. Activation of insulin-like growth factor type-1 receptor is required for H_2O_2-induced PKB phosphorylation in vascular smooth muscle cells. *Can. J. Physiol Pharmacol.* 2006; 84:777-786.
7. Azar ZM, Mehdi MZ, Srivastava AK. Insulin-like growth factor type-1 receptor transactivation in vasoactive peptide and oxidant signaling pathways in vascular smooth muscle cells. *Can. J. Physiol Pharmacol.* 2007; 85:105-111.
8. Banno Y, Ohguchi K, Matsumoto N, Koda M, Ueda M, Hara A, Dikic I, Nozawa Y. Implication of phospholipase D2 in oxidant-induced phosphoinositide 3-kinase signaling via Pyk2 activation in PC12 cells J Biol Chem., 2005; 280:16319-16324.
9. Blanc A, Pandey NR, Srivastava AK. Synchronous activation of ERK 1/2, $p38^{mapk}$ and PKB/Akt signaling by H_2O_2 in vascular smooth mucle cells: Potential involvement in vascular disease. (review) *Int. J. Mol. Med.* 2003; 11:229-234.
10. Blanc A, Pandey NR, Srivastava AK. Distinct roles of Ca2+, calmodulin, and protein kinase C in H_2O_2-induced activation of ERK1/2, p38 MAPK, and protein kinase B signaling in vascular smooth muscle cells. *Antioxid. Redox. Signal.* 2004; 6:353-366.
11. Board R, Jayson GC. Platelet-derived growth factor receptor (PDGFR): a target for anticancer therapeutics. *Drug Resist. Updat.* 2005; 8:75-83.
12. Boveris A, Chance B. The mitochondrial generation of hydrogen peroxide. General properties and effect of hyperbaric oxygen. *Biochem. J.* 1973; 134:707-716.
13. Carpenter G. The EGF receptor: a nexus for trafficking and signaling. *1 Bioessays.* 2000; 22:697-707.
14. Catarzi S, Biagioni C, Giannoni E, Favilli F, Marcucci T, Iantomasi T, Vincenzini MT. Redox regulation of platelet-derived-growth-factor-receptor: role of NADPH-oxidase and c-Src tyrosine kinase. *Biochim. Biophys. Acta.* 2005; 1745:166-175.

15. Chen K, Vita JA, Berk BC, Keaney JF Jr. c-Jun N-terminal kinase activation by hydrogen peroxide in endothelial cells involves SRC-dependent epidermal growth factor receptor transactivation. *J Biol Chem.* 2001; 276:16045-16050.
16. Daou GB, Srivastava AK. Reactive oxygen species mediate Endothelin-1-induced activation of ERK1/2, PKB, and Pyk2 signaling, as well as protein synthesis, in vascular smooth muscle cells. *Free Radic. Biol. Med.* 2004; 37:208-215.
17. Datta SR, Brunet A, Greenberg ME. Cellular survival: a play in three Akts. *Genes Dev.* 1999; 13:2905-2927.
18. Delafontaine P. Insulin-like growth factor I and its binding proteins in the cardiovascular system. 1 *Cardiovasc Res.* 1995; 30:825-834.
19. Dhalla NS, Temsah RM, Netticadan T. Role of oxidative stress in cardiovascular diseases. *J Hypertens.* 2000; 18:655-673.
20. Dolloff NG, Shulby SS, Nelson AV, Stearns ME, Johannes GJ, Thomas JD, Meucci O, Fatatis A. Bone-metastatic potential of human prostate cancer cells correlates with Akt/PKB activation by alpha platelet-derived growth factor receptor. *Oncogene.* 2005; 24:6848-6854.
21. Dreher D, Junod AF. Role of oxygen free radicals in cancer development. *Eur. J Cancer.* 1996; 32A:30-38.
22. Duan C, Bauchat JR, Hsieh T. Phosphatidylinositol 3-kinase is required for insulin-like growth factor-I-induced vascular smooth muscle cell proliferation and migration. *Circ. Res.* 2000; 86:15-23.
23. Eguchi S, Iwasaki H, Inagami T, Numaguchi K, Yamakawa T, Motley ED, Owada KM, Marumo F, Hirata Y. Involvement of PYK2 in angiotensin II signaling of vascular smooth muscle cells. *Hypertension.* 1999; 33:201-206.
24. English JM, Pearson G, Hockenberry T, Shivakumar L, White MA, Cobb MH. Contribution of the ERK5/MEK5 pathway to Ras/Raf signaling and growth control. *J Biol Chem.* 1999; 274:31588-31592.
25. Esposito F, Chirico G, Montesano GN, Posadas I, Ammendola R, Russo T, Cirino G, Cimino F. Protein kinase B activation by reactive oxygen species is independent of tyrosine kinase receptor phosphorylation and requires SRC activity. *J Biol Chem.*2003; 278:20828-20834.
26. Farese RV, Sajan MP, Standaert ML. Insulin-sensitive protein kinases (atypical protein kinase C and protein kinase B/Akt): actions and defects in obesity and type II diabetes. *Exp. Biol Med. (Maywood).* 2005; 230:593-605.
27. Finkel T. Oxygen radicals and signaling. *Curr. Opin. Cell Biol.* 1998; 10:248-253.
28. Fortuno A, Jose GS, Moreno MU, Diez J, Zalba G. Oxidative stress and vascular remodeling. *Exp. Physiol.* 2005; 90:457-462.
29. Frank GD, Motley ED, Inagami T, Eguchi S. PYK2/CAKbeta represents a redox-sensitive tyrosine kinase in vascular smooth muscle cells. *Biochem. Biophys. Res Commun.* 2000; 270:761-765.
30. Frank GD, Mifune M, Inagami T, Ohba M, Sasaki T, Higashiyama S, Dempsey PJ, Eguchi S. Distinct mechanisms of receptor and nonreceptor tyrosine kinase activation by reactive oxygen species in vascular smooth muscle cells: role of metalloprotease and protein kinase C-delta. *Mol. Cell Biol.* 2003; 23:1581-1589.
31. Fridovich I. Superoxide anion radical (O_2–.), superoxide dismutases, and related matters. *J Biol Chem.* 1997; 272:18515-18517.
32. Giorgetti S, Pelicci PG, Pelicci G, Van Obberghen E. Involvement of Src-homology/collagen (SHC) proteins in signaling through the insulin receptor and the insulin-like-growth-factor-I-receptor.1 *Eur. J Biochem.* 1994; 223:195-202.
33. Hernandez-Sanchez C, Werner H, Roberts CT Jr, Woo EJ, Hum DW, Rosenthal SM, LeRoith D. Differential regulation of insulin-like growth factor-I (IGF-I) receptor gene expression by IGF-I and basic fibroblastic growth factor.1 *J Biol Chem.* 1997; 272:4663-4670.
34. Houstis N, Rosen ED, Lander ES. Reactive oxygen species have a causal role in multiple forms of insulin resistance. *Nature.* 2006; 440:944-948.

35. Knebel A, Rahmsdorf HJ, Ullrich A, Herrlich P. Dephosphorylation of receptor tyrosine kinases as target of regulation by radiation, oxidants or alkylating agents.1 *EMBO J.* 1996;15: 5314-5325.

36. Kyosseva SV. Mitogen-activated protein kinase signaling. *Int Rev Neurobiol.* 2004; 59:201-220.

37. Lee JS, Kim SY, Kwon CH, Kim YK. EGFR-dependent ERK activation triggers hydrogen peroxide-induced apoptosis in OK renal epithelial cells. *Arch. Toxicol.* 2006; 80:337-346.

38. Lee SR, Kwon,KS, Kim SR, Rhee SG. Reversible inactivation of protein-tyrosine phosphatase 1B in A431 cells stimulated with epidermal growth factor. *J Biol Chem.* 1998; 273:15366-15372.

39. Lee SR, Yang KS, Kwon J, Lee C, Jeong W, Rhee SG. Reversible inactivation of the tumor suppressor PTEN by H_2O_2. *J Biol Chem.* 2002; 277:20336-20342.

40. Leevers SJ, Vanhaesebroeck B, Waterfield MD. Signaling through phosphoinositide 3-kinases: the lipids take centre stage. 2 *Curr. Opin. Cell Biol.* 1999; 11:219-225.

41. Le Roith D, Werner H, Beitner-Johnson D, Roberts CT Jr. Molecular and cellular aspects of the insulin-like growth factor I receptor.1 *Endocr Rev.* 1995; 16:143-163.

42. Levitzki A, Gazit A. Tyrosine kinase inhibition: an approach to drug development. *Science.* 1995; 267:1782-1788.

43. Lev S, Moreno H, Martinez R, Canoll P, Peles E, Musacchio JM, Plowman GD, Rudy B, Schlessinger J. Protein tyrosine kinase PYK2 involved in Ca(2+)-induced regulation of ion channel and MAP kinase functions. *Nature.* 1995; 376:737-745.

44. Lo YY, Wong JM, Cruz TF. Reactive oxygen species mediate cytokine activation of c-Jun NH_2-terminal kinases. *J. Biol. Chem.* 1996; 271:15703-15707.

45. Markadieu N, Crutzen R, Blero D, Erneux C, Beauwens R. Hydrogen peroxide and epidermal growth factor activate phosphatidylinositol 3-kinase and increase sodium transport in A6 cell monolayers. *Am J Physiol Renal Physiol.* 2005; 288:F1201-F1212.

46. Mehdi MZ, Pandey NR, Pandey SK, Srivastava AK. H_2O_2-induced phosphorylation of ERK1/2 and PKB requires tyrosine kinase activity of insulin receptor and c-Src. *Antioxid. Redox. Signal.* 2005; 7:1014-1020.

47. Mendelson J. Blockade of receptors for growth factors: an anticancer therapy—the fourth annual Joseph H Burchenal American Association of Cancer Research Clinical Research Award Lecture. *1 Clin Cancer Res.* 2000; 6:747-753.

48. Meng TC, Fukada T, Tonks NK. Reversible oxidation and inactivation of protein tyrosine phosphatases in vivo. *Mol. Cell.* 2002; 9:387-399.

49. Mills EM, Takeda K, Yu ZX, Ferrans V, Katagiri Y, Jiang H, Lavigne MC, Leto TL, Guroff G. Nerve growth factor treatment prevents the increase in superoxide produced by epidermal growth factor in PC12 cells. *J Biol Chem.*1998; 273:22165-22168.

50. Nicol RL, Frey N, Pearson G, Cobb M, Richardson J, Olson EN. Activated MEK5 induces serial assembly of sarcomeres and eccentric cardiac hypertrophy. *EMBO J.* 2001; 20:2757-2767.

51. Normanno N, Bianco C, Strizzi L, Mancino M, Maiello MR, De Luca A, Caponigro F, Salomon DS. The ErbB receptors and their ligands in cancer: an overview Curr. Drug Targets., 2005; 6:243-257.

52. Oldham S, Hafen E. Insulin/IGF and target of rapamycin signaling: a TOR de force in growth control. 1 *Trends Cell Biol.* 2003; 13:79-85.

53. Park SY, Avraham HK, Avraham S. RAFTK/Pyk2 activation is mediated by trans-acting autophosphorylation in a Src-independent manner. *J Biol Chem.* 2004; 279:33315-33322.

54. Parrizas M, Gazit A, Levitzki A, Wertheimer E, LeRoith D. Specific inhibition of insulin-like growth factor-1 and insulin receptor tyrosine kinase activity and biological function by tyrphostins. *Endocrinology.* 1997; 138:1427-1433.

55. Pearson G, Robinson F, Beers GT, Xu BE, Karandikar M, Berman K, Cobb MH. Mitogen-activated protein (MAP) kinase pathways: regulation and physiological functions. *Endocr Rev.*2001; 22:153-183.

56. Pollock DM, Pollock JS. Endothelin and oxidative stress in the vascular system. *Curr. Vasc. Pharmacol.* 2005; 3:365-367.

57. Prenzel N, Fischer OM, Streit S, Hart S, Ullrich A. The epidermal growth factor receptor family as a central element for cellular signal transduction and diversification. *1 Endocr Relat Cancer.* 2001; 8:11-31.

58. Purdom S, Chen QM. Epidermal growth factor receptor-dependent and -independent pathways in hydrogen peroxide-induced mitogen-activated protein kinase activation in cardiomyocytes and heart fibroblasts. *J Pharmacol Exp. Ther.* 2005; 312:1179-1186.

59. Rao GN. Hydrogen peroxide induces complex formation of SHC-Grb2-SOS with receptor tyrosine kinase and activates Ras and extracellular signal-regulated protein kinases group of mitogen-activated protein kinases. *Oncogene.* 1996; 13:713-719.

60. Rosado JA, Redondo PC, Salido GM, Gomez-Arteta E, Sage SO, Pariente JA. Hydrogen peroxide generation induces pp60src activation in human platelets: evidence for the involvement of this pathway in store-mediated calcium entry J Biol Chem., 2004; 279: 1665-1675.

61. Rosenkranz S, Kazlauskas A. Evidence for distinct signaling properties and biological responses induced by the PDGF receptor alpha and beta subtypes.1 Growth Factors, 1999: 16:201-216.

62. Rozakis-Adcock M, McGlade J, Mbamalu G, Pelicci G, Daly R, Li W, Batzer A, Thomas S, Brugge J, Pelicci PG. Association of the Shc and Grb2/Sem5 SH2-containing proteins is implicated in activation of the Ras pathway by tyrosine kinases.1 Nature, 1992; 360:689-692.

63. Sabri A, Govindarajan G, Griffin TM, Byron KL, Samarel AM, Lucchesi PA. Calcium- and protein kinase C-dependent activation of the tyrosine kinase PYK2 by angiotensin II in vascular smooth muscle. *Circ. Res.* 1998; 83:841-851.

64. Saito S, Frank GD, Mifune M, Ohba M, Utsunomiya H, Motley ED, Inagami T, Eguchi S. Ligand-independent trans-activation of the platelet-derived growth factor receptor by reactive oxygen species requires protein kinase C-delta and c-Src. *J Biol Chem.* 2002; 277:44695-44700.

65. Sasaki H, Nagura K, Ishino M, Tobioka H, Kotani K, Sasaki T. Cloning and characterization of cell adhesion kinase beta, a novel protein-tyrosine kinase of the focal adhesion kinase subfamily. 1 *J Biol Chem.* 1995; 270:21206-21219.

66. Saward L, Zahradka P. Angiotensin II activates phosphatidylinositol 3-kinase in vascular smooth muscle cells. *Circ. Res.* 1997; 81:249-257.

67. Schafer FQ Buettner GR. Redox environment of the cell as viewed through the redox state of the glutathione disulfide/glutathione couple. *Free Radic. Biol Med.* 2001; 30:1191-1212.

68. Schlessinger J. Ligand-induced, receptor-mediated dimerization and activation of EGF receptor.1 *Cell.* 2002; 110:669-672.

69. Seger R, Krebs EG. The MAPK signaling cascade. *FASEB J.*, 1995; 9:726-735.

70. Seo JH, Ahn Y, Lee SR, Yeol YC, Chung HK. The major target of the endogenously generated reactive oxygen species in response to insulin stimulation is phosphatase and tensin homolog and not phosphoinositide-3 kinase (PI-3 kinase) in the PI-3 kinase/Akt pathway. *Mol. Biol Cell.* 2005; 16:348-357.

71. Song G, Ouyang G, Bao S. The activation of Akt/PKB signaling pathway and cell survival. *J. Cell Mol. Med.* 2005; 9:59-71.

72. Sugden PH. Signalling pathways in cardiac myocyte hypertrophy. *1 Ann. Med.* 2001; 33:611-622.

73. Tabet F, Schiffrin EL, Touyz RM. Mitogen-activated protein kinase activation by hydrogen peroxide is mediated through tyrosine kinase-dependent, protein kinase C-independent pathways in vascular smooth muscle cells: upregulation in spontaneously hypertensive rats. *J Hypertens.* 2005; 23:2005-2012.

74. Tang H, Hao Q, Rutherford SA, Low B, Zhao ZJ. Inactivation of SRC family tyrosine kinases by reactive oxygen species in vivo. *J Biol Chem.* 2005; 280:23918-23925.

75. Touyz RM. Reactive oxygen species and angiotensin II signaling in vascular cells — implications in cardiovascular disease. *Braz. J Med. Biol Res.* 2004; 37:1263-1273.

76. Touyz RM, Yao G, Viel E, Amiri F, Schiffrin EL. Angiotensin II and endothelin-1 regulate MAP kinases through different redox-dependent mechanisms in human vascular smooth muscle cells. *J Hypertens.* 2004; 22:1141-1149.

77. Ushio-Fukai M, Alexander RW, Akers M, Griendling KK. p38 Mitogen-activated protein kinase is a critical component of the redox-sensitive signaling pathways activated by angiotensin II. Role in vascular smooth muscle cell hypertrophy. *J. Biol. Chem.*1998; 273:15022-15029.

78. Ushio-Fukai M, Alexander RW, Akers M, Yin Q, Fujio Y, Walsh K, Griendling KK. Reactive oxygen species mediate the activation of Akt/protein kinase B by angiotensin II in vascular smooth muscle cells. *J Biol Chem.* 1999; 274:22699-22704.

79. Van der Heide LP, Kamal A, Artola A, Gispen WH, Ramakers GM. Insulin modulates hippocampal activity-dependent synaptic plasticity in a N-methyl-d-aspartate receptor and phosphatidyl-inositol-3-kinase-dependent manner. *J Neurochem.* 2005; 94:1158-1166.

80. Wang X, McCullough KD, Franke TF, Holbrook NJ. Epidermal growth factor receptor-dependent Akt activation by oxidative stress enhances cell survival. *J Biol Chem.* 2000; 275:14624-14631.

81. White MF. The IRS-signalling system: a network of docking proteins that mediate insulin action. *Mol. Cell Biochem.* 1998; 182:3-11.

82. Whiteman EL, Cho H, Birnbaum MJ. Role of Akt/protein kinase B in metabolism Trends. *Endocrinol. Metab.* 2002; 13:444-451.

83. Yeatman TJ.A renaissance for SRC. 1 *Nat. Rev Cancer.* 2005; 4:470-480.

84. Yu H, Li X, Marchetto GS, Dy R, Hunter D, Calvo B, Dawson TL, Wilm M, Anderegg RJ, Graves LM, Earp HS. Activation of a novel calcium-dependent protein-tyrosine kinase. Correlation with c-Jun N-terminal kinase but not mitogen-activated protein kinase activation. 1 *J Biol Chem.* 1996; 271:29993-29998.

85. Zaka M, Rafi MA, Rao HZ, Luzi P, Wenger DA. Insulin-like growth factor-1 provides protection against psychosine-induced apoptosis in cultured mouse oligodendrocyte progenitor cells using primarily the PI3K/Akt pathway. *Mol. Cell Neurosci.* 2005; 30:398-407.

86. Zalba G, San Jose G, Moreno MU, Fortuno MA, Fortuno A, Beaumont FJ, Diez J. Oxidative stress in arterial hypertension: role of NAD(P)H oxidase. *Hypertension.* 2001; 38:1395-1399.

87. Zheng B, Clemmons DR. Blocking ligand occupancy of the alphaVbeta3 integrin inhibits insulin-like growth factor I signaling in vascular smooth muscle cells.1 *Proc Natl Acad Sci USA,* 1998; 95:11217-11222.

88. Zhuang S, Schnellmann RG. H_2O_2-induced transactivation of EGF receptor requires Src and mediates ERK1/2, but not Akt, activation in renal cells. *Am J Physiol Renal Physiol.* 2004; 286:F858-F865.

Adaptation Biology and Medicine (Volume 5: Health Potentials)
Editors: L. Lukyanova, N. Takeda and P.K. Singal

2. Oxidative Modulation of Angiotensin II-induced G-protein Signalling in Cardiovascular Adaptations

Yuan Li, Georgios Lappas and Madhu B. Anand-Srivastava*

Department of Physiology and Groupe de recherché sur le système nerveux autonome (GRSNA) Faculty of Medicine, University of Montreal

Abstract

Oxidative stress caused by the overproduction of reactive oxygen species (ROS) is implicated in various cardiovascular complications including hypertension. The enhanced expression of Giα proteins that was attributed to the augmented levels of angiotensin II has been reported to be the contributing factor in the pathogenesis of hypertension. We have previously reported that angiotensin II (Ang II) treatment of A10 vascular smooth muscle cells (VSMC) increased the expression of inhibitory G proteins (Gi protein) and associated adenylyl cyclase signalling which was attributed to the enhanced MAPkinase activity. Since Ang II has been shown to increase oxidative stress, we investigated the role of oxidative stress in Ang II-induced enhanced expression of Giα proteins and examined the effects of antioxidants on Ang II-induced enhanced expression of Giα proteins and associated adenylyl cyclase signalling in A10 VSMC. Ang II treatment of A10 VSMC enhanced the production of superoxide anion (O_2^-) which was attenuated towards control levels by antioxidant; diphenyleneiodonium (DPI). In addition, Ang II treatment of A10 VSMC augmented the expression of Giα-2 and Giα-3 proteins and Erk1/2 phosphorylation which was restored to control levels by DPI. Furthermore, the inhibition of forskolin (FSK)-stimulated adenylyl cyclase activity by low concentrations of GTPγS (receptor-independent Gi functions) and Ang II-, C-ANP4-23 -(NPR-C agonist) and oxotremorine-mediated inhibitions of adenylyl cyclase (receptor-dependent functions) that were augmented in Ang II-treated VSMC were also restored to control levels by antioxidant treatments. These results suggest that Ang II-induced increased oxidative stress and thereby MAPkinase activity may serve as an adaptive mechanism to enhance the levels of Gi proteins and decrease the levels of cAMP, which by increasing vascular resistance may contribute to the augmentation of blood pressure.

Keywords: Angiotensin II, G-protein, oxidative stress, MAPkinase, adenylyl cyclase, VSMC.

Introduction

Angiotensin II (Ang II), a vasoactive peptide and key component of the renin-angiotensin system, regulates a variety of physiological functions including blood pressure, cell proliferation, cell

*Address for Correspondence: Dr. Madhu B. Anand-Srivastava, Department of Physiology, Faculty of Medicine, University of Montreal, C.P. 6128, Succ. Centre-ville, Montréal, Québec, Canada, H3C 3J7 Tel: (514) 343-2091, Fax: (514) 343-2111 E-mail: madhu.anand-srivastava@umontreal.ca

differentiation and apoptosis (1-3). Ang II elicits its physiological effects by interacting with two distinct receptor subtypes designated as AT_1 and AT_2 (4). The presence of AT_1 receptor subtype has been shown in rat vascular tissues, however, a small proportion of AT_2 receptors are also present in rat aorta (5, 6). Most of the physiological effects of Ang II are mediated through the activation of AT1 receptors that are coupled to several signalling pathways including adenylyl cyclase/cAMP inhibition through Giα proteins (7-9) mitogen-activated protein kinases (MAPKs) (10-12) and phosphatidylinositol turnover (13).

The adenylyl cyclase/cAMP system is composed of three components: receptor, catalytic subunit and guanine nucleotide regulatory proteins (G proteins). The G-proteins act as transducers and in the presence of guanine nucleotides, transmit the signal from the hormone-occupied receptor to the catalytic subunit. The agonist-mediated stimulation and inhibition of adenylyl cyclase are mediated through stimulatory (Gs) and inhibitory (Gi) guanine nucleotide protein respectively (14, 15) resulting in the increased or decreased formation of cAMP respectively. G-proteins are heterotrimeric consisting of α, β and γ subunits. The α subunits bind and hydrolyze GTP and confer specificity in receptor and effector interactions. Four different isoforms of Gs have been identified which appear to be products of alternate splicing of single gene (16, 17) whereas three different isoforms of Giα (Giα-1, Giα-2, Giα-3) have been identified and are shown to be the products of three different genes (18, 19).

Alterations in the levels of Gi proteins that result in the impaired cellular functions lead to various pathological states such as hypertension. We have recently shown an increased expression of Gi proteins and Gi protein mRNA in hearts and aortas from SHR and in hearts from experimental hypertensive rats including deoxycorticosterone acetate (DOCA)-salt hypertensive rats and 1 kidney 1 clip (1K1C) rats with established hypertension (20-24). In addition, we have further shown that VSMC from SHR also exhibit enhanced levels of Giα protein as compared to WKY rats, which were restored to control levels by antioxidants (25), suggesting a role of oxidative stress in enhanced levels of Giα proteins in SHR. The levels of various vasoactive peptides, including angiotensin II (Ang II), that have been shown to be enhanced in hypertension may be responsible for the enhanced expression of Giα proteins. This was supported by studies showing that treatment of SHR with captopril, an antihypertensive drug that acts by blocking the angiotensin I-converting enzyme, decreased the blood pressure and the enhanced Gi protein levels (26). Furthermore, Ang II has been shown to increase the levels of Gi proteins in A10 smooth muscle cells (27), which was attenuated by losartan and a mitogen-activated protein (MAP)/extracellular signal-regulated (ERK) kinase inhibitor (28, 29). Ang II also increases oxidative stress by activating NADPH oxidase (30). Taken together, it may be possible that Ang II-induced enhanced expression of Giα proteins in A10 cells is also attributed to the enhanced oxidative stress. To examine this possibility, the present studies were therefore undertaken to investigate the effect of antioxidant treatment on Ang II-induced enhanced expression of Giα proteins and adenylyl cyclase signalling in A10 vascular smooth muscle cells (VSMC).

Materials and Methods

Materials- 3-Isobutyl-1-methylxanthine (IBMX) was purchased from Aldrich chemical Corporation (Milwaukee, WI). Diphenyleneiodonium (DPI), was from BDH Inc. (Toronto, Ontario, Canada). [α^{32}P]ATP was purchased from Amersham Corp. (Oakville, Ontario, Canada). Antibodies L-5, C-10 for Giα-2 and for Giα-3 and monoclonal phosphospecific-tyrosine^{204}ERK1/2, antibodies were from

Santa Cruz Biotechnology (Santa Cruz, CA, USA). Myokinase (EC 2.7.4.3) and all other chemicals used in the experiments were purchased from Sigma chemical Co. (St-Louis, MO).

Methods

Cell culture and incubation

A-10 cell line from embryonic thoracic aorta of rat was obtained from American Type Culture Collection, Rockville, MA, and USA. The cells were plated in 75 cm^2 flasks and incubated at 37°C in 95% air and 5% CO_2 humidified atmosphere in Dulbecco's modified Eagle's medium (DMEM) (with glucose, L-glutamine, and sodium bicarbonate) containing antibiotics and 10% heat-inactivated fetal bovine serum (FBS) as described previously (25-27). Confluent cell cultures were starved by incubation for 3h in DMEM without FBS at 37°C. These cells were then incubated with Ang II (10^{-7}M) for different time periods at 37°C. To examine the effect of antioxidant; DPI on Ang II-induced responses, the cells were pretreated with or without DPI for 24hr and then challenged with Ang II (10^{-7}M) for one hour. After incubation, cells were washed twice with ice-cold homogenization buffer (10 mM Tris-HCl, pH 7.5 containing 1 mM EDTA). The cells were scraped into ice-cold homogenization buffer using a rubber policeman, and collected by centrifugation at 4°C for 10 min at 600 × g. The cells were then homogenized in a Dounce homogenizer (10 strokes) and the homogenate was used for adenylyl cyclase assay and immunoblotting.

Adenylyl cyclase activity determination

Adenylyl cyclase activity was determined by measuring [^{32}P]cAMP formation from [α^{32}P]ATP, as previously described (20, 21). Briefly, the assay medium contained 50 mM glycylglycine (pH 7.5), 0.5 mM MgATP, [α-^{32} P]ATP (1.5×10^6 cpm), 5 mM $MgCl_2$ (in excess of the ATP concentration), 100 mM NaCl, 0.5 mM cAMP, 1 mM 3-isobutyl-1-methylxanthine (IBMX), 0.1 mM EGTA, 10 μM guanosine 5'-O-(3-thiotriphosphate) (GTPγS), and an ATP-regenerating system (consisting of 2 mM phosphocreatine, 0.1 mg creatine kinase/ml, and 0.1 mg myokinase/ml) in a final volume of 200 μl. Incubations were initiated by the addition of the membrane preparations (20-30 μg) to the reaction mixture, which had been thermally equilibrated for 2 min at 37°C. The reactions were conducted in triplicate for 10 min at 37°C and were terminated by the addition of 0.6 ml of 120 mM zinc acetate. cAMP was purified by coprecipitation of other nucleotides with $ZnCO_3$ by addition of 0.5 ml of 144 mM Na_2CO_3, and subsequent chromatography by the double-column system, as described by Salomon et al. (31). Under the assay conditions used, adenylyl cyclase activity was linear with respect to protein concentrations and time of incubation. Protein was determined essentially as described by Lowry et al. (32) with bovine serum albumin as standard.

Immunoblotting

Immunoblotting of G proteins and ERK1/2 was performed using specific antibodies as described previously (20, 25). After SDS-PAGE, the separated proteins were electrophoretically transferred to nitrocellulose paper (Schleicher and Schuell Inc., Keene, N.H.) with a semi-dry transblot apparatus (Bio-Rad Laboratories, Mississauga, Ontario) at 15 V for 45 min. After transfer, the membranes were washed twice in phosphate-buffered saline (PBS) and were incubated in PBS containing 8% dehydrated milk at room temperature for 2 h. The blots were then incubated with respective primary antibodies; L-5 for Giα-2 and C-10 for Giα-3, monoclonal phosphospecific Tyr^{204}, ERK1/2 antibody in PBS containing 5% dehydrated milk and 0.1% Tween-20 at room temperature for 2h. The antibody-antigen complexes were detected by horseradish peroxidase-conjugated second antibody and protein bands were visualized by enhanced-chemiluminescence Western-blotting detection reagents from Amersham as described previously (20, 25). Quantitative analysis of specific bands was performed by densitometric scanning of the autoradiographs with an enhanced laser densitometer (LKB Ultroscan XL, Pharmacia, Dorval, Que.) and quantified using gel-scan XL evaluation software (version 2.1) from Pharmacia.

Superoxide anion measurements

Basal superoxide anion production was measured using the lucigenin-enhanced chemiluminescence with low concentration (5 μM) of lucigenin as described previously (25). The cells after treatment with Ang II (1 μM) or DPI (10 μM) alone or in combination were washed in oxygenated Krebs-Hepes buffer, scraped and placed in scintillation vials containing lucigenin solution and the emitted luminescence was measured with a liquid scintillation counter (Wallac 1409; Perkin Elmer Life Sciences, St-Laurent, Quebec, Canada) for 5 minutes. The average luminescence value was estimated, the background value substracted and the result was divided by the total weight of proteins in each sample.

Statistical analysis

Results are expressed as mean ± S.E.M. Comparisons between groups were made with analysis of variance (ANOVA) in conjunction with Newman-Keuls tests. Results were considered significant at a value of $P < 0.05$.

Results

Effect of Ang II on superoxide anion production in A10 VSMC

Ang II has been shown to increase oxidative stress by activating NADPH oxidase (33). A role of Ang II in enhanced production of superoxide anion in VSMC from SHR has been reported (25). To investigate if Ang II treatment of A10 VSMC also resulted in enhanced production of superoxide anion, the levels of superoxide anion (O_2^-) were measured by lucigenin in A10 VSMC. The results shown in Fig. 1 depict that treatment of cells with Ang II for 24 hr increased the production of O_2^- by 6-fold which was significantly restored towards control levels by DPI, an inhibitor of NADPH oxidase.

Effect of DPI on Ang II-induced increased expression of Giα proteins

Since Ang II increases the ROS production, it was of interest to investigate if Ang II-induced increased ROS production in these cells contributes to the increased expression of Giα proteins. To test this, we examined the effect of DPI on Ang II-induced increased expression of Giα proteins and the results are shown in Fig. 2. Treatment of cells with Ang II (10^{-7} M) for 1 hr increased the levels of Giα-2 and Giα-3 by about 50%, which were restored to control levels by DPI, an NADPH oxidase inhibitor, suggesting the implication of ROS in Ang II-induced enhanced expression of Giα proteins in A10 VSMC.

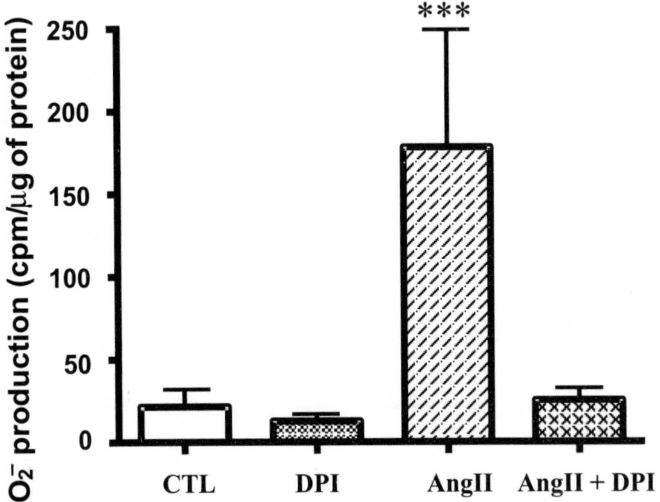

Fig. 1. **Effect of diphenyleneiodonium (DPI) on superoxide anion (O_2^-) production in A10 vascular smooth muscle cells (VSMC). A10 VSMC were pretreated without (CTL) or with DPI (1 μM) for 24 hrs and challenged with Ang II (1 μM) for one hour, and O_2^- production was determined as described in "Methods". Values are mean ± SEM of three separate experiments. *** P<0.001 vs CTL. Reproduced from (38) with permission.**

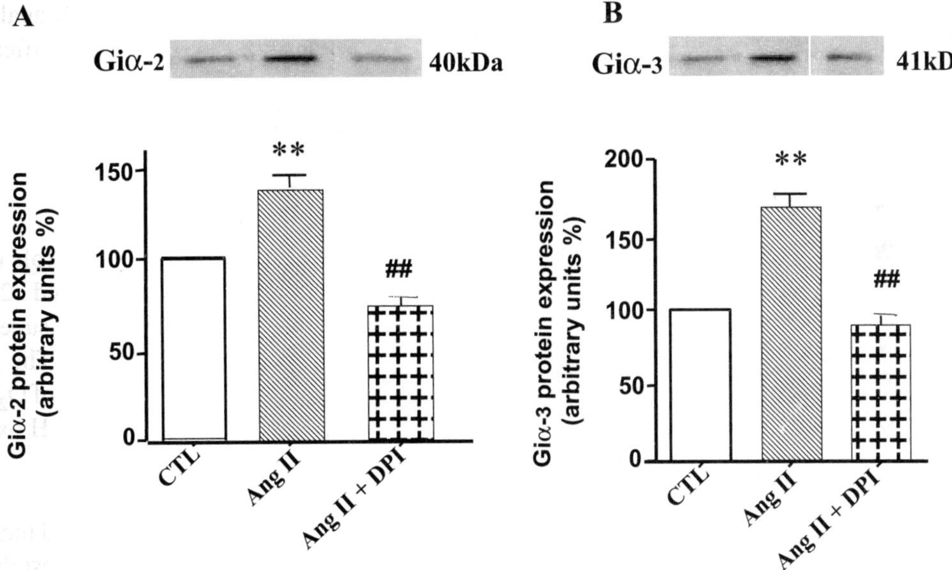

Fig. 2. **Effect of DPI on the levels of Giα-2 and Giα-3 proteins in A10 vascular smooth muscle cells (VSMC). A10 VSMC were pretreated without (CTL) or DPI (1 μM) for 24 hrs then challenged with Ang II (0.1 μM) for one hour. Membrane proteins (30 μg) were separated and transfered to nitrocellulose, which was then immunoblotted with specific antibodies against Giα-2 (A) and Giα-3 (B) as described in "Methods". Values are mean ± SEM of three separate experiments. ** P<0.01 vs Control, and ##P<0.001 vs AngII treated groups. Reproduced from (38) with permission.**

Implication of ROS in Ang II-induced enhanced Erk1/2 phosphorylation

We have previously shown that Ang II-induced enhanced expression of Giα proteins was inhibited by MEK inhibitor PD 98059 in A10 VSMC (29) suggesting the implication of MAP kinase signalling in Ang II-induced enhanced expression of Giα proteins. To investigate if Ang II could also increase the ERK1/2 phosphorylation in these cells and whether it is attributed to the increased oxidative stress, we examined the effect of DPI on Ang II-induced ERK1/2 phosphorylation in A10 cells. The results shown in Fig. 3 indicate that Ang II (10^{-7} M) increased the ERK1/2 phosphorylation by about 50% which was completely abolished by DPI treatment.

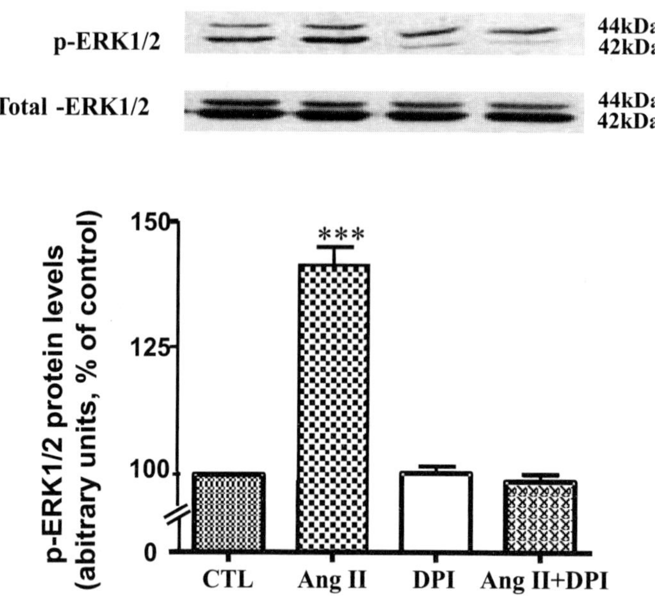

Fig. 3. Effect of diphenyleneiodonium (DPI) on Ang II-induced ERK1/2 phosphorylation in A10 vascular smooth muscle cells (VSMC). A10 VSMC were pretreated without (CTL) or with DPI (1 μM) for 24 hrs and challenged with Ang II (0.1 μM), for one hour. Membrane proteins (30 μg) were separated and transferred to nitrocellulose, which was then immunoblotted with specific antibodies against phospho-specific-Tyr^{204}-ERK1/2 (E-4) or ERK1/2 (C-14) as described in "Methods". ERK1/2 (C-14) was used to assess the loading of the protein. Data presented as the mean ± SEM of three separated experiment. *** P<0.001 vs Control. Reproduced from (38) with permission.

Effect of DPI on receptor-independent function of Giα

Since DPI restored the Ang II-induced enhanced levels of Giα-2 and Giα-3 proteins toward control levels, it was of interest to examine if the restoration of the enhanced levels of Giα proteins by DPI is also reflected in the restoration of increased Giα functions. To investigate this, the effect of DPI on receptor-independent and receptor-dependent functions was examined in Ang II-treated cells. For the receptor-independent functions of Gi, the effect of DPI was investigated on the inhibitory effect of GTPγS (10^{-12} to 10^{-8} M) on forskolin (FSK)-stimulated adenylyl cyclase activity in control and Ang II-treated A10 cells. The results shown in Fig. 4 indicate that the enhanced inhibition of FSK-stimulated adenylyl cyclase activity by different concentrations of GTPγS in Ang II-treated cells that was

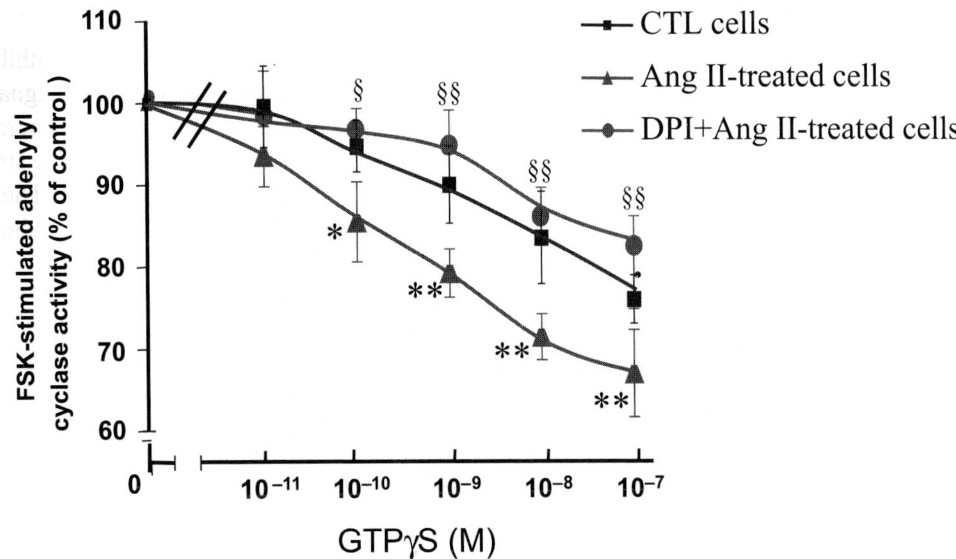

Fig. 4. Effect of diphenyleneiodonium (DPI) on GTPγS-mediated inhibition of forskolin (FSK)-stimulated adenylyl cyclase activity in control and Ang II-treated A10 vascular smooth muscle cells (VSMC). A10 VSMC were pretreated without (CTL) or with DPI (1 μM) for 24 hrs and challenged with Ang II (0.1 μM) for one hour. Membranes were prepared as described in "Methods", and adenylyl cyclase activity from these membranes was determined in the presence of 50 μM FSK alone, taken as 100%, and in the presence of various concentrations of GTPγS (10^{-12}M to 10^{-7}M). Values are mean ± SEM of three separate experiments. *$P<0.05$, **$P<0.01$ vs control group and §$P<0.05$, §§$P<0.01$ vs AngII-treated group. Reproduced from (38) with permission.

attributed to the enhanced levels of Giα proteins as compared to untreated control cells was reversed to control levels by DPI.

Effect of DPI on receptor-dependent functions

To investigate the effect of DPI on the receptor-dependent functions of Giα proteins, the effect of DPI was examined on the hormonal inhibitions of adenylyl cyclase activity in control and Ang II-treated A10 cells. For this, the effect of DPI on Ang II, C-ANP$_{4-23}$ and oxotremorine that inhibit adenylyl cyclase through Giα proteins (5, 7, 8, 34, 35) was examined on adenylyl cyclase activity in control and Ang II-treated cells. Fig. 5 shows that Ang II-, C-ANP$_{4-23}$, and oxotremorine-mediated inhibitions of adenylyl cyclase that were significantly augmented in Ang II-treated cells were restored to control levels by DPI treatment.

Discussion

We have previously reported that SHR and 1K1C hypertensive rats, which have high levels of Ang II, exhibited enhanced expression of Giα-2 and Giα-3 proteins mRNA in the heart and aorta, whereas the levels of Gsα were unaltered (20, 22, 23). The enhanced expression of Giα-2 and Giα-3 were restored toward control levels by captopril treatment (22, 23, 26), suggesting that Ang II contributes to the augmented levels of Giα proteins in SHR and 1K1C hypertensive rats. This was further supported by

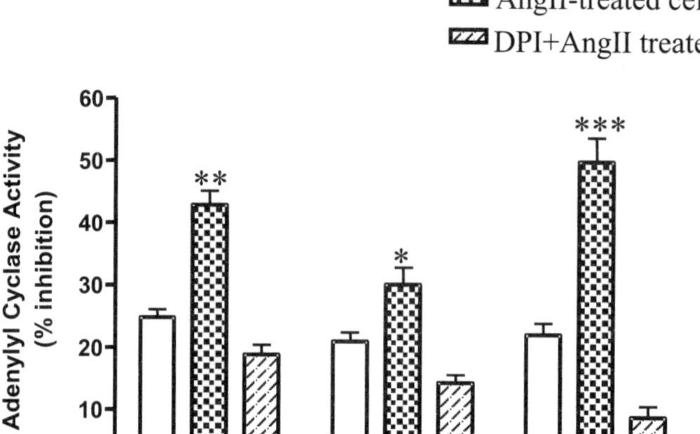

Fig. 5. Effect of diphenyleneiodonium (DPI) on hormonal inhibition of adenylyl cyclase activity in Control and Ang II-treated A10 vascular smooth muscle cells (VSMC). A10 VSMC were pretreated without (CTL) or with DPI (1 μM) for 24 hrs and challenged with Ang II (0.1 μM) for one hour. Membranes were prepared as described in "Methods", The adenylyl cyclase activity in these membranes was determined in the presence of 10 μM GTPγS alone, taken as 100% (basal) or in combination with Ang II (10 μM), C-ANP$_{4-23}$ (0.1 μM) and oxotremorine (5 μM). Values are mean±SEM of three separate experiments. *P<0.05, **P<0.01 and ***P<0.001 vs control groups. Reproduced from (38) with permission.

our recent studies demonstrating that Ang-II treatment of A10 VSMCs resulted in enhanced expression of Giα at the protein and mRNA levels which was restored to control levels by losartan (28). In addition, the Ang-II-evoked increased expression of Giα-2 and Giα-3 was inhibited by actinomycin treatment, suggesting that the Ang-II-mediated increases in Gi proteins are at the transcriptional level (27).

In the present studies, we report that A10 VSMC exposed to Ang II exhibit enhanced production of superoxide anion which contributes to the enhanced expression of Giα protein in these cells. In this regard, Ang II-induced enhanced oxidative stress has also been reported earlier in rat VSMC (20) and human internal mammary arteries (36). In addition, we have also shown recently the implication of Ang II in enhanced production of O_2^- in aortic VSMC from SHR (25).

Although a role of oxidative stress in Ang II-mediated cell signalling has been well established (37), an evidence for a direct role of oxidative stress in Ang II-mediated increased expression of Giα protein and associated adenylyl cyclase is lacking. Our results showing that antioxidant; DPI restored Ang II-evoked enhanced expression of Giα-2 and Giα-3 to control levels suggest the implication of oxidative stress in Ang II-induced enhanced levels of Giα proteins in SMC. In support of this are our earlier findings showing that enhanced oxidative stress caused by enhanced levels of Ang II contributes to the enhanced expression of Giα proteins in SHR (25). We have also shown that DPI that result in the restoration of Ang II-induced enhanced expression of Giα proteins to control levels also restored to the control levels, the enhanced inhibition of adenylyl cyclase by Ang II, C-ANP$_{4-23}$ and oxotremorine as well as GTPγS-mediated enhanced inhibition of FSK-stimulated adenylyl cyclase activity; the receptor dependent and independent functions of Giα proteins respectively.

Ang II has been shown to activate various signalling pathways including MAP kinase (9, 11, 12). We have shown previously the implication of MAP kinase signalling in Ang II-induced enhanced expression of Giα proteins in A10 VSMC (29). We also reported that the enhanced expression of ERK1/2 phosphorylation in VSMC from SHR was attenuated to control levels by antioxidants suggesting a role of oxidative stress in enhanced ERK1/2 phosphorylation (25). Thus, it can be suggested that oxidative stress by increasing MAP kinase signalling may be responsible for the enhanced expression of Giα proteins. This notion is supported by our findings showing that Ang II-induced increased ERK1/2 phosphorylation is also attenuated by DPI in A10 VSMC. Taken together, it may be suggested that Ang II-induced enhanced oxidative stress caused by O_2^- production may be responsible for the activation of MAP kinase signalling which then results in the enhanced expression of Giα proteins in VSMC.

In conclusion, we have provided the first evidence that Ang II-induced increased oxidative stress through increased MAP kinase activity may be responsible for the enhanced expression of Giα proteins in A10 VSMC. The increased expression of Giα proteins that results in decreased levels of cAMP and thereby increased vascular resistance may be one of the contributing factors in the pathogenesis of hypertension. It may thus be suggested that antioxidant treatment by decreasing oxidative stress, and thereby MAP kinase activity results in the attenuation of Ang II-induced enhanced levels of Giα proteins and increased levels of cAMP, which by decreasing vascular resistance may attenuate the development of high blood pressure.

References

1. Paul M, Bachmann J, Ganten D. The tissue renin-angiotensin system in cardiovascular disease. *Trends Cardiovasc Med* 1992; 2: 94-99.
2. Schelling P, Fischer H, Ganten H. Angiotensin and cell growth: a link to cardiovascular hypertrophy? *J Hypertens* 1991; 9: 3-15.
3. Suzuki J, Iwai M, Nakagami H, Wu L, Chen R, Sugaya T, Hamada M, Hiwada K, Horiuchi M. Role of angiotensin II-regulated apoptosis through distinct AT1 and AT2 receptors inneointimal formation. *Circulation* 2002; 106: 847-853.
4. Timmermans PB, Wong PC, Chin AT, Herblin WF, Benfield P, Carini DJ, Lee RJ, Wexler RR, Save JA, Smith RD. Angiotensin II receptors and angitoensin II receptor antagonists. *Pharmacol Rev* 1993; 45: 205-251.
5. Chang RS, Lotti VJ. Angiotensin receptor subtypes in rat, rabbit and monkey tissues: relative distribution and species dependency. *Life Sci* 1991; 49: 1485-1490.
6. Viswanathan M, Tsutsumi K, Correa FM, Saavedra JM. Changes in expression of angiotensin receptor subtypes in the rat aorta during development. *Biochem Biophys Res Commun* 1991; 179: 1361-1367.
7. Anand-Srivastava, M.B. Angiotensin II receptors negatively coupled to adenylate cyclase in rat myocardial sarcolemma. Involvement of inhibitory guanine nucleotide regulatory protein. *Biochem Pharmacol* 1989; 38: 489-496.
8. Anand-Srivastava, M.B. Angiotensin II receptors negatively coupled to adenylate cyclase in rat aorta. *Biochem Biophys Res Commun* 1983; 117: 420-428.
9. Pobiner BF, Hewlett EL, Garrison JC. Role of Ni in coupling angiotensin receptors to inhibition of adenylate cyclase in hepatocytes. *J Biol Chem* 1985; 260: 16200-16209.

10. Duff JL, Berk BC and Corson MA. Angiotensin II stimulates the pp^{44} and pp^{42} mitogen-activated protein kinases in cultured rat aortic smooth muscle cells. *Biochem Biophys Res Commun* 1992; 188: 257-264.
11. Molloy CJ, Taylor DS and Weber H. Angiotensin II stimulation of rapid protein tyrosine phosphorylation and protein kinase activation in rat aortic smooth muscle cells. *J Biol Chem* 1993; 268: 7338-7345.
12. Tsuda T, Kawahara Y, Ishida Y, Koide M, Shii K and Kokoyama M. Angiotensin II stimulates two myelin basic protein/microtubule-associated protein 2 kinases in cultured vascular smooth muscle cells. *Circ Res* 1992; 71: 620-630.
13. Griendling KK, Rittenhouse SE, Brock TA, Ekstein LS, Gimbrone MA, Alexander, RW. Sustained diacylglycerol formation from inositol phospholipids in angiotensin II-stimualted vascular smooth muscle cells. *J Biol Chem* 1986; 261: 5901-5906.
14. Robishaw JD, Smigel MD, Gilman AG. Molecular basis for two forms of the G protein that stimulates adenylate cyclase. *J Biol Chem* 1986; 261: 9587-9590.
15. Stryer L, Bourme HR. G-proteins: a family of signal transducers. *Annu Rev Cell Biol* 1986; 2: 391419.
16. Bray P, Carter A, Simons C, Guo V, Puckett C, Kamholz J, Spiegel A, Nirenberg M. Humana cDNA clones for four species of G alpa's signal transduct protein. *Proc Natl Acad Sci USA* 1987; 83: 8893-8897.
17. Jones DT, Reed RR. Molecular cloning of five GTP-binding protein cDNA species from rat olfactory neuroepithelium. *J Biol Chem* 1987; 162: 14241-14249.
18. Itoh H, Kozasa T, Nagata S, Nakamura S, Katada T, Ui M, Iwai S, Ohtsuka E, Kawasaki H, Suzuki K, Kaziro Y. Molecular cloning and sequence determination of cDNA's for alpha of the guanine nucleotide-binding proteins Gs, Gi and Go from rat brain. *Proc Natl Acad Sci USA* 1986; 83: 3776-3780.
19. Itoh H, Toyama RI, Kozasa TI, Tsukamoto TI, Matsuoka M, Kaziro Y. Presence of three distinct molecular species of G protein: a subunit structure of rat cDNA and human genomic DNAs. *J Biol Chem* 1988; 263: 6656-6664.
20. Anand-Srivastava MB. Enhanced expression of inhibitory guanine nucleotide regulatory protein in spontaneously hypertensive rats: relationship to adenylyl cyclase inhibition. *Biochem J* 1992; 288: 79-85.
21. Anand-Srivastava, M.B., de Champlain J, Thibault C. DOCA-salt hypertensive rat hearts exhibit altered expression of G-proteins. *Am J Hypertens* 1993; 6: 72-75.
22. Ge C, Garcia R, Anand-Srivastava M.B. Enhanced expression of Giα protein and adenylyl cyclase signalling in aortas from 1 kidney 1 clip hypertensive rats. *Can J Physiol Pharmacol* 2006; 84: 739-746.
23. Ge C, Garcia R, Anand-Srivastava MB. Altered expression of Gi-protein and adenylyl cyclase activity in hearts from one kidney one clip hypertensive rats: effect of captopril. *J Hypertens* 1999; 17: 1617-1626.
24. Thibault C, Anand-Srivastava MB. Altered expression of G-protein mRNA in spontaneously hypertensive rats. *FEBS Lett* 1992; 313: 160-164.
25. Lappas G, Bou Daou G and Anand-Srivastava MB. Oxidative stress contributes to the enhanced expression of Giα proteins and adenylyl cyclase signalling in vascular smooth muscle cells from spontaneously hypertensive rats. *J Hypertens* 2005; 23: 2251-2261.
26. Pandey SK, Anand-Srivastava MB. Modulation of G-protein expression by the angiotensin converting enzyme inhibitor captopril in hearts from spontaneously hypertensive rats: relationship to adenylyl cyclase. *Am J Hypertens* 1996; 9: 833-837.
27. Palaparti A, Ge C, Anand-Srivastava MB. Angiotensin II enhances the expression of Giα in A10 cells (smooth muscle): relationship with adenylyl cyclase activity. *Arch Biochem Biophys* 1999; 365: 113-122.
28. Anand-Srivastaa MB and Palaparti A. Angiotensin-II-induced enhanced expression of Gi proteins is attenuated by losartan in A10 vascular smooth muscle cells: role of AT1 receptors. *Can J Physiol Pharmacol* 2003; 81: 150-158.
29. Ge C, and Anand-Srivastava MB. Involvement of phosphatidylinositol 3-kinase and mitogen-activated protein kianse pathways in AII-mediated enhanced expression of Gi proteins in vascular smoot muscle cells. *Biochem Biophys Res Commun* 1998; 251: 570-575.

30. Griendling KK, Minieri CA, Collerenshaw JD, Alexande RW. Angiotensin II stimulates NADH and NADPH oxidase activity in cultured vascular smooth muscle cells. *Circ Res* 1994; 74: 1141-1148.

31. Salomon Y, Londos C, Rodbell M. A highly sensitive adenylyl cyclase assay. *Anal Biochem* 1974; 58: 541-548.

32. Lowry OH, Roseborough NJ, Farr AL, and Randall RJ. Protein measurement with the folin phenol reagent. *J Biol Chem.* 1951; 193: 265-275.

33. Laplant MA, Wu R, Midaoui El and de Champlain J. NADPH oxidase activation by angiotensin II is dependent on $p^{42}/^{44}$ ERK MAPkinase pathway activation in rats vascualr smooth muscle cells. *J Hypertens* 2003; 21: 927-936.

34. Anand-Srivastava MB and Cantin M. Pertussis toxin attenuates atrial natriuretic factor-mediated inhibition of adenylyl cyclase: involvement of inhibitory guanine nucleotide regulatory protein. *J Biol Chem* 1987; 262: 4931-4934.

35. Migeon JC, Thomas SL, Nathanson NM. Differential coupling of m2 and m4 muscarinic receptors to inhibition of adenylyl cyclase by Gi alpha and G(o)alpha subunits. *J Biol Chem* 1995; 270: 16070-16074.

36. Berry C, Hamilton CA, Brosman MJ, Magill, FG, Berg GA, McMurray JJ and Dominiczak AF. Investigation into the sources of superoxide in human blood levels: angiotensin II increases superoxide productors in human internal mammary arteries. *Circulation* 2000; 101: 2206-2212.

37. Touyz RM, Yao G, Viel E, Amri F and Schiffrin EL. Angiotensin II and endothelin-1 regulate MAPkinases through different redox-dependent mechanisms in human vascular smooth muscle cells. *J Hypertens* 2004; 22: 1141-1149.

38. Li Y, Lappas, G and Anand-Srivastava, M.B. Role of oxidative stress in angiotensin II-induced enhanced expression of Giα proteins and adenylyl cyclase signalling in A10 vasacular smooth muscle cells. *Am J Physiol.*, 2006. (In press)

Abbreviations

C-ANP_{4-23} [des(Glu^{18}, Ser^{19}, Glu^{20}, Leu^{21}, Gly^{22})ANP_{4-23}-NH_2]; Ang II, angiotensin II; MAPK, mitogen-activated protein kinase; MEK, mitogen extracellular signal regulated kinase; ERK1/2, extracellular signal regulated kinase; GTPγS, guanosine 5'-O-(3-thiotriphosphate); Gs, stimulatory guanine nucleotide regulatory protein; Gi, inhibitory guanine nucleotide regulatory protein; DPI, diphenyleneiodonium; VSMC, vascular smooth muscle cells.

Acknowledgements

We would like to thank Christiane Laurier for her valuable secretarial help.

Adaptation Biology and Medicine (Volume 5: Health Potentials)
Editors: L. Lukyanova, N. Takeda and P.K. Singal

3. Cardiovascular Responses to Lower Body Positive Pressure

Y. Kawai[1]*, S. Matsuo[1], H. Ohnishi[1], K. Tagawa[2] and T. Hayashi[3]

[1]Department of Adaptation Physiology, Faculty of Medicine, Tottori University, Yonago 683-8503, Japan
[2]Department of Regional Environment, Faculty of Regional Sciences, Tottori University, Tottori 680-8550, Japan
[3]Department of Applied Mathematics and Physics, Faculty of Engineering, Tottori University, Tottori 680-8550, Japan

Abstract

The purpose of the present study is firstly to elucidate the effect of lower body positive pressure (LBPP) on net ground reaction force (GRF) that is the difference between the body weight and the upward force produced by the LBPP and secondly to examine the effect of LBPP on the cardiovascular system. We hypothesized that application of LBPP decreases the net GRF and that a low level of LBPP causes minimal changes in cardiovascular parameters. Ten healthy male subjects (body weight: 64.5 ± 1.6 kg) participated in this study . During the graded application of LBPP, the net GRF decreased in a linear manner, reaching 30.9 ± 1.6 kgw at 15.0 mmHg of LBPP. There was no significant difference in heart rate, systolic blood pressure, and diastolic blood pressure between control and LBPP conditions. Calf skin blood flow measured by a laser Doppler flow meter did not change significantly during 15 mmHg of standing LBPP. These results suggest that our LBPP chamber can effectively unload lower extremities with minimal changes in local and systemic circulation.

Keywords: Lower Body Positive Pressure (LBPP), rehabilitation, body weight, systemic circulation, calf skin blood flow

Introduction

Early start of rehabilitation during the recovery period from a disease or an injury is highly recommended because immobilization causes muscle and bone atrophy in the lower extremities [Bloomfield, 1977]. Elderly patients who suffer from muscle pain or muscle weakness, however, are often unable to walk while supporting their entire body weight. They use assistant devices such as crutches, parallel bars, walkers, and suspension harnesses [Berman and Studenski, 1998], or they walk in swimming pools [Prins and Cutner, 1999] in order to partially unload their body weight. All of these, however, have some disadvantages. For example, muscle strength of the upper extremities is required for using crutches, parallel bars, and walkers, limiting the usefulness of these devices in elderly

*Address for Correspondence: Yasuaki Kawai, MD, PhD, Professor of Adaptation Physiology, Faculty of Medicine, Tottori University, 86 Nishi-cho, Yonago 683-8503 Japan. Tel: 81-859-38-6041, Fax: 81-859-38-6043,
E-mail: kawai@grape.med.tottori-u.ac.jp

patients. A risk of infection prevents a use of swimming pool for patients who underwent recent surgical operation. Thus, we developed a lower body positive pressure (LBPP) chamber which can reduce apparent body weight (net GRF) in collaboration with Dr. Hargens at University of California San Diego. The purpose of the present study is twofold: 1) To elucidate the effect of LBPP on the net GRF and 2) To examine the effect of LBPP on the cardiovascular system. We hypothesized that exposure to LBPP decreases net GRF in a linear manner and that a low level of LBPP causes minimal changes in cardiovascular parameters.

Materials and Methods

Subjects

Ten healthy male subjects, aged 27.0 ± 3.7 years, participated in this study. Informed consent was obtained from all subjects after they were fully briefed. Subjects were under normal fluid and salt intake and nobody took any medication at the time of the study.

LBPP chamber

Our LBPP chamber encloses a treadmill on which subjects can walk or run at a speed up to 15 km/h (Fig. 1). Air seal between the subject and the chamber is achieved by using a waist seal that fastens over a lip around the hole in the roof of the chamber. Air is driven into the chamber at a constant rate through a duct by means of a blower (U2V-40S; Showa Denki Co., Ltd.). The pressure inside the chamber can be adjusted to a certain level by controlling a valve through which air can go out and can be raised safely up to 80 mmHg. Subjects wore running shorts which is securely attached to the waist seal (Fig. 2a). Seal height was adjusted to the level of subject's iliac crest by changing the height of treadmill using a motor beneath the treadmill. When the pressure inside the chamber becomes higher than the

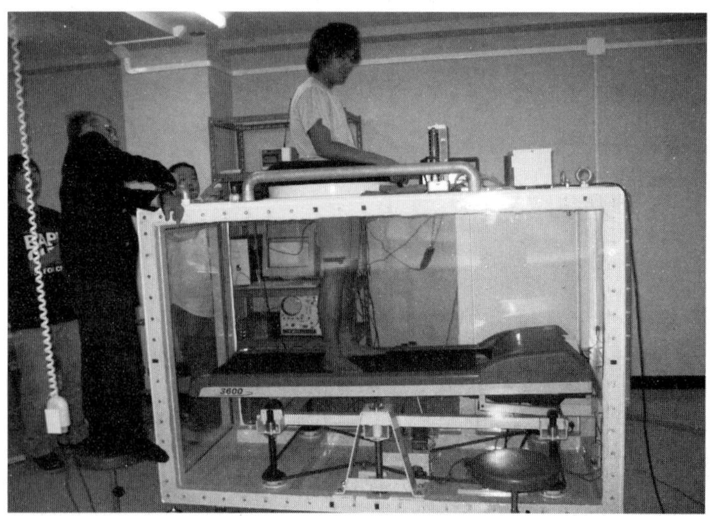

Fig. 1. Photograph of the LBPP chamber. Height of the treadmill is changed by means of a motor beneath the treadmill so that seal height is adjusted to the level of subject's iliac crest. The pressure inside the chamber can be adjusted to a certain pressure by controlling a valve through which air can go out.

atmospheric pressure, the seal bulges and pulls the subject's body upward (Fig. 2b), resulting in a decrease in the net GRF that is the difference between the body weight and the upward force produced by the LBPP. The upward force is proportional to the transmural (inside – outside) pressure multiplied by the cross-sectional area of the subject's body at the waist seal [Cutuk et al., 2006]. Therefore, a high level of the transmural pressure produced by high level of LBPP and/or the large cross-sectional area achieved by using a large waist seal are required in order to obtain a strong upward force. In the present study, we focused on the effect of low level of LBPP on cardiovascular system. Thus, we used a large size of waist seal whose diameter was 60 cm in order to minimize the level of LBPP.

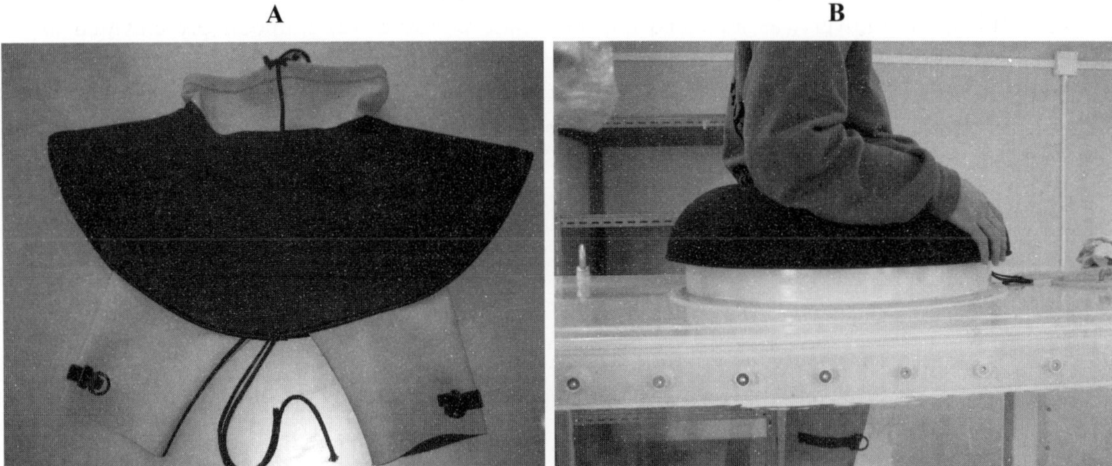

Fig. 2. (A) Photograph of the shorts with the waist seal. (B) The seal bulges and pulls the subject's body upward when the pressure inside the chamber becomes higher than the atmospheric pressure.

Measurements

The GRF was measured using a digital weighing scale. Heart rate was measured using a limb lead ECG monitor, while systolic and diastolic blood pressures were measured at the right arm by means of a manual mercury manometer. Calf skin blood flow was measured with a laser Doppler flow meter (ALF 21; Advance, Tokyo, Japan) and skin temperature was monitored at the right calf utilizing a thermistor (D226; Takara, Tokyo, Japan).

Protocol

On the experiment day, subjects stood on the weighing scale inside the LBPP chamber, and the GRF at atmospheric pressure (body weight) was measured. Then, LBPP was applied in a graded manner at 4.5, 6.0, 7.5, 9.0, 10.5, 12.0, 13.5, 15.0 mmHg, and returned back to 4.5 mmHg, and then atmospheric pressure. Net GRF was measured and recorded at each LBPP level. After the measurement of GRF, the weighing scale was removed and the treadmill height was readjusted to set the waist seal at an adequate level as described before. In the next experiment, subjects stood stationary for 5 min in the LBPP chamber at atmospheric pressure. Heart rate, systolic and diastolic blood pressures, calf skin blood flow, and calf skin temperature were measured during the last 1 minute (control). The inside pressure was then increased to 15 mmHg above the atmospheric pressure within 3 minutes. The subject continued to stand still for 5 more minutes. Data were collected again in the last 1 minute (LBPP).

Data analysis

Statisitical analyses were performed using a paired t test between control (atmospheric pressure) and LBPP. A difference in means was considered significant when $p < 0.05$. All data were expressed as mean ± SEM.

Results

Ground reaction force

In our 10 volunteers, the average body weight was 64.5 ± 1.6 kg (range: 57.1 – 73.0) at atmospheric pressure. When the blower was switched on, the inside pressure was immediately elevated to 4.5 mmHg. The net GRF was reduced by 10.0 kgw with 4.5 mmHg of LBPP. During the graded application of LBPP, the net GRF decreased in a linear manner, reaching 30.9 ± 1.6 kgw at 15.0 mmHg of LBPP (Fig. 3). When the inside pressure was reduced, the net GRF was increased again in a linear manner. The slope for the descending line of the LBPP-GRF relationship was 2.21 kgw/mmHg.

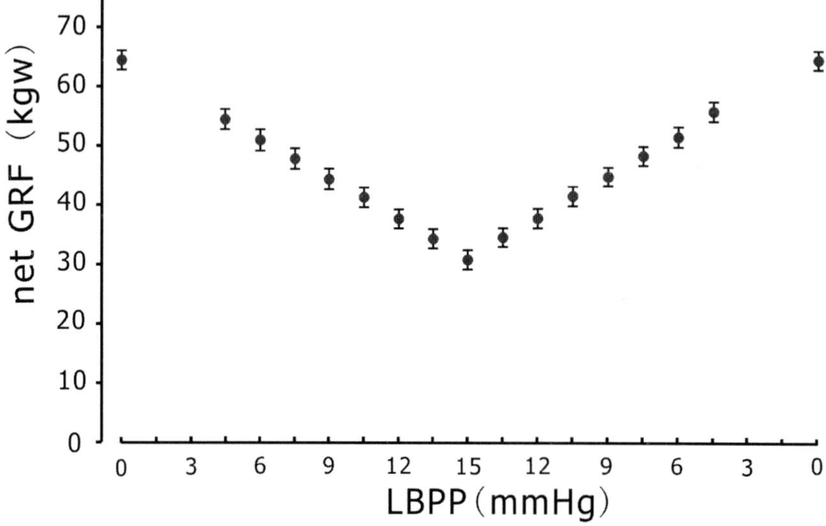

Fig. 3. Relationship between the level of LBPP and the net ground reaction force (GRF). During the graded application of LBPP, the net GRF decreased in a linear manner from 64.5 ± 1.6 kgw at atmospheric pressure to 30.9 ± 1.6 kgw at 15 mmHg of LBPP.

Heart rate and blood pressures

During the control period (atmospheric pressure), subject's heart rate was 76.0 ± 2.8 bpm, and systolic and diastolic blood pressures were 119.6 ± 3.5 mmHg and 78.0 ± 2.3 mmHg, respectively (Fig. 4). After 5 min exposure to 15 mmHg of LBPP, the heart rate, systolic, and diastolic pressures were 76.3 ± 4.1 bpm, 121.0 ± 3.5 mmHg, and 76.4 ± 3.5 mmHg, respectively. There was no significant difference in any of these between control and LBPP conditions.

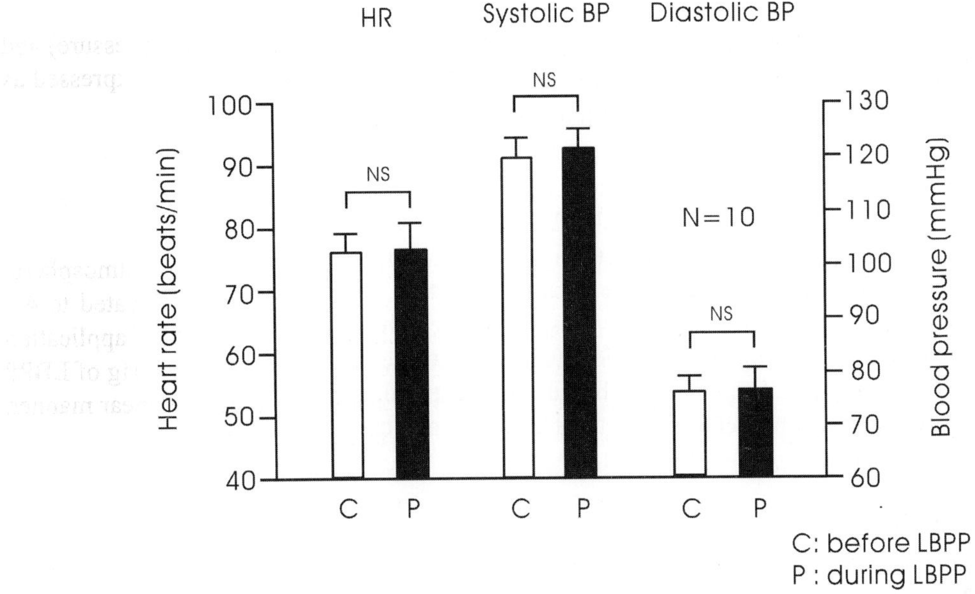

Fig. 4. Heart rate (HR), systolic blood pressure (BP), and diastolic pressure before (white column) and during LBPP (black column). There was no significant difference in any of these between the two conditions.

Calf skin blood flow and skin temperature

Calf skin blood flow during control and LBPP periods were 2.7 ± 0.2 ml/min/100 ml tissue and 2.7 ± 0.2 ml/min/100 ml tissue, respectively (Fig. 5). There was no significant difference between the two conditions. On the other hand, calf skin temperature decreased from 31.7 ± 0.3°C at atmospheric pressure to 31.1 ± 0.3°C during application of LBPP ($p < 0.01$).

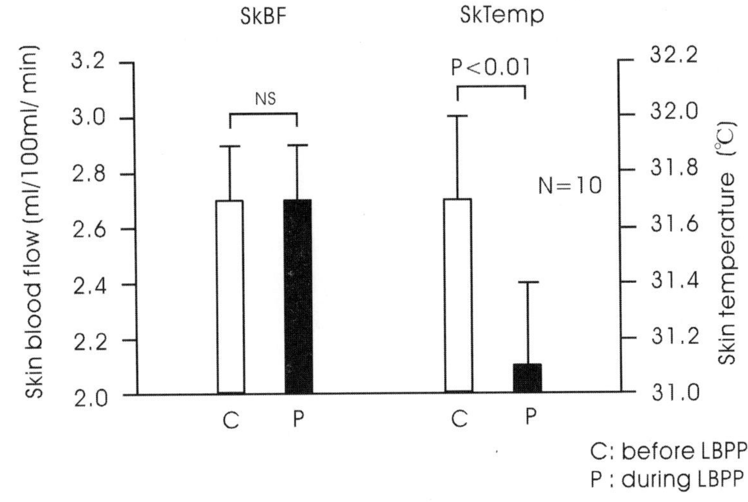

Fig. 5. Calf skin blood flow (left) and skin temperature (right) before (white column) and during LBPP (black column). Skin temperature decreased during LBPP, while skin blood flow did not change significantly.

Discussion

Our results demonstrate that net GRF decreases linearly with upright standing LBPP, indicating an inverse relationship between increasing LBPP and the net GRF (Fig. 3). Our study also confirms our hypothesis that a low level of LBPP causes minimal changes in cardiovascular parameters.

In the present study, relatively low levels of LBPP were applied. Cutuk and coworkers [2006] demonstrated that standing subject's body weight (net GRF) decreased linearly from 84 ± 5 kg at ambient pressure to 13 ± 3 kg at 50 mmHg LBPP, showing a weight reduction of 1.42 kg/mmHg. Application of 15 mmHg LBPP in our chamber, on the other hand, decreased the net GRF by 34 kg, indicating a decrease of 2.21 kg/mmHg. The greater reduction in our LBPP chamber is mainly dependent on the large size of waist seal. As previously described, the upward force produced by LBPP is proportional to the transmural (inside – outside) pressure multiplied by the cross-sectional area of the subject's body at the waist seal. Therefore, the bigger the waist seal, the greater the upward force. We speculate that our waist seal is bigger than that used by Cutuk and coworkers.

It has been shown that application of higher levels of LBPP causes an intravascular fluid shift from lower extremities to torso which in turn increases central venous pressure [Bevegard et al., 1977; Norsk et al., 1986; Shi et al., 1993], central blood volume [Mills and Robertshaw, 1983], and cardiac output [Shi et al., 1993]. On the other hand, application of LBPP increases total peripheral resistance [Fu et al., 1998; Fu et al., 2000]. The increased cardiac output and total peripheral resistance result in an elevation of mean arterial pressure in the supine subjects exposed to LBPP [Nishiyasu et al., 2000; Gallagher et al., 2001; Fu et al., 2001]. In standing subjects, on the other hand, systolic and diastolic blood pressures trended higher during 50 mmHg of LBPP than in ambient pressure but these increases of blood pressures were statistically not significant [Cutuk et al., 2006]. Our results for systolic and diastolic blood pressures are consistent with those reported by Cutuk and coworkers. The discrepancy of these results seems to be dependent on the level of LBPP and the posture difference. Fu and coworkers [2000] demonstrated that mean arterial pressure was elevated significantly by 11% at 30 mmHg LBPP but remained unchanged at low levels of LBPP. It has been also shown that increase in mean arterial pressure due to LBPP was greater in the supine than in the upright position [Nishiyasu et al., 1998].

Our data showed that heart rate did not change significantly during standing LBPP, which is different from the result observed by Cutuk and coworkers [2006] who reported that heart rate decreased from 83 ± 3 bpm at ambient pressure to 73 ± 3 bpm at 50 mmHg LBPP. The discrepancy probably reflects the difference in the level of LBPP. In their study, the decrease in heart rate became statistically significant at LBPP levels greater than 20 mmHg but not at 10 mmHg. These results suggest that standing LBPP at less than 20 mmHg has minimal effect on systemic circulation.

It is also important to observe the effect of LBPP on local circulation. Cutuk and coworkers [2006] investigated cerebrovascular responses to LBPP and showed that head capillary perfusion, blood flow velocity through the middle cerebral artery, and brain oxygenation were unchanged by the application of the 50 mmHg standing LBPP. Application of LBPP decreased calf circumstance [Tanaka et al., 2003] and muscle blood flow in the leg [Rowell et al., 1991]. On the other hand, LBPP decreased vascular resistance [Bevegard et al., 1977] and increased blood flow in the forearm [Schmedtje Jr et al., 1995; Journeay et al., 2003]. To our knowledge, however, no study has examined the effect of LBPP on skin circulation in the lower extremities. Application of positive pressure may compress veins in the skin of lower extremities, which in turn may reduce the skin blood flow and skin temperature. Our results, however, demonstrated that 15 mmHg LBPP did not decrease skin microvascular blood flow measured by a laser Doppler flow meter (Fig. 5). On the other hand, skin temperature in the calf was decreased significantly during application of LBPP. Skin temperature is determined as the difference between heat transported to the skin from the body core by skin blood flow and heat loss from the body surface. Therefore, it is a good indicator for skin blood flow if the heat loss from the body surface is constant. We monitored the air temperature inside the chamber and confirmed that it did not change significantly throughout the experiment. However, during application of LBPP, air was driven into the chamber by a blower and the same amount of air was squeezed out to keep the inside pressure constant, causing air movement around the calf. This air current increases heat loss from the body surface by convection effect, which may explain the decrease in the skin temperature of our subjects.

In conclusion, our LBPP chamber can unload lower extremities with minimal changes in heart rate, blood pressure, and local circulation in the lower extremities. The chamber can effectively provide buoyant force with relatively low level of LBPP by using a large size of waist seal.

Acknowledgment

A part of this study was technically and financially supported by Showa Denki Co., Ltd.

References

1. Berman A, Studenski S. Musculoskeletal rehabilitation. *Clin Geriatr Med.* 1998; 14: 641-665.
2. Bevegard S, Castenfors J, Lindblad LE. Effect of changes in blood volume distribution on circulatory variables and plasma renin activity in man. *Acta Physiol Scand.* 1977; 99: 237-45.
3. Bloomfield SA. Changes in musculoskeletal structure and function with prolonged bed rest. *Med Sci Sports Exerc.* 1997; 29: 197-206.
4. Cutuk AA, Groppo ER, Quigley EJ, White KW, Pedowitz RA, Hargens AR. Ambulation in stimulated fractional gravity using lower body positive pressure: Cardiovascular safety and gait analyses. *J Appl Physiol.* 2006; [Epub ahead of print]
5. Fu Q, Iwase S, Niimi Y, Kamiya A, Kawanokuchi J, Cui J, Mano T, Suzumura A. Effects of lower body positive pressure on muscle sympathetic nerve activity response to head-up tilt. *Am J Physiol.* 2001; 281: R778-R785.
6. Fu Q, Sugiyama Y, Kamiya A, Mano T. A comparison of autonomic responses in humans induced by two simulation models of weightlessness: lower body positive pressure and 6 degrees head-down tilt. *J Auton Nerv Syst.* 2000; 80: 101-107.
7. Fu Q, Sugiyama Y, Kamiya A, Shamsuzzaman AS, Mano T. Responses of muscle sympathetic nerve activity to lower body positive pressure. *Am J Physiol.* 1998; 275(4 Pt 2): H1254-H1259.
8. Gallagher KM, Fadel PJ, Smith SA, Norton KH, Querry RG, Olivencia-Yurvati A, Raven PB. Increases in intramuscular pressure raise arterial blood pressure during dynamic exercise. *J Appl Physiol.* 2001; 91: 2351-2358.
9. Journeay WS, Reardon FD, Kenny GP. Cardiovascular responses to apneic facial immersion during altered cardiac filling. *J Appl Physiol.* 2003; 94: 2249-2254.
10. Mills DE, Robertshaw D. Plasma prolactin responses to acute changes in central blood volume in man. *Horm Res.* 1983; 18: 153-159.
11. Nishiyasu T, Nagashima K, Nadel ER, Mack GW. Effects of posture on cardiovascular responses to lower body positive pressure at rest and during dynamic exercise. *J Appl Physiol.* 1998; 85: 160-167.

12. Nishiyasu T, Nagashima K, Nadel ER, Mack GW. Human cardiovascular and humoral responses to moderate muscle activation during dynamic exercise. *J Appl Physiol*. 2000; 88: 300-307.
13. Norsk P, Bonde-Petersen F, Warberg J. Influence of central venous pressure change on plasma vasopressin in humans. *J Appl Physiol*. 1986; 61: 1352-1357.
14. Prins J, Cutner D. Aquatic therapy in the rehabilitation of athletic injuries. *Clin Sports Med*. 199; 18: 447-461.
15. Rowell LB, Savage MV, Chambers J, Blackmon JR. Cardiovascular responses to graded reductions in leg perfusion in exercising humans. *Am J Physiol*. 1991; 261(5 Pt 2): H1545-H1553.
16. Schmedtje JF Jr, Gutkowska J, Taylor AA. Reciprocity of hemodynamic changes during lower body negative and positive pressure. *Aviat Space Environ Med*. 1995; 66: 346-52.
17. Shi X, Potts JT, Foresman BH, Raven PB. Carotid baroreflex responsiveness to lower body positive pressure-induced increases in central venous pressure. *Am J Physiol*. 1993; 265(3 Pt 2): H918-H922.
18. Tanaka K, Gotoh TM, Morita H, Hargens AR. R-R interval variability with lower body positive pressure assessed by wavelet packet transform. *Biol Sci Space*. 2003 Oct; 17(3): 203.

Adaptation Biology and Medicine (Volume 5: Health Potentials)
Editors: L. Lukyanova, N. Takeda and P.K. Singal

4. Nitric Oxide Synthase 3 Mediates Cardioprotection by Hyperoxia and Hyperbaria

Bernadette P. Cabigas[1] and John E. Baker[2*]

[1]Department of Hyperbaric Medicine, St. Luke's Medical Center, Milwaukee, WI 53215
[2]Division of Cardiothoracic Surgery, Medical College of Wisconsin, Milwaukee, WI, 53226

Abstract

The relative contributions of the fraction of inspired oxygen (F_IO_2) and atmospheric pressure (ATM) to cardioprotection are unknown. We determined whether the product of $F_IO_2 \times$ ATM (oxygen partial pressure) controls the extent of hyperoxic + hyperbaric-induced cardioprotection and involves activation of nitric oxide synthase (NOS). Adult Sprague Dawley rats (n=10/gp) were treated for 1 hour with (1) normoxia + normobaria (21% O_2 at 1 ATM), (2) hyperoxia + normobaria (100% O_2 at 1 ATM), (3) normoxia + hyperbaria (21% O_2 at 2 ATM) and (4) hyperoxia + hyperbaria (100% O_2 at 2 ATM). Infarct size following 25 min ischemia and 180 min reperfusion was decreased following hyperoxia + normobaria and normoxia + hyperbaria compared with normoxia + normobaria and further decreased following hyperoxia + hyperbaria treatment. L-NAME (200 μM) reversed the cardioprotective effects of hyperoxia + hyperbaria. Nitrite plus nitrate content was increased 2.2 fold in rats treated with normoxia + hyperbaria and hyperoxia + hyperbaria. NOS3 protein increased 1.2 fold and association of hsp90 with NOS3 four fold in hyperoxic + hyperbaric rats. Cardioprotection conferred by hyperoxia + hyperbaria is directly dependent on oxygen availability and mediated by NOS.

Keywords: HBO, cardioprotection, NOS, hyperoxia, hyperbaria and antiarrythmic.

Introduction

Cardiovascular disease remains a major health care problem in the United States affecting the outcome of both cardiac and non-cardiac surgery. Pharmacological or physical preconditioning protects the heart against subsequent lethal ischemia. However, there are no current therapies that have proven to directly protect the heart against the deleterious effects of ischemia in humans. Therefore other strategies to protect the heart against ischemia that can be applied clinically need to be evaluated. Pretreatment of the heart with hyperbaria (increased atmosphere pressure) prior to ischemia limits post-ischemic infarct size (1). In addition, pretreating hearts with hyperoxia (increased fraction of inspired oxygen) prior to ischemia improves post-ischemic cardiac function and reduces infarct size (2). The

*Address for Correspondence: John E. Baker, Ph.D., Division of Cardiothoracic Surgery, Medical College of Wisconsin, 8701 Watertown Plank Road, Milwaukee, WI 53226. Tel: (414) 456-8706, Fax: (414) 453-9700, E-mail: jbaker@mcw.edu

product of F_IO_2 and atmospheric pressure determines the partial pressure of oxygen in a given volume as defined in the equation:

Partial pressure of oxygen (ATM) = $F_IO_2 \times$ atmospheric pressure (ATM)

However the relative contribution of F_IO_2 and atmospheric pressure to cardioprotection and the underlying mechanisms are unknown.

Hyperbaric oxygenation (HBO) is a mainstay in the treatment of several conditions including carbon monoxide poisoning, decompression sickness and impaired oxygen delivery (3). Other benefits of clinical HBO therapy include defibrillation, enhancement of the efficacy of various drugs used in conjunction with HBO therapy, reduction of anginal attacks, relief of dyspnea, lowered cholesterol levels, normalized ECGs, antiarrhythmic effects, stabilization of impaired fat metabolism and improvement of liver function (4, 5). However repeated exposure to hyperbaria + hyperoxia (5 ATM at 100% O_2) increases sensitivity to seizures in rats (6). Nitric oxide (·NO) may be responsible for the signaling mechanism involved in brain toxicity resulting from treatment with hyperbaria + hyperoxia because NOS inhibitors are able to delay the onset of seizures. Elevations in oxygen tension at atmospheric pressures between 2.0 and 2.5 above ambient air also result in an increase in ·NO production by pulmonary endothelial cells, lungs and the cerebral cortex. Angiogenesis and downregulation of intercellular adhesion molecule (ICAM) expression caused by hyperbaric oxygen treatment have been attributed to changes in ·NO concentration (7). In human umbilical vein endothelial cells and bovine aortic endothelial cells, HBO treatment for 90 min following hypoxia/hypoglycemia abolished ICAM protein expression, polymorphonuclear adhesion and increased ·NO production (7). Hyperoxic-induced cardioprotection is mediated in part through a nuclear factor êB pathway (8) which may activate nitric oxide synthase (NOS).

Several studies indicate that ·NO plays a central role in cardioprotection (9, 10). Regulation of ·NO production from NOS occurs by multiple mechanisms including interaction with heat shock protein 90 (hsp90). Increases in hsp90 association with NOS3 play a critical role in increasing cardioprotection (11). In rat cerebral cortex exposed to (2.8 ATM) an elevation in ·NO concentration occurred after 45 minutes of exposure. The amount of NOS protein in brain remained unchanged but the association of NOS1 with hsp90 was increased two fold (12). Taken together these studies suggest hyperbaria increases ·NO production which then plays a role in ischemia/reperfusion injury in major organ systems. Heme oxygenase-1 (HO-1) belongs to the heat shock protein family. HO-1 converts heme into bilirubin iron and carbon monoxide, and is known to have cell-protective properties (13, 14). Therefore hyperoxia + hyperbaria may activate HO-1 and confer cardioprotection. However the role of NOS-hsp90 and HO-1 in mediating hyperbaric- and hyperoxic-induced cardioprotection at pressures that do not result in seizures is unknown. We hypothesized that the partial pressure of oxygen determines the extent of immediate cardioprotection with increased resistance to ischemia mediated by nitric oxide synthase and heme oxygenase.

The objectives of our study were to determine (i) whether the product of F_IO_2 and ATM determine the extent of hyperoxic- and hyperbaric-induced cardioprotection and (ii) the role of NOS and HO-1 in the mechanism by which hyperoxia and hyperbaria confers cardioprotection.

Materials and Methods

Animals

Male Sprague Dawley rats, *Rattus norvegicus*, at 8-9 weeks of age maintained on a customized 3075S rodent diet (TEKLAD, Wilmington, DE) were used in these studies. The animals were kept on a 12 hour light/dark cycle. All animal work was conducted in accordance with the Guide for the Care and Use of Laboratory Animals published by the US National Institutes of Health (NIH Publication No. 85–23, revised 1996) and approved by the Institutional Animal Care and Use at the Medical College of Wisconsin.

Cardioprotection with Hyperoxia + Hyperbaria

We performed the following experiments in random order to investigate whether the product of F_IO_2 and atmospheric pressure determines the extent of cardioprotection. Rats were exposed to hyperoxia (100% O_2) + normobaria (1 ATM), normoxia (21% O_2) + hyperbaria (2 ATM) or hyperoxia (100% O_2) + hyperbaria (2 ATM) for 1 hour. Rats exposed to normoxia (21% O_2) + normobaria (1 ATM) for 1 hour served as untreated controls. The animals were then allowed to recover in room air for 10 minutes. Resistance to myocardial ischemia was then determined using an isolated perfused heart model (15).

Normoxia + Normobaria treatment (21% O_2, 1 ATM)

All animals were placed inside the experimental hyperbaric chamber at the Medical College of Wisconsin (756.6 L volume capacity). The interior of the chamber was illuminated to maintain rats in the light portion of the light/dark cycle. Atmospheric pressure measurements were obtained each day from the National Weather Service for the city of Milwaukee, WI and the oxygen concentration of room air was measured using an oxygen sensor located in the exhaust valve of the hyperbaric chamber. Carbon dioxide levels in the chamber were measured using a catheter placed inside the hyperbaric chamber and exteriorized for measurement using a POET gas analyzer. Control animals underwent sham treatment for 1 hour in the chamber by remaining at ambient atmospheric pressure and oxygen concentration. The amount of time spent in the chamber for the control animals was the same as that for the paired treated animals. Following sham treatment animals were then removed from the chamber for a 10-minute recovery period. During this recovery period, heparin (150 U/kg) and sodium pentobarbital (50 mg/kg) were administered intraperitoneally. The chest was then opened by a median sternotomy, the heart excised and perfused.

Hyperoxia + Normobaria treatment (100% O_2, 1 ATM)

Rats were placed in the hyperbaric chamber. Once sealed, 750L of 100% oxygen was added over a six-minute period with the exhaust valve kept open to prevent any increases in chamber pressure within the chamber. After this volume was added, a final measurement of the oxygen concentration was taken to confirm hyperoxic conditions were present. All intake and exhaust valves were then closed and environmental conditions maintained for 60 minutes. After 66 minutes the door to the chamber was then opened and the animals were allowed to recover for 10 minutes.

Normoxia + Hyperbaria treatment (21% O_2, 2 ATM)

Rats were placed in the hyperbaric chamber and an additional 750L of 21% oxygen was introduced into the chamber over a 2-minute period to raise atmospheric pressure to 2 ATM. This elevated pressure was confirmed by the pressure gauge located outside the hyperbaric chamber. The intake valve was sealed with atmospheric pressure monitored every 15 minutes for 1 hour. When the pressure fell below 1.95 ATM, additional oxygen was added to the chamber *via* the intake valve to restore atmospheric pressure to 2 ATM. After 66 minutes, the chamber was decompressed to a pressure of 1.45 ATM over 35 seconds. This pressure was maintained for three minutes. A final decompression to atmospheric pressure was performed over an additional 35 seconds so that the total decompression time equaled 4 minutes and 10 seconds. The chamber door was opened and the rats were allowed to recover for 10 minutes before excision of the heart.

Hyperoxia + Hyperbaria treatment (100% F_IO_2, 2 ATM)

Rats were placed in the hyperbaric chamber and 750L of 100% oxygen was introduced into the chamber over a 2-minute period to raise atmospheric pressure to 2 ATM. Rats were then treated in an identical manner to that described for the normoxia + hyperbaria group.

Isolated heart perfusion

Isolated rat hearts were perfused retrogradely with bicarbonate buffer at constant perfusion pressure (98 mmHg) with a balloon inflated in the left ventricle. End-diastolic pressure was set to 5 mmHg and developed pressure was recorded during steady-state conditions (15). Hearts were allowed to beat spontaneously. Hearts were kept in temperature-controlled chambers to maintain myocardial temperature at 37°C during periods of perfusion and ischemia.

Infarct size determination

Following three hours of reperfusion, 1.0% (w/v) 2,3,5 Triphenyl-Tetrazolium Chloride (TTC) was delivered to the isolated heart *via* a syringe attached to the 3-way stopcock. TTC was administered at a rate of 1ml/min for 10 minutes. The left ventricle was then isolated and sliced into 1mm cross-sectional segments. The tissues were then incubated for 15 min in a 1% TTC solution in 100 mM phosphate buffer (pH 7.4) at 37°C. Tissues were then stored in 10% formaldehyde overnight and the extent infarcted myocardium was measured. The infarcted tissue was then separated from the non-infarcted tissue and measured as a percentage (% Infarct=Infarct Area/Area at Risk) × 100 using a fluorescent, digital, color camera with a 55mm lens (Nikon). Imaging analysis software (Imaging Research, St. Catharines, ON) was programmed to recognize infarcted tissue (scan area) in proportion to entire heart (total target area) (15).

Role of NOS

To determine whether cardioprotection induced by hyperoxia + hyperbaria is mediated by ·NO, isolated hearts from rats treated with normoxia + normobaria, normoxia + hyperbaria and hyperbaria + hyperoxia were perfused with L-NAME, a general NOS inhibitor (200 μM) for 15 minutes prior to 25 minutes global ischemia and 3 hours reperfusion. Resistance to myocardial ischemia was then determined.

Generation of Nitrite plus Nitrate

Hearts from rats treated with normoxia + normobaria, normoxia + hyperbaria and hyperoxia + hyperbaria were perfused with bicarbonate buffer for 15 minutes. Nitrite and nitrate content of the coronary effluent and heart tissue were determined as previously described using an ozone chemiluminescence method (15, 16).

Western analysis and immunoprecipitation studies

Western analysis and immunoprecipitation of heart homogenates from rats treated with normoxia + normobaria and hyperoxia + hyperbaria for NOS1, NOS2, NOS3, hsp90 and HO-1 were performed using the method we described previously (11). Antibodies used were polyclonal rabbit NOS1, NOS2 and NOS3 (Santa Cruz Biotechnology, Cat #SC-648, Cat #SC-651 and Cat #SC-654, monoclonal hsp90 (BD Transduction, Cat #610419) and monoclonal HO-1 (Stressgen, Cat #OSA110), respectively. For the immunoprecipitation studies we used an H32 NOS3 monoclonal antibody raised in mouse (Biomol, Plymouth Meeting, PA, Cat #SA246). The secondary antibody used for NOS3 was goat anti-rabbit IgG Horseradish peroxidase conjugate (Bio-Rad Laboratories, Cat #170-6515). The secondary antibody used for hsp90 was goat anti-mouse IgG Horseradish peroxidase conjugate (Sigma, Cat #A-8924).

Statistical analysis

Statistical analysis was performed by use of repeated-measures analysis of variance with the Greenhouse-Geisser adjustment used to correct for the inflated risk of a type I error. If significant, the Mann-Whitney test was used as a second step to confirm which groups were significantly different (15). Significance was set at $P<0.05$.

Results

Oxygen and cardioprotection studies

To investigate whether the product of F_IO_2 and ATM determines the extent of hyperbaric- and hyperoxic-induced cardioprotection, rats were treated for 1 hour with (1) normoxia + normobaria (21% O_2 at 1 ATM), (2) hyperoxia + normobaria (100% O_2 at 1 ATM), (3) normoxia + hyperbaria (21% O_2 at 2 ATM) or (4) hyperoxia + hyperbaria (100% O_2 at 2 ATM). Following the addition of oxygen to the hyperbaria chamber, carbon dioxide levels fell from 0.03% to levels that were undetectable. Over the subsequent 60 minute treatment period, carbon dioxide levels rose to a maximum value of 0.001%. We did not observe seizures in any of the four experimental groups. Immediately following treatment and recovery the heart was excised and perfused. Aerobic cardiac function prior to ischemia was unaffected by treatment with hyperoxia + normobaria, normoxia + hyperbaria and by hyperoxia + hyperbaria compared with normoxia + normobaria (heart rate = 260 ± 10 beats/min; coronary flow rate = 12 ± 2 ml/min/g; left ventricular developed pressure = 118 ± 20 mmHg). Thus there were no changes in pre-ischemic function as a consequence of hyperoxic/hyperbaric treatment that were predictive of post-ischemic recovery. Hearts were then subjected to 25 minutes global no flow ischemia followed by 3 hours of reperfusion. Recovery of left ventricular developed pressure (LVDP) and cumulate lactate dehydrogenase leakage over the first hour of reperfusion and infarct size/area at risk (% of left ventricle) at 3 hours reperfusion were used to assess resistance to ischemia. Increasing the product of $F_IO_2 \times$ ATM conferred cardioprotection in a linear manner. In untreated hearts (normoxia + normobaria), infarct size (mean ± SD, %LV) was 13 ± 2% and percent recovery of LVDP was 34 ± 7%. Treatment with hyperoxia + normobaria reduced infarct size by 31% to 9 ± 2% and increased recovery

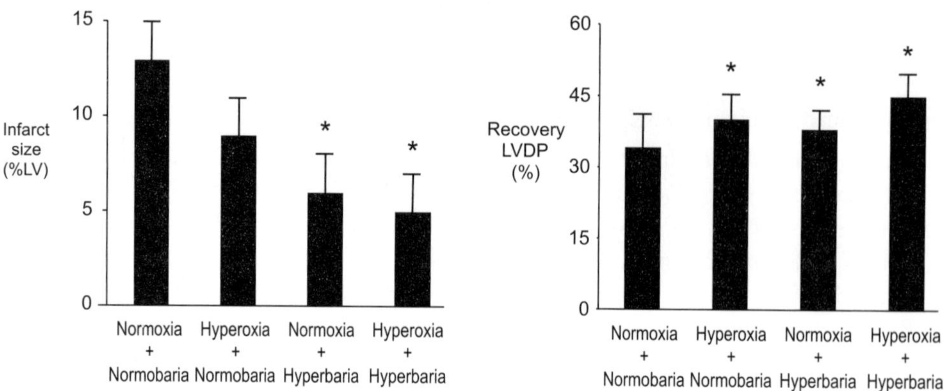

Fig. 1. Acute cardioprotection conferred by hypoxia and hyperbaria. Data are mean ± SD, n = 10 hearts/group. * = p<0.05, vs normoxia + normobaria.

of LVDP to 40 ± 11% (Figure 1). Treatment with normoxia + hyperbaria reduced infarct size by 46% to 7 ± 3% and increased recovery of LVDP to 38 ± 4% (Figure 1). Treatment with hyperoxia + hyperbaria further reduced infarct size by 62% to 5 ± 2% and increased recovery of LVDP to 45 ± 8% (Figure 1) compared with untreated rats. Coronary flow was re-established within 10-15 seconds of reperfusion and recovered to 80-90% of preischemic values in all experimental groups. We then subjected hearts to 45 minutes of ischemia to determine if hyperoxic + hyperbaric treatment protects the heart against injury from a more severe ischemic insult. In normoxic + normobaric treated rats, infarct size and recovery of LVDP were 56 ± 4% and 15 ± 5%, respectively. Hyperoxic + hyperbaric treatment decreased infarct size to 46 ± 4% and increased recovery of LVDP to 24 ± 4%. Our results suggest the product of F_IO_2 and ATM determines the extent of cardioprotection.

Hyperbaric + hyperoxic and normobaric + normoxic treated rats subjected to 25 minutes of myocardial ischemia demonstrated a low level of leakage of lactate dehydrogenase prior to ischemia (0.20 ± 0.06 IU/g and 0.15 ± 0.3 IU/g, respectively). Post-ischemic enzyme leakage was increased in both hyperbaric + hyperoxic and normobaric + normoxic groups (5.1 ± 1.3 IU/g and 5.9 ± 1.7 IU/g, respectively) compared with pre-ischemic values. However there were no differences in post-ischemic enzyme leakage between treatment groups.

Role of NOS in Oxygen-Induced Cardioprotection

To determine whether enhanced ·NO production from NOS is responsible for increased resistance to ischemia in hearts from animals treated with normoxia + hyperbaria and hyperoxia + hyperbaria, isolated hearts were perfused with a general NOS inhibitor (L-NAME, 200 μM) for 15 minutes prior to 25 minutes global ischemia. L-NAME decreased coronary flow rate from 13 ± 2 ml/min/g to 9 ± 2 ml/min/g in hearts treated with normoxia + hyperbaria and from 12 ± 2 ml/min/g to 9 ± 2 ml/min/g in hearts treated with hyperoxia + hyperbaria. Hearts were reperfused for 3 hours. L-NAME increased infarct size in hearts from rats treated with normoxia + hyperbaria (6 ± 2%) and hyperoxia + hyperbaria (11 ± 2%) to that observed in hearts from untreated (normoxia + normobaria) rats (13 ± 2%), but did not alter infarct size in hearts from normoxic + normobaric rats (11 ± 2%) (Figure 4). L-NAME

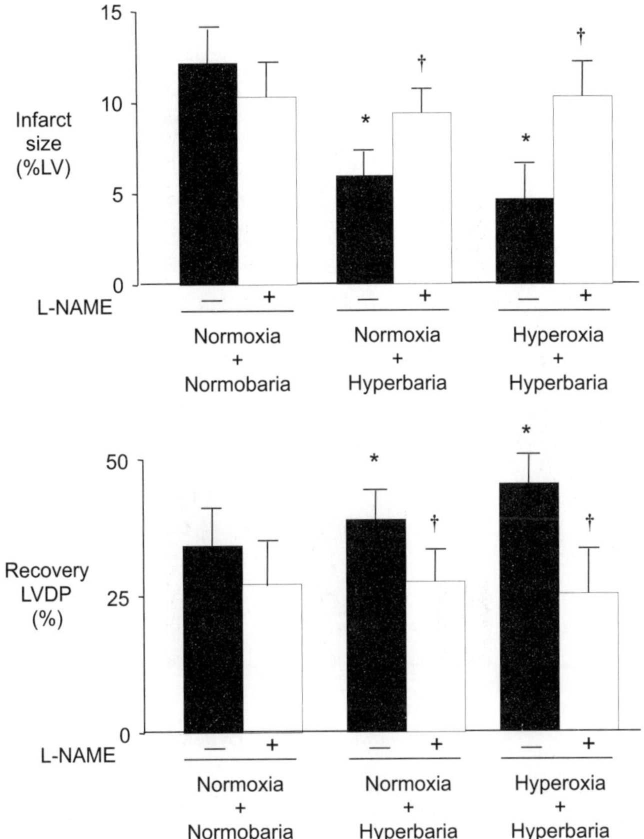

Fig. 2. Nitric oxide generation and acute cardioprotection conferred by hyperoxia + hyperbaria. Data are mean ± SD, n = 10 hearts/group. L-NAME (200 μM). * = p <0.05, vs normoxia + normobaria, † = p <0.05, with L-NAME vs without L-NAME.

decreased recovery of LVDP in hearts from rats treated with hyperoxia + hyperbaria (25 ± 8%) and normoxia + hyperbaria (26 ± 6%) but had no effect on hearts from rats treated with normoxia + normobaria (27 ± 4%) (Figure 2). Thus resistance to ischemia in hearts from rats treated with normoxia + hyperbaria and hyperoxia + hyperbaria is decreased by inhibition of NOS to levels present in hearts from untreated (normoxia + normobaria) rats.

Nitrite plus Nitrate Production

The data from the NOS inhibition studies suggest that hearts from rats treated with normoxia + hyperbaria and hyperoxia + hyperbaria produce more ·NO than hearts from untreated rats (normoxia + normobaria). To determine the extent to which normoxia + hyperbaria and hyperoxic + hyperbaric treatment increased ·NO production, hearts were analyzed for nitrite plus nitrate content and nitrite plus nitrate leakage (15). Nitrite plus nitrate levels in the coronary perfusates were approximately 0.2 μM. We found that in hearts previously treated with normoxia + hyperbaria and hyperoxia + hyperbaria

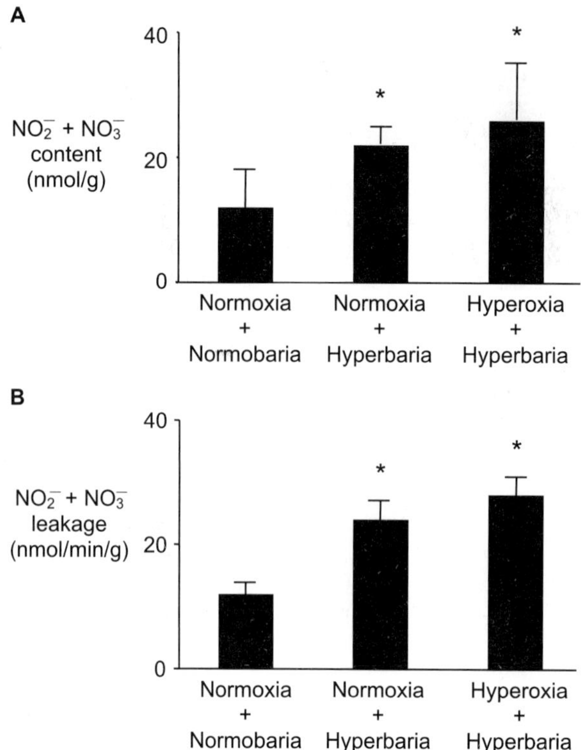

Fig. 3. Hyperoxia + hyperbaria induced $NO_2^- + NO_3^-$ formation in heart. Panel A shows $NO_2^- + NO_3^-$ content in heart tissue. Panel B shows $NO_2^- + NO_3^-$ leakage from heart into coronary effluent. Data are mean ± SD, n = 8 hearts/group. * = $p < 0.05$, normoxia + normobaria vs hyperoxia + hyperbaria.

tissue levels for nitrite plus nitrate, and leakage of nitrite plus nitrate from the heart were increased 2.2- and 2.3-fold respectively compared with normoxic + normobaric treated hearts (Figure 3). Our observation suggests that nitrite plus nitrate production by the heart is increased more than 2-fold in rats treated with normoxia + hyperbaria and hyperoxia + hyperbaria.

Protein Levels for NOS1, NOS2 and NOS3

To determine whether increased nitrite plus nitrate production in hearts treated with hyperoxia + hyperbaria was related to increased expression of NOS, we examined protein levels for NOS1, NOS2 and NOS3. Using standard Western analysis, the only NOS isoform we were able to detect was NOS3. NOS1 and NOS2 could not be detected by Western analysis. Hearts from BN/Mcw rats were used as positive controls for the Western analysis for NOS3. Treatment of rats with hyperoxia + hyperbaria increased NOS3 protein expression by 22% compared with untreated (normoxia + normobaria) rat hearts (Figure 4A). Our data suggest that the increase in ·NO production in hearts treated with hyperoxia + hyperbaria was due to increased NOS3 protein expression.

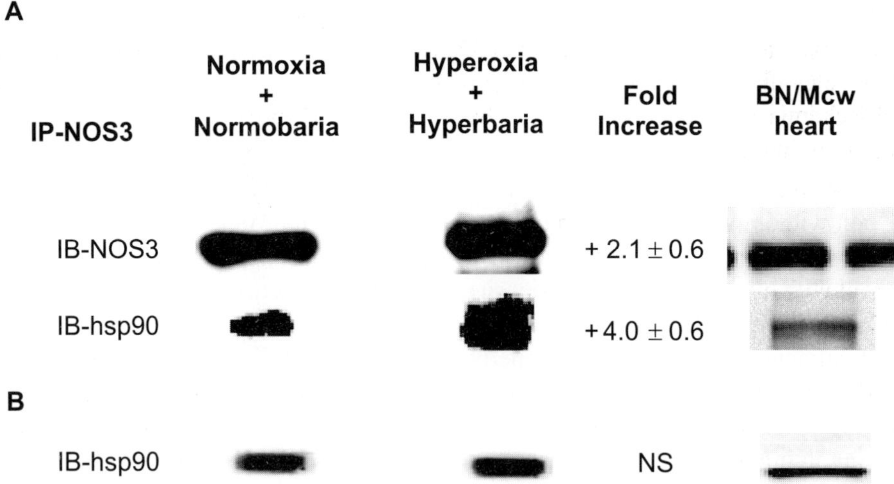

Fig. 4. Western analysis of NOS3 and hsp90 expression and association in heart homogenates from rats treated with normoxia + normobaria and hyperoxia + hyperbaria. (A) Western analysis for NOS3 content and hsp90 associated with NOS3 by immunoprecipitating NOS3 and probing for NOS3 and hsp90. (B) Western analysis for hsp90 content in homogenates of hearts from rats treated with normoxia + normobaria and hyperoxia + hyperbaria. IP, immunoprecipitation; and IB, immunoblot. Heart from BN/Mcw rat used as positive controls. N=4-6/group.

Hsp90 and its Association with NOS3 Protein

·NO production from NOS3 may be regulated by a protein-protein interaction. Since hsp90 increases ·NO generation from NOS3 (17), we then determined whether increased resistance to ischemia in hearts treated with hyperoxia + hyperbaria was related to the extent of hsp90 association with NOS3 in comparison with hearts from normoxic + normobaric-treated rats. The extent of hsp90 association with NOS3 was measured immediately prior to the onset of ischemia. Hyperoxia + hyperbaria increased the association of hsp90 with NOS3 compared with normoxic + normobaric hearts by 4-fold (Figure 4A). To determine if the increase in association of hsp90 with NOS3 was due to changes in hsp90 content, Western analysis of hsp90 in total heart homogenate was performed. Hyperoxic + hyperbaric treatment did not change the total content of hsp90 in the heart (Figure 4B). This increased association of hsp90 with NOS3 supports the notion that hsp90 helps NOS3 produce ·NO, which is responsible for the observed cardioprotection against ischemia/reperfusion injury seen in hyperoxic + hyperbaric treated rats.

Protein Levels for HO-1

The protein content of HO-1 in the heart was determined by SDS-PAGE and standard Western analysis using a monoclonal antibody specific for the protein. Purified rat HO-1 protein (Stressgen, SPP730) was used as a positive control. We did not detect the expression of HO-1 protein in either experimental group (data not shown). Thus HO-1 does not appear to play a role in cardioprotection conferred by hyperoxia + hyperbaria in our model.

Discussion

Our study demonstrates that the product of $F_IO_2 \times$ ATM determines the extent of cardioprotection conferred by elevation of the partial pressure of oxygen. Furthermore increased resistance to myocardial ischemia immediately conferred by hyperoxic + hyperbaric treatment is mediated by a mechanism involving increased ·NO production from NOS3.

During hyperoxic + hyperbaric treatment, the vascular supply of oxygen to the heart exceeds demand. Upon withdrawal from the hyperoxic + hyperbaric environment pO_2 reverts to normoxic + normobaric levels immediately. Increased resistance to myocardial ischemia is observed immediately after treatment with hyperoxia + hyperbaria, suggesting that induction of new genes are not necessary for the cardioprotective effects of hyperoxia + hyperbaria to be manifested. Indeed HO-1 was not induced by treatment with hyperoxia + hyperbaria. Therefore the myocardium must utilize existing proteins as the protective mechanism in its response to the increased partial pressure of oxygen. Our results indicate that the heart appears to adapt to short-term increases in oxygen availability by a mechanism involving an increased protein-protein interaction of hsp90 with NOS3 to increase ·NO levels, which confers increased resistance to ischemia. Increased cardioprotection could not be attributed to elevated levels of carbon dioxide as the concentration of this gas did not exceed 0.001% during treatment of rats with hyperoxia. Our findings may enhance understanding of the cardioprotective mechanisms resulting from adaptation to hyperbaria + hyperoxia in the adult heart, and may lead to the development of new strategies to protect hearts of patients undergoing surgical repair for acquired heart disease.

Our results confirm and extend previous studies of exposure of rats and mice to increased F_IO_2 to decrease infarct size and increase recovery of developed pressure following comparable periods of global ischemia in the isolated perfused heart model (2, 18, 19). In a study of the response to hyperoxic exposure between rats and mice (19), one hour of exposure to 80% or >95% oxygen resulted in greater cardioprotection as manifested by increased recovery of LVDP and a reduction of infarct size in the rat. Even 24 hours after the hyperoxic exposure, isolated rat hearts still displayed the beneficial effects of treatment. In contrast, cardioprotection in the mouse occurred after a shorter exposure time and yielded reduced beneficial effects in infarct size (35% vs 51% in rats) when compared with the rat model (19). In general, tolerance to hyperoxia is species-dependent with smaller animals less resistant to this environmental stress, which may partly explain the varying cardioprotective effects. Hyperoxia alone is an attractive mode of treatment for patients with ischemic myocardial disease in that breathing hyperoxic gas produces the same result as ischemic pre-conditioning without subjecting the patient to transient ischemic conditions, which may cause additive injury.

Our study demonstrates hyperoxia (100% O_2) alone confers cardioprotection as evidenced by a 31% reduction in post-ischemic infarct size and an 18% increase in recovery of LVDP. Furthermore the combination of hyperoxia + hyperbaria (2.0 ATM) confers increased cardioprotection with a further reduction in infarct size by 62% and an increase in recovery of LVDP by 32%. However there were no differences in LDH leakage between treated and untreated groups. Reperfusion of the heart after global ischemia leads to areas of “no reflow.” This may underestimate the extent of enzyme leakage from irreversibly injured myocardium to account for the absence of differences between groups. The underlying mechanism appears to involve ·NO as its oxidation product (nitrite plus nitrate) is increased 2 fold after hyperoxic + hyperbaric oxygen treatment compared with normoxic + normobaric controls. Also when L-NAME (a general NOS inhibitor) was administered, the observed cardioprotective effects following treatment with hyperoxia + hyperbaria were lost. In addition, NOS3 expression and NOS3 association with hsp90 was increased after 1 hour treatment with hyperoxia + hyperbaria. We suggest the mechanism for cardioprotection conferred by hyperoxic + hyperbaric treatment might involve an increase in NOS3 protein expression, which results in increased ·NO levels in the heart to confer protection against subsequent ischemia/reperfusion injury. ·NO has to interact with a cellular target for its physiological effect to be manifested. The heme site of guanylyl cyclase is one such target for ·NO and this interaction results in the conversion of guanosine-5'-triphosphate to the intracellular second messenger, cyclic guanosine-3', 5'-monophosphate (cGMP). Soluble guanylyl cyclase plays a pivotal role in the transduction of intercellular messages conveyed by ·NO. ATP-dependent potassium (K_{ATP}) channels are important mediators of protection against ischemia-reperfusion injury in myocardium (20). Thus hyperoxia + hyperbaria-induced increases in ·NO levels may activate K_{ATP} channels by a cGMP-dependent mechanism (21) to confer cardioprotection.

Under conditions of normoxia + normobaria, oxygen supply to the heart equals oxygen demand. We previously showed that under conditions of normobaric hypoxia, where oxygen supply becomes limiting despite increases in hematocrit to increase oxygen delivery, the heart increases production of ·NO from NOS3 through increased association with hsp90, which confers increased resistance to ischemia (11). Our current study reveals that under conditions of hyperoxia + hyperbaria, the heart also increases production of ·NO from NOS3 by increasing NOS3 association with hsp90 to confer increased resistance to subsequent myocardial ischemia. Taken together our studies suggest that NOS3 in the heart responds to changes in oxygen availability as an adaptive response that confers protection against a subsequent ischemic insult. Increased protein-protein association between hsp90 and NOS3 may represent a universal mechanism for controlling ·NO production from NOS3.

Thom and colleagues showed that hyperbaria causes an increase in ·NO synthesis in brain and abdominal great vessels as a stress response mechanism (12, 22). Badr and associates (23) also reported that hyperbaric oxygen therapy within 6 hours of reperfusion injury exerts a protective effect on rats with cerebral infarction. Our study extends these two previous observations to the heart and suggests a role for NOS in mediating cardioprotection conferred by hyperoxia + hyperbaria. Furthermore, our study suggests an elevation in the partial pressure of oxygen increases the association of hsp90 with NOS3 in the heart. Our findings contrast with studies in the brain by Thom *et al.* (12) who demonstrated the association between hsp90 and NOS1 was increased in brain following an elevation in the partial pressure of oxygen. However hyperbaria did not change the association of NOS3 with hsp90. We did not observe a role for hsp90 association with NOS1 in heart following hyperoxia + hyperbaria treatment.

Cardioprotection conferred by increased partial pressure of oxygen suggests this diatomic molecule acts more as a pharmaceutical rather than a support for cellular metabolism. Indeed under high pressure, oxygen has been suggested to behave as a drug (3). Coronary artery bypass surgery and coronary angioplasty remain as clinical treatments to treat myocardial ischemia. The ability to pharmacologically precondition the heart against ischemia with diatomic oxygen may offer a more practical way of harnessing protective mechanisms endogenous to the heart. Oxygen is already used in humans, so safety and toxicity profiles are established. In patients with angina pectoris, treatment with HBO and tissue plasminogen activator (TPA) alleviated myocardial infarct symptoms 6.5 hours faster than in patients who received TPA alone (5). These patients also reported relief from chest pain within

10 minutes of exposure to hyperbaria. The underlying mechanism of action may be due to increased oxygen dissolved in the plasma, which helps the damaged heart in oxygenating the body, while at the same time promotes increased collateral circulation. The potential benefit afforded to patients undergoing bypass surgery and transplantation seems to be logical and feasible since the donor or recipient hearts are available for HBO treatment prior to these surgical procedures.

All these findings demonstrate how HBO treatment is advantageous not only in attenuating ischemia/reperfusion injury in the laboratory but also in various clinical scenarios such as angina pectoris, carbon monoxide poisoning and decompression sickness. We suggest it would be worthwhile to further investigate the role of HBO treatment in increasing nitric oxide, which then leads to cardioprotection. Our findings have clinical implications for patients with ischemic heart disease (IHD) and those undergoing surgery for the treatment of IHD. However hyperbaric oxygen (HBO) treatment is not used as routinely as hyperoxia (>95% F_IO_2) in the treatment of heart disease despite the known benefits of high-pressure oxygen therapy. This may be related to lack of information and/or access to hyperbaric treatment chambers.

As delineated by Kindwall (3) and the Undersea and Hyperbaric Medical Society website (24), there exist a variety of clinical applications for HBO therapy. However in many cases HBO therapy remains clinically marginalized because of the lack of large prospective randomized outcome studies and convincing basic science to support the subcellular mechanism of action. In particular, cardiac surgery has not been an area known to take advantage of HBO therapy. By delineating the physiology, biochemistry and subcellular mechanism of the cardioprotective actions of HBO therapy in the rat heart, a natural extension of our work would eventually be clinical application in the arena of cardiac surgery in humans.

Acknowledgements

This study was supported in part by grants from the National Institutes of Health (HL54075 and HL65203) to Dr. Baker and a grant from the Undersea and Hyperbaric Medicine Society to Dr. Cabigas.

References

1. Sterling DL, Thornton JD, Swafford A, Gottlieb SF, Bishop SP, Stanley AW, Downey JM. Hyperbaric oxygen limits infarct size in ischemic rabbit myocardium in vivo. *Circulation* 1993; 88(4 Pt 1): 1931-1936.
2. Tahepold P, Valen G, Starkopf J, Kairane C, Zilmer M, Vaage J. Pretreating rats with hyperoxia attenuates ischemia-reperfusion injury of the heart. *Life Sci* 2001; 68(14): 1629-1640.
3. Kindwall E. In: *Hyperbaric Medicine Practice*: Best Publishing Company; 1995. p. 20-29.
4. Shandling AH, Ellestad MH, Hart GB, Crump R, Marlow D, Van Natta B, Messenger JC, Strauss M, Stavitsky Y. Hyperbaric oxygen and thrombolysis in myocardial infarction: the "HOT MI" pilot study. *Am Heart J* 1997; 134(3): 544-550.
5. Stavitsky Y, Shandling AH, Ellestad MH, Hart GB, Van Natta B, Messenger JC, Strauss M, Dekleva MN, Alexander JM, Mattice M, Clarke D. Hyperbaric oxygen and thrombolysis in myocardial infarction: the 'HOT MI' randomized multicenter study. *Cardiology* 1998; 90(2): 131-136.
6. Chavko M, Xing G, Keyser DO. Increased sensitivity to seizures in repeated exposures to hyperbaric oxygen: role of NOS activation. *Brain Res* 2001; 900(2): 227-233.
7. Buras JA, Stahl GL, Svoboda KK, Reenstra WR. Hyperbaric oxygen downregulates ICAM-1 expression induced by hypoxia and hypoglycemia: the role of NOS. *Am J Physiol Cell Physiol* 2000; 278(2): C292-302.

8. Tahepold P, Vaage J, Starkopf J, Valen G. Hyperoxia elicits myocardial protection through a nuclear factor kappaB-dependent mechanism in the rat heart. *J Thorac Cardiovasc Surg* 2003; 125(3): 650-660.
9. Rakhit RD, Marber MS. Nitric oxide: an emerging role in cardioprotection? *Heart* 2001; 86(4): 368-372.
10. Bolli R. Cardioprotective function of inducible nitric oxide synthase and role of nitric oxide in myocardial ischemia and preconditioning: an overview of a decade of research. *J Mol Cell Cardiol* 2001; 33(11): 1897-1918.
11. Shi Y, Baker JE, Zhang C, Tweddell JS, Su J, Pritchard KA, Jr. Chronic hypoxia increases endothelial nitric oxide synthase generation of nitric oxide by increasing heat shock protein 90 association and serine phosphorylation. *Circ Res* 2002; 91(4): 300-306.
12. Thom SR, Bhopale V, Fisher D, Manevich Y, Huang PL, Buerk DG. Stimulation of nitric oxide synthase in cerebral cortex due to elevated partial pressures of oxygen: an oxidative stress response. *J Neurobiol* 2002; 51(2): 85-100.
13. Poss KD, Tonegawa S. Reduced stress defense in heme oxygenase 1-deficient cells. *Proc Natl Acad Sci U S A* 1997; 94(20): 10925-10930.
14. Maines MD. The heme oxygenase system: a regulator of second messenger gases. *Annu Rev Pharmacol Toxicol* 1997; 37: 517-554.
15. Shi Y, Hutchins W, Su J, Siker D, Hogg N, Pritchard Jr. K, Keszler A, Tweddell J, Baker J. Delayed cardioprotection with isoflurane: role of reactive oxygen and nitrogen. *Am J Physiol* 2005; 288: H175-H184.
16. Hampl V, Walters C, Archer S. Determination of Nitric Oxide by the Chemiluminescence Reaction with Ozone. In: *Methods in Nitric Oxide Research*; 1996. p. 309-318.
17. Garcia-Cardena G, Fan R, Shah V, Sorrentino R, Cirino G, Papapetropoulos A, Sessa WC. Dynamic activation of endothelial nitric oxide synthase by Hsp90. *Nature* 1998; 392(6678): 821-824.
18. Tahepold P, Elfstrom P, Eha I, Kals J, Taal G, Talonpoika A, Valen G, Vaage J, Starkopf J. Exposure of rats to hyperoxia enhances relaxation of isolated aortic rings and reduces infarct size of isolated hearts. *Acta Physiol Scand* 2002; 175(4): 271-277.
19. Tahepold P, Ruusalepp A, Li G, Vaage J, Starkopf J, Valen G. Cardioprotection by breathing hyperoxic gas-relation to oxygen concentration and exposure time in rats and mice. *Eur J Cardiothorac Surg* 2002; 21(6): 987-994.
20. Gross GJ, Peart JN. KATP channels and myocardial preconditioning: an update. *Am J Physiol Heart Circ Physiol* 2003; 285(3): H921-930.
21. Baker JE, Contney SJ, Singh R, Kalyanaraman B, Gross GJ, Bosnjak ZJ. Nitric oxide activates the sarcolemmal K(ATP) channel in normoxic and chronically hypoxic hearts by a cyclic GMP-dependent mechanism. *J Mol Cell Cardiol* 2001; 33(2): 331-341.
22. Thom SR, Fisher D, Zhang J, Bhopale VM, Ohnishi ST, Kotake Y, Ohnishi T, Buerk DG. Stimulation of perivascular nitric oxide synthesis by oxygen. *Am J Physiol Heart Circ Physiol* 2003; 284(4): H1230-1239.
23. Badr AE, Yin W, Mychaskiw G, Zhang JH. Dual effect of HBO on cerebral infarction in MCAO rats. *Am J Physiol Regul Integr Comp Physiol* 2001; 280(3): R766-770.
24. Undersea and Hyperbaric Medical Society. In: http://www.uhms.org/.

Adaptation Biology and Medicine (Volume 5: Health Potentials)
Editors: L. Lukyanova, N. Takeda and P.K. Singal

5. Vascular Nitric Oxide: Effects of Physical Activity, Benefits for Health

Richard M. McAllister*, Sean C. Newcomer and M. Harold Laughlin

Department of Biomedical Sciences, University of Missouri, Columbia, Missouri 65211, USA

Abstract

Exercise training is known to induce several adaptations in the cardiovascular system, one of which is increased skeletal muscle blood flow at maximal exercise. Improved muscle blood flow, in turn, could in part be accounted for by augmented endothelium-dependent, nitric oxide (NO)-mediated vasodilation. Studies have indeed demonstrated that endothelium-dependent, NO-mediated dilation of conductance-type vessels is augmented after endurance exercise training; recently, this adaptation has been extended into resistance-type vessels within skeletal muscle. With the latter, however, it appears that only resistance vessels supplying muscle active during training sessions exhibit this adaptation. Increased expression of endothelial NO synthase (eNOS) appears to underlie enhanced endothelium-dependent, NO-mediated dilation of both conductance and resistance vessels. Greater eNOS expression may also underlie the preventive and/or rehabilitative effect(s) of exercise training on atherosclerosis, given that NO inhibits several steps of the atherosclerotic disease process. Thus, exercise training may induce adaptations that benefit both vasodilation and vascular health.

Keywords: Endothelium, conductance vessels, resistance vessels, nitric oxide synthase, exercise training, vasodilation, atherosclerosis

Introduction

The cardiovascular system exhibits numerous adaptations to chronic imposition of stress. One such chronic stress is endurance exercise training; that is, physical activity involving large muscles/muscle groups that is performed at moderate intensity for prolonged sessions over a period of weeks to years. We use a rat model in our laboratory, with rats running on a motor-driven treadmill at 30 m/min up a 10% grade (~70% of exercise capacity) during 1-h daily training sessions, 5 d/wk, over periods of 2-3 months. This exercise training program induces several well-established adaptations, including increased skeletal muscle mitochondrial content and left ventricular hypertrophy (2, 8).

Increased maximal cardiac output is another adaptation to endurance exercise training. A study involving dogs demonstrated that this type of exercise training led to a 30% increase in cardiac output

*Address for Correspondence: Richard M. McAllister, Ph.D., Research Associate Professor, Department of Biomedical Sciences, University of Missouri, E102 Vet Med Bldg, Columbia, Missouri 65211, USA. Phone: 573-882-3074, Fax: 573-884-6890, E-mail: mcallisterr@missouri.edu

at maximal exercise levels (9). These investigators also determined regional blood flow distribution in their dogs using radiolabelled microspheres. They reported that 80% of the additional cardiac output at maximal exercise post-training was directed to skeletal muscle (e.g., hindlimb muscles such as the gastrocnemius). Although left ventricular adaptations undoubtedly made important contributions to greater muscle blood flow post-training, a nearly 25% reduction in total peripheral resistance suggested that vascular adaptations also contributed. We have focused our research on the latter possibility, examining adaptations to endurance exercise training in both conductance and resistance vessels supplying skeletal muscle with blood flow. Much of this review will therefore summarize findings from that research. In addition, since it appears that endurance exercise training confers protection against several chronic diseases, we will briefly discuss the possibility that training may protect against the vascular disease of atherosclerosis.

Conductance Vessels

Conductance vessels are relatively large-diameter blood vessels. While their resistance to blood flow is low, they have nonetheless been considered a logical place to begin examining adaptations to exercise training. When we determined dilatory responses of aortae isolated from endurance exercise-trained rats and their sedentary (inactive) counterparts *in vitro*, a number of important findings emerged (2). First, dose-dependent relaxation responses to the endothelium-dependent agent acetylcholine (ACh), following norepinephrine-induced contraction, were greater in aortae from trained animals. In addition, both sensitivity to ACh and the maximal response to ACh were augmented with training. Second, dose-dependent responses to the endothelium-independent agent sodium nitroprusside (SNP) were not different between groups. Since SNP acts as an exogenous source of nitric oxide (NO), thereby bypassing endothelial-generated NO, the finding that SNP responses were unchanged with training suggested that the endothelium (but not vascular smooth muscle) adapted to endurance exercise training. Other experiments in this study, using the inhibitor of endothelial NO formation N^G-nitro-L-arginine methyl ester, demonstrated that the augmented relaxation response to ACh with training was primarily due to greater NO formation in the endothelium. This finding was subsequently supported by several studies (e.g., 8) that demonstrated greater expression of endothelial NO synthase (eNOS), the protein that catalyzes NO formation in aortic endothelium, in trained animals. Collectively, these findings for a conductance vessel indicate that greater endothelium-dependent dilation is a vascular adaptation that may contribute to greater skeletal muscle blood flow after a period of endurance exercise training. Nonetheless, the small contribution of conductance vessels to total peripheral resistance dictates that studies also examine the resistance vasculature.

Resistance Vessels

Resistance vessels are relatively small-diameter blood vessels. Given that their resistance to blood flow is high and therefore a primary determinant of tissue blood flow, they are considered important sites in the vasculature to examine adaptations to training. We recently conducted a study of the effects of endurance exercise training on skeletal muscle resistance vessels (8). This study involved several experimental approaches, including *in situ* determination of vasodilatory responses, *in vitro* determination of vasodilatory responses, and determination of eNOS protein expression. These different experimental approaches yielded remarkably consistent findings concerning adaptations of resistance vessels to training.

With the first approach, vasodilatory responses were determined *in situ* using the isolated perfused rat hindlimb preparation. Use of this preparation for determination of vascular function has been described in detail previously (7). Since ~80% of the rat hindlimb mass is skeletal muscle, responses to infusion of agents such as ACh largely reflect those of resistance vessels located within muscles of the hindlimb. Importantly, use of the radiolabelled microsphere technique in conjunction with the hindlimb preparation allows examination of muscle-specific blood flows during infusion of vasoactive agents. When we examined muscle-specific responses, a number of findings emerged (Fig. 1). First, in the white gastrocnemius muscle section, which consists primarily of fast glycolytic (FG) muscle fibers, no vasodilation in response to ACh occurred in either the sedentary group or the endurance exercise-trained group (Fig. 1B). In the soleus muscle, which is composed almost exclusively of slow oxidative (SO) fibers, while robust vasodilation in response to ACh occurred, this response was similar between sedentary and trained rats (Fig. 1D). In the red section of the gastrocnemius muscle, consisting largely of fast oxidative glycolytic (FOG) fibers, ACh-induced vasodilation was greater in trained than sedentary rats (Fig. 1A; 8).

These findings obtained *in situ* led us to determine dilatory responses to increases in flow, a more physiological stimulus for endothelium-dependent dilation, of resistance vessels *in vitro*. We examined

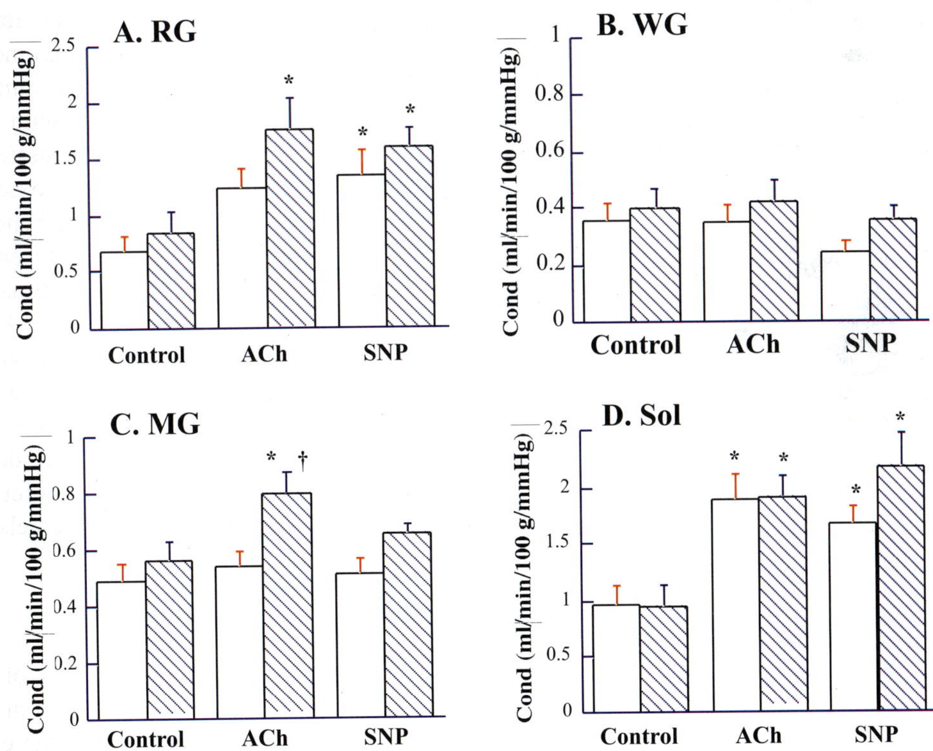

Fig. 1. Conductance (Cond; quotient of blood flow and perfusion pressure) in red section of gastrocnemius muscle (RG; Panel A), white section of gastrocnemius muscle (WG; Panel B), mixed section of gastrocnemius muscle (MG; Panel C), and soleus muscle (Sol; Panel D). Values are mean ± SE; sedentary, open bars; exercise-trained, filled bars. ACh, acetylcholine; SNP, sodium nitroprusside. * Different from control within same group, P<0.05; † different between sedentary and exercise-trained, P<0.05.

resistance vessels isolated from several hindlimb muscles/muscle sections, including the white gastrocnemius (2A arterioles), the soleus (1A arterioles), and the red gastrocnemius (2A arterioles). In each instance, the first order of resistance vessel located intramuscularly was studied. Vessels from all three muscles/muscle sections exhibited dilation in response to increases in flow (Fig. 2). Similar to responses determined *in situ*, dilatory responses to flow of 2A arterioles from red gastrocnemius of trained rats *in vitro* were greater than those of sedentary rats (Fig. 2A). Interestingly, dilation in response to SNP was also enhanced in these arterioles from trained animals, raising the possibility that enhancement of both endothelial NO formation and vascular smooth muscle sensitivity to NO contributed to augmentation of the dilatory response to flow (8).

With our third approach, we determined eNOS protein expression in a variety of resistance vessel orders from the same muscles/muscle sections in which dilatory responses were examined. For these determinations, it was necessary to pool vessels from several rats in order to have sufficient protein for Western blotting. In agreement with the above-mentioned dilatory data, eNOS expression was

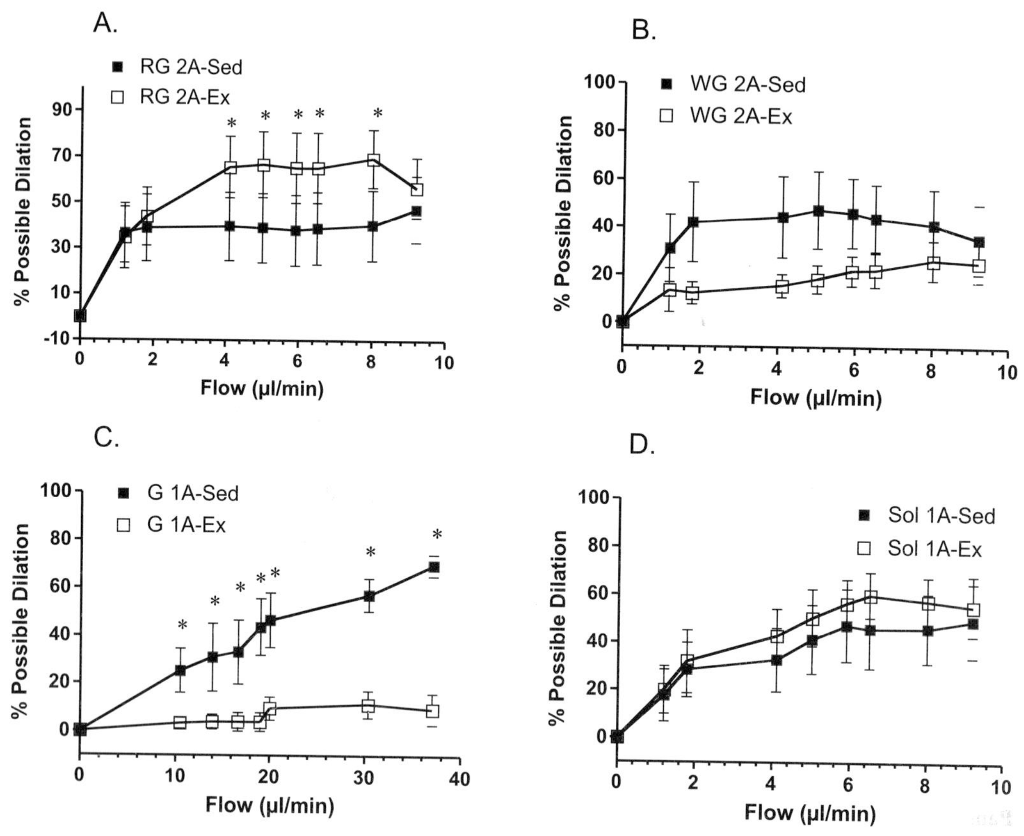

Fig. 2. Dose-dependent dilatory responses to increases in flow of RG 2A arterioles (Panel A), WG 2A arterioles (Panel B), gastrocnemius (G) 1A arterioles (Panel C), and Sol 1A arterioles (Panel D). Muscle abbreviations as per Fig. 1. Values are mean ± SE; percent possible dilation is fraction of difference in diameter between arteriole with basal tone and without tone (in Ca^{2+}-free perfusate). * Different between sedentary (Sed) and exercise-trained (Ex), P<0.05.

augmented significantly by endurance exercise training only in vessels (2A, 4A, and 5A arterioles) from the red gastrocnemius muscle section (Fig. 3A). No resistance vessels from either the white gastrocnemius (Fig. 3A) or soleus muscles (Fig. 4A) exhibited greater eNOS expression in trained rats compared with their sedentary counterparts. Interestingly, expression of superoxide dismutase-1 (SOD-1), the Cu/Zn-dependent isoform of SOD-1, was only increased in the gastrocnemius feed artery (Figs. 3B, 4B; 8). This suggests that increased NO production via eNOS, but not reduced NO scavenging by free radicals such as superoxide as a consequence of greater antioxidant action by SOD-1, is solely responsible for increased endothelium-dependent vasodilation.

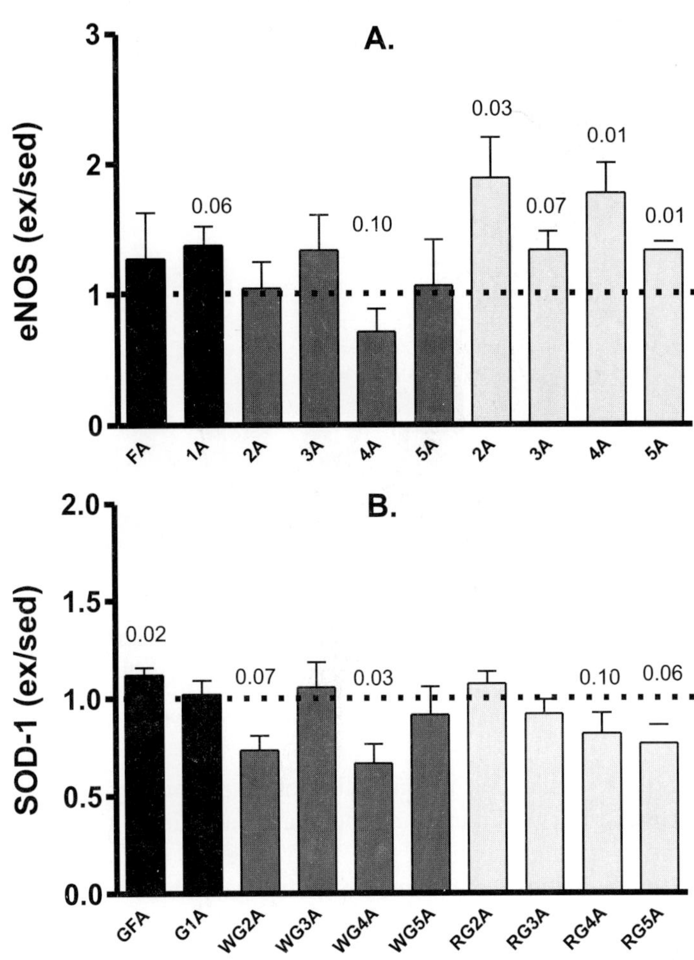

Fig. 3. Panel A. eNOS protein expression in gastrocnemius (G) resistance vessels. GFA, G feed artery; other abbreviations as per Figs. 1 and 2. Values are mean ± SE; height of bars represents ratio of exercise-trained (Ex) to sedentary (Sed). Ex is significantly greater than Sed ($P<0.05$) for RG 2A, 4A, and 5A arterioles; borderline significance ($0.05<P<0.10$) also indicated. Panel B. SOD-1 protein expression in G resistance vessels. Abbreviations as per Panel A. Note that Ex is significantly greater than Sed in only GFA.

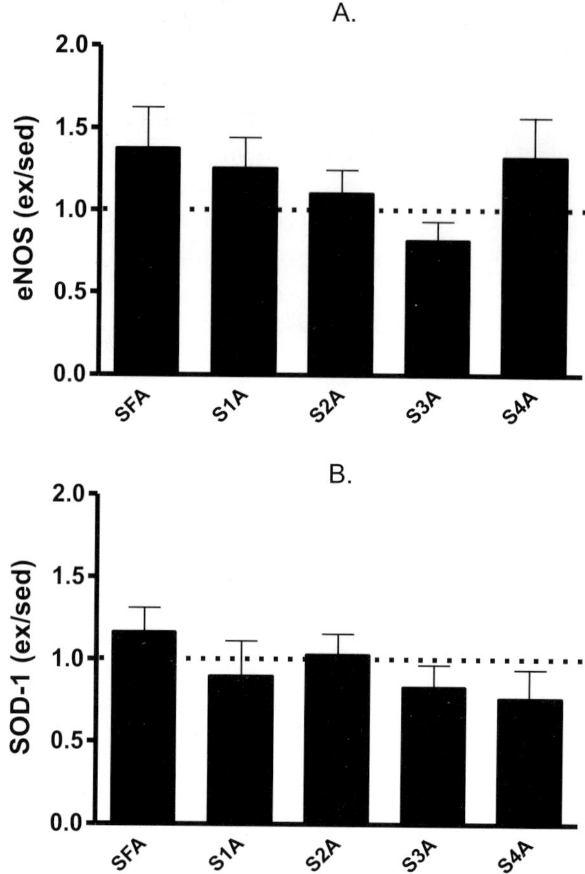

Fig. 4. Panel A. eNOS protein expression in soleus (S) resistance vessels. SFA, soleus feed artery. Values are mean ± SE; height of bars represents ratio of exercise-trained (Ex) to sedentary (Sed). Ex is not significantly greater than Sed ($P<0.05$) in any arteriolar order. Panel B. SOD-1 protein expression in S resistance vessels. Abbreviations as per Panel A. Note that Ex is not significantly greater than Sed in any arteriolar order.

The consistency of our data, obtained using three different experimental approaches, suggests that endurance exercise training primarily induces endothelial adaptations in vessels that experience increased blood flow during training sessions. The FOG-type red gastrocnemius muscle section exhibits increased blood flow (~3-fold over resting level) during treadmill running at 30 m/min, while flows to the FG-type white gastrocnemius and SO-type soleus muscles increase only slightly (soleus) or not at all (white gastrocnemius; 5). These different blood flow responses to treadmill running at a speed of 30 m/min likely reflect differences in muscle fiber recruitment and metabolism (c.f., 5). We have also investigated effects of sprint exercise training (repeated short bursts of faster treadmill running at 60 m/min, separated by rest intervals). This type of training also increases recruitment of, as well as blood flow to, FG-type white gastrocnemius fibers (5). Sprint training was found to enhance both ACh-induced dilatory responses in *vitro* and eNOS protein expression in certain resistance vessels of the white gastrocnemius muscle section (6). Collectively, these findings suggest that magnitude of the increase in blood flow during exercise training sessions is a key determinant of vascular adaptations to exercise training.

Effects of Endurance Exercise Training on Atherosclerosis

Available human data (e.g., 10) suggest that endurance exercise training prevents the development and/or progression of atherosclerosis. Several steps of the atherosclerotic disease process, including leukocyte adhesion to the endothelium, vascular smooth muscle cell migration into the subendothelial space, and platelet aggregation in the vascular lumen (c.f., 1), are inhibited by NO (c.f., 3). Since, as detailed above, exercise training augments endothelial NO formation, this may be an adaptation underlying the protective effect of exercise against atherosclerosis.

Concerning leukocyte adhesion, a recent study by Yang and colleagues (11) examined expression of P-selectin and vascular cell adhesion molecule-1 (VCAM-1), proteins that recruit leukocytes to the endothelium. These leukocytes may subsequently enter the subendothelial space, differentiate into macrophages, and become cholesterol-laden foam cells, a key constituent of an atherosclerotic lesion. Cholesterol-fed rabbits in this study expressed P-selection and VCAM-1 in aortic vascular endothelium, but exercise training on a treadmill blunted dietary cholesterol-induced expression of these proteins. Importantly, intimal thickness (an index of extent of atherosclerosis) was reduced in parallel with reduced expression of these proteins.

A study involving rats examined the anti-atherosclerotic effects of swimming training on vascular smooth muscle cell migration and platelet aggregation (4). In this study, carotid arteries were subjected to balloon angioplasty, a procedure known to induce smooth muscle migration and proliferation. Number of smooth muscle cells in the subendothelial space, an index of smooth muscle migration, was reduced in swim-trained rats compared with their sedentary counterparts. Extent of atherosclerosis was reduced in parallel. Another beneficial adaptation to exercise training was a reduction in platelet aggregation in blood sampled from these rats.

These two studies require verification, but nonetheless suggest that increased NO formation in the vascular endothelium, a predicted outcome of greater eNOS expression with endurance exercise training, results in inhibition of one or more steps of atherosclerosis. This adaptation may underlie the preventive effect of exercise training in healthy individuals, as well as the rehabilitative effect of training in patients with cardiovascular disease (10).

Acknowledgements

The authors acknowledge the important technical contributions to their studies cited in this review by Ivelisse Albarracin, Beth Degarmo, Molly Edmonds, Ann Melloh, Tara Smith, and Pam Thorne. Also acknowledged are our valued collaborators, Drs. Jeffrey Jasperse and Elmer Price.

This work was supported by funds from the National Institutes of Health, the American Heart Association, and the American Physiological Society.

References

1. Berliner JA, Navab M, Fogelman AM, Frank JS, Demer LL, Edwards PA, Watson AD, Lusis AJ. Atherosclerosis: basic mechanisms. Oxidation, inflammation, and genetics. Circulation. 1995; 91: 2488-2496.

2. Delp MD, McAllister RM, Laughlin MH. Exercise training alters endothelium-dependent vasoreactivity of rat abdominal aorta. J Appl Physiol. 1993; 75: 1354-1363.
3. Gewaltig MT, Kojda G. Vasoprotection by nitric oxide: mechanisms and therapeutic potential. Cardiovasc Res. 2002; 55: 250-260.
4. Indolfi C, Torella D, Coppola C, Curcio A, Rodriguez F, Bilancio A, Leccia A, Arcucci O, Falco M, Leosco D, Chiariello M. Physical training increases eNOS vascular expression and activity and reduces restenosis after balloon angioplasty or arterial stenting in rats. Circ Res. 2002; 91: 1190-1197.
5. Laughlin MH, Armstrong RB. Muscular blood flow distribution patterns as a function of running speed in rats. Am J Physiol Heart Circ Physiol. 1982; 243: H296-H306.
6. Laughlin MH, Woodman CR, Schrage WG, Gute D, Price EM. Interval sprint training enhances endothelial function and eNOS content in some arteries that perfuse white gastrocnemius muscle. J Appl Physiol. 2004; 96: 233-244.
7. McAllister RM. Endothelium-dependent vasodilation in different rat hindlimb skeletal muscles. J Appl Physiol. 2003; 94: 1777-1784.
8. McAllister RM, Jasperse JL, Laughlin MH. Nonuniform effects of endurance exercise training on vasodilation in rat skeletal muscle. J Appl Physiol. 2005; 98: 753-761.
9. Musch TI, Haidet GC, Ordway GA, Longhurst JC, Mitchell JH. Training effects on regional blood flow response to maximal exercise in foxhounds. J Appl Physiol. 1987; 62: 1724-1732.
10. Myers J, Prakash M, Froelicher V, Do D, Partington S, Atwood JE. Exercise capacity and mortality among men referred for exercise testing. New Engl J Med. 2002; 346: 793-801.
11. Yang A-L, Jen CJ, Chen H-i. Effects of high-cholesterol diet and parallel exercise training on the vascular function of rabbit aortas: a time course study. J Appl Physiol. 2003; 95: 1194-1200.

Adaptation Biology and Medicine (Volume 5: Health Potentials)
Editors: L. Lukyanova, N. Takeda and P.K. Singal

6. Intermittent Hypoxic Conditioning of Canine Myocardium: Robust Protection Against Ischemia-reperfusion Injury

Robert T. Mallet[1]*, Myoung-Gwi Ryou[1], Eugenia B. Manukhina[2] and H. Fred Downey[1]

[1]Department of Integrative Physiology, University of North Texas Health Science Center, Fort Worth, Texas, USA,
[2]Institute of General Pathology and Pathophysiology, Moscow, Russia

Abbreviations

AAR: area at risk; FIO_2: fraction of inspired O_2; IHC: intermittent hypoxia conditioning; IS: infarct size; LAD: left anterior descending coronary artery; LV: left ventricular mass; NAC: *N*-acetylcysteine; NOS: nitric oxide synthase; ROS: reactive oxygen species; SOD: superoxide dismutase; VF: ventricular fibrillation; VT: ventricular tachycardia

Summary

Recognition that sojourns to high altitude decrease incidence and severity of heart disease[1,2] has prompted extensive research to develop hypoxic conditioning as a means of treating and preventing cardiovascular disease.[3-6] Empirical evidence indicates that intermittent hypoxia conditioning (IHC) programs lessen incidence and severity of angina and arrhythmias in patients with coronary artery disease.[7-10] These findings suggest that IHC may provide a powerful, noninvasive intervention to prevent myocardial ischemic injury in patients. However, implementation of IHC as a cardioprotective therapy requires rigorous preclinical studies to challenge and optimize the conditioning program before it can be applied to human subjects for prophylaxis of myocardial infarction and arrhythmias. Moreover, the mechanisms responsible for IHC-induced cardioprotection have not been delineated. This article summarizes published information on hypoxia conditioning of myocardium, and presents recent studies in the authors' laboratories that provide important insights on the cellular and molecular mechanisms of this clinically relevant cardioprotective phenomenon.

Hypoxia Conditioning and Myocardial Ischemic Tolerance

Numerous investigations in small mammals have examined the impact of chronic hypoxia conditioning programs on myocardial infarction, arrhythmias, and contractile dysfunction during acute ischemia and following reperfusion. Table 1 presents a sampling of these hypoxia programs.[11-21] In rats, hypoxia bouts of several h/d for 3-6 wk decreased infarction following acute coronary occlusion-reperfusion by

*Address for Correspondence: Robert T. Mallet, Ph.D., Department of Integrative Physiology, University of North Texas Health Science Center, 3500 Camp Bowie Boulevard, Fort Worth, TX 76107-2699 USA. Ph: 817-735-2260, E-mail: malletr@hsc.unt.edu

Table 1. Studies in small mammals of hypoxia conditioning-induced cardioprotection against ischemia-reperfusion injury.

Citation	Species	FIO_2 (%)	h/d	Days	Outcome
Asemu *et al.*, 1999[11]	rat	11*	4	25-30	64% reduction in ventricular tachyarrhythmias
Baker *et al.*, 1999[12]	rabbit†	12	24	9	Improved post-ischemic LV contractile recovery
Eells *et al.*, 2000[13]	rabbit†	12	24	7-10	Improved post-ischemic LV contractile recovery
Forkel *et al.*, 2004[14]	rat	10.5	24	14	Improved post-ischemic RV but not LV function
Manukhina *et al.*, 2000[15]	rat	11*	5	40	Less aortic hypotension during coronary occlusion
Neckár *et al.*, 2002[16]	rat	11*	8	24-32	15% smaller infarct; increased post-ischemic dP/dt
Neckár *et al.*, 2004[17]	rat	11*	8	24-32	31% smaller infarct
Zhong *et al.*, 2000[18]	rat	11*	6	14-42	Decreased cardiac arrhythmias
Zhong *et al.*, 2002[19]	rat	11*	6	28-42	Improved recovery of LV systolic, diastolic function
Zhu *et al.*, 2003[20]	rat†	11*	6	42	Delayed contracture; improved LV systolic recovery
Zhu *et al.*, 2004[21]	rat†	11*	6	60	Improved LV recovery; decreased LDH release

LDH: lactate dehydrogenase; LV: left ventricular; RV: right ventricular; dP/dt: rate of change of LV pressure. *:These studies used hypobaric hypoxia; FIO_2 values are normobaric equivalents of the reduced atmospheric O_2 content. †Neonates.

15-30%[16,17,22] and lowered the incidence of ventricular tachycardia and fibrillation during ischemia and reperfusion.[8,11,18,23,24] Hypoxia conditioning enhances post-ischemic contractile recovery,[12,13,19-21] although one study reported improvement only in right ventricle.[14] Even single bouts of hypoxia protected rodent myocardium from ischemia-induced infarction and contractile dysfunction 24 h later.[25,26] In these protocols, the animals were exposed to continuous hypoxia, or to single daily hypoxia bouts of several h, without intermittent reoxygenation during hypoxia. The ability of multiple, brief hypoxic exposures, with intervening normoxia, to prevent myocardial infarction was unknown.

Intermittent hypoxia conditioning prevents myocardial ischemia-reperfusion injury

Recently, we conducted the first investigation of hypoxia-induced cardioprotection in a large mammal, the dog.[27] Moreover, this study was the first to test whether a program of multiple, brief hypoxia cycles could prevent ischemia-reperfusion injury in an experimental animal. Mongrel dogs were subjected to a 20 day program of intermittent hypoxia conditioning (IHC) by breathing hypoxic atmospheres in an acrylic chamber.[27] This IHC program is modeled after small clinical trials in which IHC decreased cardiac arrhythmias and improved exercise tolerance in patients with coronary artery disease.[7-9] Daily conditioning sessions consisted of 5-8 cycles of moderately intense hypoxia (10-9.5% FIO_2), each 5-10 min, with intervening 4 min reoxygenation periods. The target fraction of inspired O_2 (FIO_2) was attained within 90 s by introducing N_2 gas into the chamber, and abrupt reoxygenation was achieved by opening the chamber to room air. The number, duration and intensity of the hypoxia cycles were progressively increased from 5 cycles of 5 min exposure to 10% FIO_2 on day 1 to 7 cycles of 10 min exposure to 9.5% FIO_2 on days 11-20. The dogs exhibited no discomfort or distress during the sessions. Sham control dogs completed the same 20 d program, except compressed air (21% O_2) instead of N_2 was introduced into the chamber. Nonconditioned control dogs were neither hypoxia- nor sham-conditioned. During each hypoxia cycle, systemic arterial PO_2 fell from *c.* 100 mm Hg to 45-50 mm Hg, and O_2 content fell by 25-30%; both variables quickly recovered during reoxygenation.[28] Arterial PCO_2 fell and pH increased due to hyperventilation during hypoxia.

Coronary occlusion-reperfusion experiments were conducted on the day following completion of the IHC or sham conditioning program. The heart was exposed via left lateral thoracotomy under pentobarbital anesthesia. The left anterior descending coronary artery (LAD) was dissected distal to its first major diagonal branch and a silk suture was placed around the artery, forming a snare. The LAD was occluded for 60 min by tightening the snare, then reperfused by releasing the snare. Collateral blood flow to the ischemic territory was measured with ^{46}Sc-radiolabeled microspheres injected at 30 min occlusion.[29] Cardiac arrhythmias were monitored by lead II electrocardiography, and analyzed and scored according to the Lambeth conventions.[30] In the event of ventricular fibrillation, cardioversion was achieved by applying 10 J DC countershocks to the epicardium. After 5 h reperfusion, the ischemic area at risk was demarcated by perfusing non-ischemic myocardium with Evans blue dye while simultaneously perfusing at-risk myocardium with 0.9% NaCl.[31] The left ventricle was sectioned and incubated with 2,3,5-triphenyl tetrazolium chloride to differentiate infarcted from viable myocardium within the area at risk.[32] Masses of infarcted, viable ischemic and non-ischemic myocardium were measured by planimetry.[31] Myocardial ischemic injury was assessed from the percentage of ischemic myocardium that infarcted.

The ischemic fraction of the total left ventricular mass (AAR/LV) was similar in non-conditioned, sham-conditioned and IHC dogs, and comparable fractions of the ischemic tissue, nearly 40%, infarcted in sham-conditioned and acute dogs (Figure 1). Remarkably, only 1.1 ± 0.3% of the ischemic

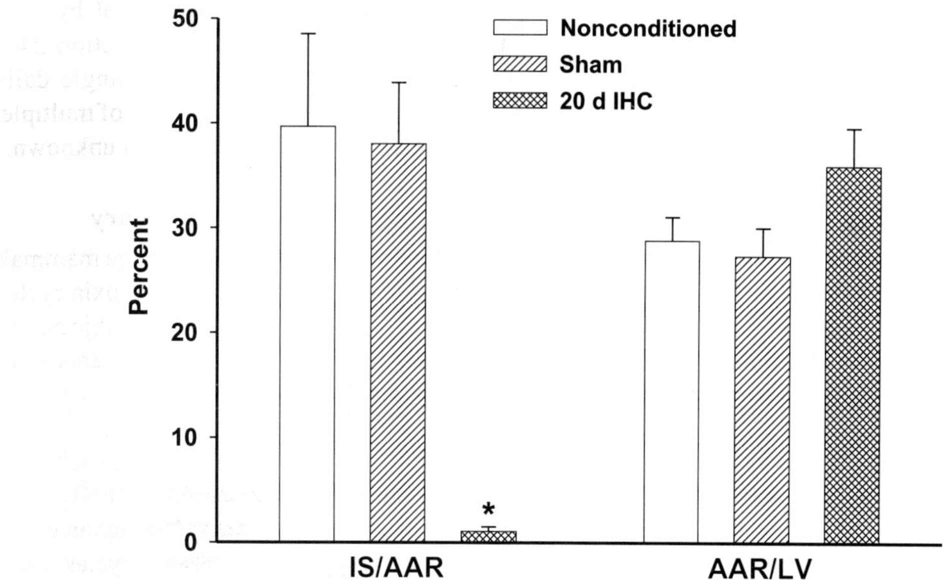

Fig. 1. ***Intermittent, normobaric hypoxia conditioning (IHC) prevents myocardial infarction.*** **Masses of myocardial infarct (infarct size; IS), ischemic area at risk (AAR), and left ventricular free wall + septum (LV) were measured after 1 h occlusion and 5 h reperfusion of the left anterior descending coronary artery (LAD). IS/AAR and AAR/LV are expressed as percentages. Values are means ± SEM from 7 acute non-conditioned dogs (open bars), 10 sham-conditioned dogs (hatched bars) and 9 dogs hypoxia-conditioned for 20 consecutive days (IHC; cross-hatched bars). *P < 0.001 *vs.* both non-hypoxic groups. Data in this figure and Figures 2-5 are from reference 28 and are presented with the generous consent of Springer Science and Business Media.**

myocardium infarcted in the IHC dogs. IHC also mitigated ventricular arrhythmias, which were especially severe during the first 30 min reperfusion in non-IHC dogs. Isolated ventricular premature beats occurred in all of the nonconditioned and sham-conditioned dogs, but in only 63% of the IHC dogs. More strikingly, 14 of the 17 non-IHC dogs suffered ventricular tachycardia and/or ventricular fibrillation following reperfusion, but none of the 12 IHC dogs experienced these dangerous tachyarrhythmias. Accordingly, arrhythmia scores[24,28,30] in the IHC dogs were less than 20% of scores in the non-IHC controls (Figure 2). Collectively, these results demonstrate that powerful cardioprotection was produced by the IHC program.

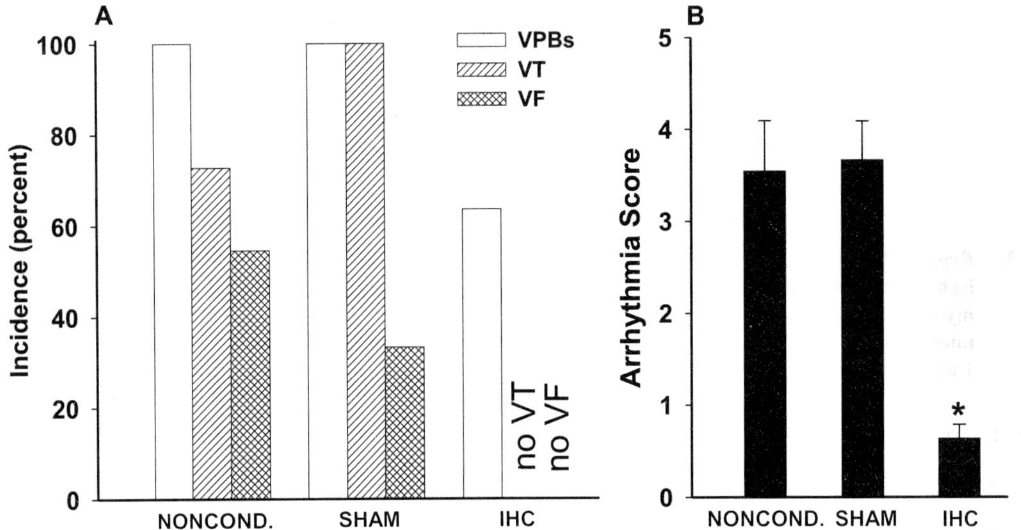

Fig. 2. ***Incidences of ventricular arrhythmias during LAD reperfusion (panel A) and arrhythmia score (panel B).*** **Ventricular arrhythmias during the first 30 min LAD reperfusion were monitored by standard limb lead II electrocardiography, and analyzed according to the Lambeth conventions.[30] Panel A: The numbers of experiments in which the arrhythmias occurred (open bars: isolated ventricular premature beats; hatched bars: ventricular tachycardia [VT]; cross-hatched bars: ventricular fibrillation [VF]) are expressed as percentages of the total number of experiments in each group. Panel B: Arrhythmia scores from the experiments in Panel A. Values are means ± SEM. *P < 0.001 *vs.* the other groups.**

Prolonged hypoxia stimulates angiogenesis in myocardium,[19,33] which could increase nutritive collateral blood flow to the ischemic territory and, thus, minimize ischemic injury. However, the current IHC program did not increase collateral blood flow to ischemic myocardium, nor perfusion of remote, nonischemic tissue (Figure 3). Chronic hypoxia conditioning programs in rats sharply increase hematocrit,[14,34,35] which would improve delivery of oxygen to ischemic myocardium. However, blood viscosity also increases, which increases cardiac work and energy demand. In contrast, the IHC program altered neither systemic arterial hematocrit nor hemoglobin content in dogs (Table 2). Thus, the IHC program imparted robust cardioprotection without increasing collateral flow, or O_2-carrying capacity of arterial blood, and without the detrimental effect of increased viscosity.

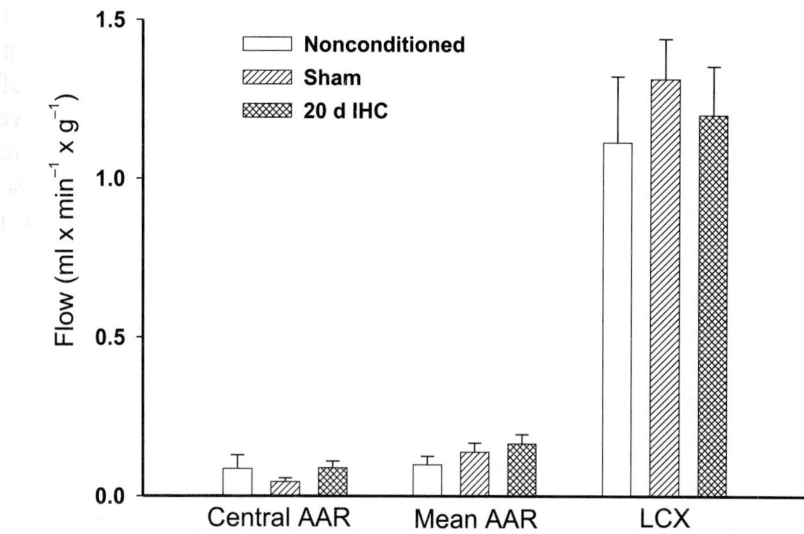

Fig. 3. ***Regional myocardial blood flow during coronary artery occlusion.*** **Blood flow to the central portion of the ischemic area at risk (Central AAR), mean flow to the area at risk (Mean AAR) and flow to non-ischemic myocardium perfused by the left circumflex coronary artery (LCX) were measured from distributions of microspheres injected at 30 min LAD occlusion. Values are means ± SEM from the same groups as in Figures 1 and 2. No statistically significant differences between groups were detected.**

Table 2. **Arterial hematocrit, hemoglobin and O_2 content measured at conclusion of intermittent hypoxia or sham conditioning programs.**

	Nonconditioned	Sham	IHC
N	11	6	12
Hematocrit, %	40 ± 2	42 ± 1	38 ± 1
Hemoglobin content, g/dl	13.5 ± 0.4	13.5 ± 0.4	12.6 ± 0.3
O_2 content, ml/dl	17.9 ± 1.0	18.1 ± 0.6	16.6 ± 0.8

Hematocrit and hemoglobin and O_2 contents were measured in systemic arterial blood samples collected before LAD occlusion-reperfusion protocols. Values are means ± SEM. No statistically significant differences between the groups were detected by ANOVA. Hematocrit and hemoglobin data are from reference 28 and are reproduced with the kind permission of Springer Science and Business Media.

Progressive development of cardioprotection over multiple IHC sessions

Qualitatively, the reduction in ischemic injury one day after the IHC program resembles the cardioprotection afforded by the delayed 'second window' of ischemic preconditioning, in which significant cardioprotection is manifest 1-3 days after a brief, nonlethal coronary occlusion (reference 36 and references therein), although IHC cardioprotection appears to be much more extensive than that afforded by delayed preconditioning. It could be argued that IHC is a preconditioning-like phenomenon and is mediated by similar mechanisms. Delayed preconditioning cardioprotection is evoked by a single ischemic event.[36] However, a single bout of IHC (7 cycles of 9.5% FIO_2; 10 min/cycle) did not protect canine myocardium from infarction and arrhythmias following 60 min LAD occlusion imposed the next day (Figure 4). In these experiments, infarcts (36 ± 8% of area at risk) were

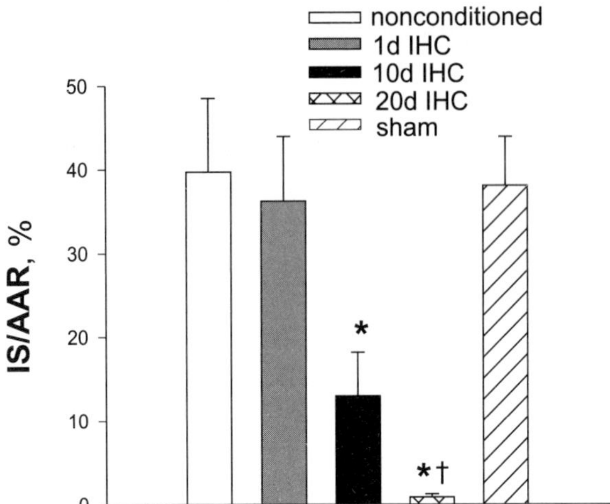

Fig. 4. ***Incomplete cardioprotection against infarct imparted by abbreviated hypoxia conditioning programs.*** **Infarct size (IS) and area at risk (AAR) were measured in dogs conditioned by abbreviated 1 and 10 day intermittent hypoxia programs (1d IHC and 10d IHC, respectively), then subjected to 1 hour occlusion and 5 h reperfusion of the LAD. Infarcts in the 1d IHC group (gray bar) were similar to those in non-hypoxic naïve and sham dogs. Infarcts in the 10d IHC group (black bar) were smaller than those of the non-hypoxic dogs, but larger than those of dogs hypoxia-conditioned for 20 d (cross-hatched bar). *P < 0.05 *vs.* nonconditioned and sham; †P < 0.01 *vs.* 10d IHC.**

as large as those of normoxic controls, and arrhythmia score (3.5 ± 0.5) also approximated control values.

To further define the evolving nature of the cardioprotection produced by IHC, we assessed cardioprotection after the first 10 d of the IHC program. Infarct size in these dogs (13 ± 5% of area at risk) was one-third that of normoxic control infarcts, but greater than the minute infarcts of 20-d IHC dogs (Figure 4). Thus, multiple sessions of IHC are necessary to impart cardioprotection, and approximately 20 d are required for protection to be fully manifest.[28] This is very unlike acute or delayed ischemic preconditioning[36] and suggests that different mechanisms are responsible for IHC-induced cardioprotection.

The gradual development of cardioprotection over multiple hypoxia sessions strongly suggests that IHC protects myocardium from ischemic injury by modifying gene expression, leading to progressive buildup of cardioprotective proteins. Indeed, hypoxia-reoxygenation of isolated cardiomyocytes and intact hearts activates genes encoding an array of potentially cardioprotective proteins[37,38] including the heat shock proteins Hsp-70[18,39,40] and Hsp-32, *a.k.a.* heme oxygenase,[33,40-42] antioxidant enzymes superoxide dismutase,[40,43] catalase[21,40] and glutathione peroxidase,[40,44] glucose 6-phosphate dehydrogenase and 6-phosphogluconate dehydrogenase of the NADPH-generating hexose monophosphate pathway,[45] subunits of AP-1 transcription factor[46,47] and a novel, anti-apoptotic, oxygen-regulated protein.[48]

β_1-Adrenergic activity mediates IHC-induced cardioprotection

Hypoxemia elicits chemoreceptor-mediated β-adrenergic activation of the myocardium, which increases cardiac output to restore O_2 delivery to tissues.[49,50] Cyclic hypoxia-reoxygenation provokes catecholamine release from intrinsic cardiac adrenergic cells by augmenting cytosolic $[Ca^{2+}]$ transients.[51] Thus, intermittent reoxygenation may intensify β-adrenergic activation of hypoxic myocardium.

In several studies,[52-55] β-adrenergic activation has been found to be cardioprotective. For example, norepinephrine pretreatment suppressed ventricular tachycardia and ventricular fibrillation, and increased survival of dogs subjected to LAD occlusion-reperfusion.[56] Multiple intravenous isoproterenol injections in conscious rabbits decreased ST segment displacement and lowered left ventricular end-diastolic and end-systolic pressures during α + β-adrenergic stress imposed 24 or 48 h later.[57] Catecholamines activate potentially cardioprotective genes, including hypoxia-responsive genes.[58,59]

To test the role of β-adrenergic activity in IHC cardioprotection, dogs were administered the β_1-selective adrenoceptor antagonist metoprolol (1 mg/kg orally at 12-h intervals) throughout the IHC program.[28] This regimen selectively antagonizes β_1-adrenoceptors in dogs, but the receptor antagonism completely subsides within 24 h of discontinuing treatment.[60] Metoprolol selectively blunted tachycardia, mediated by β_1-adrenoceptors, but didn't alter β_2-adrenoceptor-mediated vasodilation and hypotension following isoproterenol injections.[28]

β_1-Adrenoceptor blockade nearly abrogated IHC-induced cardioprotection (Figure 5).[28] Metoprolol treatment during the IHC program enlarged infarct size from 1.1 ± 0.3% to 27 ± 3% of area at risk (Figure 5A), and arrhythmia score increased from 0.6 ± 0.1 in untreated to 2.8 ± 0.7 in metoprolol-treated IHC dogs (Figure 5B). Metoprolol increased neither infarct size (Figure 5A) nor arrhythmia score (Figure 5B) in sham-conditioned dogs,[33] so its blunting of cardioprotection in IHC dogs was not due to a direct toxic effect of the β-blocker. These results demonstrate a crucial role of β_1-adrenoceptor activation in the cardioprotective mechanism of IHC. Thus, episodic β-adrenergic activity during IHC contributes to progressive development of myocardial resistance to ischemia.

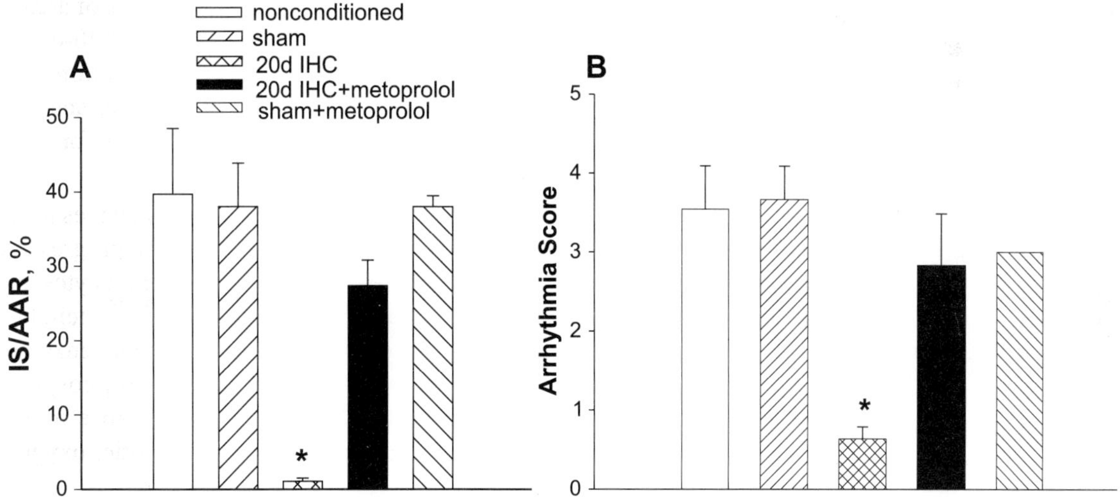

Fig. 5. ***Competitive antagonism of β_1-adrenoceptors during IHC regimen abrogates cardioprotection.*** ***P < 0.001 *vs.* all other groups. Abbreviations as in Figures 1 and 2. Arrhythmia score = 3 in all 4 sham + metoprolol experiments, so SEM = 0 in this group.**

Reactive oxygen species and cardioprotection

The metoprolol studies indicate that β_1-adrenergic activity is *required* for IHC-induced cardioprotection, but do not prove that β_1-adrenergic activity alone is *sufficient* for cardioprotection. Systemic hypoxia-reoxygenation is a complex metabolic and physiological challenge to the organism, and β_1-adrenergic signaling pathways may collaborate with additional mechanisms to produce the adapted, cardioprotected phenotype. Specifically, we propose that reactive oxygen species (ROS), including superoxide ($\cdot O_2^-$), hydrogen peroxide (H_2O_2) and their metabolites, are generated during cyclic hypoxia-reoxygenation and evoke cardioprotection by modifying gene expression. $\cdot O_2^-$ [61,62] and H_2O_2[42,61,63] function as signaling molecules to activate cytoprotective genes, including heat shock proteins[42,61,62] and antioxidant enzymes.[63] ROS are generated primarily upon reoxygenation of hypoxic tissue.[38,62] During hypoxia, electrons accumulate in the mitochondrial respiratory chain. When O_2 is reintroduced, some of these electrons are directly transferred to O_2, generating $\cdot O_2^-$. Hypoxia-induced adrenergic activity may be another source of ROS. For example, norepinephrine induced $\cdot O_2^-$ formation in rabbit hearts.[64] In rat ventricular myocytes, isoproterenol stimulation generated ROS,[65] and pharmacological mimetics of the antioxidant enzyme superoxide dismutase (SOD) prevented norepinephrine-enhancement of Mn-SOD gene expression and protein synthesis.[59] Reoxygenation was necessary for robust activation of gene expression in hypoxic myocardium[40] and for activation of the redox-sensitive transcription factors NF-κB and AP-1 in human dermal microvascular endothelial cells.[66]

To investigate whether ROS generated by hypoxia-reoxygenation evoked cardioprotection, we modified the IHC protocol to limit ROS formation. In one set of experiments, four dogs received daily hypoxia exposures of the same total duration and intensity as in the IHC program, but without intervening reoxygenations; consequently, reoxygenation-induced ROS formation would have occurred only once per session, at the end of the sustained hypoxic bout. Although the total duration and intensity of hypoxia were identical to the IHC program, this continuous hypoxia conditioning program did not prevent infarction or arrhythmias (Figure 6), in striking contrast to IHC. Thus, intermittent reoxygenations are essential for IHC to produce cardioprotection.

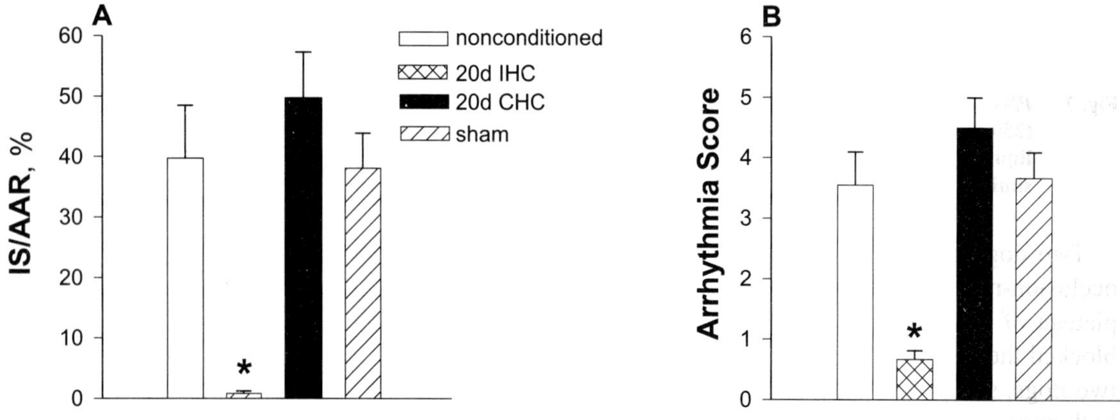

Fig. 6. ***IHC cardioprotection requires intermittent reoxygenations.*** **When hypoxia bouts were uninterrupted by 4 min reoxygenations (continuous hypoxia conditioning: CHC), hypoxia conditioning failed to protect myocardium from infarction (panel A) or arrhythmias (panel B). *P < 0.001 *vs.* all other groups.**

If ROS orchestrate IHC-induced cardioprotective adaptations, the presence of pharmacological antioxidants in the circulation during the hypoxia sessions could interfere with development of cardioprotection. To test this proposal, *N*-acetylcysteine (NAC), a broad-spectrum, membrane-permeable aminothiol antioxidant, was added to the chow diet consumed each morning of the IHC program, at a dose (250 mg/kg body wt) sufficient to abrogate oxidative stress produced by inspiratory resistive breathing in dogs.[67] NAC detoxifies a host of ROS and reactive nitrogen species including $\cdot O_2^-$,[68] H_2O_2,[42,68] $\cdot OH$,[69] nitric oxide[70] and peroxynitrite.[71-73]

Preliminary experiments were conducted to define the time-course of plasma NAC concentration after NAC intake (Figure 7). The antioxidant was detectable in systemic venous plasma 30 min after intake, and rose to a plateau at 2-3 h before subsiding. NAC was completely cleared from the plasma 24 h after intake, so it would neither accumulate in the plasma as the IHC program progresses, nor exert any residual antioxidant effects during the subsequent coronary occlusion-reperfusion experiments.

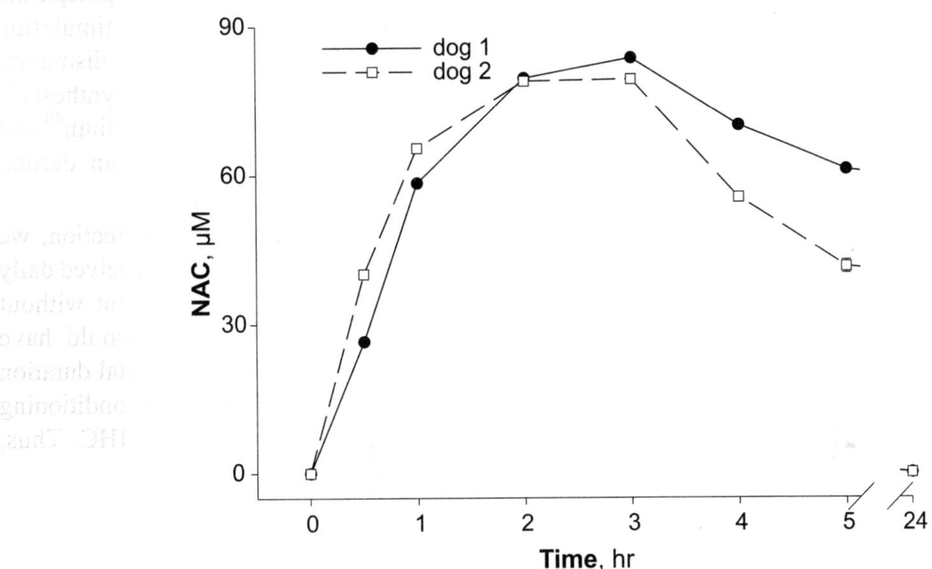

Fig. 7. ***Plasma* N-*acetylcysteine (NAC) after oral NAC intake.* Two different dogs consumed chow containing NAC (250 mg/kg body wt). NAC concentrations in systemic venous plasma were measured by high performance liquid chromatography.[114] Maximal NAC concentrations were attained at 2-3 h post-intake, and were similar in the two dogs. NAC had cleared from the plasma by 24 h post-intake.**

Two dogs have completed the IHC program with daily NAC consumption, followed by LAD occlusion-reperfusion protocol. IHC sessions commenced 2 h after NAC intake to coincide with the plateau of plasma NAC concentration (Figure 7). Antioxidant treatment during the IHC program blocked the cardioprotection. Twenty-eight and 41% of the ischemic myocardium infarcted in these two dogs, similar to values in non-hypoxic controls. Moreover, ventricular tachycardia occurred in both dogs and ventricular fibrillation in one dog during reperfusion. Although additional experiments must be conducted to confirm these early results, these two experiments support the proposal that ROS, generated during intermittent reoxygenation, orchestrate IHC-induced cardioprotective adaptations.

Collectively, these results indicate that both ROS and β-adrenergic activity are pivotal contributors to IHC-induced cardioprotection. Figure 8 presents a hypothetical mechanism incorporating both of these factors. We propose that β_1-adrenergic signaling and ROS impart cardioprotection by evoking gene expression and synthesis of cardioprotective proteins. As discussed above, ROS may be generated as a direct result of the pro-oxidant effects of hypoxia-reoxygenation on the mitochondrial respiratory chain, and/or by β_1-adrenergic stimulation of cardiomyocytes. In the latter case, β_1-adrenergic antagonism may abrogate cardioprotection at least in part by suppressing ROS formation. The dual requirement of β_1-adrenergic activity and ROS is concordant with the 'cardioprotective threshold' concept of Yellon and Downey,[36] wherein two or more stimuli must be combined to afford protection, and removal of only one stimulus is enough to block the protection as observed in this study.

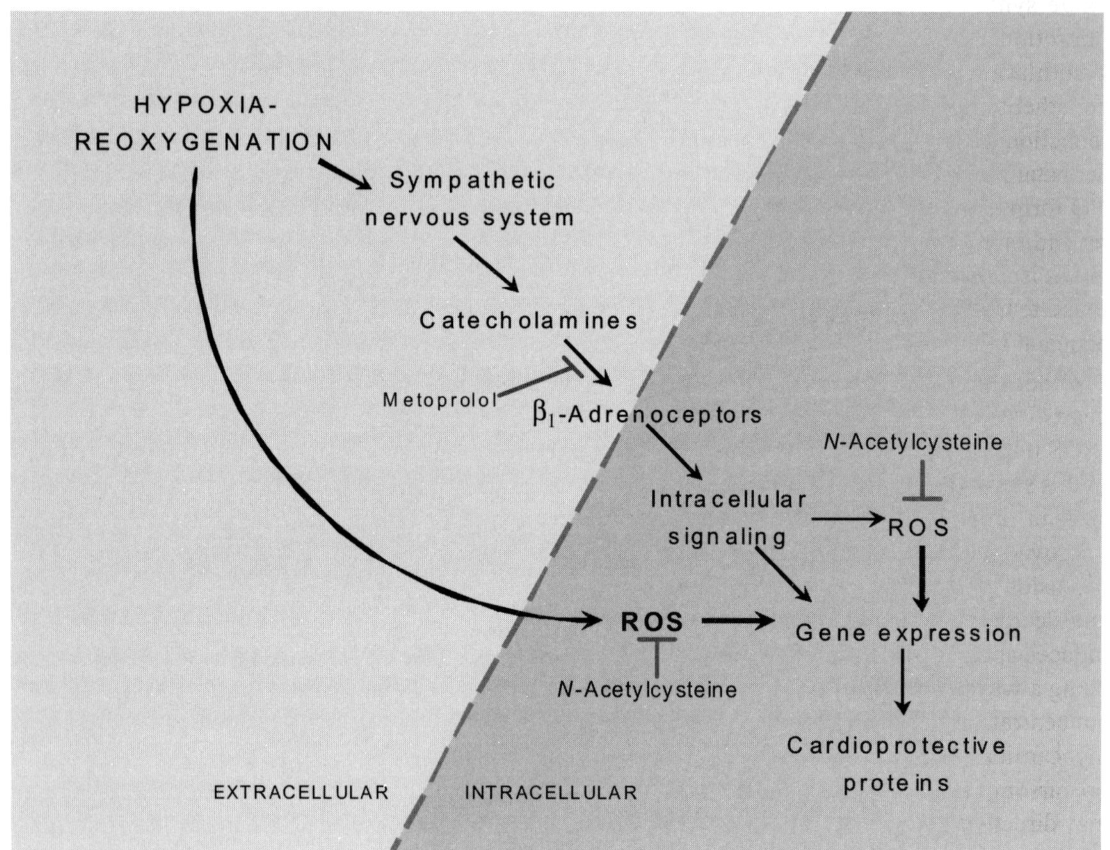

Fig. 8. ***Proposed mechanisms of IHC-induced cardioprotection mediated by sympathetic activity and reactive oxygen species (ROS).*** **Inhibition of these mechanisms by β_1-adrenoceptor antagonist metoprolol and antioxidant *N*-acetylcycteine is indicated. For simplicity, the scheme omits intracellular mediators and effectors of signaling cascades.**

The progressive development of cardioprotection over the 20-d IHC program, and the pivotal contributions of intermittent reoxygenation, β_1-adrenergic activity and ROS, strongly suggest that cardioprotection is imparted by *de novo* synthesis of cardioprotective proteins. Although the protective proteins haven't been identified in the current model, the roster of hypoxia-, β-adrenergic- and ROS-responsive genes offers many candidates, including heat shock proteins, antioxidant enzymes, erythropoietin,[26] glucose transporters and glycolytic enzymes, and transcription factors.[38,74]

Suppression of nitric oxide synthesis: a cardioprotective mechanism?

Whether NO is protective or harmful to ischemic and post-ischemic myocardium is a matter of debate, with evidence supporting either alternative.[75] Numerous studies have documented salutary effects of NO, particularly in non-preconditioned myocardium (reference 76 and references therein), and nitric oxide synthase (NOS) inhibitors reportedly exacerbate ischemia-reperfusion-induced myocardial infarction[77,78] and lethal arrhythmias.[79] NO's cardioprotective actions may include coronary vasodilation,[80] suppression of inflammatory neutrophil infiltration,[81] preservation of coronary endothelial function,[82] enhancement of myofilament Ca^{2+} sensitivity and contractile efficiency,[83] and reduction of myocardial O_2 demand.[84,85] Continuous hypoxia or prolonged intermittent hypoxia in neonatal rabbit and adult rat hearts evokes NOS expression, which increases myocardial capacity for NO formation. Thus, hypoxia increases NOS content[12,14,25,86-90] and abundance of its mRNA[12,86,87] in rat and rabbit myocardium. Inhibitors of the inducible NOS (iNOS) isoform increased infarct size of hearts isolated from hypoxia-conditioned mice and subjected to global ischemia/reperfusion.[25] NO was found to mediate hypoxic conditioning against post-ischemic contractile stunning of isolated, perfused neonatal rabbit heart.[12,91]

Other studies reveal cardiotoxic actions of NO. NOS inhibition lessened infarction in *in situ* rabbit hearts.[92] Wang *et al.*[90] detected intense accumulation of inducible NOS (iNOS) protein and increased iNOS mRNA expression in ischemic myocardium during 24 h coronary occlusion of *in situ* rat hearts. In this study, the selective iNOS inhibitors aminoguanidine and *S*-methylisothiourea *decreased* infarct size, in proportion to the extent of iNOS inhibition. Aminoguanidine also reduced infarction and improved recovery of contractile performance in *in situ* rabbit hearts following 30 min coronary artery occlusion.[93] The nonselective NOS inhibitor *L*-NAME improved post-ischemic contractile recovery and decreased arrhythmias in non-preconditioned rat hearts.[94] NO scavengers prevented cytokine-induced apoptosis in cultured rat ventricular cardiomyocytes.[95] These studies suggest that NO, far from being always cardioprotective, may have contributed to myocardial ischemic injury. Indeed, high NO concentrations are cytotoxic.[75,92,96,97] During oxidative stress, as with reperfusion of ischemic myocardium, NO rapidly and irreversibly condenses with $\cdot O_2^-$[98] to generate peroxynitrite ($ONOO^-$), the precursor of a host of highly reactive, cytotoxic ROS and reactive nitrogen species.[75] Furthermore, NO may directly attack catalytic Fe·S-centers of respiratory chain components and the Krebs cycle enzyme aconitase, and thereby disrupt ATP production.[92,99-101] Moreover, excess NO inactivates creatine kinase in isolated rabbit hearts[102] and the antioxidant enzymes glutathione peroxidase, catalase and Mn-SOD in cultured rat glial cells.[103] Ca^{2+} activates endothelial NOS (eNOS),[104] the principal constitutive NOS isoform in cardiomyocytes. Intracellular Ca^{2+} overload upon reperfusion has been implicated in the pathogenesis of myocardial infarction and arrhythmias. Ca^{2+}-activated eNOS could exacerbate myocardial reperfusion injury by producing toxic amounts of NO, some of which may have been produced in cardiomyocytes.

Intermittent hypoxia conditioning suppressed NOS content and activity in guinea-pig atria[105] and mouse brain,[106] and dampened expression of eNOS in vascular endothelial cells.[107,108] These reports raise the possibility that IHC, in contrast to chronic hypoxia, could suppress myocardial NOS. Accordingly, we tested the hypothesis that IHC lowers myocardial NOS activity and thereby dampens NO formation upon LAD reperfusion.[109] NOS activity was measured (colorimetric assay) in left and right ventricular myocardium of IHC and non-hypoxic control dogs. Hypoxic conditioning lowered NOS activity by ~60% in the left circumflex and LAD perfusion territories of the left ventricular free wall, and also in the right ventricle (Figure 9). Immunoblotting revealed a parallel decrease in eNOS content in LAD-perfused myocardium of IHC *vs.* control dogs.[109]

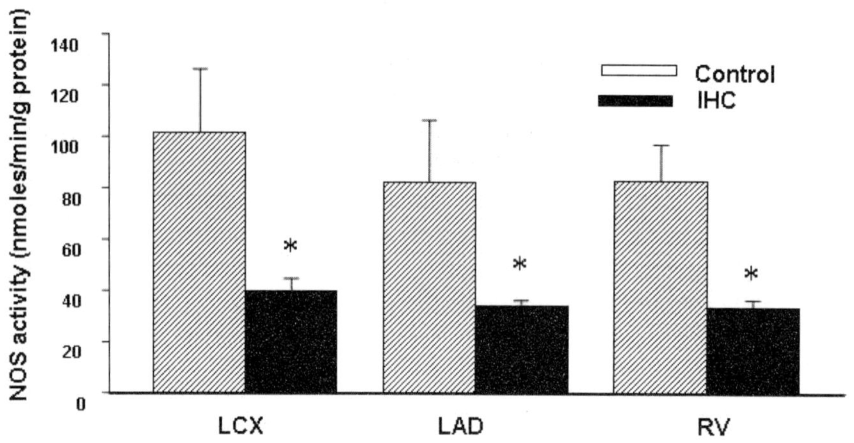

Fig. 9. ***Nitric oxide synthase (NOS) activities in left and right ventricular myocardium.*** **NOS activity was measured in left ventricular myocardium from the left circumflex (LCX) and left anterior descending (LAD) coronary artery perfusion territories, and in right ventricular myocardium. Values are means ± SEM from 7 non-hypoxic control and 6 IHC dogs. *P < 0.05 *vs.* control.**

Next, the hypothesis that the IHC-induced suppression of NOS decreased explosive NO formation during LAD reperfusion was tested. Myocardial release of nitrite (NO_2^-), a stable product of NO metabolism, indexed NO formation.[110,111] A sharp burst of NO_2^- release occurred in non-hypoxic controls, which peaked at 2 min reperfusion and then quickly subsided. No such burst occurred in the IHC dogs. Consequently, IHC lowered cumulative NO_2^- release during the first 5 min reperfusion by 70% (Figure 10A), commensurate with the reductions in NOS activity (Figure 9). However, cumulative LAD flow during the same period was unaltered by IHC (Figure 10B). Thus, the reduced NO formation did not attenuate reactive hyperemia, suggesting that NO is not an important contributor to coronary vasodilation during reperfusion. Moreover, decreased NO formation in reperfused myocardium of IHC dogs could not be ascribed to decreased vascular shear stress, an important physiological stimulus of eNOS.[112]

How might IHC suppress NOS expression? We propose that repeated, intermittent episodes of normobaric hypoxia and reoxygenation could provoke formation of moderate, sub-toxic amounts of NO in canine ventricular myocardium. This NO, acting alone or in collaboration with ROS formed upon reoxygenation, could suppresses expression of eNOS[15,113] and, thus, progressively lower

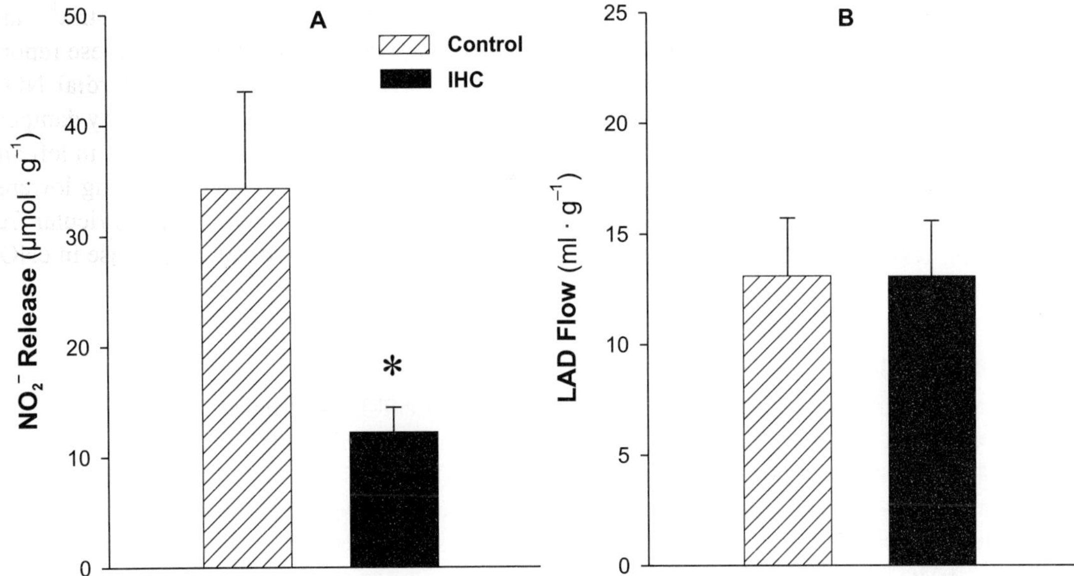

Fig. 10. ***Cumulative left ventricular myocardial NO_2^- release (panel A) and LAD flow (panel B) during the first 5 min of LAD reperfusion.*** **Means ± SEM from 5 experiments per group. *P < 0.05 *vs.* control.**

myocardial NOS activity over the course of the conditioning program. By dampening the lethal NO burst upon release of coronary occlusion, suppression of eNOS could ameliorate myocardial injury. The residual NOS activity in the hypoxia-conditioned myocardium may be sufficient to produce moderate, cytoprotective amounts of NO upon reperfusion.

Future Directions

The studies described herein demonstrate that intermittent, normobaric hypoxia conditioning of dogs imparts robust cardioprotection against ischemia-reperfusion injury. Intermittent reoxygenation of the dogs during hypoxia sessions is essential to evoke the protection. Myocardial resistance to ischemia-reperfusion injury developed gradually over the course of the 20-d IHC program. Neither collateral blood flow, hematocrit, nor hemoglobin content were altered by IHC. Although the mechanisms of this remarkable cardioprotection have not been fully delineated, formation of reactive O_2 metabolites and β-adrenergic activation of the myocardium during cyclic hypoxia-reoxygenation appear to be pivotal to the developing protection. IHC-induced suppression of NOS activity may limit ischemic injury by dampening the toxic burst of NO formation at the onset of coronary reperfusion.

Intermittent hypoxia conditioning appears to constitute a novel, noninvasive paradigm to increase myocardial resistance to ischemic injury, and could potentially be used for prophylaxis of myocardial infarction and arrhythmias in individuals at risk of myocardial ischemia. However, several practical issues remain to be resolved in preclinical studies. The persistence of the cardioprotection, *i.e.* the rate at which the protection subsides once IHC is discontinued, must be defined. Another practical issue is whether IHC is effective against coronary occlusions exceeding 1 h, since victims of heart attack are often slow to respond to the symptoms, and travel time to the coronary unit can be prolonged. It is uncertain if less intense hypoxia is capable of imparting significant cardioprotection in this model. Importantly, the current findings suggest that persons undergoing treatment with β-adrenergic antagonists or antioxidants might not benefit from IHC. Also, it must be emphasized that IHC is a strategy that is cardioprotective only if implemented in advance of an ischemic event.

Important questions remain regarding the cardioprotective mechanisms of IHC. The specific ROS spawned by hypoxia-reoxygenation and the cardioprotective proteins produced in response to ROS and/or β-adrenergic activity have not been identified. Although myocardial eNOS content was suppressed, the response of the other NOS isoforms to IHC has not been determined, and the targets of excess NO and its toxic metabolites are unknown.

Acknowledgements

This work was supported by grants AT-003598, HL-064785 and HL-071684 from the US National Institutes of Health, grant 047-014-016 from the Netherlands Organization for Scientific Research, grant 03-04-49065 from the Russian Foundation for Basic Research, and faculty research grant no. 67703 from the University of North Texas Health Science Center (UNTHSC). MGR was supported by a fellowship from the UNTHSC Graduate School of Biomedical Sciences. Portions of this work were completed by MGR in partial fulfillment of the requirements for the Master of Science degree at UNTHSC. Arti B. Sharma, M.B.B.S., Ph.D., E. Marty Knott, D.O., Ph.D., Linda L. Howard, Janneth Guarin, Joseph Menefee, and Kevin Oguayo assisted with conditioning of the dogs. Arthur G. Williams, Jr. provided expert assistance with LAD occlusion-reperfusion experiments and measurements of infarct size. Jie Sun, B.S. measured *N*-acetylcysteine and helped prepare the figures. Animal experimentation was approved by the Institutional Animal Care and Use Committee of UNTHSC, and was conducted in accordance with the *Guide to the Care and Use of Laboratory Animals* (NIH publication no. 85-23, revised 1996).

References

1. Hurtado A. Some clinical aspects of life at high altitudes. Ann Intern Med 1960; 53: 247-258.
2. Poupa O, Krofta K, Procházka J, Turek Z. Acclimatization to simulated high altitude and acute cardiac necrosis. Fed Proc 1966; 25: 1243-1246.
3. Powell FL, Garcia N. Physiological effects of intermittent hypoxia. High Alt Med Biol 2000; 1: 125-136.
4. Levine BD. Intermittent hypoxic training: fact and fancy. High Alt Med Biol 2002; 3: 177-193.
5. Tin'kov AN, Aksenov VA. Effects of intermittent hypobaric hypoxia on blood lipid concentrations in male coronary heart disease patients. High Alt Med Biol 2002; 3: 277-282.
6. Serebrovskaya TV. Intermittent hypoxia research in the former Soviet Union and the Commonwealth of Independent States: history and review of the concept and selected applications. High Alt Med Biol 2002; 3: 205-221.
7. Burtscher M, Pachinger O, Ehrenbourg I, Mitterbauer G, Faulhaber M, Pühringer R, Tkatchouk E. Intermittent hypoxia increases exercise tolerance in elderly men with and without coronary artery disease. Intl J Cardiol 2004; 96: 247-254.
8. Meerson FZ, Kots IaI, Volovich VB, Aleshin IA, Tverdokhlib VP. Positive effect of the adaptation to periodic anoxia in patients with idiopathic arrhythmia in the atmosphere exposure chamber. Kardiologiia 1989; 29: 53-56.
9. Ehrenbourg IV, Gorbatchenkov AA. Clinical characteristics of angina pectoris patients in relation to interval hypoxic training effect. Hypox Med J 1993; 2/93: 10-13.

10. Lyamina NP. Changes in plasma β-endorphine in patients with coronary heart disease exposed to psychoemotional load and under the influence of adaptation to hypoxia. Hypox Med J 1995; 2/95: 17-18.
11. Asemu G, Papoušek F, Oštádal B, Kolár F. Adaptation to high altitude hypoxia protects the rat heart against ischemia-induced arrhythmias. Involvement of mitochondrial K_{ATP} channel. J Mol Cell Cardiol 1999; 31: 1821-1831.
12. Baker JE, Holman P, Kalyanaraman B, Griffith OW, Pritchard KA Jr. Adaptation to chronic hypoxia confers tolerance to subsequent myocardial ischemia by increased nitric oxide production. Ann NY Acad Sci 1999; 874: 236-253.
13. Eells JT, Henry MM, Gross GJ, Baker JE. Increased mitochondrial K_{ATP} channel activity during chronic myocardial hypoxia: is cardioprotection mediated by improved bioenergetics? Circ Res 2000; 87: 915-921.
14. Forkel J, Chen X, Wandinger S, Keser F, Duschin A, Schwanke U, Frede S, Massoudy P, Schulz R, Heusch G. Responses of chronically hypoxic rat hearts to ischemia: K_{ATP} channel blockade does not abolish increased RV tolerance to ischemia. Am J Physiol Heart Circ Physiol 2004; 286: H545-H551.
15. Manukhina EB, Mashina SYu, Smirin BV, Lyamina NP, Senchikhin VN, Vanin AF, Malyshev IYu. Role of nitric oxide in adaptation to hypoxia and adaptive defense. Physiol Res 2000; 49: 89-97.
16. Neckár J, Szárszoi O, Koten L, Papousek F, Ost'ádal B, Grover GJ, Kolár F. Effects of mitochondrial K_{ATP} modulators on cardioprotection induced by chronic high altitude hypoxia in rats. Cardiovasc Res 2002; 55: 567-575.
17. Neckár J, Ost'ádal B, Kolár F. Myocardial infarct size-limiting effect of chronic hypoxia persists for five weeks of normoxic recovery. Physiol Res 2004; 52: 621-628.
18. Zhong N, Zhang Y, Fang QZ, Zhou ZN. Intermittent hypoxia exposure-induced heat-shock protein 70 expression increases resistance of rat heart to ischemic injury. Acta Pharmacol Sin 2000; 21: 467-472.
19. Zhong N, Zhang Y, Zhu H-F, Wang J-C, Fang Q-Z, Zhou Z-N. Myocardial capillary angiogenesis and coronary flow in ischemia tolerance rat by adaptation to intermittent high altitude hypoxia. Acta Pharmacol Sin 2002; 23: 305-310.
20. Zhu H-F, Dong J-W, Zhu W-Z, Ding H-L, Zhou Z-N. ATP-dependent potassium channels involved in the cardiac protection induced by intermittent hypoxia against ischemia/reperfusion injury. Life Sci 2003; 73: 1275-1287.
21. Zhu W-Z, Dong J-W, Ding H-L, Yang H-T, Zhou Z-N. Postnatal development in intermittent hypoxia enhances resistance to myocardial ischemia/reperfusion in male rats. Eur J Apply Physiol 2004; 91: 716-722.
22. Kolár F, Oštádal B. Molecular mechanisms of cardiac protection by adaptation to chronic hypoxia. Physiol Res 2004; 53 (suppl 1): S3-S13.
23. Meerson FZ, Ustinova EE, Orlova EH. Prevention and elimination of heart arrhythmias by adaptation to intermittent high altitude hypoxia. Clin Cardiol 1987; 10: 783-789.
24. Asemu G, Neckár J, Szárszoi O, Papoušek F, Oštádal B, Kolár F. Effects of adaptation to intermittent high altitude hypoxia on ischemic ventricular arrhythmias in rats. Physiol Res 2000; 49: 597-606.
25. Xi L, Tekin D, Gursoy E, Salloum F, Levasseur JE, Kukreja RC. Evidence that NOS2 acts as a trigger and mediator of late preconditioning induced by acute systemic hypoxia. Am J Physiol Heart Circ Physiol 2002; 283: H5-H12.
26. Cai Z, Manalo DJ, Wei G, Rodriguez ER, Fox-Talbot K, Lu H, Zweier JL, Semenza GL. Hearts from rodents exposed to intermittent hypoxia or erythropoietin are protected against ischemia-reperfusion injury. Circulation 2003; 108: 79-85.
27. Zong P, Setty S, Sun W, Martinez R, Tune JD, Ehrenburg IV, Tkatchouk EN, Mallet RT, Downey HF. Intermittent hypoxic training protects canine myocardium from infarction. Exp Biol Med 2004; 229: 806-812.
28. Mallet RT, Ryou M-G, Williams AG Jr., Howard L, Downey HF. b_1-Adrenergic receptor antagonism abrogates cardioprotective effects of intermittent hypoxia. Basic Res Cardiol 2006; 101: 436-446.

29. Heymann MA, Payne BD, Hoffman JIE, Rudolph AM. Blood flow measurements with radionuclide-labeled particles. Prog Cardiovasc Dis 1997; 20: 55-79.
30. Walker MJA, Curtis MJ, Hearse DJ, Campbell RWF, Janse MJ, Yellon DM, Cobbe SM, Coker SJ, Harness JB, Harron DWG, Higgins AJ, Julian DG, Lab MJ, Manning AS, Northover BJ, Parratt JR, Riemersma RA, Riva E, Russell DC, Sheridan DJ, Winslow E, Woodward B. The Lambeth conventions: guidelines for the study of arrhythmias in ischaemia, infarction, and reperfusion. Cardiovasc Res 1988; 22: 447-455.
31. Shizukuda Y, Mallet RT, Lee SC, Downey HF. Hypoxic preconditioning of ischaemic canine myocardium. Cardiovasc Res 1992; 26: 534-542.
32. Nachlas MM, Shnitka TK. Macroscopic identification of early myocardial infarcts by alterations in dehydrogenase activity. Am J Pathol 1963; 42: 379-405.
33. Deindl E, Kolár F, Neubauer E, Vogel S, Schaper W, Oštádal B. Effect of intermittent high altitude hypoxia on gene expression in rat heart and lung. Physiol Res 2003; 52: 147-157.
34. Daneshrad Z, Garcia-Riera MP, Verdys M, Rossi A. Differential responses to chronic hypoxia and dietary restriction of aerobic capacity and enzyme levels in the rat myocardium. Mol Cell Biochem 2000; 210: 159-166.
35. Morel S, Milano G, Ludunge KM, Corno AF, Samaja M, Fleury S, Bonny C, Kappenberger L, von Segesser LK, Vassalli G. Brief reoxygenation episodes during chronic hypoxia enhance posthypoxic recovery of LV function. Basic Res Cardiol 2006; 101: 336-345.
36. Yellon DM, Downey JM. Preconditioning the myocardium: from cellular physiology to clinical cardiology. Physiol Rev 2003; 83: 1113-1151.
37. Meerson FZ, Malyshev IYu, Zamotrinsky AV. Differences in adaptive stabilization of structures in response to stress and hypoxia relate with the accumulation of hsp70 isoforms. Mol Cell Biochem 1992; 111: 87-95.
38. Prabhakar NR. Oxygen sensing during intermittent hypoxia: cellular and molecular mechanisms. J Appl Physiol 2001; 90: 1986-1994.
39. Iwaki K, Chi S-H, Dillmann WH, Mestril R. Induction of HSP70 in cultured rat neonatal cardiomyocytes by hypoxia and metabolic stress. Circulation 1993; 87: 2023-2032.
40. Piacentini L, Karliner JS. Altered gene expression during hypoxia and reoxygenation of the heart. Pharmacol Therapeut 1999; 83: 21-37.
41. Lee PJ, Jiang B-H, Chin BY, Iyer NV, Alam J, Semenza GL, Choi AMK. Hypoxia-inducible factor-1 mediates transcriptional activation of the heme oxygenase-1 gene in response to hypoxia. J Biol Chem 1997; 272: 5375-5381.
42. Borger DR, Essig DA. Induction of HSP 32 gene in hypoxic cardiomyocytes is attenuated by treatment with N-acetyl-L-cysteine. Am J Physiol Heart Circ Physiol 1998; 274: H965-H973.
43. Zhou X, Zhai X, Ashraf M. Direct evidence that initial oxidative stress triggered by preconditioning contributes to second window of protection by endogenous antioxidant enzyme in myocytes. Circulation 1996; 93: 1177-1184.
44. Nakanishi K, Tajima F, Nakamura A, Yagura S, Ookawara T, Yamashita H, Suzuki K, Taniguchi N, Ohno H. Effects of hypobaric hypoxia on antioxidant enzymes in rats. J Physiol 1995; 489.3: 869-876.
45. Irlbeck M, Iwai T, Lerner T, Zimmer H-G. Effects of angiotensin II receptor blockade on hypoxia-induced right ventricular hypertrophy in rats. J Mol Cell Cardiol 1997; 29: 2931-2939.
46. Adhikary G, Kline D, Yuan G, Kumar GK, Simonson MS, Cherniack NS, Prabhakar NR. Gene regulation during intermittent hypoxia: evidence for the involvement of reactive oxygen species. In: *Frontiers in Modeling and Control of Breathing* (Poon C-S, Kazemi H, eds). New York: Kluwer/Plenum, 2001, pp. 297-302.
47. Yuan G, Adhikary G, McCormick AA, Holcroft JJ, Kumar GK, Prabhakar NR. Role of oxidative stress in intermittent hypoxia-induced immediate early gene activation in rat PC12 cells. J Physiol 2004; 557.3: 773-783.

48. Aleshin AN, Sawa Y, Kitagawa-Sakakida S, Bando Y, Ono M, Memon IA, Tohyama M, Ogawa S, Matsuda H. 150-kDa oxygen-regulated protein attenuates myocardial ischemia-reperfusion injury in rat heart. J Mol Cell Cardiol 2005; 38: 517-525.

49. Vogel JA, Hartley LH, Cruz JC, Hogan RP. Cardiac output during exercise in sea-level residents at sea level and high altitude. J Appl Physiol 1974; 36: 169-172.

50. Mazzeo RS, Bender PR, Brooks GA, Butterfield GE, Groves BM, Sutton JR, Wolfel EE, Reeves JT. Arterial catecholamine responses during exercise with acute and chronic high-altitude exposure. Am J Physiol Endocrinol Metab 1991; 261: E419-E424.

51. Huang M-H, Bahl JJ, Wu Y, Hu F, Larson DF, Roeske WR, Ewy GA. Neuroendocrine properties of intrinsic cardiac adrenergic cells in fetal rat heart. Am J Physiol Heart Circ Physiol 2005; 288: H497-H503.

52. Asimakis GK, Inners-McBride K, Conti VR, Yang CJ. Transient β adrenergic stimulation can precondition the rat heart against postischaemic contractile dysfunction. Cardiovasc Res 1994; 28: 1726-1734.

53. Miyawaki H, Ashraf M. Isoproterenol mimics calcium preconditioning-induced protection against ischemia. Am J Physiol Heart Circ Physiol 1997; 272: H927-H936.

54. Nasa Y, Yabe K, Takeo S. β-adrenoceptor stimulation-mediated preconditioning-like cardioprotection in perfused rat hearts. J Cardiovasc Pharmacol 1997;29:436-443.

55. Frances C, Nazeyrollas P, Prevost A, Moreau F, Pisani J, Davani S, Kantelip JP, Millart H. Role of β_1- and β_2-adrenoceptor subtypes in preconditioning against myocardial dysfunction after ischemia and reperfusion. J Cardiovasc Pharmacol 2003; 41: 396-405.

56. Végh A, Parratt JR. Noradrenaline, infused locally, reduces arrhythmia severity during coronary artery occlusion in anesthetized dogs. Cardiovasc Res 2002; 55: 53-63.

57. Kovanecz I, Papp JG, Szerkeres L. Long-term ischaemic preconditioning of the heart induced by repeated β-adrenergic stress. Acta Physiol Hung 1996; 84: 297-298.

58. Meng X, Brown JM, Ao L, Banerjee A, Harken AH. Norepinephrine induces cardiac heat shock protein 70 and delayed cardioprotection in the rat through α_1 receptors. Cardiovasc Res 1996; 32: 374-383

59. Amin JK, Xiao L, Pimental DR, Pagano PJ, Singh K, Sawyer DB, Colucci WS. Reactive oxygen species mediate alpha-adrenergic receptor-stimulated hypertrophy in adult rat ventricular myocytes. J Mol Cell Cardiol 2001; 33: 131-139.

60. Panchal AR, Stanley WC, Kerner J, Sabbah HN. β-receptor blockade decreases carnitine palmitoyl transferase I activity in dogs with heart failure. J Card Fail 1998; 4: 121-126.

61. Schoeniger LO, Curtis W, Esnaola NF, Beck SC, Gardner T, Buchman TG. Myocardial heat shock gene expression in pigs is dependent on superoxide anion generated at reperfusion. Shock 1994; 1: 31-35.

62. Kukreja RC, Kontos MC, Loesser KE, Batra SK, Qian YZ, Gbur CJ Jr., Naseem SA, Jesse RL, Hess ML. Am J Physiol Heart Circ Physiol 1994; 267: H2213-H2219.

63. Csonka C, Pataki T, Kovacs P, Müller SL, Schroeter ML, Tosaki A, Blasig IE. Effects of oxidative stress on the expression of antioxidative defense enzymes in spontaneously hypertensive rat hearts. Free Radic Biol Med 2000; 29: 612-619.

64. Rump AF, Schierholz J, Klaus W. Studies on the cardiotoxicity of noradrenaline in isolated rabbit hearts. Arzneimittelforschung 2002; 52: 543-551.

65. Persoon-Rothert M, van der Valk-Kokshoorn EJ, Egas-Kenniphaas JM, Mauve I, van der Laarse A. Isoproterenol-induced cytotoxicity in neonatal rat heart cell cultures is mediated by free radical formation. J Mol Cell Cardiol 1989; 21: 1285-1291.

66. Natarajan R, Fisher BJ, Jones DG, Ghosh S, Fowler AA 3rd. Reoxygenating microvascular endothelium exhibits temporal dissociation of NF-κB and AP-1 activation. Free Radic Biol Med 2002; 32: 1033-1045.

67. Barreiro E, Sanchez D, Galdiz JB, Hussain SN, Gea J. N-acetylcysteine increases manganese superoxide dismutase activity in septic rat diaphragms. Eur Respir J 2005; 26: 1032-1039.

68. Genet S, Kale RK, Baquer NZ. Effects of free radicals on cytosolic creatine kinase activities and protection by antioxidant enzymes and sulfhydryl compounds. Mol Cell Biochem 2000; 210: 23-28.
69. Brunet J, Boily M-J, Cordeau S, Des Rosiers C. Effects of N-acetylcysteine in the rat heart reperfused after low-flow ischemia: evidence for a direct scavenging of hydroxyl radicals and a nitric oxide-dependent increase in coronary flow. Free Radic Biol Med 1995; 19: 627-638.
70. Foresti R, Clark JE, Green CJ, Motterlini R. Thiol compounds interact with nitric oxide in regulating heme oxygenase-1 induction in endothelial cells. J Biol Chem 1997; 272: 18411-18417.
71. Kondo H, Takahashi M, Niki E. Peroxynitrite-induced hemolysisof human erythrocytes and its inhibition by antioxidants. FEBS Lett 1997; 413: 236-238.
72. Koppal T, Drake J, Butterfield DA. In vivo modulation of rodent glutathione and its role in peroxynitrite-induced neocortical synaptosomal membrane protein damage. Biochim Biophys Acta 1999; 1453: 407-411.
73. Cabassi A, Dumont EC, Girouard H, Bouchard JF, Le Jossec M, Lamontagne D, Besner JG, de Champlain J. Effects of chronic N-acetylcysteine treatment on the actions of peroxynitrite on aortic vascular reactivity in hypertensive rats. J Hypertens 2001; 19: 1233-1244.
74. Wenger RH. Cellular adaptation to hypoxia: O_2-sensing protein hydroxylases, hypoxia-inducible transcription factors, and O_2-regulated gene expression. FASEB J 2002; 16: 1151-1162.
75. Manukhina EB, Downey HF, Mallet RT. Role of nitric oxide in cardiovascular adaptation to intermittent hypoxia. Exp Biol Med 2006; 231: 343-365.
76. Bolli R. Cardioprotective function of inducible nitric oxide synthase and role of nitric oxide in myocardial ischemia and preconditioning: an overview of a decade of research. J Mol Cell Cardiol 2001; 33: 1897-1918.
77. Williams MW, Taft CS, Ramnauth S, Zhao ZQ, Vinten-Johansen J. Endogenous nitric oxide (NO) protects against ischaemia-reperfusion injury in the rabbit. Cardiovasc Res 1995; 30: 79-86.
78. Masini E, Salvemini D, Ndisang JF, Gai P, Berni L, Moncini M, Bianchi S, Mannaioni PF. Cardioprotective activity of endogenous and exogenous nitric oxide on ischaemia reperfusion injury in isolated guinea pig hearts. Inflamm Res 1999; 48: 561-568.
79. Pabla R, Curtis MJ. Endogenous protection against reperfusion-induced ventricular fibrillation: role of neuronal versus non-neuronal sources of nitric oxide and species dependence in the rat versus rabbit isolated heart. J Mol Cell Cardiol 1996; 28: 2097-2110.
80. Gutterman DD, Miura H, Liu Y. Redox modulation of vascular tone: focus of potassium channel mechanisms of dilation. Arterioscler Thromb Vasc Biol 2005; 25: 671-678.
81. Pabla R, Buda AJ, Flynn DM, Blesse SA, Shin AM, Curtis MJ, Lefer DJ. Nitric oxide attenuates neutrophil-mediated myocardial contractile dysfunction after ischemia and reperfusion. Circ Res 1996; 78: 65-72.
82. Weyrich AS, Ma XL, Lefer AM. The role of L-arginine in ameliorating reperfusion injury after myocardial ischemia. Circulation 1992; 86: 279-288.
83. Heusch G, Post H, Michel MC, Kelm M, Schulz R. Endogenous nitric oxide and myocardial adaptation to ischemia. Circ Res 2000; 87: 146-152.
84. Shen W, Xu X, Ochoa M, Zhao G, Wolin MS, Hintze TH. Role of nitric oxide in the regulation of oxygen consumption in conscious dogs. Circ Res 1994; 75: 1086-1095.
85. Setty S, Sun W, Martinez R, Downey HF, Tune JD. Alpha-adrenoceptor-mediated coronary vasoconstriction is augmented during exercise in experimental diabetes mellitus. J Appl Physiol 2004; 97: 431-438.
86. Felaco M, Grilli A, Gorbunov N, Di Napoli P, De Lutiis MA, Di Giulio C, Taccardi AA, Barsotti A, Barbacane RC, Reale M, Conti P. Endothelial NOS expression and ischemia-reperfusion in isolated working rat heart from hypoxic and hyperoxic conditions. Biochim Biophys Acta 2000; 1524: 203-211.
87. Grilli A, DeLutiis MA, Patruno A, Speranza L, Cataldi A, Centurione L, Taccardi AA, Di Napoli P, De Caterina R, Barbacane R, Conti P, Felaco M. Effect of chronic hypoxia on inducible nitric oxide synthase expression in rat myocardial tissue. Exp Biol Med 2003; 228: 935-942.

88. Rouet-Benzineb P, Eddahibi S, Raffestin B, Laplace M, Depond S, Adnot S, Crozatier B. Induction of cardiac nitric oxide synthase 2 in rats exposed to chronic hypoxia. J Mol Cell Cardiol 1999; 31: 1697-1708.
89. Shi Y, Pritchard KA Jr., Holman P, Rafiee P, Griffith OW, Kalyanaraman B, Baker JE. Chronic myocardial hypoxia increases nitric oxide synthase and decreases caveolin-3. Free Radic Biol Med 2000; 29: 695-703.
90. Wang X, Kevil CG, Gladwin MT, Lefer DJ. Cytoprotective effects of nitrite during in vivo ischemia-reperfusion of the heart and liver. J Clin Invest 2005; 115: 1232-1240.
91. Baker JE, Holman P, Kalyanaraman B, Pritchard KA Jr. Adaptation of hearts to chronic hypoxia increases tolerance to subsequent ischemia by increased nitric oxide production. Adv Exp Med Biol 1998; 454: 203-217.
92. Patel VC, Yellon DM, Singh KJ, Neild GH, Woolfson RG. Inhibition of nitric oxide limits infarct size in the in situ rabbit heart. Biochem Biophys Res Commun 1993; 194: 234-238.
93. Wildhirt SM, Schulze C, Conrad N, Kornberg A, Horstman D, Reichart B. Aminoguanidine inhibits inducible NOS and reverses cardiac dysfunction late after ischemia and reperfusion – implications for iNOS-mediated myocardial stunning. Thorac Cardiovasc Surg 1999; 47: 137-143.
94. Andelová E, Barteková M, Pancza D, Styk J, Ravingerová T. The role of NO in ischemia/reperfusion injury in isolated rat heart. Gen Physiol Biophys 2005; 24: 411-426.
95. Arstall MA, Sawyer DB, Fukazawa R, Kelly RA. Cytokine-mediated apoptosis in cardiac myocytes. Circ Res 1999; 85: 829-840.
96. Beckman JS, Beckman TW, Chen J, Marshall PA, Freeman BA. Apparent hydroxyl radical production by peroxynitrite: implications for endothelial injury from nitric oxide and superoxide. Proc Natl Acad Sci USA 1990; 87: 1620-1624.
97. Malyshev IYu, Manukhina EB, Mikoyan VD, Kubrina LN, Vanin AF. Nitric oxide is involved in heat-induced HSP70 accumulation. FEBS Lett 1995; 370: 159-162.
98. Beckman JS, Koppenol WH. Nitric oxide, superoxide, and peroxynitrite: the good, the bad, and the ugly. Am J Physiol Cell Physiol 1996; 271: C1424-C1437.
99. Geng Y, Hansson GK, Holme E. Interferon-gamma and tumor necrosis factor synergize to induce nitric oxide production and inhibit mitochondrial respiration in vascular smooth muscle cells. Circ Res 1992; 71: 1268-1276.
100. Takeuchi K, McGowan FX, Danh H-C, Glynn P, Simplaceanu E, del Nido PJ. Direct detrimental effects of L-arginine upon ischemia-reperfusion injury to myocardium. J Mol Cell Cardiol 1995; 27: 1405-1414.
101. Yuyuma K, Yamamoto H, Nishizaki I, Kato T, Sora I, Yamamoto T. Caspase-independent cell death by low concentrations of nitric oxide in PC12 cells: involvement of cytochrome c oxidase inhibition and the production of reactive oxygen species in mitochondria. J Neurosci Res 2003; 73: 351-363.
102. Gross WL, Bak MI, Ingwall JS, Arstall MA, Smith TW, Balligand J-L, Kelly RA. Nitric oxide inhibits creatine kinase and regulates rat heart contractile reserve. Proc Natl Acad Sci USA 1996; 93: 5604-5609.
103. Dobashi K, Pahan K, Chahal A, Singh I. Modulation of endogenous antioxidant enzymes by nitric oxide in rat C6 glial cells. J Neurochem 1997; 68: 1896-1903.
104. Andrew PJ, Mayer B. Enzymatic function of nitric oxide synthases. Cardiovasc Res 1999; 43: 521-531.
105. Mohan RM, Golding S, Paterson DJ. Intermittent hypoxia improves atrial tolerance to subsequent anoxia and reduces stress protein expression. Acta Physiol Scand 2001; 172: 89-95.
106. Lu G-W, Liu H-Y. Downregulation of nitric oxide in the brain of mice during their hypoxic preconditioning. J Appl Physiol 2001; 91: 1193-1198.
107. McQuillan LP, Leung GK, Marsden PA, Kostyk SK, Kourembanas S. Hypoxia inhibits expression of eNOS via transcriptional and posttranslational mechanisms. Am J Physiol Heart Circ Physiol 1994; 267: H1921-H1927.
108. Liao JK, Zulueta JJ, Yu F-S, Peng H-B, Cote CG, Hassoun PM. Regulation of bovine endothelial constitutive nitric oxide synthase by oxygen. J Clin Invest 1995; 96: 2661-2666.

109. Ryou M-G, Zong P, Sun J, Manukhina EB, Downey HF, Mallet RT. Hypoxic conditioning suppresses cytotoxic nitric oxide production upon myocardial reperfusion. FASEB J 2006; 20: A793.

110. Lauer T, Preik M, Rassaf T, Strauer BE, Deussen A, Feelisch M, Kelm M. Plasma nitrite rather than nitrate reflects regional endothelial nitric oxide synthase activity but lacks intrinsic vasodilator action. Proc Natl Acad Sci USA 2001; 98: 12814-12819.

111. Kleinbongard P, Dejam A, Lauer T, Rassaf T, Schindler A, Picker O, Scheeren T, Gödecke A, Schrader J, Schulz R, Heusch G, Schaub G, Bryan N, Feelisch M, Kelm M. Plasma nitrite reflects constitutive nitric oxide synthase activity in mammals. Free Radic Biol Med 2003; 35: 790-796.

112. Maiorana A, O-Driscoll G, Taylor R, Green D. Exercise and the nitric oxide vasodilator system. Sports Med 2003; 33: 1013-1035.

113. Grumbach IM, Chen W, Mertens SA, Harrison DG. A negative feedback mechanism involving nitric oxide and nuclear factor kappa-B modulates endothelial nitric oxide synthase transcription. J Mol Cell Cardiol 2005; 39: 595-603.

114. Medved I, Brown MJ, Bjorksten AR, Leppik JA, Sostaric S, McKenna MJ. N-acetylcysteine infusion alters blood redox status but not time to fatigue during intense exercise in humans. J Appl Physiol 2003; 94: 1572-1582.

Adaptation Biology and Medicine (Volume 5: Health Potentials)
Editors: L. Lukyanova, N. Takeda and P.K. Singal

7. Acute Heat Shock Treatment Enhances Cardiac Tolerance to Ischemia/ reperfusion-induced Damage

Ger J. van der Vusse[1]*, Lászlo Ligeti[2], Tamás Ivanics[2], Richard N. Cornelussen[1] and Luc H.E.H. Snoeckx[1]

[1]Dept. of Physiology, Cardiovascular Research Institute Maastricht, Maastricht University, P.O. Box 616, 6200 MD Maastricht, The Netherlands
[2]Institute of Human Physiology and Clinical Experimental Research, Semmelweis University, Budapest, Hungary

Keywords: Heat Shock, arachidonic acid, necrosis and coronary angioplasty

Summary

Cessation of blood flow to myocardial tissue (cardiac ischemia) eventually results in cardiac muscle cell death and loss of myocardial mechanical function. Timely restoration of blood flow is required to prevent irreversible damage inflicted upon the heart. Pharmacological treatment is often applied to mitigate ischemia/reperfusion-induced injury. Prompted by the discovery of various families of heat shock proteins (HSP), either constitutively expressed or induced by a variety of stress-full events, including acute heat treatment of the body, the potentially beneficial effects of acute heat shock on cardiac ischemia/reperfusion-induced injury has been extensively studied. In our laboratory we have explored the effect of induction of one of the heat shock proteins, i.e. the inducible form of HSP70, on the deleterious effect of ischemia/reperfusion of rat myocardium, with special attention to hypertrophied and/or aged hearts.

To this end, anesthetized male Lewis rats were placed on a temperature-controlled heating pad which maintained body temperature at 42°C for 15 minutes. Biochemical analysis revealed a substantial increase in cardiac HSP70 at the protein level six hours after heat treatment.

Animals were routinely sacrificed 24 hours after heat shock. The hearts were extirpated and mounted on a perfusion apparatus, perfused with a crystalline buffer and allowed to eject fluid from the left ventricle (working mode).

*Address for Correspondence: Ger J. van der Vusse, Ph.D., Department of Physiology, Cardiovascular Research Institute Maastricht, Maastricht University, P.O. Box 616, 6200 MD Maastricht, The Netherlands. E-mail: Vandervusse@fys.unimaas.nl

The hearts were subjected to 30-45 min total ischemia followed by reperfusion. Post-ischemic hemodynamic recovery (cardiac output, CO) increased from 58% to 77% (compared to pre-ischemic values) in adult (4 months of age) rat hearts when the animals were pretreated by acute heat shock. This beneficial effect was relatively more pronounced when the hearts were made hypertrophic by aortic banding: post-ischemic CO increased from 14% to 53% due to heat shock treatment. In aged (17 months) animals the post-ischemic recovery of CO amounted to 3% and 20% in sham-operated control and heat-shocked animals, respectively. These observations indicate that the protective effect of heat shock pretreatment was well preserved under conditions of cardiac hypertrophy and ageing.

Subsequent studies have shown that heat shock mitigated the post-ischemic cardiac accumulation of poly-unsaturated fatty acids, including arachidonic acid, strongly suggesting a causative role of HSP's in preventing membrane phospholipid degradation and providing an explanation for the decline in release of cytoplasmic proteins during reperfusion after heat pretreatment. We also observed that heat shock prevented the increase in end-diastolic cytoplasmic calcium ion concentration in the post-ischemic phase, which is in line with the improved functional recovery of the heat-pretreated hearts.

The collective findings show the powerful protective effect of heat pretreatment on the ischemic/ reperfused normal, hypertrophied or aged heart. The clinical potential of this intervention remains to be explored.

Introduction

Impaired blood flow through the coronary arteries (ischemia) results in insufficient supply of O_2 to the cardiac muscle cells to maintain proper electro-mechanical activity. If the unbalance between oxygen supply and demand is not restored in due time, the affected cardiomyocytes will become damaged and get lost by necrotic and/or apoptotic processes. Evidence is accumulating that reperfusion itself can augment the ischemia-induced damage. The precise mechanism underlying reperfusion-induced injury is incompletely understood: calcium-overload, osmotic swelling as well as enhanced production of oxygen free radicals have been proposed as potentially injuring events.

In the clinical setting, surgical and pharmacological measures have been developed to investigate the ischemic insult or to restore myocardial blood flow before a critical period of time has been surpassed.

Examples of surgical measures are acute aorta coronary bypass surgery and percutaneous coronary angioplasty (PTCA). Pharmaceutical interventions include, among others, coronary vasodilators, calcium entry blockers, and enzymes capable of dissolving blood clots in the coronary vessels. Although these protective interventions have proven their effectiveness in the past, there is still room for further improvement of measures aiming at the prevention of cardiac ischemia/reperfusion-induced damage.

An alternative approach to cardiac protection is the mobilization of intrinsic properties of the cardiomyocytes to protect themselves against the deleterious effects of ischemia and reperfusion by environmental stress. Since the discovery of stress proteins, i.e., heat-shock proteins (HSP), their role in intrinsic cellular protective pathways has been intensively studied. Experimental studies have disclosed at least 23 members of the HSP super family (1, 2). Evidence has been provided that, among others, the inducible form of HSP70 is playing an important role in the intrinsic property of the heart to mitigate the deleterious effect of shortage of oxygen and subsequent reperfusion (3, 4).

In our laboratory we have investigated the effect of total body heat treatment of rats on post-ischemic functioning of the heart (5, 6). Body temperature was raised to 42°C for 15 min and 24 hours later the heart was removed from the body, perfused ex vivo and subsequently subjected to 30-45 min ischemia, followed by reperfusion. The effect on post-ischemic cardiac performance was monitored by measuring hemodynamic parameters such as cardiac output, left ventricular pressure and coronary flow.

In a subset of experiments hearts were freeze-clamped at various time points, i.e., just prior to ischemia, at the end of the ischemic period and at the end of reperfusion for analyzing the tissue content of unesterified arachidonic acid (7). Since almost all cardiac arachidonic acid is incorporated in membrane phospholipids an increase in the unesterified content of this poly-unsaturated fatty acid reflects enhanced membrane degradation (8). Finally the systolic and diastolic cytoplasmic calcium ion levels were measured with an epifluorescence technique to investigate whether the protective effect of heat pretreatment is related to preservation of adequate calcium handling during the cardiac cycle (9).

It should be noted that the majority of studies on cardiac ischemia and reperfusion and the protective measures thereon are performed on hearts of relatively young and healthy animals. To facilitate extrapolation of the animal findings to the clinical setting we have extended our studies to hearts of aged animals and rats subjected to aortic constriction to induce pressure-overload cardiac hypertrophy (10). Since elderly people and patients suffering from cardiac hypertrophy are more prone to an ischemic insult it was felt that the latter two animal models are highly relevant for exploring the potentially beneficial influence of heat shock on mitigating or preventing ischemia and reperfusion-induced cardiac damage.

Materials and Methods

Animal preparation

The experiments were routinely performed on 14-15 weeks old male Lewis rats (11). Aged hearts were obtained from 16-17 months old male Lewis rats (10). All experiments were approved by the Institutional Animal Care and Use Committee of the Maastricht University.

Aortic banding protocol: Pressure-overload cardiac hypertrophy was induced by banding of the aorta, 8 weeks before sacrificing the animals at the age of 14-15 weeks as described by ref. 10. In sham-operated animals the aorta was only dissected free.

Heat shock protocol: Rats were anesthetized with Hypnorm ® and placed on a temperature-controlled heating-pad until body temperature, which was continuously monitored rectally, reached 42°C. The animals were kept at this temperature for 15 min (11). Control animals were only anesthetized.

Isolated heart perfusion: Rats were sacrificed by cervical dislocation under light Hypnorm ® anesthesia 24 h after heat shock or sham treatment. The heart was immediately removed from the body, chilled in a modified Tyrode solution (11) and mounted on a perfusion apparatus. Retrograde perfusion was immediately started at a diastolic aortic pressure of 60 mmHg, unless otherwise indicated. The perfusion medium containing (in mM) NaCl (130.0), KCl (5.6), $CaCl_2$ (2.2), $MgCl_2$ (1.0), NaH_2PO_4 (1.2), $NaHCO_3$ (28.6) and D(+) glucose (11.1) was continuously gassed with 95% O_2 and 5% CO_2. After a stabilization period of 10 min, the left atrium was connected to a second cannula allowing the heart to eject fluid through the aortic valve ("working heart model"). The left atrial filling pressure was set at 10 mmHg.

Ischemia was introduced by cross-clamping the left atrial and aortic cannulas. After the period of no-flow ischemia, hearts were reperfused by removing the clamps on the left atrial and aortic cannulas (11). The duration of ischemia and reperfusion lasted 45 min, unless otherwise indicated.

Hemodynamic variables were measured as described earlier (11). Coronary effluent was collected to monitor the loss of cytoplasmic enzymes, lactate dehydrogenase and creatine kinase, being markers of irreversible damage of the cardiomyocyte plasmalemma.

In a subset of experiments normotensive animals at the age of 14-15 weeks were sacrificed 30 min, 3 h, 6 h and 24 h after heat treatment. Hearts were snap-frozen in liquid nitrogen and tissue was stored at −80ºC for analysis of the protein content of the inducible form of HSP70. HSP70 content was measured with an immuno-chemical technique as described earlier (6).

Cardiac tissue obtained 24 h after heat shock was also used to measure the content of unesterified arachidonic acid. The content of this poly-unsaturated long-chain fatty acid was measured with gas chromatography as described before (7).

The effect of heat treatment on left ventricular calcium handling was monitored in a separate set of experiments in Langendorff-mode perfused hearts (9). The hearts, obtained from 14-15 week old rats, were perfused with a modified Krebs-Henseleit medium containing (in mM) NaCl (118.0), KCl (4.3), $NaHCO_3$ (25.0), $MgSO_4$ (1.2), KH_2PO_4 (1.2), Na-EDTA (2.0), glucose (11.0) and Na-pyruvate (5.0) continuously gassed with 95% O_2 and 5% CO_2. Left ventricular pressure was measured with a latex balloon inserted via the mitral valve orifice into the left ventricle. After a stabilization period of 30 min the hearts were loaded with Indo 1-AM (250 μg), solubilized in a mixture of 0.25 ml DMSO and 0.08 g pluronic F-127T and added to the perfusion medium as described in detail before (9). Indo-1 was excited at 340 nm. Epicardial fluorescence emission was recorded at 400 nm (Ca^{2+} - bound dye) and 506 nm (Ca^{2+} - free dye). From these signals the changes in cytoplasmic Ca^{2+} concentration during the systolic and diastolic phase were estimated.

Statistics

All data were expressed as means ± SD. Statistical analyses was carried out using one-way ANOVA followed by Fischer's post hoc test or Mann-Whitney U test to evaluate differences between groups.

Values of $p \leq 0.05$ were considered to be statistically significant.

Results

Time course of HSP70 protein expression

In control hearts the content of the inducible form of HSP70 amounted to approximately 0.3 mg per gram total left ventricular protein (Fig. 1). After heat treatment the HSP70 content significantly increased, showing a 3 fold increase at 6 h after heat shock. Thereafter the HSP70 content leveled off. At 24 h after total body heat treatment the cardiac HSP70 content was still on the order of 0.9 mg per gram total protein (6).

Post-ischemic functional recovery of normotensive hearts

The effect of heat treatment on ischemia/reperfusion tolerance was monitored in hearts of 14-15 week old rats, subjected to 45 min ischemia, followed by 45 min reperfusion.

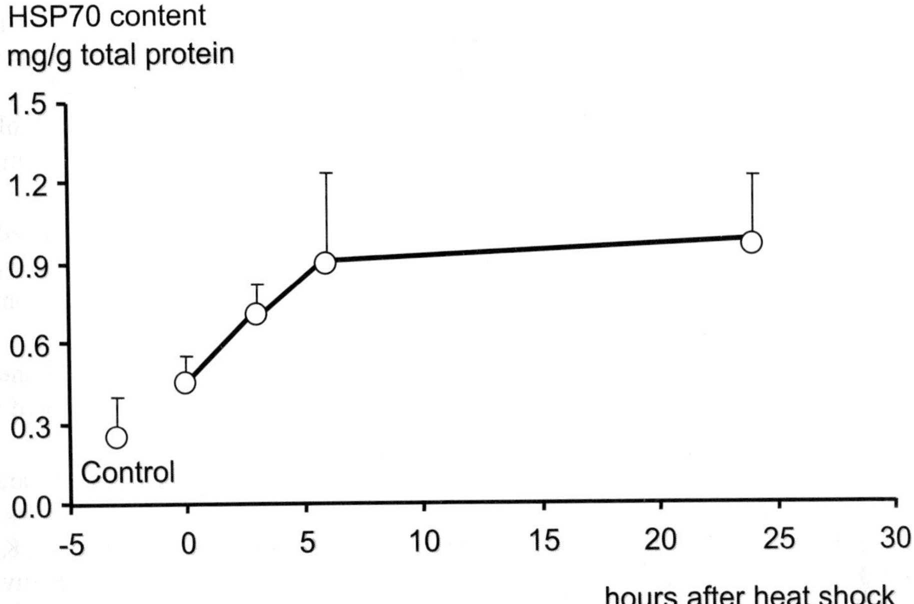

Fig. 1. Time course of the cardiac expression of the inducible form of HSP70. The cardiac protein content of HSP70 was measured just prior to heat treatment and at 30 min, 3 h, 6 h, and 24 h after total body heat shock. The protein content was expressed as mg per gram total cardiac protein. The values after 3, 6 and 24 h post-heat treatment were significantly different from control (after ref. 6).

The recovery of the hemodynamic parameters, cardiac output, left ventricular developed pressure and contractility (first derivative of developed pressure) is shown in Figure 2. The data were normalized to pre-ischemic values. The findings clearly show that all indices for cardiac performance after restoration of flow were improved when the animals were subjected to total body heat stress 24 h prior to the ischemic insult. The highest percentage improvement of function was seen in cardiac output. It should be noted that the values of the hemodynamic parameters did not differ between the heat-shocked group and the sham animals prior to the ischemic insult with the exception of coronary flow. As indicated in Figure 3 heat treatment had a significant effect on coronary flow before the onset of ischemia. On the average coronary flow was increased by 20%. Post-ischemic coronary flow in both heat-shocked and sham hearts were not different from the corresponding pre-ischemic values (Fig. 3).

Post-ischemic hemodynamic recovery of hypertrophied and aged hearts 24 h after heat treatment

When hearts of young adult rats were made hypertrophic 8 weeks prior to sacrifice, the functional outcome of the ex-vivo perfused hypertrophied heart subjected to 45 min ischemia and 45 min reperfusion was substantially less than the recovery observed in normotensive hearts (c.f. Fig. 2 and Fig. 4). Relatively, the protective effect of heat treatment was more pronounced in hypertrophied heart than in normotensive hearts.

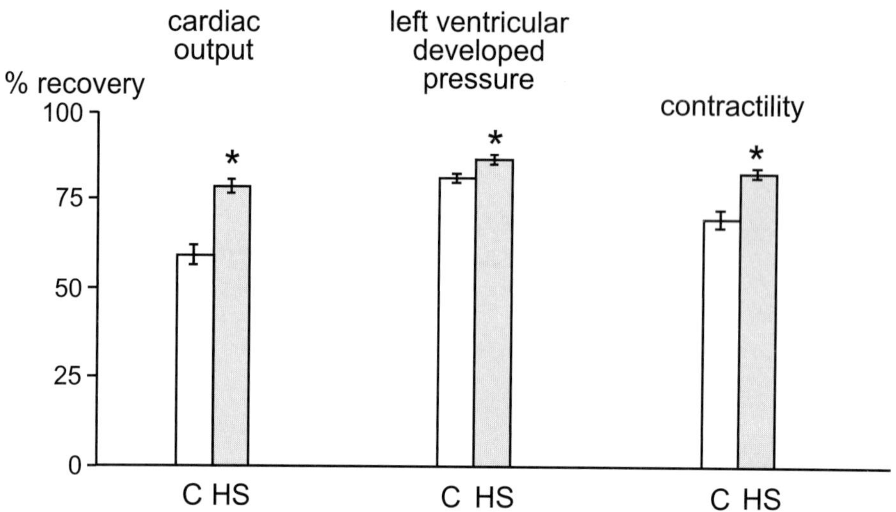

Fig. 2. Post-ischemic functional recovery of ex vivo rat hearts. Normotensive rat hearts, 14-15 weeks of age, were subjected to 45 min ischemia, followed by 45 min reperfusion. End-diastolic aortic pressure was set at 60 mmHg (11). The end-reperfusion values of cardiac output, left ventricular developed pressure (i.e. difference between peak systolic and end-diastolic left ventricular pressure) and contractility (left ventricular dp/dt max) were normalized on the corresponding pre-ischemic values and expressed as percentage recovery. C and HS refer to control sham and heat-shocked hearts, respectively. Heat shock was applied 24 h before sacrifying the animals. Asterisk indicates statistically different.

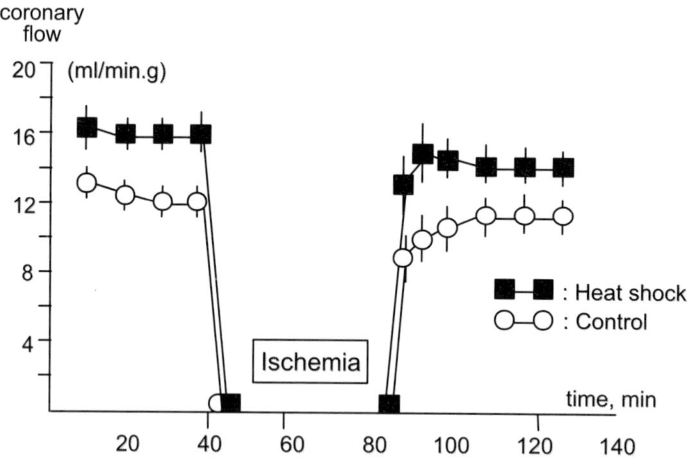

Fig. 3. Recovery of coronary flow 24 h after heat treatment. Coronary flow was measured in ex vivo perfused hearts obtained from 14-15 weeks old normotensive rats. Ischemia lasted for 45 min. Thereafter the hearts were reperfused for 45 min. End-diastolic aortic pressure prior to and after ischemia was set at 60 mm Hg (11). At all time points before ischemia and during the reperfusion phase coronary flow in the heat-shocked hearts was significantly higher than in the sham group. The post-ischemic values did not significantly differ from corresponding pre-ischemic values.

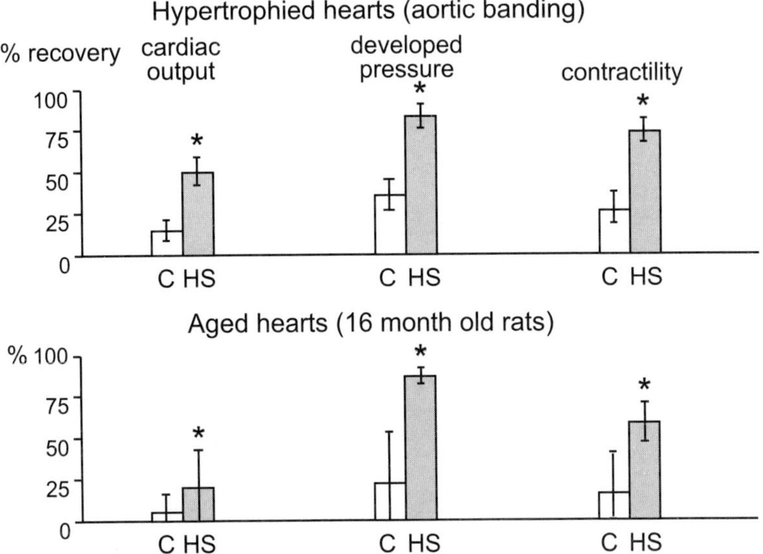

Fig. 4. Post-ischemic functional recovery of ex vivo perfused hearts of hypertensive or aged rats. In the upper panel the findings of hypertrophied hearts of 14-15 week old rats are shown (after Cornelussen et al., 1994). Hearts were made hypertrophic by constriction of the aorta 8 weeks prior to sacrifice at the age of 14-15 weeks. The duration of ischemia was 45 min. The hearts were reperfused thereafter for 45 min. In the lower panel the effect of heat shock of aged rats is shown. The animals were sacrificed at the age of 15-16 months, 24 h after heat treatment. The duration of ischemia was 30 min, followed by 45 min reperfusion (10). The end-diastolic aortic pressure was set at 75 mm Hg. See for further details legends to Fig. 2.

Figure 4 also shows that aged hearts are very vulnerable to a period of 30 min ischemia, followed by 45 min reperfusion. The percentage recovery showed a high degree of variation between the different aged hearts. On the average the degree of recovery was much smaller than that seen in normotensive hearts of young animals (Fig. 2). Heat shock treatment had a highly significant beneficial effect on the post-ischemic recovery of hearts of 16-17 months old rats (Fig. 4, lower panel).

Release of cytoplasmic marker enzymes

Analysis of the release of two cytoplasmic cardiac enzymes, creatine kinase and lactate dehydrogenase, showed that heat treatment of young, normotensive rats had a mitigating effect on the post-ischemic release of these marker enzymes from heart ex vivo subjected to 45 min ischemia followed by 45 min reperfusion. On the average a 20-25% decline in released amounts of enzymes could be observed in the heat-treated group (Fig. 5).

Accumulation of unesterified arachidonic acid in ischemic and reperfused hearts, 24 h after heat treatment

Figure 6 displays the cardiac content of the poly-unsaturated long-chain arachidonic acid prior to ischemia, at the end of 45 min ischemia and at the end of the 45 min reperfusion period. In control sham hearts the arachidonic acid content significantly increased during the ischemic episode. After

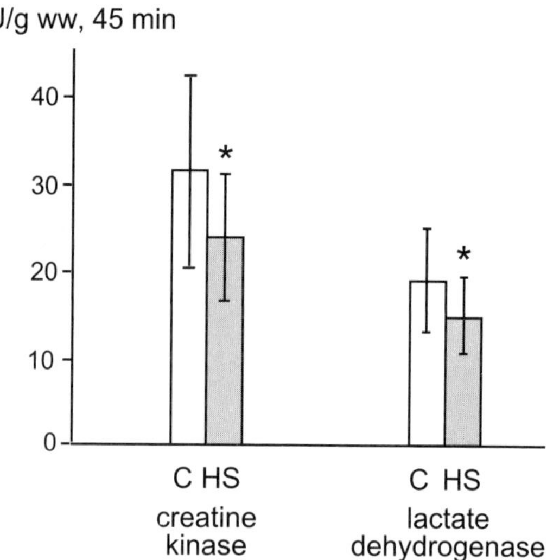

Fig. 5. Release of cytoplasmic proteins from ischemia/reperfusion-damaged hearts after 24 h of heat treatment. Hearts of normotensive rats, 14-15 weeks of age, were ex vivo subjected to 45 min ischemia, followed by 45 min reperfusion. End-diastolic aortic pressure was set at 100 mm Hg. Enzyme activity was measured in samples obtained during the reperfusion phase. The values represent the total amount of enzyme released during 45 min reperfusion. Asterisk indicates significant differences between heat-shocked (HS) and control sham (C) hearts.

restoration of flow the tissue arachidonic acid content continued to rise. Heat pretreatment did not affect pre-ischemic levels of cardiac arachidonic acid (Fig. 6). Also the accumulation of unesterified arachidonic acid during the ischemic insult was similar in sham and heat-shocked hearts. The most striking difference was observed in the reperfused hearts. In contrast to sham hearts, the cardiac content of arachidonic acid did not further increase in the heat-shocked group. Data indicated even a tendency to decline during the reperfusion phase (Fig. 6).

Effect of heat shock on calcium handling in the ischemic-reperfused heart

In Figure 7 diastolic cytoplasmic free Ca^{2+} [Ca_i^{2+}] in ex vivo perfused hearts are shown. Prior to ischemia the diastolic [Ca_i^{2+}] were similar in sham and heat-shocked rat hearts. During the ischemic episode, the [Ca_i^{2+}] monotonously increased in both control and heat-shocked hearts. Due to the relatively large standard deviations the differences between control and heat-shocked hearts did in general not reach the level of significance. In contrast, after restoration of flow [Ca_i^{2+}] precipitously increased further in control hearts, while diastolic [Ca_i^{2+}] in heat-shocked hearts leveled off to values not significantly different from corresponding pre-ischemic levels.

Figure 8 shows that pre-ischemic systolic cytoplasmic [Ca_i^{2+}] levels were similar in control and heat-shocked hearts. During the ischemic episode cardiac contraction came to a complete halt so that systolic [Ca_i^{2+}] could not be measured. Following restoration of coronary flow systolic [Ca_i^{2+}] showed a substantial increase in control hearts. In striking contrast, the post-ischemic systolic Ca^{2+} levels in heat-shocked hearts were statistically not different from corresponding pre-ischemic values (Fig. 8).

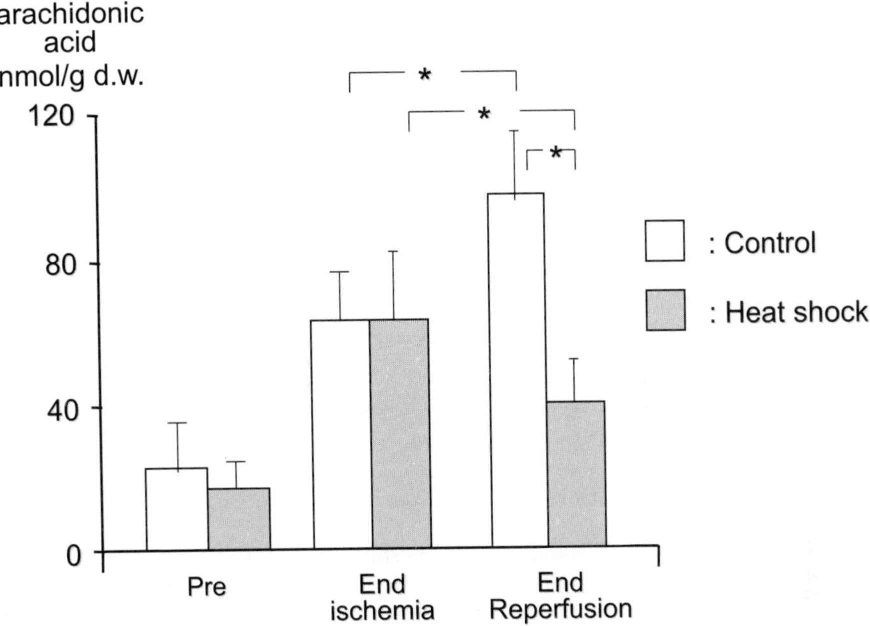

Fig. 6. Accumulation of unesterified arachidonic acid in ischemic and reperfused rat hearts, 24 h after heat treatment. Heart of normotensive rats, 14 weeks of age, were subjected to 45 min of ischemia followed by 45 min reperfusion. Hearts were freeze-clamped just prior to ischemia (pre), and at the end of ischemia and reperfusion, respectively. Arachidonic levels are expressed as nmol per gram dry weight of tissue. Asterisk indicates significant differences between groups of hearts (after refs. 7, 17).

Discussion

A common cause of cardiac muscle cell damage eventually resulting in cardiac mechanical failure and in irreversible loss of contractile tissue is prolonged underperfusion of the heart by impaired blood flow through the coronary arteries. Restoration of flow before a critical time interval has been surpassed prevents the induction of irreversible damage. In the clinical setting, invasive techniques such as acute aorta-coronary bypass surgery and percutaneous coronary angioplasty in combination with the positioning of stents in the narrowed region of the coronary vessel have proven their effectiveness during the past decade. Pharmacological interventions, aiming at reducing the energy demand of the underperfused heart or improving coronary flow, have also shown to exert their beneficial effects on cardiac performance (12). Despite the progress made to protect the heart against ischemia-induced damage, the search has been continued to further enhancing the protection of the jeopardized myocardium against the deleterious effects of flow deprivation. Since evidence is accumulating that in addition to ischemia-induced damage, the affected muscle cells may also be injured by reperfusion itself the need for additional protective measures is widely recognized.

In our laboratory we have sought to increase the intrinsic capability of the heart to mitigate the vulnerability against ischemia/reperfusion-induced damage. To this end experiments were performed on rats pretreated with total body heat shock one day prior to the ischemic insult. A time lapse of 24 h was chosen since previous studies had shown that this kind of stress intervention only showed the late window of protection present between 20 and 72 h after the heat treatment (13).

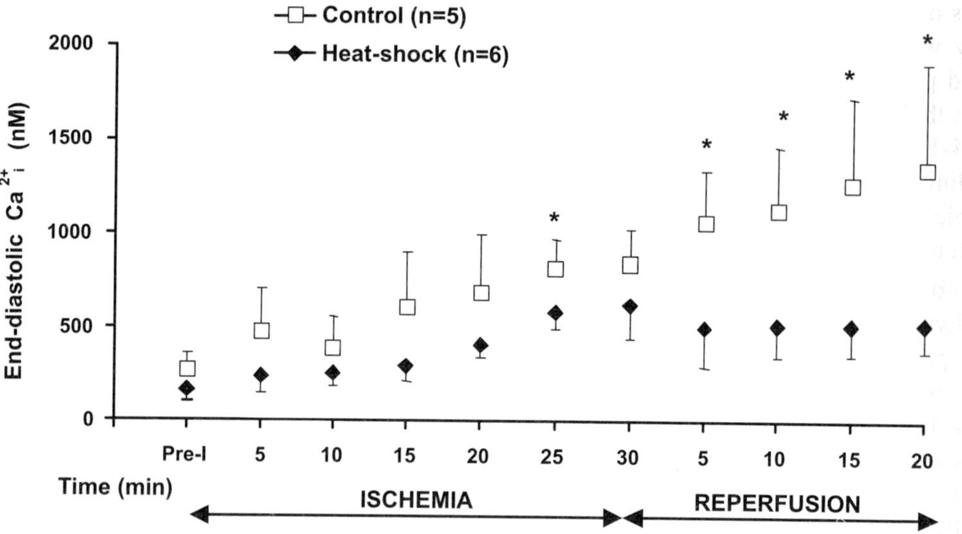

Fig. 7. End-diastolic cytoplasmic Ca^{2+} levels in ex vivo perfused hearts of normotensive young adult rats prior to ischemia, during ischemia and during the reperfusion phase, following 24 h of heat treatment. Hearts were perfused according to Langendorff. Ischemia lasted 30 min and was followed by 20 min reperfusion. End-diastolic aortic pressure was set at 70 mm Hg. End-diastolic $[Ca_i^{2+}]$ was estimated with an epifluorescence technique. Asterisk indicates significant differences between control sham and heat-shocked hearts (after ref. 9).

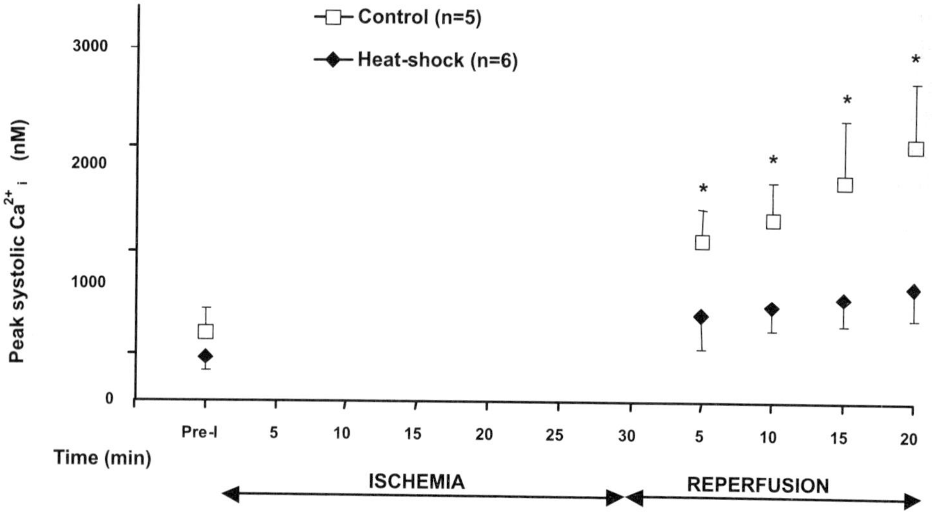

Fig. 8. Systolic $[Ca_i^{2+}]$ in ex vivo perfused hearts of normotensive young adult rats prior to ischemia and during reperfusion, 24 h after heat treatment. Hearts were perfused according to Langendorff. Ischemia lasted 30 min and was followed by 20 min reperfusion. End-diastolic aortic pressure was set at 70 mm Hg. Systolic $[Ca_i^{2+}]$ was estimated with an epifluorescence technique. Asterisk indicates significant differences between control sham and heat-shocked hearts (after ref. 9).

Hearts of heat-treated rats were ex vivo subjected to ischemia and reperfusion and the functional recovery was monitored. The collected findings indicate that heat treatment indeed effectively improved post-ischemic pump function of the transiently ischemic heart. It should be emphasized, however, that these experiments were performed on hearts of healthy, relatively young rats, i.e., at the age of 14-15 weeks. These experimental conditions hamper extrapolation of the experimental findings to the clinical setting, since most patients who are suffering from ischemic heart disease are more vulnerable to shortage of coronary flow due to cardiac hypertrophy or are of advanced age with a heart more often affected by coronary artherosclerosis than hearts of young human beings.

Therefore we extended our studies to animals either at the age of 15-17 months or rats in which cardiac hypertrophy was introduced by aorta constriction. When the ex vivo perfused hearts of these two groups of animals were subjected to transient ischemia the functional outcome was found to be more depressed than that of hearts of relatively young animals. Heat treatment of aged rats and rats suffering from cardiac hypertrophy also showed the beneficial effect of this intervention on post-ischemic mechanical performance. This finding indicates that activation of the intrinsic capability to withstand ischemia/reperfusion-induced injury is still present in diseased or aged hearts, an observation with potential clinical ramifications.

The cause of ischemia/reperfusion-induced damage of the heart is multifactoral in nature as is the intrinsic protective capability (1, 14). In our studies we have explored a variety of processes that could explain the beneficial effect of heat treatment.

Impaired recovery of pump function after an ischemic insult may be caused by lack of appropriate restoration of flow during the reperfusion phase, among others, due to coronary vessel insufficiency or cardiomyocyte hypercontraction raising the vascular resistance above physiological levels. Our findings, however, indicate that prevention of a persistent underperfusion of the myocardium is not a likely cause of heat-shock induced protection. Both in sham and heat-shocked hearts coronary flow recovered to pre-ischemic levels. It should be noted, however, that heat shock did affect coronary flow as shown in the hearts prior to ischemia. Before the ischemic insult the coronary flow was significantly increased without a notable change in mechanical performance. The causes and consequences of this increase are incompletely understood, but the finding points most likely to an effect on the vascular endothelium or smooth muscle cells resulting in an apparently luxurious coronary flow.

The beneficial effect may also be caused by a decline in number of irreversible damaged muscle cells after the ischemic-reperfusion insult, resulting in a decrease in infarct size (15). It is of interest to note that the cardiac release of two cytoplasmic enzymes, lactate dehydrogenase and creatine kinase, after reinstallation of coronary flow, was found to be significantly reduced in the heat-shocked group. This finding strongly suggests that heat treatment stabilizes a membrane damaging process ultimately leading to irreversible loss of cell membrane integrity. This notion is supported by the observation that heat-shocked hearts showed a less pronounced accumulation of unesterified arachidonic acid at the end of the reperfusion phase. Under normal circumstances arachidonic acid is almost exclusively incorporated in membrane phospholipids (8). Enhanced tissue levels of this poly-unsaturated long-chain fatty acid are a sensitive index of increased degradation of cellular membrane phospholipids. The fact that in both sham and heat-shocked hearts arachidonic acid accumulated to the same extent during the ischemic episode and that the mitigating effect was only visible after the reinstallation of flow suggests that the protective effect of heat shock relates in particular to deleterious processes set into motion after reperfusion of the coronary vessels. This intriguing possibility is further substantiated by the effect of heat pre-treatment on calcium handling in the ischemic and reperfused heart.

Our findings indicate that during the ischemic insult the cytoplasmic Ca^{2+} starts to rise. This increase may be caused by an enhanced permeability of the sarcolemma for Ca^{2+} and by impaired activity of SERCA2a. Although the precise mechanism is unknown, it is worthy to note that heat shock pretreatment barely affected the rise in cytoplasmic Ca^{2+} during the ischemic insult. In contrast, the most profound effect was observed after restoration of coronary flow. While the cytoplasmic calcium ion concentration further increased, both during systole and diastole, in the post-ischemic control heart, a completely different picture emerged after restoration of flow in the heat-shocked heart. Heat pretreatment showed to be highly effective in preventing Ca_i^{2+} accumulation after reperfusion. The lack of post-ischemic increase in $[Ca_i^{2+}]$ may on the one hand explain the reduced arachidonic acid accumulation in the reperfused heart by mitigating the activity of Ca^{2+} - sensitive phospholipase, on the other the better preserved mechanical performance of the heat-shocked post-ischemic heart due to improved intracellular calcium ion handling. In this respect, it is worthy to note that mouse hearts lacking functional HSP70 show abnormal Ca^{2+} handling: a delayed Ca^{2+} decline, decreased sarcoplasmic reticulum Ca^{2+} content and low SERCA2a activity (16).

In addition to improving calcium homeostasis in the reperfused heart, heat pretreatment may also intervene with deleterious processes such as enhanced production of oxygen free radicals and swelling of post-ischemic cardiac muscle cells due to osmosis, known to occur readily after restoration of flow (15).

Circumstantial evidence indicates that heat treatment exerts its beneficial effect on the ischemic/reperfused heart via enhanced expression of heat shock proteins (HSP). At present a number of HSP families have been identified of which the HSP70 family was found to be a likely candidate for increasing the capability of the cardiac muscle cell to withstand an ischemia-reperfusion insult. Of the HSP70 family the inducible form of HSP70, i.e., HSP72, does play most likely an important role in the beneficial action of heat shock. The time course of the enhanced expression of the inducible form of HSP70 clearly showed that 6 h after the intervention the expression reached its maximal level. Since the protective effect is fully displayed only 20 h after heat treatment, the combined data leads to the conclusion that HSP70 exerts its effect not by its mere existence, but via other, as yet, incompletely elucidated processes. An important feature of such a process must be that a certain amount of time is required to fully develop its protective capacity mediated by increased intracellular levels of HSP70. In the past it was assumed that HSP70 acted as chaperone of proteins that should not become degraded during the ischemic period. Although this option can not be dismissed, it is, however, difficult to reconcile with the observation that the pre-ischemic functioning of the heart is totally unaffected by heat pretreatment. Interaction of HSP70 with pivotal proteins is supposed to influence the activity of the latter also under normoxic conditions, unless binding to the proteins to be protected occurs at epitopes not involved in their biological action.

Recent studies with genetically engineered mice suggest that HSP70 may interact with proteins partaking in signaling pathways that might become only activated during ischemia/reperfusion and eventually lead to enhanced necrosis and apoptosis of the affected cardiomyocytes (16). To delineate the precise mechanism underlying the protective effects of heat-shock on the ischemic/reperfused heart, further detailed investigations are obviously required.

Finally the question remains as to whether heat treatment bears clinical significance. Under conditions that an ischemic episode can be expected such as during aorta-coronary bypass surgery or PTCA, activation of the intrinsic capacity of the heart to protect itself against the deleterious effect of shortage of flow and obligatory reperfusion may be an additive measure to improve post-ischemic mechanical performance. In this respect locally applied hyperthermia might be a future development. The innate protective mechanism could also be activated by specific agents mimicking heat shock by enhancing the intracellular HSP70 content or by interfering directly with important intermediates in the sequel of events leading to the heat-shock mediated protection against ischemia/reperfusion. Promising data has been recently published by Taniguchi and coworkers, indicating that pretreatment of rats with PPARγ-ligands enhances the cardiac tolerance against ischemia/reperfusion-induced damage by increased expression of the inducible form of HSP70 (18).

Acknowledgements

The help of Vivian Schellings and Theo Roemen in preparing the manuscript is greatly appreciated.

References

1. Snoeckx LHEH, Cornelussen RN, van Nieuwenhoven FA, Reneman RS, Van der Vusse GJ. The role of stress proteins in cardiovascular pathophysiology. Physiol. Rev., 81: 1461-1497, 2001.
2. Cornelussen RN, Vanagt WYR, Prinzen FW, Snoeckx LHEH. Proteins involved in salvage of the myocardium. Adv Exp Med Biol 543: 277-292, 2003.
3. Donnelly TJ, Sievers RE, Vissern FL, Welch WJ, Wolfe CL. Heat shock protein induction in rat hearts. A role for improved myocardial salvage after ischemia and reperfusion? Circulation 85: 769-78, 1992.
4. Currie RW, Tanguay RM, Kingma JG Jr. Heat-shock response and limitation of tissue necrosis during occlusion/reperfusion in rabbit hearts. Circulation 87: 963-71, 1993.
5. Cornelussen R, Van der Vusse GJ, Reneman RS, Snoeckx LHEH. Inability of the heat-shocked heart to adjust its preischemic and postischemic performance to variable loading conditions. J Mol Cell Cardiol 28: 291-298, 1996.
6. Cornelussen R, Garnier AV, Van der Vusse GJ, Reneman RS, Snoeckx LHEH. Biphasic effect of heat stress pretreatment on ischemic tolerance of isolated hearts. J Mol Cell Cardiol 30: 365-372, 1998.
7. Cornelussen RN, Van der Vusse GJ, Roemen, THM, Snoeckx LHEH. Heat pretreatment differentially affects cardiac fatty acid accumulation during ischemia and postischemic reperfusion. Am J Physiol 280: H1736-H1743, 2001.
8. Van der Vusse GJ, Glatz JFC, Stam H, Reneman RS. Fatty acid homeostasis in the normoxic and ischemic heart. Physiol. Rev. 72: 881-940, 1992.
9. Szenczi O, Kemecsei P, Miklos Z, Ligeti L, Snoeckx LH, van Riel NA, Op den Buijs J, Van der Vusse GJ, Ivanics T. In vivo heat shock preconditioning mitigates calcium overload during ischaemia/reperfusion in the isolated, perfused rat heart. Pflügers Arch 449: 518-25, 2005.
10. Cornelussen R, Reneman RS, Van der Vusse GJ, Snoeckx LHEH. Heat shock protects aged hypertrophied and non-hypertrophied rat hearts against ischemic damage Am J Physiol 273: H1333-H1341, 1997.
11. Cornelussen R, Spiering W, Webers JHG, Reneman RS, Van der Vusse GJ, Snoeckx LHEH. Heat-shock improves the ischemia tolerance of the hypertrophied rat heart. Am J Physiol 267: 1941-1947, 1994.
12. Stanley WC, Marzilli M. Metabolic therapy in the treatment of ischaemic heart disease: the pharmacology of trimetazidine. Fund Clin Pharmacol 17: 133-145, 2003.
13. Qiau YZ, Shipley JB, Levasseur JE, Kukreja RC. Dissociation of heat shock proteins expression with ischemic tolerance by whole body hyperthermia in rat heart. J Mol Cell Cardiol 30: 1163-1172, 1998.

14. Mitchell MB, Winter CB, Banerjee A, Harken AH. The relationship between ischemia-reperfusion injury, myocardial stunning and cardiac preconditioning. Surg Gynecol Obstet 177: 97-114, 1993.
15. Currie RW, Karmazyn M, Kloc M, Mailer K. Heat-shock response is associated with enhanced post-ischemic ventricular recovery. Circ Res 63: 543-549, 1988.
16. Kim YK, Suarez J, Hu Y, McDonough PM, Boer C, Dix DJ, Dillmann WH. Deletion of the inducible 70-kDa heat shock protein genes in mice impairs cardiac contractile function and calcium handling associated with hypertrophy. Circulation 113: 2589-2597, 2006.
17. Van der Vusse GJ, Cornelussen RN, Roemen THM, Snoeckx LHEH. Heat stress pretreatment mitigates postischemic arachidonic acid accumulation in rat heart. Mol Cell Biochem 185: 205-211, 1998.
18. Taniguchi Y, Ooie T, Takahasi N, Shinohara T, Nakagawa M, Yonemochi H, Hara M, Yoshimatsu H, Saikawa T. Pioglitazone but not glibenclamide improves cardiac expression of heat shock protein 72 and tolerance against ischemia/reperfusion injury in the heredity insulin-resistant rat. Diabetes 55: 2371-2378, 2006.

Adaptation Biology and Medicine (Volume 5: Health Potentials)
Editors: L. Lukyanova, N. Takeda and P.K. Singal

8. Adaptation of the Heart to Toxic Action of Adriamycin

V.I. Kapelko*, V.L. Lakomkin and V.G. Tsyplenkova

Russian Cardiological Research Complex, Moscow

Summary

The results of our study have shown that even a single adriamycin introduction leaves a deep trace. The earliest effects of adriamycin are increased coronary vessel tone, and decreased ventricular pressure development combined with slowed myocardial relaxation. Most probably, these effects are explained by binding of adriamycin to endothelial NOS and ryanodine-sensitive protein in sarcoplasmic reticulum membranes, respectively. Oxidative modification of these molecules due to adriamycin-induced cellular ROS increase may be regarded as alternative explanation.

Keywords: Superoxide dismutase, Adriamycin, nitric oxide synthase and reactive oxygen species

Introduction

Anthracycline antibiotic adriamycin (doxorubicin), being one of the most effective antitumoral preparations, renders prominent cardiotoxic action with the subsequent development of cardiomyopathy at prolonged application. The majority of researchers believe that toxic action of adriamycin is realized by means of activation of free radical oxidation (1-3). Mitochondria represent basic cellular target of adriamycin (4-6). It has been shown that addition of adriamycin (40-160 μM) to cardiomyocytes amplify the oxidative reactions close to mitochondria (4). An increased susceptibility of those in the myocardium in comparison with mitochondria in the liver is caused by the presence of cardiolipin in cardiomyocyte mitochondrial membranes which possess increased affinity to adriamycin (7) and of external NADH dehydrogenase catalyzing auto-oxidation of adriamycin semiquinones (8). Mitochondrial dysfunction has been observed even at nanomolar concentration of adriamycin (5) resulting in rapid superoxide formation with the subsequent exit of cytochrome c especially in the presence of aglycone derivatives of adriamycin (6).

The inevitably occurring alterations of myocardial contractility depend on a doze of adriamycin, animal species and heart rate. The contractility of rat myocardium, that is highly dependent on the amount of Ca^{++} liberated from sarcoplasmic reticulum, is usually reduced (10-12) even under influence of small dozes of adriamycin due to diminished Ca^{++} stores in the reticulum (13,14). Along with this alteration, damages of other membranes, cytoskeleton structure, extracellular matrix and disturbed

*Address for Correspondence: 121552, 3rd Cherepkovskaya Str. 15A, Moscow, Fax (495) 414-66-99, E-mail: v.kapelko@cardio.ru

function of the nucleus are also observed, especially at higher dozes of adriamycin [see reviews 3,9,15,16].

In view of such multimodal toxic action of adriamycin, it is important to find out what structures are injured and how reparation of these damages proceeds. The latter aspect till now has attracted a rather little attention. Therefore, in the given work we studied the contractile function of the heart and a tone of coronary vessels, their resistivity to oxidative stress, and also the activity of antioxidant enzymes in the myocardium. It is known that cardiomyocytes with preliminary reduced SOD activity exhibited increased superoxide radical production (17), while various antioxidants or iron chelators prevent toxic action of adriamycin (18-20).

Design of the Study

Male Wistar rats weighing 340-380 g have been used. All experiments were performed in the isolated heart. Rats were anesthetized with urethane (1.7 g/kg b.w.); the hearts were perfused at 37°C through the aorta with modified Krebs-Henseleit solution containing 11 mM glucose and saturated with carbogen (95% O2 and 5% CO_2). The perfusion was carried out with perfusion pump at controlled rate. The pressure in latex balloon introduced into the left ventricular cavity was continuously monitored. All studies described in this manuscript were performed in accordance with institutional guidelines for animal research.

Adriamycin (doxorubicin, "Ferrein", Russia) was either introduced into coronary perfusate by infusion pump to maintain a stable concentration in the perfusate or injected intraperitoneally at a doze of 2.2 mg/kg. Half of adriamycin-injected rats were taken for the study 2 hours after injection, and the other half at 2-3 weeks later. At the end of physiological experiments the hearts were taken for morphological and biochemical analysis. Details of the methods have been indicated elsewhere (21).

All experiments have been subdivided in three groups. The first group is based on the results of adriamycin infusion into coronary vessels of the isolated heart and data obtained represent early effect of adriamycin. The other two groups are based on the results of adriamycin injection *in vivo*, the second being hearts taken in 2 hours represents acute effect of adriamycin and the third group taken at 2-3 weeks after injection represents postponed effects of adriamycin.

Early effects of adriamycin

Adriamycin introduction in the perfusate of the isolated heart was performed at middle perfusion rate, approximately 10 ml/min/g and was accompanied by changes in cardiac contractile function and perfusion pressure. Minimal functional changes were observed at application of 1 μM adriamycin while 3 μM significantly affected both cardiac function and perfusion pressure. The latter steadily rose at the start of 3 μM adriamycin introduction and its rise continued till the end of adriamycin introduction (Fig. 1). The cardiac contractile function represented by RPP (rate-pressure product = developed pressure multiplied by heart rate) decreased insignificantly after 1 μM adriamycin introduction, and by the end of adriamycin introduction only developed pressure and the rate of maximal pressure fall, –dP/dt, were significantly reduced by $8 \pm 2\%$ and $13 \pm 5\%$, respectively. The 3 μM adriamycin introduction was followed by RPP depletion that became prominent after 15 min (Fig. 2). A depletion of developed pressure was more prominent than heart rate. The rates of maximal pressure development and fall have been reduced similarly to developed pressure. Minimal diastolic pressure was raised only at 3 μM adriamycin by 41 mmHg and its degree and time course closely corresponded to perfusion pressure rise.

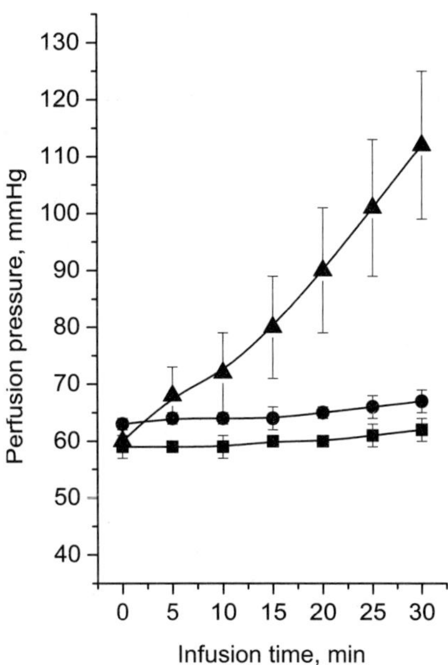

Fig. 1. Effects of 30-min adriamycin infusion into coronary vessels of the isolated rat heart on the perfusion pressure (mmHg). Groups: control (no infusion, circles), adriamycin 1 μM (squares), adriamycin 3 μM (triangles). Values are presented as M ± m.

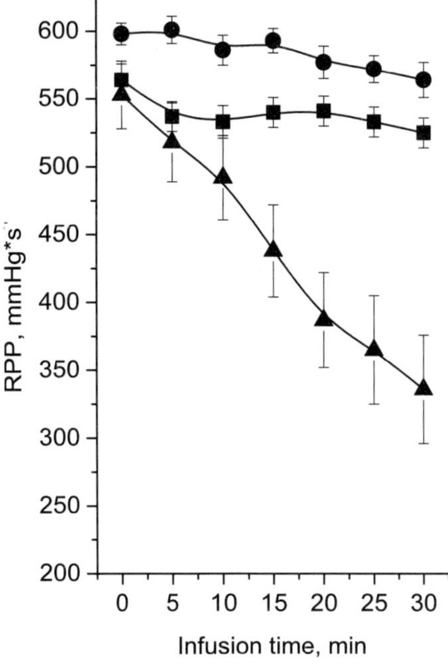

Fig. 2. Effects of 30-min adriamycin infusion into coronary vessels of the isolated rat heart on RPP (rate-pressure product, mmHg*s^{-1}). Legends as in Fig. 1.

In the course of experiment the functional load elicited by increased perfusion rate was applied both before and after adriamycin introduction. The perfusion rate was increased twice in 3 steps at 3-min intervals. As a result, developed pressure steadily increased (Gregg phenomenon) and was accompanied by increased perfusion pressure (Fig. 3). This functional load showed prominent depression of developed pressure rise after 1 μM adriamycin introduction. Further, a fall of developed pressure even at increased perfusion rate was observed after 3 μM adriamycin introduction (Fig. 4). The clear difference between effects of two adriamycin concentrations was also observed when the response of perfusion pressure was compared (Fig. 3). Only minimal increase in this parameter was seen after 1 μM adriamycin concentration as compared to the initial load while perfusion pressure was greatly increased after injecting 3 μM adriamycin.

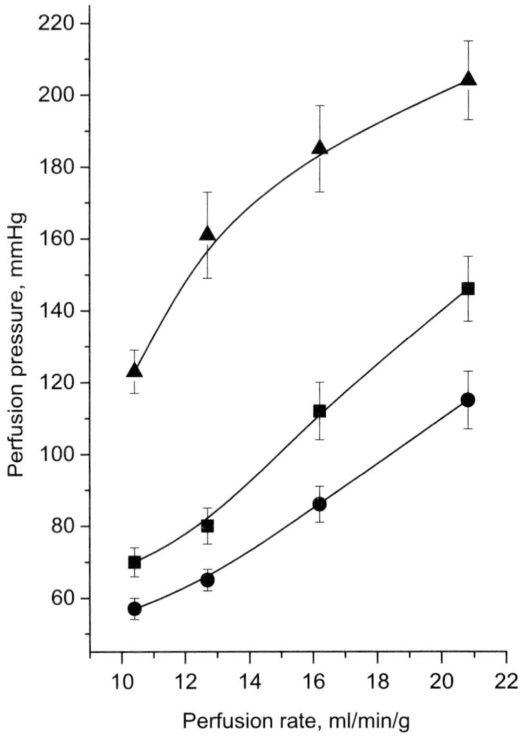

Fig. 3. Dependence of perfusion pressure of the isolated rat heart on perfusion rate after adriamycin infusion and withdrawal. Legends as in Fig. 1.

These results have shown that among functional parameters of the heart, the most sensitive to adriamycin action are perfusion and developed pressures and –dP/dt. Adriamycin-induced increment of perfusion pressure, which in conditions of stabilized perfusion rate is the indicator of coronary vessels resistance, has been reported earlier (10-12). Our data show that this rise seems to be one of the earliest manifestations of the toxic adriamycin action on the myocardium manifesting both at adriamycin introduction and after its removal when perfusion rate was increased. The latter phenomenon is usually prevented *in vivo* due to increased activity of endothelial NOS resulted in the relaxation of smooth muscle cells under Nitric Oxide (NO) influence. There are some findings to

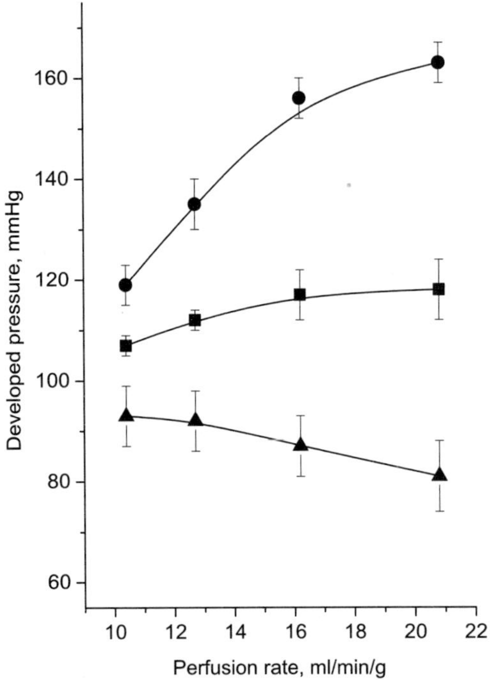

Fig. 4. Dependence of developed pressure of the isolated rat heart on perfusion rate after adriamycin infusion and withdrawal. Legends as in Fig. 1.

believe that eNOS activity can be the initial link of oxidative stress action on the heart. Adriamycin was reported to cause a dose-dependent inhibition of eNOS activity (22) followed by decreased NO production but increased superoxide and H_2O_2 production (23). The affinity of adriamycin to eNOS is much higher than to other myocardial reductases (22) that may explain earlier responsiveness of coronary vessels to adriamycin as compared to the myocardial contractile function. The functional consequence of increased coronary vascular resistance is an elevation of hydrostatic pressure in the intercellular space which, in turn, directly affects diastolic pressure in the same direction.

Negative inotropic action of adriamycin manifested at functional load even under low adriamycin dose most likely is caused by binding of its molecule to proteins involved in Ca^{++} transport. Adriamycin has high affinity to ryanodine-sensitive protein in sarcoplasmic reticulum membranes [24], reduces Ca^{++} stores in SR (14) and by binding to cardiolipin in mitochondrial internal membranes increases its permeability followed by increased Ca^{++} efflux from mitochondria (25). This may result in a rise in sarcoplasmic Ca^{++} level. The two-fold elevation of its level in cardiomyocytes was detected at 50 μM adriamycin concentration (26). Adriamycin at low concentrations did not influence Ca^{++} transients in the sarcoplasm of skinned fibres and rate or extent of shortening but even at nanomolar range of concentrations, permanently elongated time to peak shortening and time of relaxation (27). Our data showing more profound decrease in –dP/dt as compared to +dP/dt are in accordance with these findings.

Acute effects of adriamycin

The contractile function of the heart isolated two hours after adriamycin injection *in vivo* was similar to control group both at middle and high perfusion rate. In the latter, heart rate was 4.4 ± 0.1 Hz in both groups and developed pressure 189 ± 3 mmHg and 199 ± 9 mmHg in the control and adriamycin-treated group, respectively. Maximum RPP level was similar in both groups. Only perfusion pressure in adriamycin-treated group was increased in comparison with the control group by 12% at middle perfusion rate and by 26% at high perfusion rate (120 ± 6 mmHg vs. 95 ± 3 mmHg, $p < 0.01$). This effect of adriamycin injection *in vivo* was similar to that found after direct adriamycin introduction into the heart while the effect on the contractile function was not observed. The action of adriamycin injected *in vivo* was four times longer and the dose injected was comparable to that introduced in the isolated heart. However, peak concentration in the blood should be rapidly decreased due to binding of adriamycin with various tissues and the real concentration in the coronary vessels can be several times less during two-hour period before heart isolation. In this case, elevated tonus of coronary vessels of the isolated heart also evidences higher sensitivity of coronary vessels as compared to cardiomyocytes to adriamycin.

It's meaningful to compare these results with the study in aqueorin-loaded isolated rat heart (28) in which three consecutive injections of adriamycin, 2 mg/kg, were made in 5 days. In the study performed just after last adriamycin injection, developed pressure was unchanged and sarcoplasmic Ca^{++} level was increased at varied perfusion rate (28). Moreover, these hearts could maintain higher systolic Ca^{++} level in conditions of lower perfusate Ca^{++} concentration or lower perfusion pressure. It seems that calcium activation of the contractile apparatus became less efficient in these adriamycin-treated hearts. In similar cumulative dosage of adriamycin, 7.5 mg/kg, a profound fall in the force development of the isolated papillary muscle was found (29). Probably, similar elevation of cytosolic Ca^{++} takes place in smooth muscle cells. If so, then myosin light chain kinases (MLCK) content is decreased (30) and probably its activity may be a simple regulatory phenomenon preventing overcontraction of these cells.

The scheme of this series of experiments included the determination of myocardial antioxidant protection by means of introduction of hydrogen peroxide (100 μM), an inductor of free radical oxidation, in the perfusate (21). The H_2O_2 was infused with Sage infusion pump at adjustable rate maintaining constant H_2O_2 concentration. An introduction of H_2O_2 was accompanied by steadily decreased developed pressure and RPP, and its steepness was appreciably more prominent in 2-hour adriamycin-treated group since 20-th minute of H_2O_2 infusion (Fig. 5). Heart rate at H_2O_2 action practically did not change in both groups. The positive correlation between initial level of perfusion pressure and the degree of developed pressure reduction under the influence of H_2O_2 was found out in the control group ($r = 0.61$) as well as in 2-hour adriamycin-treated group ($r = 0.65$).

The ultrastructural changes in the myocardium of both groups were similar and mostly caused by oxidative stress. Small perivascular edema and expansion of contacts between some cardiomyocytes were most typical alterations. In a small amount of cardiomyocytes, the marked sarcoplasmic edema and glycogen disappearance were noted. In some mitochondria with increased matrix density, the dense arrow-like inclusions were seen. Mitochondria, as a rule, were isolated, but sometimes clusters were found in which mitochondria were interconnected by specific intermitochondrial contacts in adriamycin-treated myocardium.

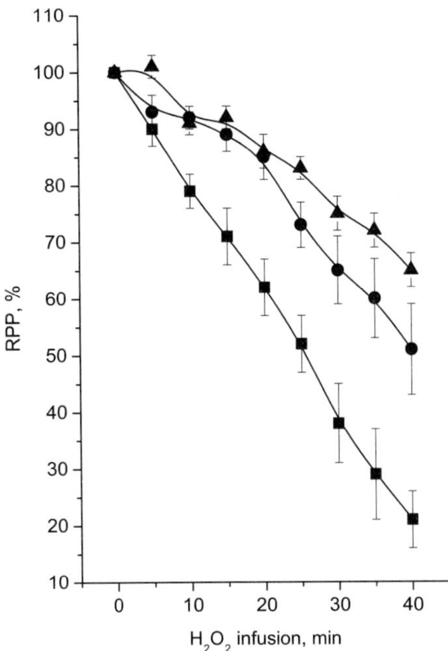

Fig. 5. Effect of H_2O_2 infusion on RPP (in percents to value before infusion). Control (no infusion, circles), adriamycin, 2 hours (squares), adriamycin, 2-3 weeks (triangles).

The biochemical analysis of myocardial samples taken after the end of experiments has shown an increased MDA concentration in adriamycin-treated group by 23% (2.49 ± 0.08 vs. 20.2 ± 0.08 nM/mg of protein, $p<0.05$). The activity of three antioxidant enzymes (SOD, GSH-Px and catalase) remained unaltered.

Thus, adriamycin-treated hearts with practically unchanged contractile function showed higher susceptibility to toxic action of H_2O_2. The most plausible explanation of this finding, established for the first time, may be the common basis for the action of adriamycin and H_2O_2 on cellular Ca^{++} transport. It is known that ROS in micromolar concentrations block calmodulin, inhibit ryanodine receptor, reduce Ca^{++}-ATPase activity of SR and cause massive Ca^{++} efflux from the reticulum (31, 32). It is natural to assume that the amount of Ca^{++} activating myofibrils decreases under H_2O_2 influence to greater extent in cardiomyocytes with previously disturbed Ca^{++} transport under adriamycin influence because further elevation of H_2O_2 concentration may exceed the ability of cellular GSH-Px and catalase to neutralize H_2O_2.

Elevation of ROS in the acute period of adriamycin action is also supported by increased MDA level that has also been reported earlier (33-35) with the peak of ROS generation observed after 2 hours (35). The activity of basic antioxidant enzymes in our study did not differ from control values that coincided with results by Li et al (33), who also found stable activity of GSH-Px, catalase and MnSOD at a period of 1-24 hours after adriamycin introduction at a similar doze, 2.5 mg/kg. Probably, the level of adriamycin-induced oxidative stress was not so high to overload antioxidant defense system and this suggestion is corresponded to our data showing only mildly increased lipid peroxidation. A high stability of antioxidant enzymes to toxic adriamycin action is supported by the observation that

GSH-Px activity can remain constant even after much higher dose of adriamycin, 15 mg/kg (36). The only enzyme that showed reduced activity during the first day after adriamycin introduction was Cu/Zn SOD (33). The action of SOD for prevention of toxic action of adriamycin has been evidenced in isolated cardiomyocytes. Cells with reduced SOD activity were less resistant to adriamycin action (17) while preliminary application of SOD or GSH-Px prevented the development of apoptosis of embryonic cardiomyocytes subjected to adriamycin (37).

It is noteworthy that even at profoundly decreased myocardial function after H_2O_2 application, the ultrastructural changes were rather mild and seemed more likely be caused by the action of hydrogen peroxide (38) than adriamycin. The meaning of mitochondria clusters formed soon after adriamycin injection is intriguing. Probably, the formation of such clusters may preserve centrally located mitochondria from damage by ROS acting in the periphery of a cluster. An absence of visible morphological alterations was, however, associated with some destruction of intracellular matrix as has been found in our previous work performed at similar adriamycin dosage and term (30). Reduced contents of some cytoskeleton and matrix proteins as well as MLCK and KRP protein were revealed. Nevertheless, these mild alterations seemed to be compensated and did not disturb myocardial contractile function, at least in isovolumic mode.

Delayed effects of adriamycin

In 2-3 weeks after 2.2 mg/kg adriamycin introduction, the myocardial contractile indices (developed pressure, +dP/dt and RPP) were similar to the control group; however, relaxation parameters were significantly altered. In this regard, -dP/dt was lower (control 3400 ± 94 vs. adriamycin 3040 ± 108 mmHg/s, $p<0.05$), and the relaxation index (maximal −dP/dt divided by developed pressure) was lower (control 17.5 ± 0.5 vs. adriamycin 15.0 ± 0.5 s-1, $p<0.01$), while diastolic pressure was increased (control 2 ± 1 vs. adriamycin 7 ± 1 mmHg, $p<0.05$). The perfusion pressure at high perfusion rate was kept at the increased level, control 95 ± 3 vs. adriamycin 115 ± 4 mmHg ($p<0.01$).

Antioxidant status of the myocardium in this series was also studied by introduction of H_2O_2 and determination of MDA concentration and activity of antioxidant enzymes. Hydrogen peroxide introduction was accompanied by similar RPP depletion as in the control group; however, at some points RPP values in late adriamycin-treated group were even higher (Fig. 5). An increased MDA concentration in adriamycin-treated group was preserved (2.71 ± 0.05 vs. 2.02 ± 0.08 nM/mg of protein, + 34%, $p < 0.05$) while the activity of antioxidant enzymes remained unaltered.

A prolongation of time interval between adriamycin injection and the myocardial study facilitates reparation or even supercompensation of acute alterations. So, the resistivity of the contractile function to oxidative stress was completely restored or even became slightly higher, despite increased MDA level. At this time the activity of antioxidant enzymes was on the stable level that coincided with data by Li et al. (19). In another study the injection of a similar doze of adriamycin within 4 weeks was accompanied by increased level of antioxidant enzymes (39) that allows to consider this fact as an adaptive phenomenon. In recently performed study with the same dose of adriamycin, the full restoration of previously decreased content of desmin, vinkulin and fibronectin was combined with increased content of tubulin and KRP protein as well as increased density of collagens I and III types (30).

At the same time, some alterations seen before became more pronounced, namely, slowed myocardial relaxation. This finding along with increased level of initial diastolic pressure suggests decreased and incomplete Ca^{++} extraction from myofibrils. At this time, the systolic Ca^{++} level remained increased [28] that may at least partly be due to increased Ca^{++} entry into cardiomyocytes. This suggestion is supported by the observation that even at 4-10 weeks after adriamycin introduction a stable elevation of mRNA for sarcolemmal Ca^{++} channel was detected (40). Decreased Ca^{++} extraction from the contractile apparatus of cardiomyocytes and smooth muscle cells may result not only from direct action of adriamycin on Ca^{++} transport proteins but also from changes in their function caused by altered membrane environment of these structures.

This suggestion has arised from our previous data in rat hearts studied at 2-3 weeks after 3 adriamycin injections, of 2 mg/kg every other day. The ultrastructural study revealed sparse cardiomyocyte alterations mostly associated with membrane structures (11). Alterations of nuclei were observed most frequently (8.6%) and alterations in membranes of mitochondria (5.9%) and the SR (4.2%) were slightly less frequent. Some nuclei contained mitochondria, different slices showing various phases of their penetration (30). This phenomenon was also seen in patients with cardiac pathology (41). The biochemical study of mitochondria in skinned fibers demonstrated unchanged maximal level of oxygen consumption, but respiratory index in adriamycin-treated group was lower by 30%. A decrease (by 35%) in apparent affinity constant for ADP evidenced an increased permeability of mitochondrial membranes (42). It might be noted that these biochemical data relate to rather large part of the left ventricle and thus represent majority of cardiomyocytes.

Such alterations of mitochondria may suggest a disturbance in energy production. However, when isolated hearts were perfused in NMR-spectrometer, nor ATP and creatine phosphate content or isovolumic pressure development were altered in hearts of adriamycin-treated group (43). As maximal energy demand in the isovolumic mode is lower in comparison with the heart performing pump function, a special series of experiments has been done in so-called working heart (11). Maximal cardiac work at increased aortic resistance was 31% higher in hearts of adriamycin-treated group and this was mainly due to increased coronary flow.

The latter result were unexpected being contradictory to acute negative inotropic and vascular constricting effects of adriamycin. Also, in the present study an increased vascular resistance was detected both in acute and postponed phases of adriamycin action. The reason for this discrepancy may relate to differences in the mode of study of vascular resistance. In the present study, when perfusion rate has been increased, the coronary vessel tone mainly depended on shear stress-induced activation of endothelial NOS. The functional power of this enzyme in the intact heart is sufficient to almost prevent a rise in coronary vessel tone, but in adriamycin-treated heart with disturbed endothelial NOS function, this response may be markedly diminished resulting in raised perfusion pressure. In the former study, when aortic resistance was increased, the Bayliss response dominated manifesting increased contraction of smooth muscle cells that prevented coronary overflow. In adriamycin-treated hearts with stable depletion of the activity and/or the content of MLCK (30), the contractile response of vascular cells is expected to be weakened with resulting rise in coronary flow. Moreover, metabolic autoregulation through adenosine released from cardiomyocytes may be enhanced in adriamycin-treated hearts. Thus, increased maximal cardiac work in pumping mode may rather reflect increased coronary flow rate than pressure development.

These data suggest that secondary or postponed alterations induced by adriamycin introduction may be due to continuing adriamycin incorporation to myocardial structures. In addition to its direct action on protein molecules to which adriamycin has high affinity, it also binds to cardiolipin (7) representing one of most abundant components of various membrane protein environment. As a result, the function of many proteins such as ATP/ADP translocase or Na^+-K^+-ATPase and so on may be disturbed making the basis for well-known cumulative effects of adriamycin.

Conclusion

Endothelial cells represent the first barrier on the way of adriamycin molecule into the myocardium and therefore, are exposed to higher adriamycin concentration in comparison to other adriamycin-sensitive proteins. A disturbance in endothelial NOS function shifts the balance between relaxing and constricting influences on smooth muscle cells to the latter with resulting rise in coronary vessel tone. The inhibition of endothelial NOS seems to be stable and observed depletion of MLCK content or activity in smooth muscle cells may represent adaptational alteration to counteract the diminished relaxing influence from the endothelium.

A disturbance in function of ryanodine-sensitive protein in sarcoplasmic reticulum membranes results both in diminished activation of the contractile apparatus especially at increased functional load and slowed Ca^{++} uptake from myofibrils. To maintain sufficient level of Ca^{++} activation, an additional Ca^{++} entry from intercellular space is required, and stable elevation of mRNA for sarcolemmal Ca^{++} channel seems to represent other adaptational alteration. However, when cellular ROS level is elevated due to hydrogen peroxide action the additional disturbance in Ca^{++} transport proteins results in profound fall of cardiac contractile function.

During several weeks between adriamycin injection and myocardial study, adriamycin can steadily accumulate in the myocardium and, in addition to acute adriamycin effects, other manifestations of its action appear, namely, increased permeability of sarcoplasmic reticulum, mitochondrial and nuclear membranes. These alterations most probably are explained by adriamycin binding to cardiolipin and can be compensated at least under the action of middle cumulative doses. An increment of this dose may exceed cellular ability to compensate these alterations and well-known cumulative toxic effects of adriamycin can appear leading to the cardiomyopathy.

References

1. Olson RD, Mushlin PS. Doxorubicin cardiotoxicity: analysis of prevailing hypotheses. FASEB J. 1990; 4: 3076-3086.
2. Gille L, Nohl H. Analyses of the molecular mechanism of adriamycin-induced cardiotoxicity. Free Radic Biol Med 1997; 23: 775-782.
3. Singal PK, Iliskovic N, Li T, Kumar D. Adriamycin cardiomyopathy: pathophysiology and prevention. FASEB J. 1997; 11: 931-936.
4. Sarvazyan N. Visualization of doxorubicin-induced oxidative stress in isolated cardiac myocytes. Am J Physiol 1996; 271[5 Pt 2]: H2079-85.
5. Green PS, Leeuwenburgh C. Mitochondrial dysfunction is an early indicator of doxorubicin-induced apoptosis. Biochim Biophys Acta 2002; 1588 [1]: 94-101.
6. Clementi ME, Giardina B, Di Stasio E, Mordente A, Misiti F. Doxorubicin-derived metabolites induce release of cytochrome C and inhibition of respiration on cardiac isolated mitochondria. Anticancer Res. 2003; 23[3B]: 2445-50.

7. Goormaghtigh E, Ruysschaert JM. Anthracycline glycoside-membrane interactions. Bioch Bioph Acta 1984; 779: 271-288.
8. Nohl H, Gille L, Staniek K. The exogenous NadriamycinH dehydrogenase of heart mitochondria is the key enzyme responsible for selective cardiotoxicity of anthracyclines. Z Naturforsch [C]. 1998; 53[3-4]: 279-85.
9. Wallace KB. Doxorubicin-induced cardiac mitochondrionopathy. Pharmacol Toxicol 2003; 93[3]: 105-15.
10. Pelikan PC, Weisfeldt ML, Jacobus WE, Miceli MV, Bulkley BH, Gerstenblith G. Acute doxorubicin cardiotoxicity: functional, metabolic, and morphologic alterations in the isolated, perfused rat heart. J Cardiovasc Pharmacol 1986; 8[5]: 1058-66.
11. Kapelko VI, Tsyplenkova VG, Khatkevich AN, Beskrovnova NN. Morphological and functional estimation of acute and protracted cardiomyocyte alterations caused by adriamycin in varied doses. Exp Clin Cardiol 1999; 4 [1]: 35-42.
12. Joyeux M, Godin-Ribuot D, Faure P, Demenge P, Ribuot C. Heat stress protects against electrophysiological damages induced by acute doxorubicin exposure in isolated rat hearts. Cardiovasc Drugs Ther 2001; 15[3]: 219-24.
13. Kim DH, Landry III, AB, Lee YS, Katz AM. Doxorubicin-induced calcium release from sarcoplasmic reticulum vesicles. J Mol Cell Cardiol 1989; 21: 433-436.
14. Matsushita T, Okamato M, Toyama J, Kodama I, Ito S, Fukutomi T, Suzuki S, Itoh M. Adriamycin causes dual inotropic effects through complex modulation of myocardial Ca^{2+} handling. Jpn Circ J. 2000; 64[1]: 65-71.
15. Singal PK, Li T, Kumar D, Danelisen I, Iliskovic N. Adriamycin-induced heart failure: mechanism and modulation. Mol Cell Biochem 2000; 207: 77-86.
16. Kapelko VI. Myocardial fibrosis in cardiomyopathy: the possible crucial role in the dilative form. Exp Clin Cardiol 2001; 6 [1]: 41-49.
17. Sarvazyan NA, Askari A, Huang WH. Effects of doxorubicin on cardiomyocytes with reduced level of superoxide dismutase. Life Sci. 1995; 57[10]: 1003-10.
18. Rajagopalan S, Politi PM, Sinha BK, Myers CE. Adriamycin-induced free radical formation in the perfused rat heart: implications for cardiotoxicity. Cancer Res. 1988; 48[17]: 4766-9.
19. Li T, Danelisen I, Bello-Klein A, Singal PK. Effects of probucol on changes of antioxidant enzymes in adriamycin-induced cardiomyopathy in rats. Cardiovasc Res 2000; 46[3]: 523-30.
20. Kang YJ. The antioxidant function of metallothionein in the heart. Proc Soc Exp Biol Med 1999; 222 [3]: 263-273.
21. Lakomkin VL, Konovalova GG, Kalenikova EI, Zabbarova IV, Kaminnyi AI, Tihaze AK, Lankin VZ, Ruuge EK, Kapelko VI. Changes in antioxidant status of myocardium during oxidative stress under the influence of coenzyme Q_{10}. Biochemistry [Moscow] 2005; 70[1]: 79-84.
22. Vasquez-Vivar J, Martasek P, Hogg N, Masters BS, Pritchard KA Jr, Kalyanaraman B. Endothelial nitric oxide synthase-dependent superoxide generation from adriamycin. Biochemistry 1997; 36[38]: 11293-7.
23. Kalivendi SV, Kotamraju S, Zhao H, Joseph J, Kalyanaraman B. Doxorubicin-induced apoptosis is associated with increased transcription of endothelial nitric-oxide synthase. Effect of antiapoptotic antioxidants and calcium. J. Biol. Chem. 2001; 276: 47266-76.
24. Saeki K, Obi I, Ogiku N, Shigekawa M, Imagawa T, Matsumoto T. Doxorubicin directly binds to the cardiac-type ryanodine receptor. Life Sci 2002; 70[20]: 2377-89.
25. Sokolove PM. Interactions of adriamycin aglycones with mitochondria may mediate adriamycin cardiotoxicity. Int.J.Biochem 1994; 26: 1341-1350.
26. Kusuoka H, Futaki S, Koretsune Y, Kitabatake A, Suga H, Kamada T, Inoue M. Alterations of intracellular calcium homeostasis and myocardial energetics in acute adriamycin-induced heart failure J. Cardiovasc. Pharmacol. 1991; 18: 437-444.
27. Wold LE, Aberle NS 2nd, Ren J. Doxorubicin induces cardiomyocyte dysfunction via a p38 MAP kinase-dependent oxidative stress mechanism. Cancer Detect Prev. 2005; 29[3]: 294-299.

28. Kapelko VI, Williams CP, Gutstein DE, Morgan JP. Abnormal myocardial calcium handling in the early stage of adriamycin cardiomyopathy. Arch Physiol Biochem 1996; 104: 185-191.
29. Singal PK, Forbes M, Sperelakis N. Occurence of intramitochondrial Ca^{++} granules in hypertrophied heart exposed to adriamycin. Can J Physiol Pharmacol 1984; 62: 1239-1244.
30. Dudnakova TV, Lakomkin VL, Tsyplenkova VG, Shekhonin BV, Shirinsky VP, Kapelko VI. Alterations in myocardial cytoskeletal and regulatory protein expression following a single doxorubicin injection. J Cardiovasc Pharmacol. 2003; 41[5]: 788-94.
31. Okabe E, Tsujimoto Y, Kobayashi Y. Calmodulin and cyclic adriamycinP-ribose interaction in Ca^{2+} signaling related to cardiac sarcoplasmic reticulum: superoxide anion radical-triggered Ca^{2+} release. Antioxid Redox Signal. 2000; 2[1]: 47-54.
32. Goldhaber JI, Liu E. Excitation-contraction coupling in single guinea pig ventricular myocytes. J Physiol [London] 1994; 477: 135-147.
33. Li T, Danelisen I, Singal PK. Early changes in myocardial antioxidant enzymes in rats treated with adriamycin. Mol Cell Biochem 2002; 232[1-2]: 19-26.
34. Hrelia S, Fiorentini D, Maraldi T, Angeloni C, Bordoni A, Biagi PL, Hakim G. Doxorubicin induces early lipid peroxidation associated with changes in glucose transport in cultured cardiomyocytes. Biochim Biophys Acta. 2002; 1567[1-2]: 150-6.
35. Luo X, Evrovsky Y, Cole D, Trines J, Benson LN, Lehotay DC. Doxorubicin-induced acute changes in cytotoxic aldehydes, antioxidant status and cardiac function in the rat. Biochim Biophys Acta 1997; 1360: 45-52.
36. Yin X, Wu H, Chen Y, Kang YJ. Induction of antioxidants by adriamycin in mouse heart. Biochem Pharmacol. 1998; 56: 87-93.
37. Kalivendi SV, Konorev EA, Cunningham S, Vanamala SK, Kaji EH, Joseph J, Kalyanaraman B. Doxorubicin activates nuclear factor of activated T-lymphocytes and Fas ligand transcription: role of mitochondrial reactive oxygen species and calcium. Biochem J. 2005; 389[Pt 2]: 527-39.
38. Lakomkin VL, Konovalova GG, Kalenikova EI, Zabbarova IV, Tsyplenkova VG, Tihaze AK, Lankin VZ, Ruuge EK, Kapelko VI. Protection of rat myocardium by coenzyme Q during oxidative stress induced by hydrogen peroxide. Biochemistry [Moscow] 2004; 69 [5]: 520-526.
39. Dziegiel P, Murawska-Cialowicz E, Jethon Z, Januszewska L, Podhorska-Okolow M, Surowiak P, Zawadzki M, Rabczynski J, Zabel M. Melatonin stimulates the activity of protective antioxidative enzymes in myocardial cells of rats in the course of doxorubicin intoxication. J Pineal Res. 2003; 35[3]: 183-7.
40. Boucek RJ, Jr, Miracle A, Anderson M, Engelman R, Atkinson J, Dodd DA. Persistent effects of doxorubicin on cardiac gene expression. J Mol Cell Cardiol 1999; 31: 1435-1446.
41. Engedal H, Jensen H, Saetersdal TS. Ultrastructure of abnormal membrane inclusions in nuclei of human myocardial cells. Br Heart J 1977; 39: 145-151.
42. Kapelko V, Khatkevich A, Benevolensky D, Belikova Y. Cardiac pump function and mitochondrial respiration at early doxorubicin treatment and withdrawal. Exp Clin Cardiol 1999; 4[3]: 147-151.
43. Kapelko VI, Khatkevich AN, Dvoriantsev SN, Ruuge EK, Belikova IuO, Benevolensky DS. Contractile function and energetic metabolism of the heart at the early stage of adriamycin cardiomyopathy (in Russian). Kardiologiya (Mos) 1997, 37 [2]: 31-35.

Adaptation Biology and Medicine (Volume 5: Health Potentials)
Editors: L. Lukyanova, N. Takeda and P.K. Singal

9. The Activation of Mitochondrial ATP-dependent Potassium Channel is One of the Mechanisms of Adaptation to Hypoxia

G.D. Mironova*

Russian Academy of Sciences, Moscow

Abstract

In this review the structural organization, properties and functional role of ATP-dependent potassium channel, found recently in mitochondria will be discussed. The role of this channel in heart and brain protection against ischemic injuries is widely discussed now. According to the recent data the activation of this channel is one of the key mechanisms of organism adaptation to ischemia (Garlid *et al.*, 1997; Oldenburg *et al.*, 2003; O'Rourke, 2004).

The classical adaptation to ischemia is usually formed by brief episodes of coronary artery occlusion (5 minutes at once or 2-3 times in 5 minutes) (Murry *et al.*, 1986; Kitigawa *et al.*, 1990) or brief decrease in oxygen partial pressure in inhaled air before the subsequent prolonged hypoxia (Zhu *et al.*, 2003). Such procedure leads to significant decrease of the size of myocardial infarct zone, which occurs as a result of subsequent prolonged occlusion (Kloner *et al.*, 1998). This phenomenon was named "ischemic preconditioning". The K_{ATP} channels inhibitors were found to abolish the positive effect of such adaptation, whereas the activators of the channel imitate it (Gross *et al.*, 1992; Auchampach *et al.*, 1992). Thus the channel plays an important role in adaptation of the organism to ischemia.

The acute phase of preconditioning lasts for only 1-2 hours. However in 24 hours, the second window of tolerance occurs (Yellon *et al.*, 1998; Patle *et al.*, 2005). This lasts for 3-4 days and sometimes to a month (Carrol and Yellow, 2000). The phenomenon was named the delayed tolerance of myocardium to ischemia, and it is this phenomenon that lies, probably, in the base of intermittent hypoxic therapy widely used in clinic. Activation of mitochondrial ATP-dependent potassium channel is on the basis of the first and the second phases of tolerance since the specific inhibitors of the channel abolish both phases of tolerance (Wang *et al.*, 2001). Protein synthesis inhibitors have no effect on the first phase but inhibit the second phase of tolerance. It is thought that the opening of mitoK_{ATP} on the second phase can be connected with activation of protein synthesis and/or with modification of functional activity of the channel on post-translational level (Petrishchev *et al.*, 2001).

Phenomenon of ischemic myocardial adaptation is considered to be the most effective way to decrease the size of myocardial infarction zone. Cell mechanisms of the adaptation are intensively studied in many laboratories and clinics all over the world (Gross *et al.*, 1992; Garlid *et al.*, 1997; Carreira *et al.*, 2005; Murata *et al.*, 2001). The protective effect is supposed to be three-staged and includes different

*Address for Correspondence: G.D. Mironova, Institute of Theoretical and Experimental Biophysics, Russian Academy of Sciences, Pushchino, Moscow region, 142290 Russia, E-mail: mironova@iteb.ru

extracellular trigger mechanisms, and intracellular signal mechanisms. However, independently of the intra- or extracellular mechanisms involved in adaptation the final effecter of this process is the mitoK_{ATP} channel (Grover and Garlid, 2000; Petrishchev *et al.*, 2001; Palet *et al.* 2005).

Keywords: Hypoxia, pre-conditioning, sulphonylurea, glibenclamide, cromakalin, and hydroxydecanoic acid

Structural Properties and Regulation of the ATP-dependent Potassium Channels

There are two types of K_{ATP} channels in the cell: sarcoplasmic (or cell) and mitochondrial. Cell channel was found earlier than mitochondrial and was studied in more detail. This channel consists of two subunits: channel-forming (m.w. 47-49 kDa) which forms the channel pore, and regulatory with molecular weight in the range of 60-140 kDa (Aguilar-Bryan *et al.*, 1995). The regulatory subunit binds modulators, and particularly sulphonylurea, and called SUR subunit. The channel subunit forms a channel pore in the membrane which possess inwardly rectifying properties. In other words, it is capable to transport potassium ions mainly in one direction (Yokoshiki *et al.*, 1998). Therefore the channel-forming subunit was called KIR subunit. ATP in physiological concentrations inhibits the channel (Sellers *et al.*, 1992) binding mostly with the channel-forming subunit. The regulatory subunit increases affinity of the channel-forming subunit with ATP more than on an order and adds to the channel sensitivity to modulators (Tucker *et al.*, 1997).

In 1991 using the patch-clamp technique Japanese researchers identified ATP-dependent potassium channel in the inner mitochondrial membrane (mitoK_{ATP}) (Inoue *et al.*, 1991). This channel was also inwardly rectifying the sensitivity to ATP. It should be noted that, in our laboratory, 10 years before, the selective potassium channel (Fig. 1A), was isolated from the inner mitochondrial membrane by using the water- ethanol extraction (Mironova *et al.*, 1981). The protein obtained formed inwardly rectifying channel in artificial membranes (Fig. 1B). Later in our joint research work with professor Garlid (USA), the protein with similar properties was isolated from inner mitochondrial membrane by using the detergent extraction and it was found that this channel was ATP-sensitive (Paucek *et al.*, 1992).

Being reconstituted into BLM, the water- ethanol extracted protein formed potassium-selective channels with elementary conductivity 10 pS in the presence of 100 mM KCl (Fig. 1A). It corresponds with the data obtained on the native membrane by patch-clamp technique when the conductivity of a single channel was 9.7 pS. The channel was inhibited by ATP (Fig. 2) and activated by GDP (Mironova *et al.*, 1996, 1997, 1999).

Data on the effects of different modulators indicate community and not identity of cell and mitochondrial potassium channels (O'Rourke, 2004). Those channels have some common regulators. Thus, glibenclamide, the sulphonylurea preparation, inhibits both channels whereas cromakalim and pinacidil activate them. However, there are some differences in both channels regulation. For example, 5-hydroxydecanoic acid (5-HD) is capable to inhibit only mitoK_{ATP} and has no effect on cellK_{ATP} (Sato *et al.*, 2000). At the same time, HMR-1098 shows an opposite effect being the specific inhibitor of cell channel (Jung *et al.*, 2000). Diazoxide is 10 times more effective opener of mitochondrial than of sarcoplasmic channel, and therefore is used as specific mitoK_{ATP} activator (Garlid *et al.*, 1997). UDP is also much more effective as the opener of mitoK_{ATP} (1,64), than cellK_{ATP} (Alekseev *et al.*, 1998).

The study of the properties of isolated and reconstituted into BLM mitoK_{ATP} and of the channel in native membrane allowed us to propose the model of mitoK_{ATP} (Fig. 3). According to this model, mitoK_{ATP} as the cellK_{ATP} is heteromultimer and consists of the channel-forming and regulatory

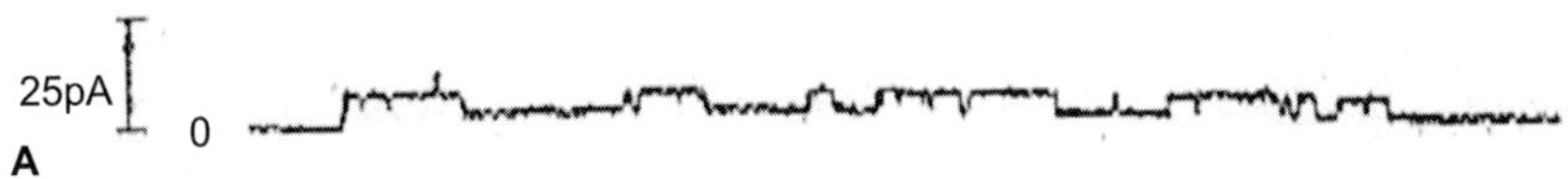

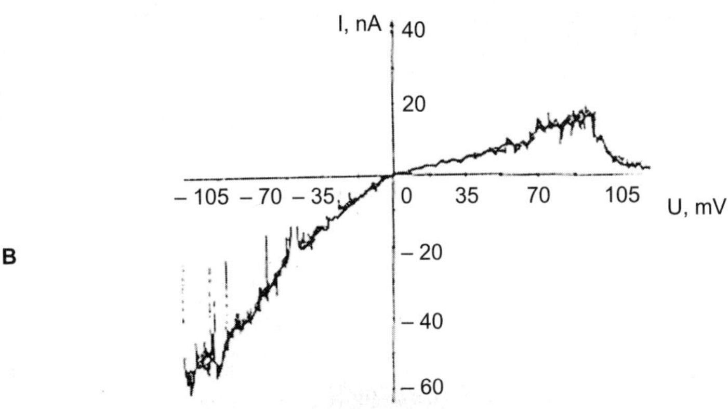

Fig. 1. **Channel activity of the 55 kDa protein reconstituted into BLM (A) Single channel record. The protein forms the potassium-selective channels with the elementary conductivity of 10 pS in the presence of 100 mM KCl. (B) Current-voltage dependence of the channel in symmetric 100 mM KCI. The protein forms inwardly rectifying channel in artificial membrane.**

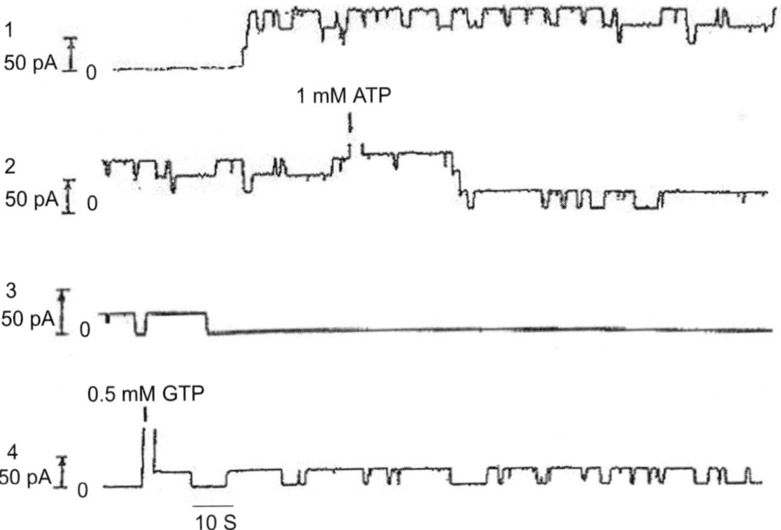

Fig. 2. **The electrical properties of mitoK$_{ATP}$. BLM modified by mitochondrial 55 kDa protein. The 2-4 are the continuation of the 1.**

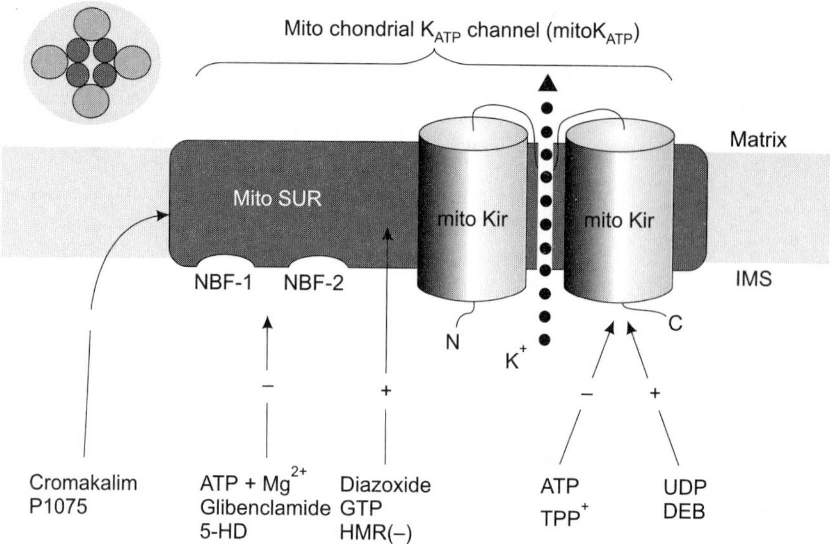

Fig. 3. The model of the interaction of regulatory ligands with mitoK_{ATP} structure. This model was proposed on the basis of the study of the properties of isolated mitoKATP reconstituted into BLM and of the channel in native mitochondria. According to the model mitochondrial channel is heteromultimer and consists of the channel-forming and regulatory subunits. 55 kDa protein is the channel-forming subunit.

subunits and 55kDa protein is the channel-forming subunit (Mironova *et al.*, 2004). The channel formed by the 55kDa protein is not sensitive to whole channel modulators and has rectifying properties. Thus, on the analogy of cellK_{ATP} this subunit was called mitoKIR. In intact mitochondria, ATP-dependent potassium transport is sensitive to sulphonylurea and other modulators. Consequently, we proposed the presence of regulatory subunit in inner mitochondrial membrane. This subunit makes the channel sensitive to sulphonylureas and other modulators and was named mitoSUR (Mironova *et al.*, 2004).

To confirm directly that mitochondrial 55kDa protein is the channel-forming subunit of the whole mitoK_{ATP} the polyclonal antibodies against this protein were obtained. Using these antibodies the inhibitory analysis was carried out. ABs obtained showed the inhibition of ATP-dependent K^+ transport in intact mitochondria (Fig. 4A) without any effect on parameters of mitochondrial respiration (Fig. 4B). Hence, such inhibition was specific and we can assert that 55 kDa protein isolated from rat liver mitochondria is the channel-forming subunit of ATP-sensitive potassium channel. According the other recent research the structure of the channel-forming subunit of mitoK_{ATP} from brain mitochondria is similar to sarcoplasmic KIR6.1 (Lacza *et al.*, 2003).

The Redox State of Active Groups of the Channel Protein in Regulation of the MitoK_{ATP}

To understand the mechanism of cardioprotective action of mitoK_{ATP} activators, the properties of the channel were studied. It was found that functioning of mito- and cellK_{ATP} depends on redox-state of active groups of protein-channels. We showed that redox agents modulate mitoK_{ATP} activity (Grigoriev

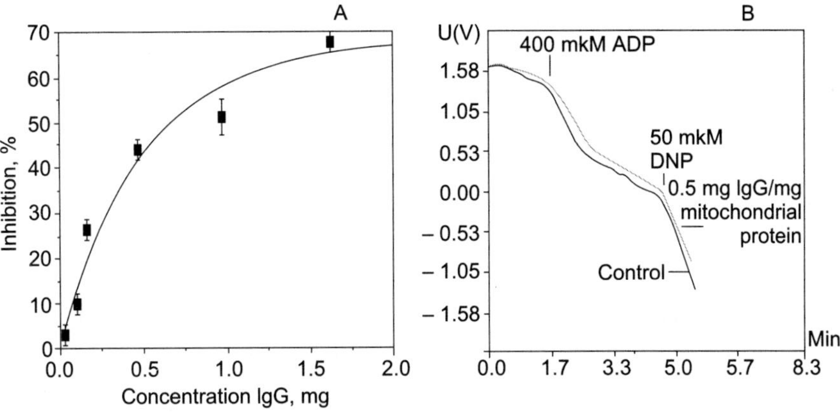

Fig. 4. Inhibition of ATP-dependent potassium transport in mitochondria by the polyclonal antibodies against 55 kDa protein. (A) Influence of the ABs on ATP-dependent K+ influx into mitochondria. (B) Influence of the ABs on the mitochondria respiration.

et al., 1999). As it is shown in the figure (Fig. 5A), electron donor, p-dimethylaminoethylbenzoate (DEB) activates mitoK$_{ATP}$ reconstituted into BLM and is localized in the native membrane (Fig. 5B). In experiments on the native membrane, the blockers, glibenclamide and 5-HD, abolish the effect of DEB. At the same time DEB has no effect on the cellK$_{ATP}$ in cardiomyocytes (Fig. 5C).

The redox agents affect, probably, the SH-groups of the channel. The channel being activated by dithiotrietol (DTT) was demonstrated to lose the potassium selectivity (Grigoriev *et al.*, 1999). Hence, we made a conclusion about the presence of functionally active SH-groups on gate and selective channel filter as in case of the bacterial potassium channel (Meury and Kepes, 1982). Later, it was found that mercury containing preparation, n-ethylmalienemide, is also shown to block mitoK$_{ATP}$ (Coetzee *et al.,* 1995). This inhibitory effect was reversed by thiol-reducing agent (DTT).

In the cardiomyocytes, the channel openers were shown to induce the reactive oxygen species (ROS) generation (Forbes *et al.*, 2001; Carroll *et al.*, 2001; Oldenburg *et al.*, 2003). On the other hand, experiments on submitochondrial particles (SMP) reconstituted into BLM demonstrated that addition of the xanthine/xanthine-oxidase system of generation of superoxide anion radical, into incubation medium leads to the channel activation during the first minute of incubation (Zhang *et al.,* 2003). Under existing conditions, probability of mitoK$_{ATP}$ opening increased almost three times. Subsequent addition of 5-HD, glibenclamide or SH-agent, N-ethylenmaleimide, into incubation medium completely abolished the activating effect of superoxide anion. Superoxide also blocked the inhibitory effect of ATP. Consequently the channel is regulated by positive feedback. Data obtained in experiment with N-ethylenmaleimide allow to suggest that oxygen radical effect directly on sulfide groups of the channel.

The decrease of the ROS production during activation of mitoK$_{ATP}$ was recently observed in heart, liver and brain mitochondria (Ferranti *et al.,* 2003). Authors discussed, in detail, the discrepancy in results obtained in the intact cells and mitochondria and, suggested that mitoK$_{ATP}$ opening in cells may not directly enhance mitochondrial ROS generation, but may activate other intracellular pathway responsible for ROS production. It should be mentioned that during reperfusion, the activation of mitoK$_{ATP}$ leads to decrease in the ROS production (Pain *et al.,* 2000; Zweier *et al.*, 1987). In any case,

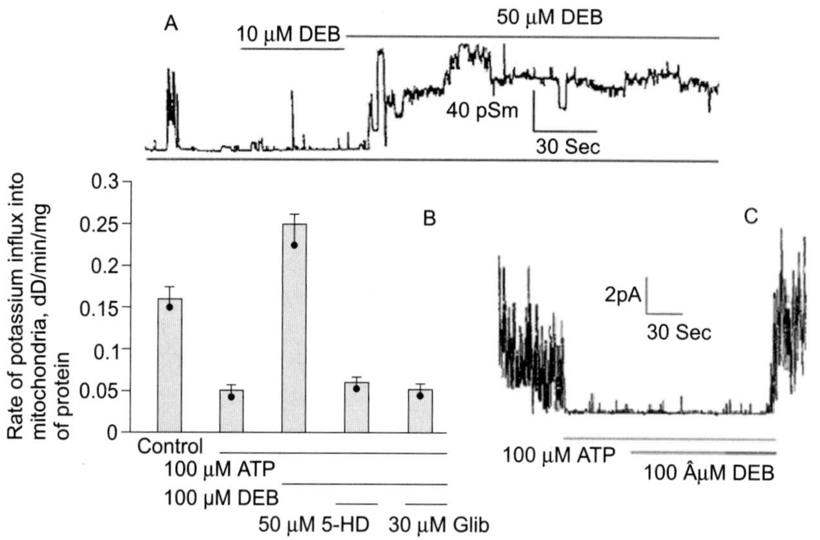

Fig. 5. Modulation of mitoKATP channel by redox agent diethylaminoethylbenzoate (DEB). Activation of mitoK_{ATP} reconstituted into BLM (A) and in intact mitochondria (B) and absence of DEB effect on cellK_{ATP} (C).

it is generally acknowledged that ROS play an important role in producing lethal cell injury associated with cardiac ischemia and reperfusion (McCord, 1988; Das *et al.*, 1994).

Another universal factor of metabolism regulation is nitrogen oxide (NO). It was found that NO generation stimulated by donor NO-SNAP (S-nitrozo-N-acetyl-DL-penicillamine) causes dissociation of mitochondrial respiration, which can be observed by increase in FAD fluorescence (Sasaki *et al.*, 2000). The fluorescence is abolished by mitoK_{ATP} inhibitor. In the presence of NO-synthase inhibitor, cardioprotective effect stimulated by diazoxide did not manifest (Ockaili *et al.*, 2001). Activating effect of NO on mitoK_{ATP} is supposed to be realized through protein kinase C activation (Krenz *et al.*, 2002). Thus, redox agents like SH-, ROS, and NO regulate the channel activity.

Cardioprotective Properties of the ATP-dependent Potassium Channel Activators

Earlier, it was supposed that cellK_{ATP} opening leads to protection of myocardium against ischemic injuries (Auchampach *et al.*, 1992, Oldenburg *et al.*, 2003). However, further research showed that it is mitoK_{ATP} channel that plays more significant role in the system of anti-ischemic protection of myocardium (Garlid *et al.*, 1997; Sato *et al.*, 2000).

As it was noted above, in the base of ischemic preconditioning phenomenon mitoK_{ATP} activation takes place. The fact that specific inhibitor of this channel abolishes the tolerance effect, and specific mitoK_{ATP} opener, diazoxide, imitates this effect proves this suggestion. In the last 5 years, a lot of other evidences about the role of channel in cardioprotection appeared. They were obtained mainly using the channel activators. However, the mechanism of the tolerance development still remains unclear (O'Rourke et al., 2004; Ardehali *et al.*, 2005; Carreira *et.al.*, 2005).

Nevertheless, any activation of potassium transport in mitochondria leads to myocardial protection against ischemia. Activation of other potassium channels, for example, Ca^{2+}-activated K^+ channel, also prevents ischemic injuries (Xu *et al.,* 2002). This fact supports the idea that K^+ influx into mitochondria and concomitant swelling of the organelles underlies cardioprotective effect. However, the ischemic preconditioning is abolished by the inhibitors of mitoK_{ATP} (Garlid *et. al.,* 1997). Consequently, any potassium influx could provide adaptation to hypoxia, but it is the activation of mitoK_{ATP} that is the key point of the preconditioning.

Cardioprotective effect of mitoK_{ATP} activators brings up a question about application of these pharmacological agents in medicine. It is known, however, that use of medical products is often accompanied by nonspecific side effects. On the other hand, as far as during preconditioning the activation of mitoK_{ATP} is observed, it is logically to suppose that at this moment some metabolic channel activators appear in the cell.

According to this suggestion, the search of natural activators of mitoK_{ATP} has been carried out in our laboratory for the last few years. To solve this problem, the effect of different nucleotides and phosphonucleotides on the channel was studied. Experiments on the isolated and reconstituted into BLM channel-forming subunit showed that, mainly, diphosphonucleotides possess the channel-activating properties. However, the effectiveness of their action was different. GDP, for example, activated the channel at concentration of 0.5 mM (Fig. 6) that corresponds with the data obtained on intact mitochondria by patch-clamp technique (Bednarczyk *et al.,* 2005).

The most effective mitoK_{ATP} activator was uridine-5'-diphosphate (UDP) (Fig. 7). Micromoles concentrations of this nucleotide reactivated reconstituted into BLM and inhibited by ATP mitochondrial K_{ATP} channel (Fig. 7A). Studies were also carried out on intact mitochondria where both subunits of the channel are presented. In intact mitochondria, UDP in concentration of 30 ìM reactivated ATP-sensitive K^+ transport inhibited by ATP (Fig. 7B). Effect of UDP is specific since the

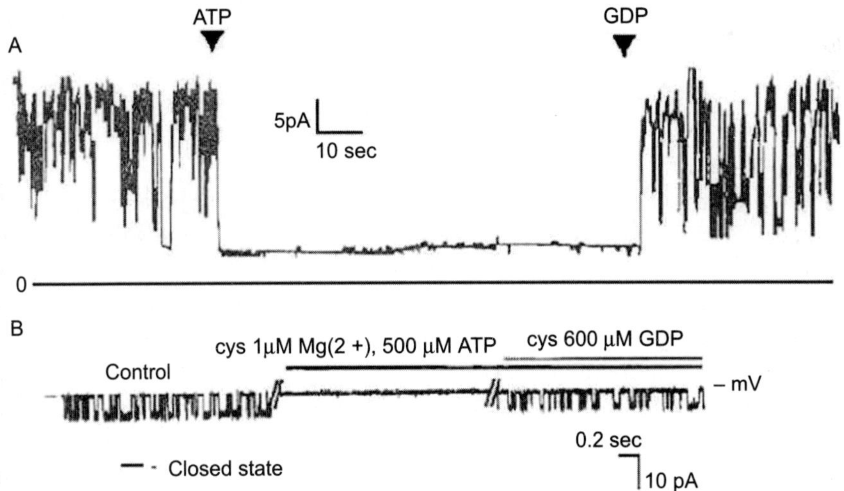

Fig. 6. Influence of ATP and GDP on the channel-forming subunit (55 kDa protein) in BLM (A) and on the whole channel in cardiomyocytes (B).
A. ATP and GDP concentrations were 2mM and 0.5 mM, respectively.
B. From Bednarczyk P. *et al.*, FEBS. 2005; 579: 1625-1632.

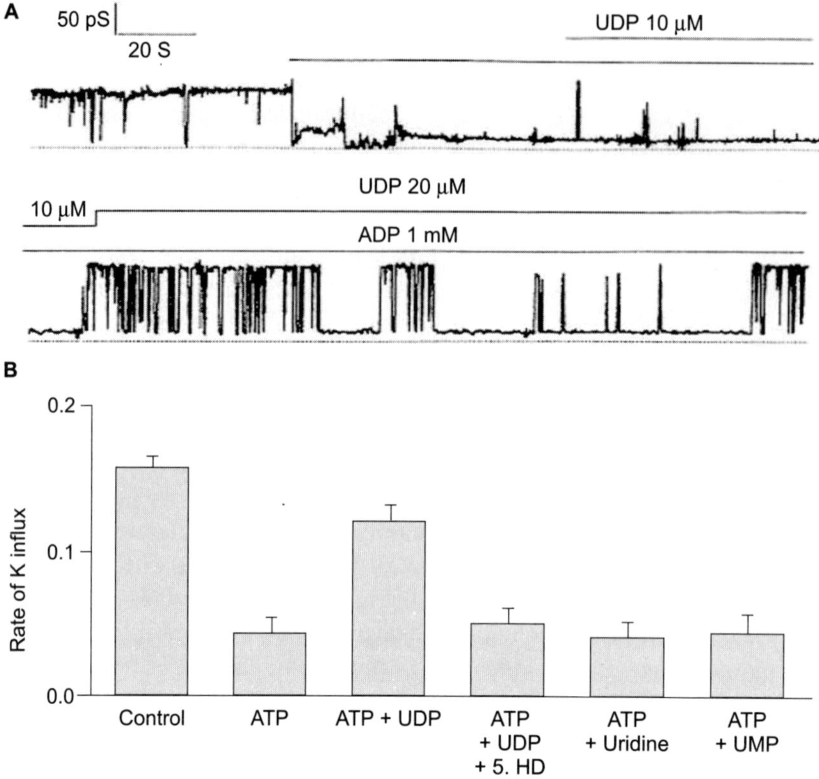

Fig. 7. UDP activation of mitoK-ATP channel reconstituted into BLM (A) and in intact mitochondria (B). In experiments on intact mitochondria, concentrations of ATP, UDP and 5-HD were 100 μM, 30 μM and 200 μM correspondingly.

addition of mitoK$_{ATP}$ blocker, 5-HD, abolishes reactivation observed. It should be noted that UTP and UDP precursors: uridine and UMP did not show such effect (Fig. 7B). Hence, UDP in the cell can serve as the channel activator.

In native conditions, mitoK$_{ATP}$ channel is usually closed since ATP concentration in the cell significantly exceeds the value necessary for the channel inhibition. As far as ATP concentration in the cell is stable value the presence of natural activator is necessary for opening and regulation of the channel. UDP can play role of such activator. Consequently, UDP concentration in the cell can regulate extent of the channel opening. According to literature, the rate of consumption of UDP precursors (uridine) from blood flow by cardiomyocytes increases during myocardial ischemia (Aussedat *et al.*, 1984). This process can promote UDP accumulation in the heart, which favors to mitoK$_{ATP}$ activation.

It is well known that UDP cannot penetrate into the cell but can be synthesized there from uridine and UMP (Matsushita and Fanburg, 1970). Thus, we examined cardioprotective action using uridine and UMP, which unlike UDP can penetrate into the cell and form UDP there. To examine the cardioprotective effect of the preparations, the model of acute myocardial ischemia of the rat left ventricle was used (namely, 60 min occlusion of the left coronary artery (LCA) without following

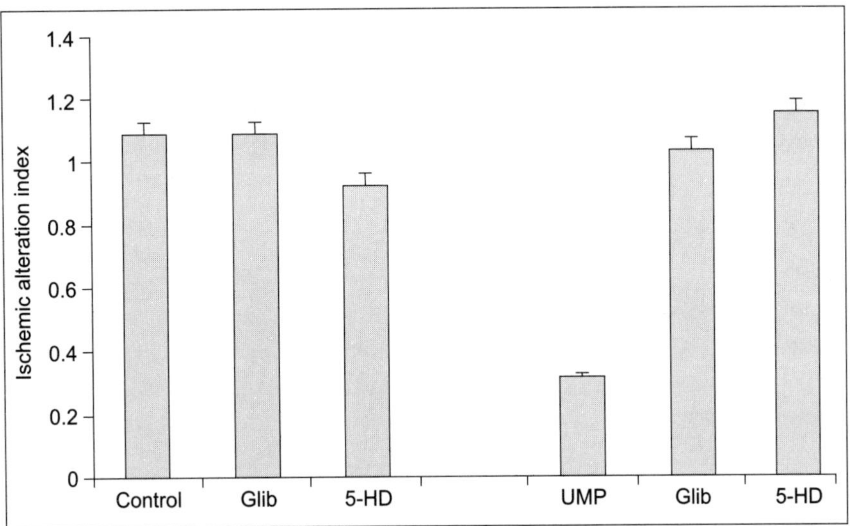

Fig. 8. Influence of UMP on the ischemic alteration index (IAI) 60 min after left coronary artery occlusion. UMP decreases the infarct zone size almost 4 times.

reperfusion). The work was carried out in collaboration with Drs Krylova and Sapronov from Saint Petersburg Institute of Experimental Medicine.

Uridine and UMP were shown to possess clear cardioprotective properties. To estimate these properties the change in index of ischemic alteration (Krylova *et al.*, 2006) and T-wave amplitude on ECG were measured. Both preparations decrease several times the index of ischemic alteration that estimates the infarct zone size. As one can see from the Fig. 8, UMP decreases the infarct zone size almost 4 times. K_{ATP} channel inhibitors – glibenclamide and 5-HD reverse the positive effect of UMP. The blockers themselves have no effect on infarct zone development. Uridine possesses the same property (Krylova *et al.,* 2006). Since 5-HD is the specific inhibitor of mitoK_{ATP}, the fact that this preparation is capable of blocking the action of uridine and UMP indicates the supreme role of mitochondrial K_{ATP} channel in heart protection against ischemic injuries.

Besides the anti-ischemic action uridine and UMP demonstrated anti-arrhythmic effect. The number of rhythm disturbances like ventricular tachycardia and premature ventricular beats were evaluated (Fig. 9). UMP injection decreased by 9 and 4 times, respectively, the duration of ventricular tachycardia and the number of premature ventricular beat. However, it should be noted that glibenclamide completely abolished the positive effect of uridine and UMP. The effect of the specific mitoK_{ATP} inhibitor 5-HD on rhythmic disorders was less manifested. On basis of the data obtained, we supposed the antiarrhythmic action of the preparations caused both ATP-dependent potassium channels but mainly by opening of cellK_{ATP} rather than mitoK_{ATP}.

The data obtained are in agreement with that of Japanese study of the female sex hormone β-estradiol (Tsai *et al.*, 2002). The hormone was shown to decrease the size of the infarct zone by 3 times. The specific mitoK_{ATP} inhibitor, 5-HD, abolishes the hormone effect whereas the cellK_{ATP} inhibitor, HMR-1098 does not block the positive effect of β-estradiol. The male hormone testosterone possesses the same property as β-estradiol (Fikret *et al.*, 2004). Consequently, the hormone status of the

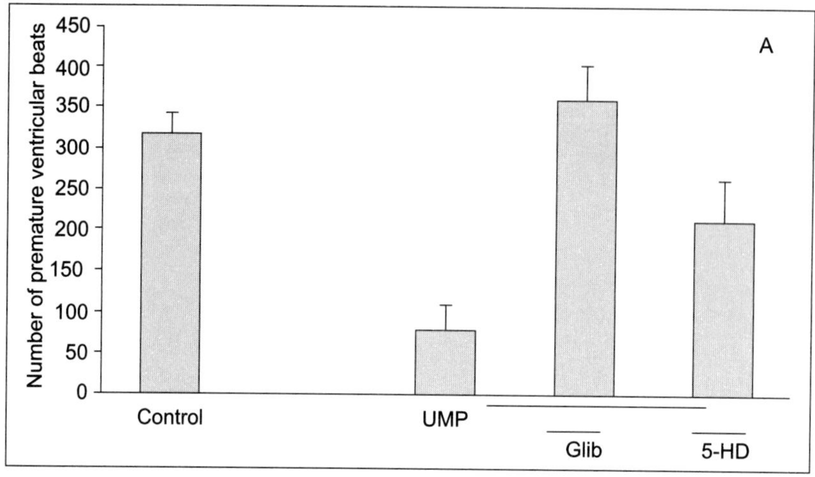

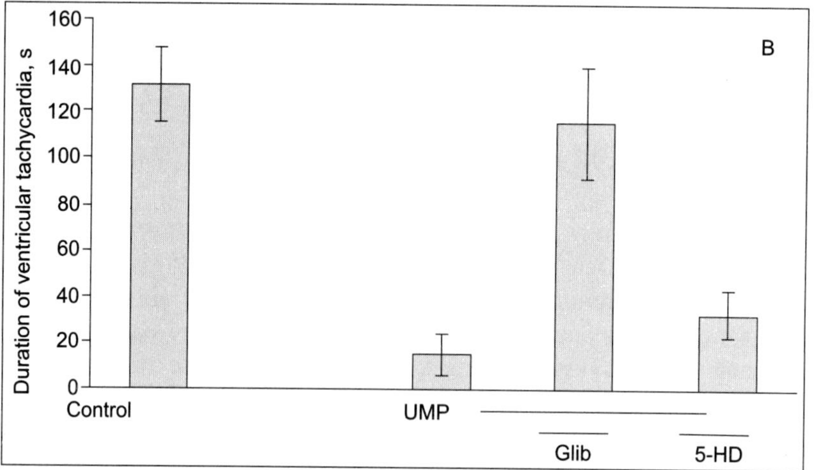

Fig. 9. Influence of UMP on rhythm disturbances. UMP injection decreased the number of premature ventricular beats 4 times (A) and the duration of ventricular tachycardia 9 times (B) Glibenclamide completely abolished the positive effect of uridine and UMP. The effect of 5-HD was less manifested.

organism can determine the work of mitoK_{ATP}. Thus, the change in hormone composition during aging can be a reason of high probability of infarct development among seniors.

We studied the mitoK_{ATP} activity of Vister rats of different age. The results are presented in the Figure 10. The activity of the channel from liver of 8 month old rats is 5 times lower than of 1 month old animal. The same tendency is observed in the heart. The age of 8 months of Vister rats corresponds to that of 50 years in human. The decrease in the activity and, probably, in the amount of mitoK_{ATP} during aging or change in hormone composition can be one of the reasons of myocardial infarction origin

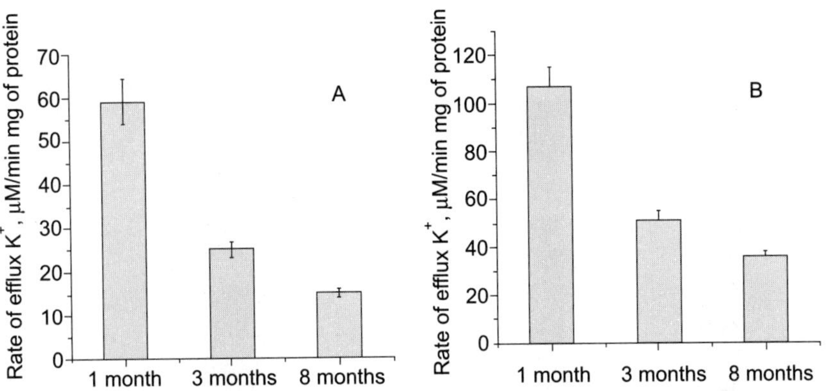

Fig. 10. Rate of ATP-dependent potassium transport in intact mitochondria of rat liver (A) and rat heart (B) of different age groups of rats.

The Role of mitoK_{ATP} Channel in Adaptation to Intermittent Hypoxia

In spite of the fact that the role of the mitoK_{ATP} in protection of myocardium and other tissues against ischemic injuries is evident, the reasons underlying this effect, as was mentioned above, still remains unclear. To clarify this question we studied the parameters of respiration and ATP-dependent potassium transport in mitochondria of rats possessing different resistance to ischemia (Lukyanova, 1997), and adapted to hypoxia. The adaptation was carried out by the intermittent normobaric hypoxia (IH) procedure (Zhu *et al.*, 2003). It is known, that the K_{ATP} channels being induced by IH contribute to the cardiac protection against ischemia/reperfusion injury (Zhu *et al.*, 2003). In our joint research with professor Lukyanova, the male Vistar rats low resisted to hypoxia were inhaled the air with twice-lower oxygen concentration for 5 minutes 6 times a day with the intervals of 3 minutes. During the intervals, the animals inhaled the air with the normal oxygen concentration. The procedure was repeated during 12 days.

We found that adaptation to intermittent normobaric hypoxia results in respiration coupling that manifests in decrease of the respiratory rates in all metabolic states (Fig. 11A), increase of respiratory control and shortening of ADP phosphorylation time (Fig. 11B). The rates of ATP-dependent

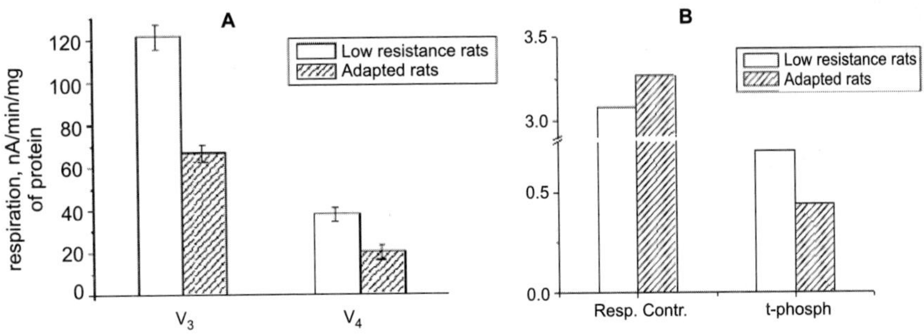

Fig. 11. Effect of intermittent hypoxia of low resisted rats on the mitochondrial respiration and oxidative phosphorylation.

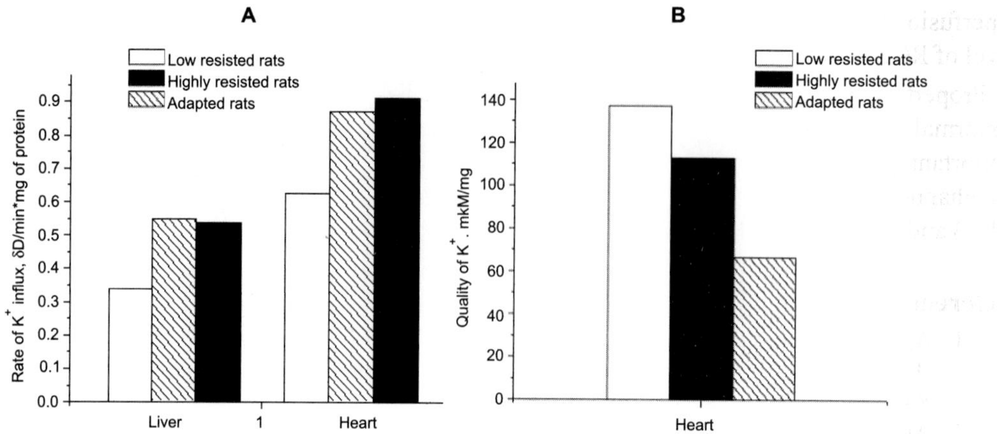

Fig. 12. Parameters of mitochondrial potassium transport in animals with different resistance to hypoxia and adapted to hypoxia.

potassium influx in rats low and highly resisted to hypoxia and low-resisted animals after adaptation have been compared. It was found that the rate is higher in adapted and highly-resisted animals to hypoxia animals (Fig. 12A). Moreover, inhibition constant for ATP is two fold less during adaptation (data not shown). Interestingly, the potassium concentration in mitochondria of highly-resisted to hypoxia and adapted animals did not significantly change and even decreased (Fig. 12B). It means that mitochondria during adaptation did not swell, and even contracted. This indicates that adaptation leads to intensification of K^+/H^+-exchanger activity in mitochondria, which regulates potassium efflux. The recycling of potassium in mitochondria maintains the definite mitochondrial volume, and, probably, necessary for adaptation of animal to hypoxia. Thus, the decrease in activity of K^+/H^+-exchanger might be the main reason of high-amplitude swelling and following injury of mitochondria during ischemia. Possibly, during animal adaptation to hypoxia not only the system of ATP-dependent K^+ influx into mitochondria but also the system of K^+ efflux (K^+/H^+-exchanger) are functionally significant.

During hypoxia the lack of oxygen leads to respiratory chain carriers reduction. According to literature, the reduction of the carriers localized on the complex I and III segments of the respiratory chain lead to the increase in reactive oxygen species (ROS) production (Kaplan-Bresler, 1965; Ferrannti *et al.*, 2003). As it was recently found, the major site of ROS generation is protein-associated ubisemiquinones which are spin-coupled with iron-sulfer cluster N2 (Ohnishi *et al.*, 2005). Under physiological conditions, mitochondria are known to convert a few percent of consumed oxygen into ROS (Lenaz *et al.*, 2002). As we observed earlier, small amount of ROS is needed for functioning of the respiratory chain (Kondrashova and Mironova, 1971).

Increase of the abnormal ROS production can be partially eliminated by activation of the potassium cycle. Our data indirectly demonstrated that during the adaptation not only influx but also potassium efflux from mitochondria are activated. The activation of potassium cycle accelerates the electrons flow along the respiratory chain, decreases a respiratory chain carriers reduction and leads to decrease in ROS production. This putative mechanism can explain the positive role of mitoK_{ATP} activation during reperfusion. It is well known that ROS are the main damaging factors during hypoxia/

reperfusion (McCord, 1988; Das *et al.*, 1994) and opening of the mitoK_{ATP} results in a decrease in the level of ROS generation during reperfusion phase (Zweier *et al.*,1987, Ozcan *et al,* 2002).

Properties of mitoK_{ATP} activators to prevent ATP degradation (Murry *et al.*, 1990) and to decrease abnormal ROS production (Zweier *et al.*, 1987, Ozcan *et al*, 2002) make the channel physiologically important not only during ischemia but also during other pathological states. According to literature, the channel can play a crucial role in different physiological (Fedotcheva *et al.*, 1985; Kloner *et al.*, 1998) and pathological (Kloner *et al.*, 1998; O'Rourke, 2004;) states.

References

1. Aguilar-Bryan L, Nichols C, Wechsler SW, Clement JPt, Boyd AE, Gonzalez G, Herrara-Sosa H, Nguy K, Bryan J, Nelson DA. Cloning of the beta cell high-affinity sulfonylurea receptor: a regulator of insulin secretion. *Science.* 1995; 268: 423-426.
2. Alekseev AE, Brady PA, Terzic A. Ligand-insensitive state of cardiac ATP-sensitive K^+ channels. Basis for channel opening. *J. Gen. Physiol.*, 1998; 111: 381-394.
3. Ardehali H, O'Rourke B. Mitochondrial K_{ATP} channel in cell survival and death. *J.Mol. Cell. Cardiol.*, 2005; 39: 7-16.
4. Auchampach J, Grover G, Gross G. Blockade of ischaemic preconditioning in dogs by the novel ATP dependent potassium channel antagonist sodium 5-hydroxidecanoate. *Cardiovasc. Res.* 1992; 26: 1054-1062.
5. Aussedat J, Ray A, and Rossi A. Uridine Incorporation in normal and ischaemic/perfused rat heart. *Mol. Physiol.*, 1984; 6: 247-56.
6. Bednarczyk P, Dolovy K, Szewczyk A. Matrix Mg^{2+} regulates mitochondrial ATP-dependent potassium channel from heart. *FEBS Letters.* 2005; 579: 1625-1632.
7. Carreira R, Facundo H, Kowaltowski A. Mitochondriual K+ transport and cardiac protection during ischemia/ reperfusion. *Braz. J. Med. Biol. Res.* 2005; 38(3): 345-352.
8. Carroll R. and Yellon D. Delay cardioprotection in human cardiomyocyte-derived cell line: the role of the adenosine, p38MAP kinase and mitochondrial K_{ATP}. *Basic Res.Cardiol.* 2000; 95: 243-249.
9. Carroll R, Gant VA, Yellon DM. Mitochondrial K_{ATP} channel opening protects a human atrial-derived cell line by a mechanism involving free radical generation. *Cardiovasc Res.* 2001; 51: 691-700.
10. Coetzee WA, Nakamura TY, Faivre J-F. Effects of thiol-modifying agents on K_{ATP} channels in guinea pig ventricular cells. *Am. J. Physiol. Heart Circ. Physiol.* 1995; 269: 1625-1633.
11. Das DK. Cellular biochemical and molecular aspects of reperfusion injury. *Ann. N.Y. Acad. Sci.* 1994; 723: 116-127.
12. Fedotcheva NI, Sharichev AA, Mironova GD, Kondrasheva MN. Inhibition of succinate oxidation and K^+ transport in mitochondria during hibernation. *Comp. Biochem. Physiol.* 1985; 82B: 191-195.
13. Ferranti RF, de Silva MM, Kowaltowski AJ. Vitochondrial ATP-sensitive K^+ channel opening decreases reactive oxygen species generation. *FEBS Letters* 2003; 536: 51-55.
14. Fikret E, Guido M., Gassanov N. Francisco R, Uta H. Testosterone induces cytoprotection by activating ATP-sensitive K^+ channel in the cardiac mitochondrial inner membrane. *Circulation 2004*; 110(19): 3100-3107.
15. Forbes RA, Stenberger C, Murphy E. Diazoxide-induced cardioprotection requires signaling through a redox-sensitive mechanism. *Circ.Res.* 2001; 88: 802-809.
16. Garlid KD, Paucek P, Yarov-Yarovoy V. Murray ND, D'Alonzo AJ, Lodge NJ, Smith MA and Grover GJ. Cardioprotective effect of diazoxide and its interaction with mitochondrial ATP-sensitive K^+ channels: possible mechanism of cardioprotection. *Circ. Res.* 1997; 81: 1072-1082.
17. Grigoriev S, Skarga YY, Mironova GD, Marinov BS. Regulation of mitochondrial K_{ATP} channel by redox agents. *Biochim. Biophys. Acta.*, 1999; 1410(1): 91-96.

18. Gross G, O'Rourke S, Pelc L, Warltier D. Myocardial and endothelial dysfunction after multiple, brief coronary occlusions: role of oxygen radicals. *Am J Physiol.* 1992; 263(6 Pt 2): 1703-1709.
19. Grover G. and Garlid K. ATP-sensitive potassium channels: a review of their cardioprotective pharmacology. *J.Mol.Cell Cardiol.*, 2000; 32: 677-695.
20. Inoue I, Nagase, Kishi K. and Higuti T. ATP-sensitive K^+ channel in the mitochondrial inner membrane. *Nature* 1991; 352: 244-247.
21. Jung O, Englert HC, Jung W. et al. The KATP channel blocker HMR 1883 does not abolish the benefit of ischemic preconditioning on myocardial infarct mass in anesthetized rabbits. *Arch Pharmacol.* 2000; 361: 445-451.
22. Kaplan-Bresler AE. Metmyoglobin oxidation during electron transport reactions in mitochondria. *J. General physiol.* 1965; 48: 685-697.
23. Kitigawa K, Matsumoto M, Tagaya M, Hata R, Ueda H, Ninobe M, Handa N, Fukunaga R, Kimura K, Mikoshiba K, Kamada T. "Ischemic tolerance" phenomenon found in the brain. *Brain Res.* 1990; 528: 21-24.
24. Kloner RA, Bolli R, Marban E, Reinlib L, Braunwald E. Medical and cellular implications of stunning, hibernation and preconditioning: an NHLBI workshop (special report). *Circulation.* 1998; 97(18): 1848-1867.
25. Kondrashova MN. and Mironova GD. Oxygen requirement in the phosphorilating od ADP under cyanide block conditions. Biokhimya, 1971; 36: 729-731.
26. Krenz M, Oldenburg O, Wimpee H. et al. Opening of ATP-sensitive potassium channels causes generation of free radicals in vascular smooth muscle cells. *Basic Res. Cardiol.* 2002; 97: 365-373.
27. Krylova IB, Kachaeva EV, Rodionova OM, Negoda AE, Evdokimova NR, Balina MI, Sapronov NS, Mironova GD. The cardioprotective effect of uridine and uridine-5'-monophosphate: the role of the mitochondrial ATP-dependent potassium channel. *Exp. Gerontol.* 2006; 41: 697-703.
28. Lacza Z, Snipes J, Kis B, Szabo C, Grover G, Busija D. Investigation of the subunit composition and the pharmacology of the mitochondrial ATP-dependent K^+ channel in the brain. *Brain Res.* 2003; 994: 27-36.
29. Lenaz G, Bovona C, D'Aurelio M, Fato R, Formiggini G, Genova ML, Giuliano G, Pich M, Paolucci U, Castelli G, and Ventura B. Role of mitochondria in oxidative stress and aging, *Ann. N.Y. Acad. Sci.* 2002; 959: 199-213.
30. Lukyanova LD. Bioenergetics hypoxia: mechanisms and ways of corrections. *Bul. Exp. Bio.l and Med. (Rus.)* 1997; 124(9); 244-254.
31. McCord JM, Free radicals and myocardial ischemia: overview and outlook. *Free Radic Biol. Med.* 1988; 4: 9-14.
32. Matsushita S. and Fanburg B. Pyrimidine nucleotide synthesis in the normal and hypertrophying rat heart. Relative importance of the de novo and "salvage" pathways. *Circ. Res.*, 1970; 27: 415-28.
33. Meury J.and Kepes A. Glutathione and the gated potassium channels of *Escherichia coli.*
34. *EMBO J.* 1982; 1: 339-343.
35. Mironova G, Fedotcheva N, Makarov P, Pronevich L, Mironov G. Protein from beef heart mitochondria inducing the channel potassium conductivity of bilayer lipid membranes. *Biofizika,* 1981; 26 (3): 451-457.
36. Mironova G, Skarga Yu, Grigoriev S, Yarov-Yarovoi V, Alexandrov A, Kolomytkin O. The ATP-dependent potassium channel from rat liver mitochondria. 1. Isolation, purification, and reconstitution in a bilayer lipid membrane. *Membr. And Cell Biol.*, 1996; 10(4): 429-437.
37. Mironova G, Grigoriev S, Skarga Yu, Negoda A, Kolomytkin O. ATP-dependent potassium channel from rat liver mitochondria: inhibitory analysis, channel clusterization. *Membr Cell Biol.,* 1997; 10(5): 583-91.
38. Mironova GD, Skarga Yu, Grigoriev SM, Negoda AE, Kolomytkin OV, Marinov BS. Reconstitution of the mitochondrial ATP-dependent potassium channel into bilayer lipid membrane. *J. Bioenerg. Biomembr.,* 1999; 31(2): 157-161.
39. Mironova G, Negoda A, Marinov B, Paucek P, Costa A, Grigoriev S, Skarga Yu, Garlid K. Functional distinctions between the mitochondrial ATP-dependent K^+ channel (mito K_{ATP}) and its inward rectifier subunit (mitoKIR) *J. Biol. Chem.*, 2004; 279(31): 32562-32568.

40. Murata M, Akao, O'Rourke B, Marban E. Mitochondrial ATP-sensitive potassium channels attenuate matrix Ca^{2+} overload during simulated ischemia and reperfusion: possible mechanism of cardioprotection. *Circ. Res.* 2001; 89: 891-898.

41. Murry C, Jennings., Reimer K. Preconditioning with ischemia: a delay of lethal cell injury in ischemic myocardium. *Circulation,* 1986; 74: 1124-1136.

42. Ockaili R, Bhargava P, Kukreja R. Chemical preconditioning with 3-nitropropionic acid in hearts: role of mitochondrial K_{ATP} channel. Am. J. Physiol. Heart Circ. Physiol. 2001; 280: 2406–2411.

43. Oldenburg O, Cohen MF, Downey JM. Mitochondrial K_{ATP} channels in preconditioning. *J Mol Cell Cardiol.* 2003; 35: 569-575.

44. Ohnishi ST, Ohnishi T, Muranaka S, Fujita H, Kimura H, Uemura K, Yoshida K, Utsumi K. A possible site of superoxide generation in the complex I segment of rat heart mitochondria. *J. Bioenerg. Biomembr.* 2005; 37: 1-15.

45. Ozcan C, Bienergraeber V, Dzeja PP., Terzic A. Potassium channel opens protect cardiac mitochondria by attenuating oxidant stress at reoxigenation. *Am. J. Physiol. Heart Circ. Physiol.* 2002; 282: H531-H539.

46. Pain T, Yang X, Critz S, Yue Y, Nakano A, Liu G, Heusch G, Cohen M, Downey J. Opening of mitochondrial K_{ATP} channels triggers the preconditioned state by generating free radicals. *Circ.Res.* 2000; 87: 460-466.

47. Patel HH, GrossER, Peart JN, Hsu AK, Gross GJ. Sarcolemmal KATP channel triggers delayed ischemic preconditioning in rats. Sarcolemmal KATP channel triggers delayed ischemic preconditioning in rats. *Am J Physiol Heart Circ Physiol.* 2005; 288: H445-H447.

48. Paucek P, Mironova G, Mahdi F, Beavis F, Woldegiorgis G, Garlid KD. Reconstruction and partial purification of the glibenclamide-sensitive, ATP-sensitive K^+ channel from rat liver and beef heart mitochondria. *J. Biol. Chem.,* 1992; 267(36): 26062-26069.

49. Petrishchev NN, Shliakhto EV, Vlasov TD, Galagudza MM. Myocardial ischemic preconditioning: pathophysiological mechanisms and prospects of clinical application (a literature review) *Ross. Fiziol. Zh. Im IM Sechenova* 2001; 87: 688-705.

50. O'Rourke B. Evidence for mitochondrial K^+ channels and their role in cardioprotection. *Circ.Res.,* 2004; 94: 420-432.

51. Sasaki N, Sato T, Ohler A, O'Rourke B, Marban E. Activation of mitochondrial ATP-dependent potassium channels by nitric oxide. *Circulation.* 2000; 101: 439-445.

52. Sellers AJ, Boden PR, Ashford ML. Lack of effect of potassium channel openers on ATP-modulated potassium channels recorded from rat ventromedial hypothalamic neurons. *Br.J.Pharmacol.* 1992; 107, 1068-1074.

53. Sato T, Sasaki N, Seharaseyon J, O'Rourke B, Marbán E. Selective pharmacological agents implicate mitochondrial but not sarcolemmal KATP channels in ischemic cardioprotection. *Circulation.* 2000; 101: 2418-2423.

54. Tsai C, Su S, Chou T, Lee T. Differential effects of sarcolemmal and mitochondrial K_{ATP} channels activated by 17b-estradiol on reperfusion arrhythmias and infarct sizes in canine hearts. *J. Pharmacol. Exper. Therap.* 2002; 301: 234-240.

55. Tucker SJ, Gribble FM, Zhao C, Trapp S. and Ashcroft FM. Truncation of Kir6.2 produces ATP-sensitive K^+ channels in the absence of the sulphonylurea receptor. *Nature* 1997; 387: 179-183.

56. Wang Y, Kudo M, Xu M, Ayub A, Ashraf M. Mitochondrial K_{ATP} channel as an end effecter of cardioprotection during late preconditioning: triggering role of nitric oxide. *J.Mol.Cell Cardiol.* 2001; 33: 2037-2046.

57. Xu W, Liu Y, Wang S, McDonald T, Van Eyk JE, Sidor A, O'Rurke B. Cytoprotective role of Ca^{2+}-activated K^+ channels in cardiac inner mitochondrial membrane. *Science* 2002; 298: 1029-1033.

58. Yellon D, Baxter GF, Garcia-Dorado D, Heusch G, Sumeray MS. Ischemic preconditioning: present position and future directions. *Cardiovasc. Res.* 1998; 37(1): 21-33.

59. Yokoshiki H, Sunagawa M, Seki T, Sperelakis N. ATP-sensitive K+ channels in pancreatic, cardiac, and vascular smooth muscle cells. *Am. J. Physiol*. 1998; 274: C25-C37.

60. Zhang D, Chen Y, Campbell W. Zou A, Gross G, Li P. Characteristics and suproxide-induced activation of reconstituted myocardial mitochondrial ATP-sensitive potassium channel. *Circ. Res*. 2001; 89: 1177-1173.

61. Zhu HF, Dong JW, Zhu WZ, Ding HL, Zhou ZN. ATP-dependent potassium channels involved in the cardic protection induced by intermittent hypoxia against ischemia/reperfusion injury. *Life Sci*. 2003; 1275-1287.

62. Zweier J, Flaherty J, Weisfeldt M. Direct measurement of free radical generation following reperfusion of ischemic myocardium. *Natl Acad Sci USA*. 1987; 84: 1404-1407.

Adaptation Biology and Medicine (Volume 5: Health Potentials)
Editors: L. Lukyanova, N. Takeda and P.K. Singal

10. Characteristics of Cholinergic and Cognitive Functions in Rats with Individual Resistance to the Cerebral Ischemia

E.I. Zakharova, Z.I. Storojeva[1], E.L. Germanova, M.Y. Monakov, A.T. Proshin[1], A.M. Dudchenko and L.D. Lukyanova

Summary

The investigation of the neuronal mechanisms of the individual sensitivity to chronic cerebral ischemia was performed on rats with low (LR) and high (HR) resistance to oxygen deficiency. The cerebral ischemia was obtained by permanent occlusion of bilateral common carotid arteries. Functional state of cholinergic systems was estimated by investigation of activity of membrane-bound (m) and water-soluble (c) isoforms of choline acetyltransferase (ChAT) and acetylcholinesterase and investigation of the level of m- and c-proteins in the synaptic subfractions isolated from the neocortex and the hippocampus. The early response of cholinergic interneurons both in the neocortex and the hippocampus (the subfractions of «heavy» synaptosomes) to ischemic exposure (3 hour) was investigated. These responses of the cholinergic systems were correlated with preservation of the synaptic pool after 7-10 days of ischemia, involving the state of the cholinergic synapses from the forebrain nuclei (the subfractions of «light» synaptosomes). The resistance to ischemic effects was found in the neocortex in HR rats and in the hippocampus in LR rats. Experiments at 7-9 days after occlusion with "Object recognition" and Morris water maze models confirmed that neocortex-dependent cognition was impaired in LR rats and hippocampus/neocortex-dependent cognition was impaired in LR and HR rats. Behavior and ChAT activity were correlated in the neocortex and/or hippocampal synaptic subfractions of ischemic rats. It seems possible that regional neuronal regulation plays an important role in the resistance of neocortex and hippocampus to cerebral ischemia and that cholinergic interneurons are involved into these regional mechanisms. The effects on the resistance to acute hypobaric hypoxia of muscarinic (M) and nicotinic (N) cholinergic receptors (AChR) reagents were tested. It was revealed that N-AChR non-α7 and α7 subunits antagonists mecamilamine and methyllycaconitine (MLA) enhanced resistance to hypoxia in LR rats and it was found broad effective concentration diapason (3.9×10^{-3} –3.9 and (0.14-1.4) $\times 10^{-3}$ nmoles respectively, icv). According to previous literature, functional N-AChRs of the neocortex and the hippocampus are localized on interneurons. The effects on the resistance to the chronic cerebral ischemia of N-AChR reagents were studied. Mecamilamine, MLA and lobeline were injected i.p. (3.9, 1.4 and 4.3 nmoles per kg) once a day for three days after the occlusion and the same scheme of injection was used for mecamilamine before the occlusion. Behavioral studies 1-2 weeks post-occlusion showed a significant role of N-AChRs in regulation of neocortical and hippocampal functions both in LR and HR rats. A positive influence of mecamylamine on the learning was documented in "object recognition" models and on the spatial

[1]Address for Correspondence: Institute of General Pathology and Pathophysiology, RAMS, P.K. Anokhin Institute of Normal Physiology, RAMS, Moscow, Russia, E-mail: iszakharov@yandex.ru

(non-cued) learning in Morris water maze model in ischemic LR and HR rats. These effects were antagonistic in LR and HR rats and were depended on the manner of administration of the substance (before or after the occlusion). In the Morris water maze model, mecamylamine did not take part in the spatial learning of intact non-ischemic rats. Moreover, a strong influence on the spatial learning in ischemic LR and HR rats was revealed. These results show that the interneurons, which express the non-α7 subunits, are involved in mechanisms of resistance to ischemia of the neocortex and the hippocampus. Thus, we believe that these results of pharmacological experiments confirm our supposition that regional neuronal regulation plays an important role in the resistance of neocortex and hippocampus to cerebral ischemia.

Keywords: Cerebral ischemia, neocortex, hippocampus, synaptic subfractions, membrane-bound and water-soluble isoforms of choline acetyltransferase and acetylcholinesterase, neocortex-dependent and hippocampus/neocortex-dependent cognition, interneurons, nicotinic non-α7 and α7 receptors, mecamilamine, methyllycaconitine, lobeline

Introduction

The brain is structurally the most sophisticated organ in the mammalian body, with its functional principle based on a continuous interaction between its constituent nerve elements. Since the neuronal regulation is designed for providing a rapid response of the body to external actions, neuronal regulation may also be especially important for the brain in the extreme conditions.

Our study was aimed at evaluating the function of cholinergic systems in the neocortex and the hippocampus under the conditions of experimental chronic brain ischemia to reveal the role of the systems in the ischemic pathology and correction of ischemic damage.

The interest to cholinergic afferents in the hippocampus and the neocortex has originated from the data on selective sensitivity of hippocampal and neocortical pyramidal neurons to cerebral ischemia. Current physiology regards hippocampal pyramidal neurons as a morphological substrate of memory. It is well known that cholinergic afferents are modulators of pyramidal neuron activity, on the one hand, and an obligatory component of neocortical and hippocampal cognitive function, on the other hand.

During the past 20 years, the implication of the hippocampal and neorcortical cholinergic system in ischemic pathology has been clearly demonstrated. This includes, first, early activation of the system in response to a stimulus, which occurs already within the first hours and does not fall behind the glutamatergic response [18, 21, 23, 35]. Secondly, this includes a correlation between development of subsequent cholinergic dysfunction and death of pyramids [17, 36, 52] as well as between cholinergic dysfunction and cognitive disorders in animals [18, 33, 51, 52, 53]. Previous research suggests a leading role of cholinergic afferent dysfunction in development of ischemic pathology [17, 35, 37].

It should be emphasized, both the neocortex and the hippocampus are supplied from two sources of cholinergic effects based on immunological data (Fig. 1) including: (1) potent neuronal projections from the forebrain nuclei basalis in the neocortex (Meynert nuclei in primates and human) and projections from the forebrain medial septal nuclei and vertical limb diagonal band (Broca) nuclei in the hippocampus and, (2) interneurons [1, 3, 9, 13, 16, 38, 43].

We have earlier shown for both the neocortex and the hippocampus that synapses of cholinergic neurons of different groups can be isolated to different synaptosomal fractions during preparation in the density sucrose gradient. Synapses (presynapse parts of synapses with its junction complexes) of cholinergic projections from forebrain nuclei collect primarily in fractions of so-called "light"

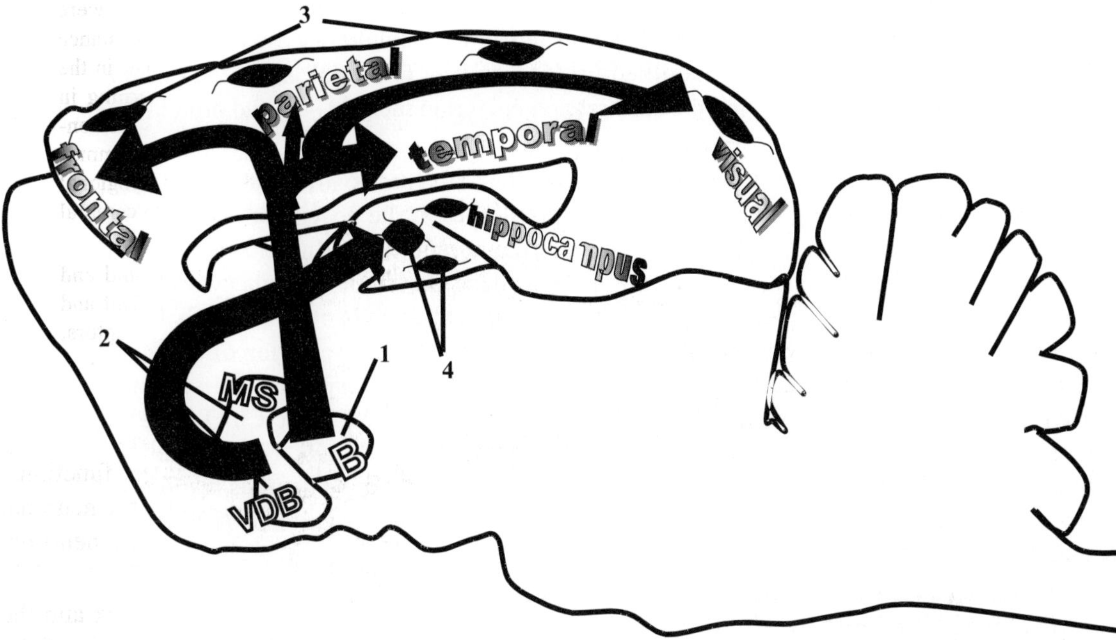

Fig. 1. Arrangement of cholinergic systems in the neocortex and the hippocampus. 1- projections from the forebrain nuclei basalis in the neocortex (B); 2 - projections from the forebrain medial septal nuclei (MS) and vertical limb diagonal band Broca nuclei (VDB) in the hippocampus; 3 and 4 - interneurons of the neocortex and the hippocampus, respectively.

synaptosomes while synapses of cholinergic interneurons collect in fractions of "heavy synaptosomes" [30, 54].

Differential examination of the two cholinergic systems in the neocortex and the hippocampus and of related cognitive functions in the conditions of cerebral ischemia has led us to an understanding of the important role of regional neuronal mechanisms in the resistance of neocortex and hippocampus to ischemic impacts.

Materials and Methods

Experiments were carried out on mongrel white, sexually mature male rats. Rats were pre-tested in the hypobaric conditions ("elevation" to 11,000 m) to select low- (LR) and high-resistance (HR) rats to acute hypoxia. Survival time was less than 3 min for LR rats and more than 10 min for HR rats. It has been shown earlier that the division of animals by their resistance to the oxygen level is based on different regulatory mechanisms [25]. The division of rats by that feature, which was used in the present study, allowed determination of certain regularities in the individual response to cerebral ischemia.

Following testing, rats were kept in the normobaric conditions for a month, and then they were taken to the experiment. Global chronic ischemia was produced by permanent bilateral occlusion of common carotid arteries. Rats were anaesthetized with Nembutal (40 mg/kg IP). Sham operated animals underwent a similar surgery, excluding the ligation.

Cholinergic function was evaluated by activities of the acetylcholine (Ach) synthesizing enzyme choline acetyltransferase (ChAT), the Ach degrading enzyme acetylcholine esterase (AChE) using the F. Fonnum radiometric method [11] modified by [54], and by universal parameters of the synaptic function, Na/K-ATPase activity using the substrate regeneration method [41] and protein levels by the Lowry method [24].

Biochemical parameters were measured in subfractions of synaptic membranes and synaptoplasm isolated from fractions of "light" and "heavy" synaptosomes in the neocortex and the hippocampus using preparative methods [30, 54]. Correspondingly, membrane-bound (m) enzyme forms and levels of membrane proteins were measured in synaptic membrane subfractions while cytosolic (c) enzyme forms and levels of water-soluble proteins were measured in synaptoplasm sub fractions. The differential study of m- and c-parameters permitted a more unbiased evaluation of the functional and metabolic state of cholinergic synapses since the synaptic membranes and the synaptoplasm are fundamentally different cell compartments in the functional respect. M- and c-ChAT have been shown to display different *in vitro* dependence on cell metabolites [8, 44, 44, 46, 47]. In addition, we have regarded the coupled decreases in m- and c-ChAT activities (cholinergic neurons markers) and levels of m- and c-proteins in ischemic rats as an evidence for changed amount of synapses as their degradation is secondary to neurodegeneration.

Cholinergic influences on neuronal muscarinic acetylcholine receptors (M-AChRs) were studied using the M-AChR agonist arecoline hydrochloride and the M-AChR antagonist atropine sulfate and also the M-AChR subtype selective antagonists pirenzepine dihydrochloride (M1 subtype), gallamine triethiodide (M2 subtype), and *para*-fluoro-hexahidro-sila-difenidol (*p*-F-HHSiD) (M3 subtype) (all from "RBI"). Cholinergic effects on neuronal nicotinic acetylcholine receptors (N-AChRs) were studied using the N-AChR agonist lobeline hydrochloride, the selective N-AChR α7 subunit antagonist methyllycaconitine citrate (MLA, "Tocris") and the selective non-α7 subunits antagonist mecamylamine hydrochloride ("RBI").

Cholinoreactive substances were dissolved in sterile isotonic saline (*p*-F-HHSiD - in distilled water) and rats in control groups received saline or distilled water. Substances were administered by intraperitonial (i.p.) or intracerebroventricular injections (i.c.v). Standard stereotaxic methods were used in the case of i.c.v injections. The solutions of substances were injected into right lateral ventricle through cannula at the rate of 0,5 -1 μl an min using the syringe pump (model 355, "Sage instruments", USA). Coordinates of cannula placement (B -1.5, L 2.0, H 3.5) were established by the Paxinos a. Watson stereotaxic atlas [39].

Behavior was studied in cued and non-cued (spatial) models of learning in the Morris water maze [28] and the "object recognition" test [31, 32].

Data were statistically treated using the nonparametric exact Fisher's test and the Spearman's rank correlation test.

Results and Discussion

Initially we found that interneurons in both the neocortex and the hippocampus represented correlates for the individual sensitivity of rats to ischemia. First, the interneurons (but not projection neurons) displayed the obligatory early response (in 3 hours) to ischemia in both the neocortex and the hippocampus. Four different types of the cholinergic response were identified: activation of the

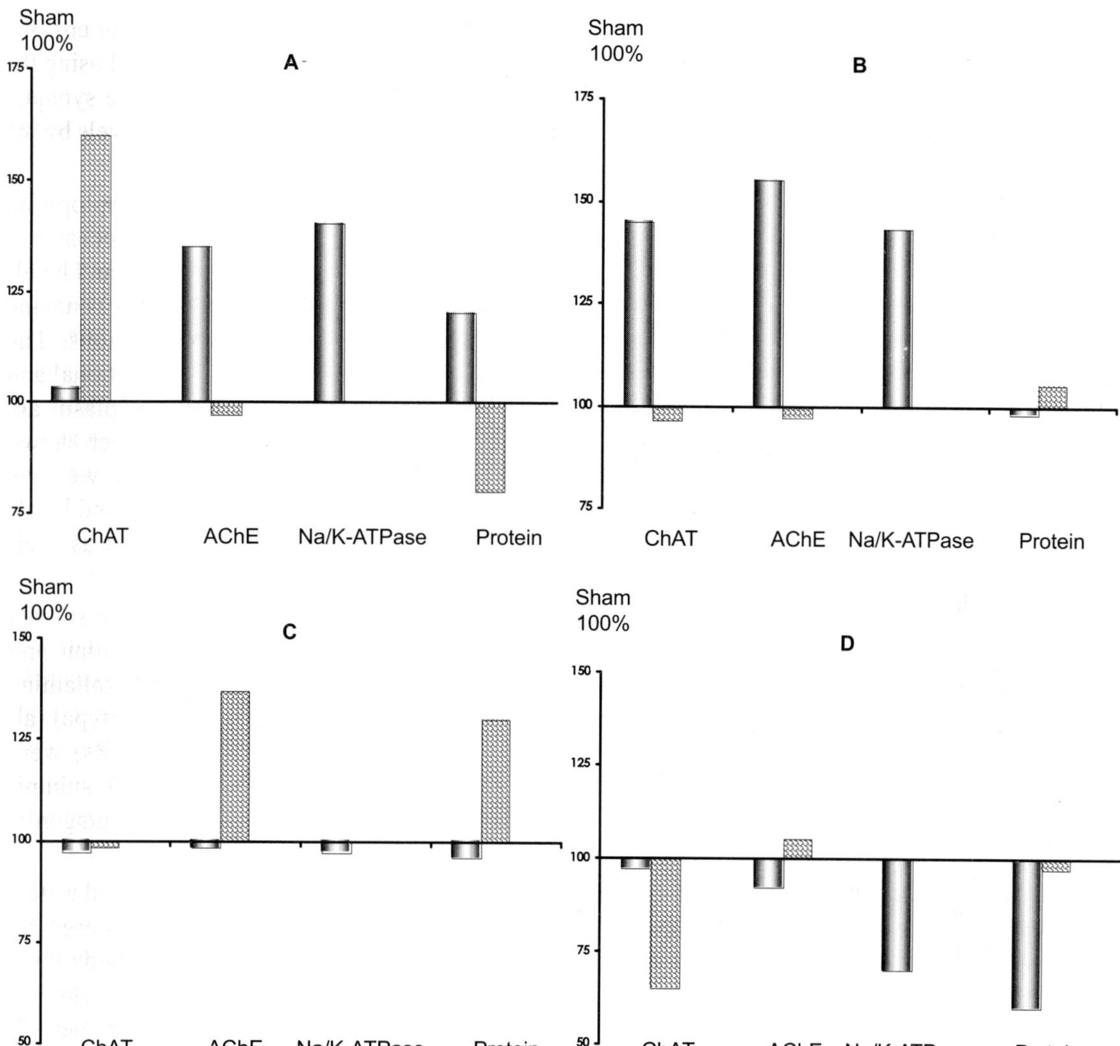

Fig. 2. Four metabolic pathways for rearrangement of the cholinergic function in response to acute hypoxia/ ischemia exposures. For each pair of bars: left bar, synaptic membrane subfraction, right bar, synaptoplasm subfraction (for Na/K-ATPase is absent); all values are presented as percentage of control values and they reflect maximum observed changes; A, activation of the cholinergic function involving c-ChAT; B, activation of the cholinergic function involving m-ChAT; C, light inhibition of the cholinergic function involving c-AChE; D, inhibition of the cholinergic function involving c-ChAT (cholinergic disfunction).
Note: as a rule c-ChAT and m-ChAT activation was accompanied by m-AChE and Na/K-ATPase activation and by growing of synaptic membrane proteins in the case of c-ChAT strong activation –there are signs of activation of the cholinergic function; c-AChE activation was accompanied by accumulation of cytosolic proteins – there are sings of prevalence metabolism over catabolism; c-ChAT inhibition was accompanied by Na/K-ATPase inhibition and destruction of synaptic membranes in the case of c-ChAT strong inhibition - there are sings of cholinergic disfunction.

cholinergic function involving c-ChAT or m-ChAT; inhibition of the cholinergic function by the mechanism of c-AChE activation or c-ChAT inactivation (Fig. 2). Changes in the other biochemical parameters confirmed these functional states.

Second, we observed conformity between the type or early interneuron response and the condition of synaptic pool after 1-2 weeks of ischemia (Fig. 3). While the activation involving c-ChAT was observed in early ischemia, the synaptic pool was preserved in late ischemia (Fig. 3A, 3-C). Other responses of interneurons were associated with late disintegration of cholinergic or non-cholinergic synapses.

In this process, low-resistance (LR) rats displayed the early c-ChAT activation and had an intact synaptic pool in the hippocampus in late ischemia while the neocortex was damaged. In high-resistance (HR) rats, at the same time, the early positive activation was observed in the neocortex, which remained intact in late ischemia, while the hiðpocampus was damaged. These data suggest that the neocortex and the hippocampus were characterized by individual sensitivity to the ischemic exposure. The hypothesis implied that LR and HR rats might differ by the profile of cognitive dysfunction in ischemic damage. Therefore the next step included behavioral experiments.

Some up-to-date behavioral tests allow distinguishing between neocortex-dependent and hippocampus-dependent animal behavior. We choose four behavioral models. The first two models of cued and non-cued behavior were developed for Morris water maze. In the cued behavioral model, the location of hidden platform is determined by a cue. According to previous literature, this function is considered as a neocortex-dependent and hippocampus-independent one. In the non-cued (spatial) behavioral model, the location of hidden platform determines the capability of animal for spatial orientation. This function depends on both the hippocampus and the neocortex [5, 14, 28]. Both models permit assessment of both working memory (average from 2d-4th trials of first session) and long-term memory (first trial of second session) in animals.

The other two models were "object recognition" tests. The tests evaluate the ratio between the «object exploration» time for familiar and new objects by using the formula, which was elaborated by the models authors. When the interval between two presentations of the objects was 5 minutes, the memory trace retention was considered as neocortex-dependent and when the interval was 30 minutes, the retention was considered as neocortex- and hippocampus-dependent [14, 31, 32].

Behavior was studied at the critical term of 7-10 days of chronic ischemia. The following was revealed:

- In the "object recognition" models, the neocortex-dependent memory was disordered in LR and not disordered in HR rats; the hippocampus/neocortex dependent memory was impaired in both LR and HR rats (Fig. 4);
- In the water maze learning models, the neocortex-dependent cued learning was impaired only in LR rats and was undisturbed in HR rats while the hippocampus/neocortex dependent spatial learning was disturbed in both LR and HR rats (Fig. 5).

Then we traced the correlation between water maze behavior and individual ischemic responses of the neocortex and the hippocampus by the ChAT activity as a marker of cholinergic synapses.

Studying behavior of sham-operated rats showed that both HR and LR rats lacked a correlation between cholinergic systems and long-term memory in both models of learning (Table). Cholinergic systems selectively correlated with the working memory only. In the spatial learning model, c-ChAT

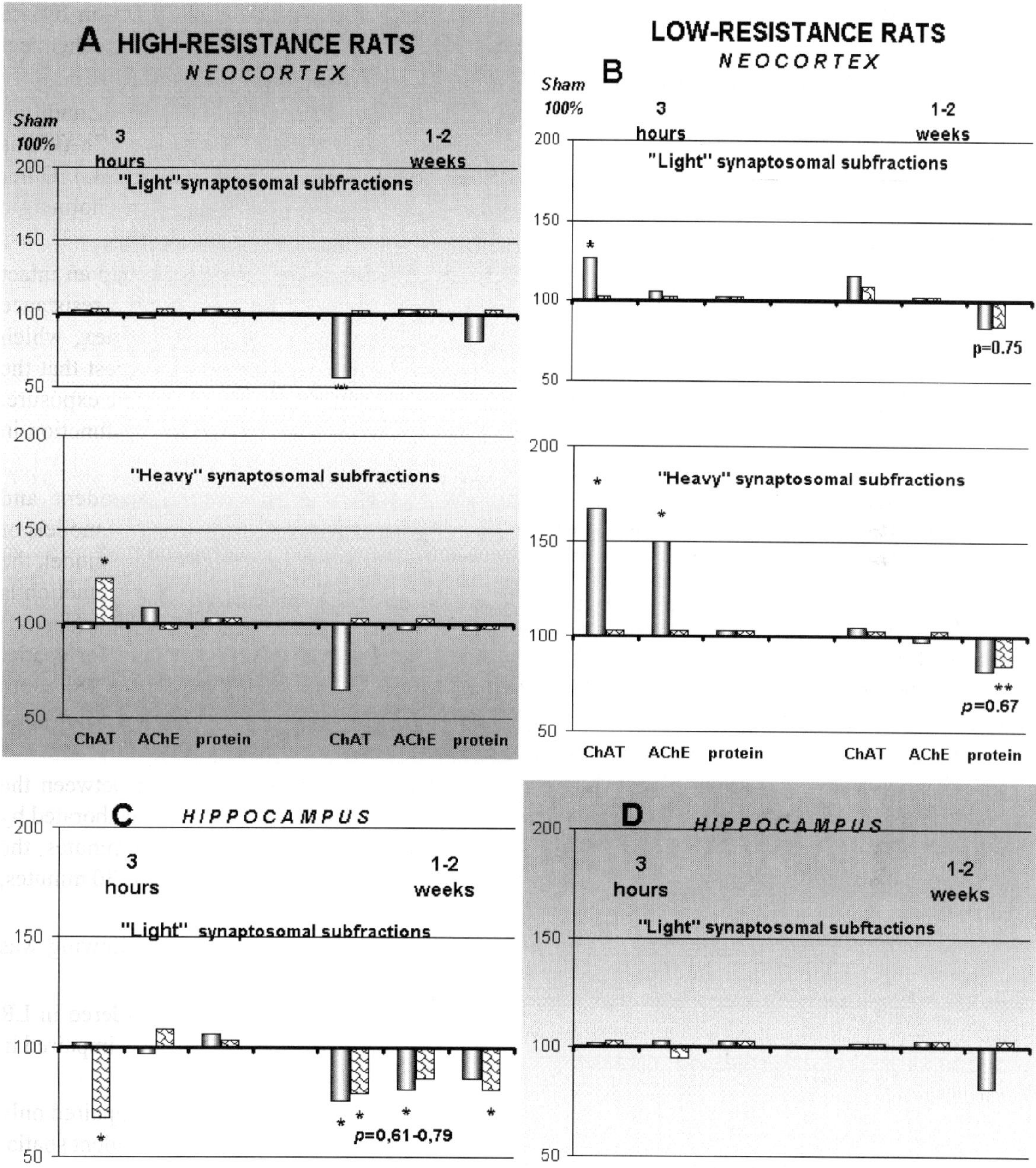
A HIGH-RESISTANCE RATS
NEOCORTEX
Sham 100%
3 hours
1-2 weeks
"Light"synaptosomal subfractions
"Heavy" synaptosomal subfractions
ChAT AChE protein ChAT AChE protein
LOW-RESISTANCE RATS
NEOCORTEX
B
p=0.75
p=0.67
C HIPPOCAMPUS
"Light" synaptosomal subfractions
p=0,61-0,79
D HIPPOCAMPUS
"Light" synaptosomal subftactions

Fig. Contd

Fig. Contd

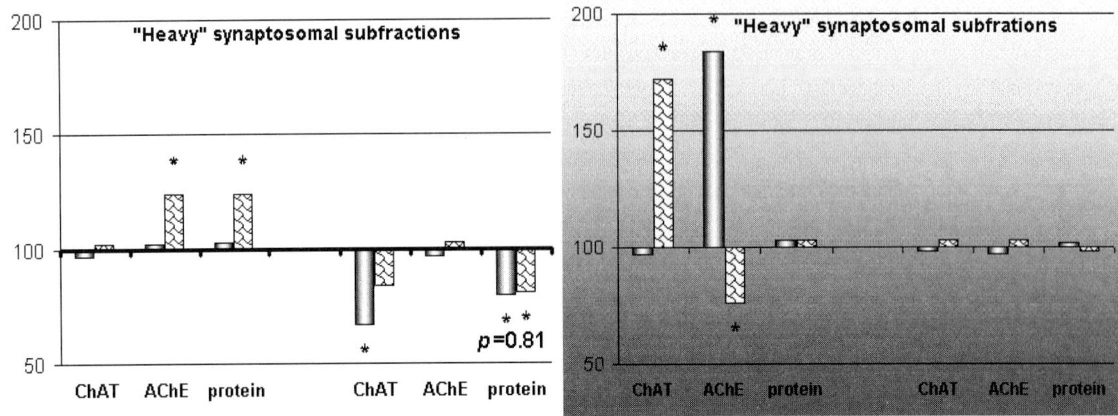

Fig. 3. Effect of the global, chronic cerebral ischemia on ChAT, AChE activity and protein content in the " light" and "heavy" neocortical and hippocampal synaptosomes subfractions of HR rats (left column of figures) and LR rats (right column of figures) 3 hours and 1-2 weeks postocclusion.

For each pair of bars: left bar, synaptic membrane subfraction; right bar, synaptoplasm subfraction; all values are presented as percentage of values in sham operated rat groups; *, **, significant differences from control (p<0.05 and p<0.01, respectively) as estimated using the exact Fisher's test; p, values of significant correlations as estimated using Spirman's test; quantity of animals in experimental groups was 5-8.

Note: A and D, the neocortex of HR-rats and the hippocampus of LR rats: c-ChAT activation was observed in the interneuron synapses (the "heavy" synaptosomes fractions) in early ischemia (3 hour), projection synapses was intact in early ischemia (the "light" synaptosomes fractions), the synaptic pool was preserved in late ischemia; B, the neocortex of LR rats: m-ChAT activation was observed in the interneuron synapses in early ischemia, projection synapses was not intact in early ischemia, disintegration of non-cholinergic synapses had place in late ischemia (correlation between decrement of m- and c-protein levels both in the "light" and in the "heavy" synaptosomes fractions, n = 10, p < 0, 05); C, the hippocampus of HR-rats: c-AChE activation was observed in the interneuron synapses, projection synapses was not intact in early ischemia (the cholinergic disfunction), disintegration of cholinergic synapses had place in late ischemia (correlation between decrement of m- and c-protein levels, between decrement of m- and c-ChAT activity, correlation between decrement of m- ChAT activity and m-protein levels and between decrement of c- ChAT activity and c-protein levels, n = 11, p < 0, 05).

activities of hippocampal interneurons correlated with parameters of working memory. In the cued learning model, which is neocortex-dependent according to literature, activities of m- and c-ChAT projection systems correlated with parameters of working memory in both the neocortex and the hippocampus.

The dependence of behavior on corresponding cholinergic systems was confirmed on ischemic rats. Below are some examples (Figs. 6 and 7).

Working memory was disturbed in LR rats when one of two cholinergic systems important for this function was disordered:

Fig. 6-A. Cholinergic dysfunction in hippocampal projection synapses (neocortical projections are intact); cued behavior was disturbed.

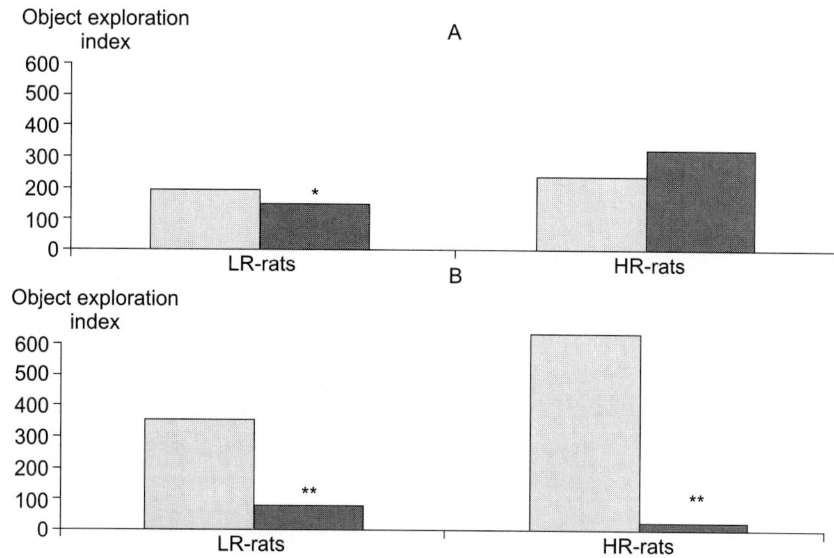

Fig. 4. Behavior of LR and HR rats in the "Object recognition" models after 7-10 days of the chronic ischemia. For each pair of bars: left bar, sham operated rats; right bar, ischemic rats; *, **, significant differences from control ($p < 0.05$ and $p < 0.01$, respectively); quantity of animals was 8-12 in both LR and HR rats groups. Note: for the "object recognition" model, the higher recognition index value the better memory; A, the cerebral ischemia impaired neocortex-dependent memory only in LR rats; B, the cerebral ischemia impaired hippocampus/neocortex -dependent memory both in LR and in HR rats.

Table 1. Correlation between behavior in Morris water maze and ChAT activity (n=14)

	cortex		**hippocampus**	
	Forebrain nucleus projections (membrane/ cytosolic ChAT)	**Interneurons (membrane/ cytosolic ChAT)**	**Forebrain nucleus projections (membrane/cytosolic ChAT)**	**Interneurons (membrane/cytosolic ChAT)**
Cue Working memory	**–0.60*/–0.62* p<0.05**	–0.53/–0.41	**0.83****/–0.11 **p<0.01**	0.35/0.12
Cue Long-term memory	0.01/0.08	–0.12/0.28	0.05/–0.06	0.14/0.04
Spacious Working memory	0.32/–0.10	–0.22/0.15	–0.19/0.12	**–0.55***/–0.21 **p<0.05**
Spacious Long-term memory	–0.08/–0.08	–0.05/–0.13	–0.12/–0.11	–0.04/–0.30

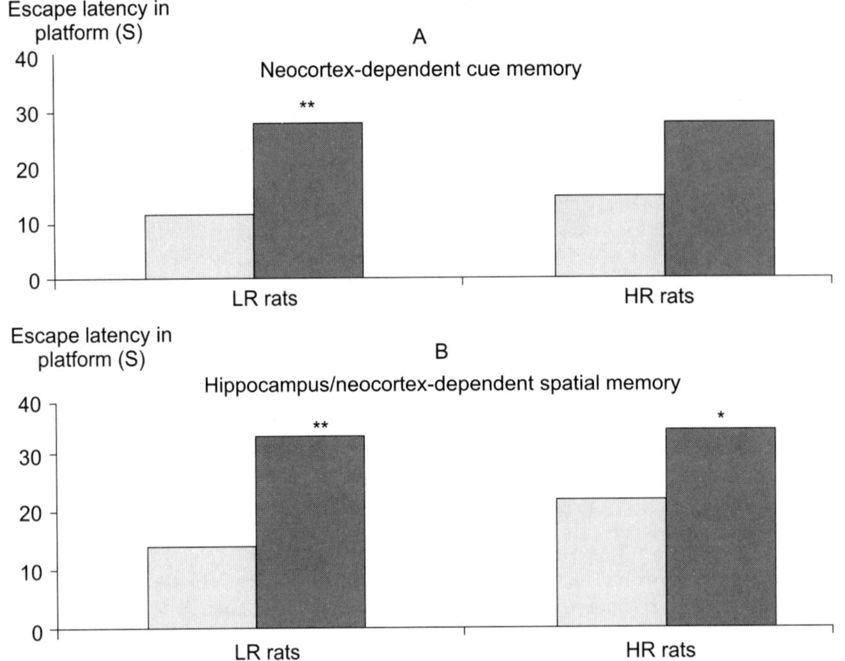

Fig. 5. Behavior of LR and HR rats in the Morris water maze cued and noncued (spatial) models after 7-10 days of the chronic ischemia. Abbreviations the same as on Fig. 4; quantity of animals was 15-16 in LR rats groups and 10 in HR rats groups.

Note: for this models, the shorter time of reaching the platform the better memory; for simplify of demonstration on this figure the data presented as means of work memory and long-term memory values because its had same results; A, the cerebral ischemia impaired neocortex-dependent cued memory only in LR rats; B, the cerebral ischemia impaired hippocampus/neocortex -dependent noncue (spatial) memory both in LR rats and in HR rats.

Fig. 6-B. Degeneration of a considerable part of projection synapses in the neocortex with no considerable disturbances of hippocampal projection function; cued behavior was disturbed.

Fig. 6-C. Both projection systems were intact; working memory was also undisturbed.

Fig. 6-D. Degeneration of synapses in hippocampal interneurons (the other cholinergic systems are intact); spatial working memory was disturbed.

The last example (Fig. 6-D) presented on Figure 7-A shows all neocortical and hippocampal cholinergic systems as compared with data for the same batch of rats but in later ischemia (Fig. 7-B). In one month LR rats already did not display any significant disorders of the hippocampal interneuron function. Consistently, they also had no significant disorders of spatial behavior despite the damage in both neocortical cholinergic systems.

In our opinion, the present behavioral study confirmed on the whole the existence of individual resistance of the neocortex and the hippocampus. Together with the previous data on the early ischemia response of cholinergic interneurons, the present results suggested that regional regulation might play an important role in the resistance of the neocortex and the hippocampus and that mechanism of the regulation involved cholinergic interneurons.

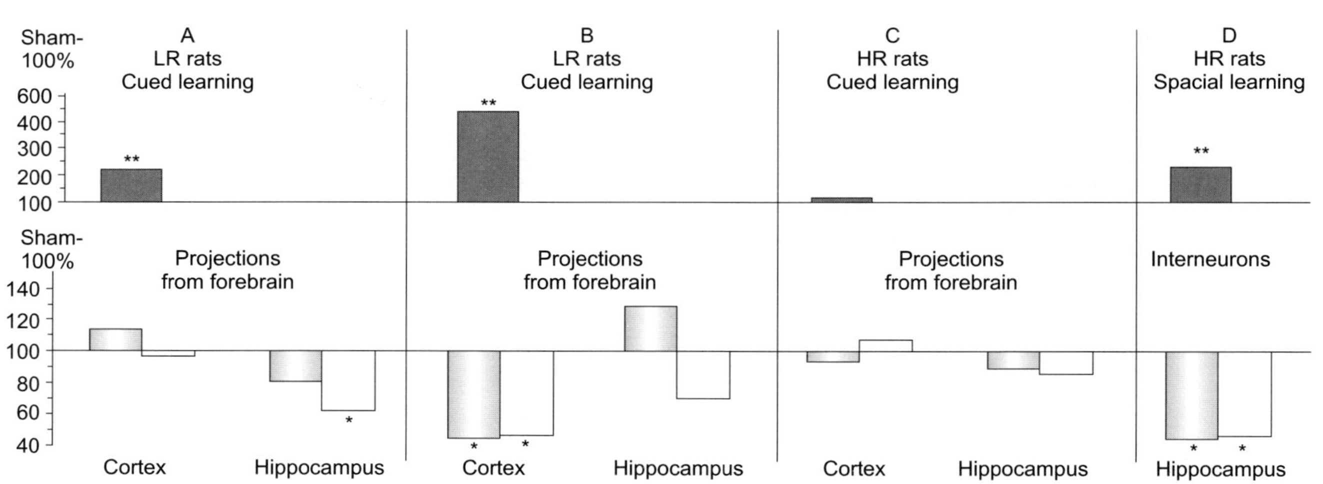

Fig. 6. Working memory in the Morris water maze models and ChAT activity 1-2 weeks post-occlusion. All values are presented as percentage of values in sham operated groups; for sectors A-D: upper bars, working memory; ground bars, ChAT activity; for each ground pair of bars: left bar, synaptic membrane subfraction; right bar, synaptoplasm subfraction; only important for corresponding cognitive functions cholinergic systems ChAT activity data are demonstrated in this figure; *, **, significant differences from control ($p < 0.05$ and $p < 0.01$, respectively); quantity of animals in sham operated groups was 8-12 and in operated groups was 3.

Note: A, hippocampal projection synapses degeneration enhanced correlation between m- and c-ChAT comparatively with correlation in sham operated group (p = 0,74, n = 15, p < 0,01 and p = 0,34, n = 12, p > 0,05 respectively); D, the same as A with hippocampal interneurons synapses degeneration (p = 0,72, n = 15, p < 0,01 and p = 0,64, n = 12, p < 0,05 respectively).

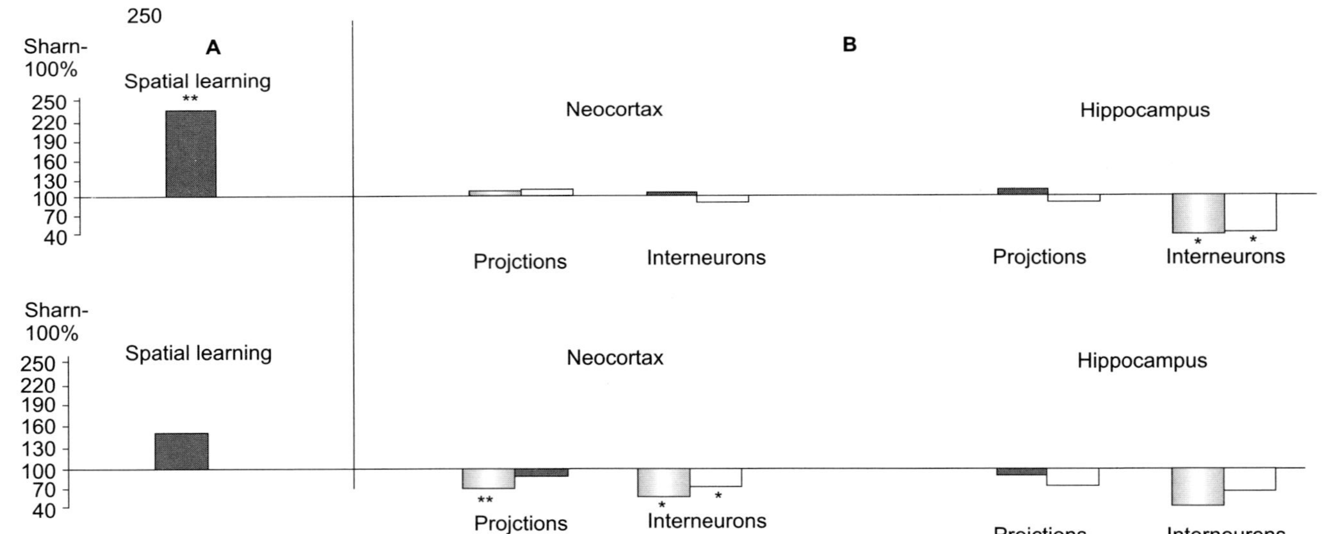

Fig. 7. **The spatial working memory in the Morris water maze model in different late periods of chronic ischemia. A, the spatial working memory and B, ChAT activity in the all synaptosomes subftactions of neocortex and hippocampus of LR rats; upper row of bars, 1-2 weeks postocclusion; ground row of bars, 4-5 weeks postocclusion; quantity of animals in sham operated groups was 12 and in ischemic groups was 3 for A and 7 for B; another abbreviations the same as on Fig. 6.**

We tested our hypothesis by modulating cholinergic effects using cholinergic agonists and antagonists. Current electrophysiology has accumulated much data that functional N-AChRs of the neocortex and the hippocampus are localized on interneurons [4, 6, 12, 19, 40, 50]. According the data of a pharmacological profile of neocortical N-AChRs is represented mainly by non-α7 subtypes (principally by α4β2subtype) and hippocampal ones – mainly by α7 subtype [10, 26, 27, 29, 40, 45, 50].

We have tested both N-AChR and also M-AChR subtypes. Their effects on resistance to acute hypobaric hypoxia were studied. LR and HR rats were used in repeated "elevation" to 11,000 m after keeping in the normobaric conditions for a month as in the cases of ischemic experiments (see methods). Substances were injected i.c.v, one time 20-30 min before test (60 min - for pirenzepine). Part of substances (arecoline, atropin, lobeline, mecamilamine, MLA) were injected also i.p. in doses above i.c.v ones on 10^3 (per kg). In this case, the effects were similar with the effects were obtained by i.c.v manner. A positive influence during surviving time was received after inhibition of both N-AChR subtypes and M2-AChR in LR rats (Fig. 8). Gallamine exerted the most pronounced effect (19 times) on LR rats; however, the effect was evident at doses, which provoked fatal convulsions in both LR and

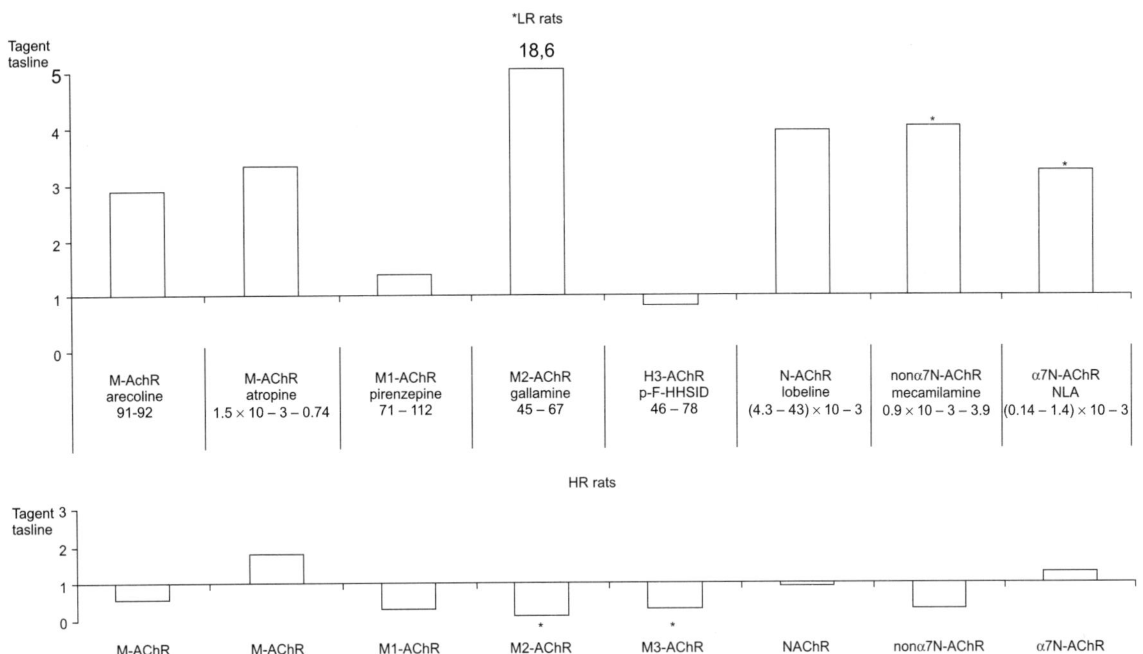

Fig. 8. Cholinergic reagents effects on resistance to acute hypobaric hypoxia in LR and HR rats. All values of survive time in hypobaric hypoxia are presented as regards to same values of saline or distilled water rats used as a control; upper row of bars, LR rats; ground row of bars, HR rats; all substances were injected icv one time: effective doses, which are presented in this figure in nmoles, were consisted in 0,5-1μl infusion volume (p-F-HHSiD – in 3μl); *, significant differences from control ($p < 0.05$); quantity respectively of LR and HR rats was 8-7 in control groups, 10-8 in arecoline groups, 13-8 in atropine groups, 5-2 in pirenzepine groups, 2-2 in gallamine groups, 8-3 in p-F-HHSiD groups, 5-6 in lobeline groups, 18-6 in mecamilamine groups and 6-8 in MLA groups.

HR rats. Therefore, we chose only N-AChR regulators for studies on the model of cerebral ischemia because they were nontoxic and exerted effects in a broad range of concentrations.

In the model of chronic cerebral ischemia, all N-AChR regulators (N-AChR agonist lobeline, non-α7-AChR antagonist mecamylamine and α7-AChR antagonist MLA) or saline were injected i.p. three times on the early days critical for ischemia; the 1st injection was made within 3 hours postsurgery and subsequent injections were made with 1-day intervals. Mecamylamine was also injected preventively three times with 1-day intervals. The last injection was made on the day of surgery 30 min before the narcosis. One time doses were 4.3 for lobeline, 3.9 for mecamylamine and 1.4 for MLA (in nmoles per kg).

Notably, it appeared within the first weeks after occlusion that N-AChRs influenced the vital capacity of ischemic rats as evaluated by parameters of weight loss, visual pathology and death rate. The preventive mecamylamine treatment was found to improve ischemia tolerance and abolished mortality in LR rats (n=6) but worsened these parameters in HR rats resulting in death of all HR rat group (n=3) within 10 days. And vice versa, the post-ischemic mecamylamine injection increased ischemic injury in LR rats (n=12) but at the same time it left unchanged vital capacity parameters in HR ischemic rats (n=9). The post-ischemic MLA treatment caused death in LR rats within one day (4 rats). Finally, the post-ischemic treatment with the N-AChR agonist lobeline left unchanged vital capacity parameters in ischemic rats of both LR and HR groups.

Behavioral studies in the late ischemic period, 1-2 weeks post-occlusion, showed the following:

In the "object recognition" model, neocortex-dependent memory was disturbed in LR rats and was undisturbed in HR rats (Fig. 9-A). The mecamylamine pretreatment exerted a protective and positive effects on memory of ischemic LR rats (HR rats were died). On the contrary, the post-occlusion treatment of ischemic rats with mecamylamine increased ischemic injury of behavior in LR rats and was ineffective in HR rats. The post-occlusion treatment of ischemic rats with the agonist lobeline was ineffective in LR and HR rats.

The hippocampus/neocortex-dependent memory was disturbed in ischemic LR and still more in HR rats (Fig. 9-B). The mecamylamine treatment before and after occlusion did not influence memory of LR rats. The post-occlusion mecamylamine treatment exerted a beneficial effect on memory of ischemic HR rats. The post-occlusion treatment with the agonist lobeline was ineffective in LR and HR rats.

In the model of spatial learning in Morris water maze, the cerebral ischemia undisturbed working memory in LR and impaired in HR rats (Fig. 10-A). The mecamylamine pre-treatment was ineffective in ischemic LR rats. The post-occlusion mecamylamine treatment exerted a negative effect on working memory in LR rats and a protective effect in HR rats. The post-occlusion treatment with the agonist lobeline induced a negative effect on working memory in both LR and HR rats.

Ischemia impaired the long-term memory in the spatial learning in LR and HR rats (Fig. 10-B). The mecamylamine treatment before and after occlusion exerted a protective effect on the long-term memory in ischemic LR rats. The post-occlusion mecamylamine treatment had a protective effect on long-term memory in HR rats. The post-occlusion treatment with the agonist lobeline negatively influenced long-term memory in both LR and HR rats.

This pharmacological experiment revealed two groups of data:

- Influence on resistance of rats to hypobaric hypoxia and cerebral ischemia;
- Influence on behavior of ischemic rats.

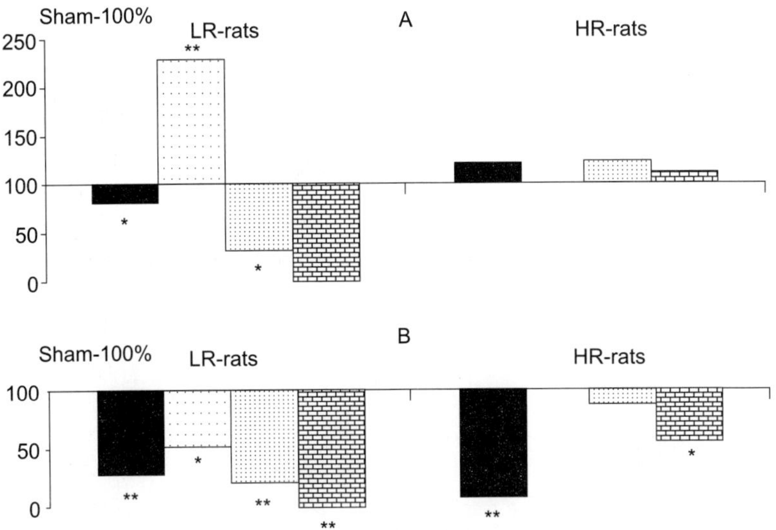

Fig. 9. Behavior in "Object recognition" models 1-2 weeks post-occlusion: effects of N-AChR reagents. All values are presented as percentage of values in sham operated groups (n = 8-12); A, neocortex-dependent memory, B, hippocampus/neocortex-dependent memory; in each group of bars: first, dark bar - ischemia and saline (n = 8-10), second, dotted bar – ischemia and non α7 N-AChR antagonist mecamylamine injected before occlusion (n = 5), third, carpeted bar - ischemia and non α7 N-AChR antagonist mecamylamine injected after occlusion (n = 5-8), fourth, brick bar – common N-AChR agonist lobeline injected after occlusion (n = 3-4); *, **, significant differences from control ($p < 0.05$ and $p < 0.01$, respectively).

The first group of data shows a significant part of N-AChRs in regulation of vital capacity functions in conditions of oxygen deficiency. Influence on vital capacity parameters of ischemic rats was very strong and broad, from positive up to lethal, in dependence of the manner of injection of the substance. It seems to suppose these mechanisms in respiratory or other centers of autonomic nervous system. This data need more detailed analysis and researches. It's necessary to note, that correlation between vital and leaning capacity of rat was absent. More over in the group of experimental rats we regularly record exhausted animals with good cognitive capacity and animals in good physiological conditions but with impaired learning capacity.

The second group of data shows a significant part of N-AChRs in regulation of neocortical and hippocampal functions. Effects of the non-α7-AChR antagonist mecamylamine depended on the manner of administration of the substance. As a rule the results were antagonistic in LR and HR rats and were revealed only by its comparative study. The mechanisms of regulation of functions by the non-α7-AChRs are, obviously, different in LR and HR rats. Perhaps, it's based on the differences in resistance of the neocortex and the hippocampus to ischemia in LR and HR rats as was showed above. In fact, mecamylamine didn't display positive effects if ischemia didn't impaired the learning. It was in the case of neocortex-dependent memory in HR rats (see Fig.9-A) i.e. in the rats with the preserved neocortex in ischemia and in the case of hippocampus/ neocortex-dependent memory in LR rats (see Fig.10-A) i.e. in the rats with the preserved hippocampus and may be the preserved neocortex also (see Fig.6-C) in ischemia.

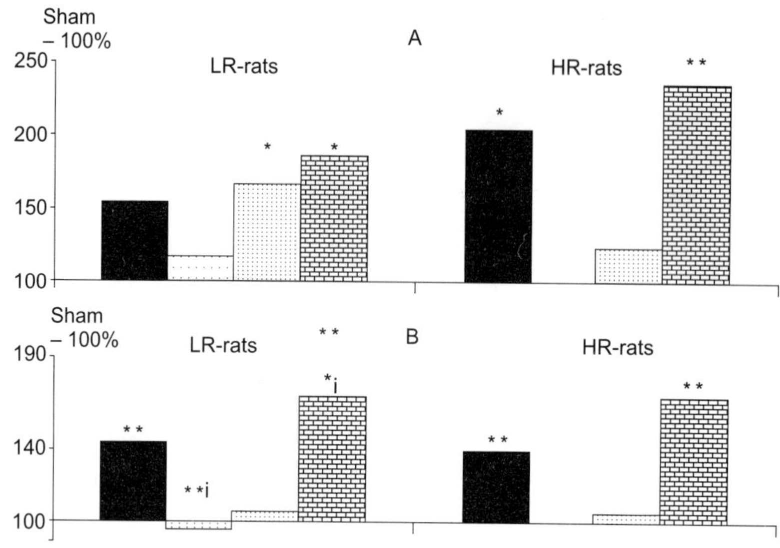

Fig. 10. The spatial behavior in Morris water maze 1-2 weeks post-occlusion: effects of N-AChR reagents. All values are presented as percentage of values in sham operated groups (n = 7-12); A, working memory, B, long term memory; in each group of bar: first, dark bar - ischemia and saline (n = 9-13), second, dotted bar – ischemia and non α7 N-AChR antagonist mecamylamine injected before occlusion (n = 4), third, carpeted bar - ischemia and non α7 N-AChR antagonist mecamylamine injected after occlusion (n = 5-6), fourth, brick bar – common N-AChR agonist lobeline injected after occlusion (n = 3-4); *, **, significant differences from control ($p < 0.05$ and $p < 0.01$, respectively); *i, **i, significant differences from ischemia ($p < 0.05$ and $p < 0.01$, respectively).

Nevertheless the fact of involving N-AChRs of both LR and HR rats in the regulative mechanisms indicates that interneurons, - at least of non-cholinergic interneurons, - were involved in these regulations in both rat groups. There is still not enough information to answer the question if the neocortical and hippocampal interneurons take part in mechanisms of resistance to ischemia or they take part in mechanisms of cognitions only.

In order to clear this result we conducted a pilot experiment on a group of intact, non-ischemic rats to compare effects of the non-α7-AChR antagonist mecamylamine and the α7-AChR antagonist MLA. Antagonists were injected into the rats in same doses and by the same schedule as for operated rats: IP, three times, the 1^{sr} injection was immediately after the end of the narcosis. The rats were also tested 1-2 weeks after the last injection (Fig. 11).

In the "object recognition" model, neocortex-dependent memory was impaired by blocking the non-α7-AChRs and was enhanced by blocking the α7-AChRs (Fig. 11-A). Blocking both the non-α7-AChRs and the α7-AChRs impaired hippocampus/neocortex-dependent memory. This result clearly reveals the involvement of N-AChRs in forming of both forms of memory. They are in good accordance with literature data concerning leading role of non-α7-AChRs in the neocortical function and of α7-AChRs in the hippocampus ones. However, the role of the interneurons in the regional mechanisms of resistance to ischemia of the neocortex and the hippocampus is still opened. What's more, improvement of neocortex-dependent memory by blocking the α7-AChRs indicates cortical-hippocampal interactions in realization of this cognitive function.

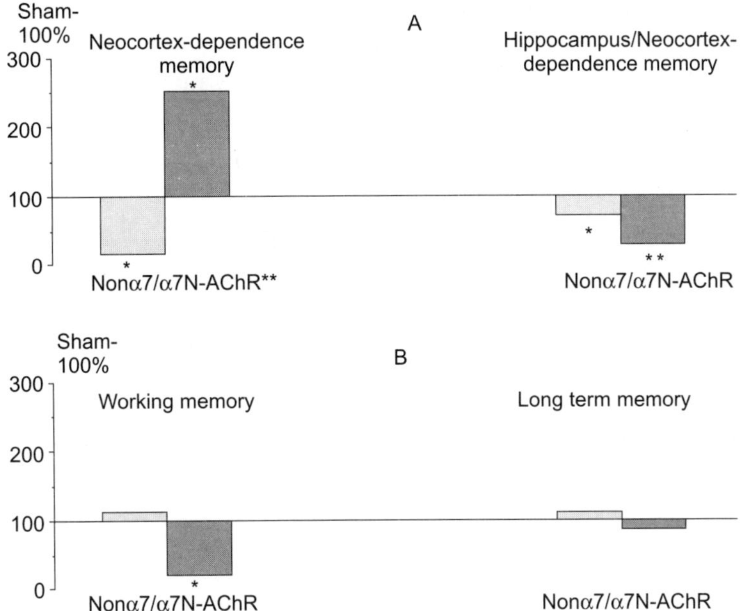

Fig. 11. Effects of N-AChR specific antagonists mecamylamine and MLA on behavior in intact rats. All values are presented as percentage of values in sham operated groups (n = 7-12); A, memory in "Object recognition" models, B, the spatial memory in Morris water maze model; in each pair of bars: left, dotted on light background bar – non α7 N-AChR antagonist mecamylamine (n = 3-4), right, dotted on dark background bar - α7 N-AChR antagonist MLA (n = 2); *, **, significant differences from control ($p < 0.05$ and $p < 0.01$, respectively).

Involvement of N-AChRs, - at least of the non-α7 subunits, - in mechanisms of resistance was revealed on the model of spatial learning in Morris water maze. In this model the working memory was enhanced by blocking the α7-AChRs and was unchanged by blocking the non α7-AChRs (Fig. 11- B). The long-term memory was unchanged by blocking both the α7-AChRs and the non α7-AChRs. These data are in good accordance with our results, that hippocampal interneurons in the intact rats takes part in the spatial working memory (but not both cholinergic neocortical systems), and that no one of the cholinergic systems of the neocortex or the hippocampus takes part in the long-term memory (see Table 1).

So it was obtained that the non-α7 subunits did not take part in the learning (on the model of spatial learning) and it was revealed strong influence on the both forms of the spatial memory in ischemic LR and HR rats through the non-α7-AChRs. These results show that the interneurons, which express the non-α7 subunits, are involved in mechanisms of resistance of the neocortex and the hippocampus. Correction of behavior by mecamylamine in the case of spatial learning was a result of adaptation (or disadaptation) of the neocortex and the hippocampus to ischemia.

Summing up the results of pharmacological experiments, we believe that the above results support one part of our hypotheses that regional neuronal regulation plays an important role in the resistance of neocortex and hippocampus to cerebral ischemia. The interneurons may receive nicotinic stimulation as from the cholinergic projections and the cholinergic interneurons. So, concerning of involvement of the cholinergic interneurons in neuronal mechanisms of regional regulation, we have only our biochemical data about early response of the cholinergic interneurons on ischemia.

Different effects of mecamylamine on ischemic LR and HR rats and different resistance of the neocortex and the hippocampus may be connected. Recent researches show for the α4β2-AChRs and for the α7-AChRs that regulation of functions by N-AChRs in the neocortex differs with the regulations in the hippocampus [22, 34, 42].

The importance of regional neuronal mechanisms in the control of brain trophic functions is discussed in previous literature [2, 7, 20, 48, 49]. However, we have found a relationship with the problem of individual resistance of neocortex and hippocampus to extreme conditions in only one paper, the review by Williamina A. Himvich [15]. In this review she postulated that individual anatomic structures of the brain should be regarded as individual organs. Such view requires a corresponding insight into the extent of autonomic nature of mechanisms for cerebral macrostructure resistance to pathological impacts.

References

1. Anderson KJ., Gibbs RB, Salvaterra PM, Cotman CW, Ultrastructural characterization of identified cholinergic neurons transplanted to the hippocampal formation of the rat. J Comp Neurol. 1986; 249(2): 279-292.
2. Arneris SP, Iadecola C, Underwood MD, Donald JR. Local cholinergic mechanisms participate in the increase in cortical cerebral blood flow elicited by electrical stimulation of the fastigial nucleus in rat. Brain Res. 1987; 411: 212-225.
3. Batchelor PE, Armstrong DM, Blaker SN, Gage FH. Nerve growth factor receptor and choline acetyltransferase colocalization in neurons within the rat forebrain: response to fimbria-fornix transection. J Comp Neurol. 1989; 284(2): 187-204.
4. Buhler AV, Dunwiddle TV. α7 nicotinic acetylcholine receptors on GABAergic interneurons evoke dendritic and somatic inhibition of hippocampal neurons. J Neurophysiol. 2002; 87: 548-557.
5. Carli M, Luschi R, Samanin R. Dose-related impairment of spatial learning by intrahippocampal scopolamine: antagonism by ondansetron, a 5-HT3 receptor antagonist. Behav Brain Res. 1997; 82(2): 185-194.
6. Christophe E, Roebuck A, Staiger JF, Lavery DJ, Charpak S, Audinat E. Two types of nicotinic receptors mediate an excitation of neocortical layer I interneurons. Neurophysiol. 2002; 88(3): 1318-1327.
7. Dauphin F, Lacombe P, Sercombe R, Hamel E, Seylaz J. Hypercapnia and stimulation of the substantia innominata increase rat frontal cortical blood flow by different cholinergic mechanisms. Brain Res. 1991; 553(1): 75-83.
8. Dobransky T, Rylett RJ. Functional regulation of choline acetyltransferase by phosphorylation. Neurochem Res. 2003; 28(3-4): 537-542.
9. Eckenstein FP, Baughman RW, Quinn J. An anatomical study of cholinergic innervation in rat cerebral cortex. Neuroscience. 1988; 25(2): 457-474.
10. Fabian-Fine R, Skehel P, Errington ML, Davies HA, Sher E, Steward MG, Fine A. Ultrastructural distribution of the alpha7 nicotinic acetylcholine receptor. J Neurosci. 2001; 10 (13): 7993-8003.
11. Fonnum F. Radiochemical microassays for the determination of choline acetyltransferase and acetylcholinesterase activities. Biochem. J. 1969; 115. P.465-472.
12. Frazier CJ, Srtowbridge BW, Papke RL. Nicotinic receptors on local circuit neurons in dentate gyrus: potential role in regulation of granule cell excitability. J Neurophysiol. 2003; 89: 3018-3028.
13. Frotscher M, Vida I, Bender R. Evidence for the existence of non-GABAergic, cholinergic interneurons in the rodent hippocampus. // Neuroscience. 2000; 96(1): 27-31.

14. Gould TG, Rowl WB, Heman KL, Meshes PA, Yuong PA, Rose FM, Bickford PC. Effects of hippocampal lesions on pattern of motor learning in the rat. Brain Res Bull. 2002; 58(6): 581-586.
15. Himwich WA. Biological foundations of psychiatry. Develop. Neurobiol.1976; 2: 591-632.
16. Houser CK, Crawford GD, Salvaterra PM, Vaughn JE. Immunocytochemical localization of choline acetyltransferase in rat cerebral cortex: a study of cholinergic neurons and synapses. J Compar Neurol. 1985; 234(1): 17-34.
17. Ishimaru H, Takahashi A, Ikarashi Y, Maruiama Y. NGF delays rather then prevents the colinergic terminal and delayed neuronal death in the hippocampus after ischemia. Brain Res.1998; 789(2): 194-200.
18. Iwasaki K, Kitamura Y, Ohgami Y, Mishima K, Fujiwara M. The disruption of spatial and changes in brain amino acid, monoamine and acetylcholine in tars with transient cerebral ischemia. Brain Res. 1996; 709(2): 163-172.
19. Ji D, Dani JA. Inhibition and disinhibition of pyramidal neurons by activation of nicotinic receptors on hippocampal interneurons. J Neurophysiol. 2000; 83(5): 2682-2690.
20. Krimer LS, Christopher Muly III E, Williams GV, Goldman-Rakic PS. Dopaminergic regulation of cerebral cortical microcirculation. Nat Neurosci. 1998; 1(4): 286-289.
21. Kumagae Y, Matsui Y. Output, tissue levels, and synthesis of acetylcholine during and after transient forebrain ischemia in the rat. J Neurochem. 1991; 56(4): 1169-1173.
22. Li XD, Arias E, Jonnala RR, Mruthinti S, Buccafusco JJ. Effects amyloid peptides on the increase in TrkA receptor expression induced by nicotine *in vitro* and *in vivo*. J Mol Neurosci. 2005; 27(3): 325-336.
23. Lipton P. Ischemic cell death in brain neurons. Physiol. Rev. 1999; 79(4): 1431-1568.
24. Lowry OH, Rosenbrough NJ, Farr AL, Randall R. Protein measurement with Folin phenol reagen. J Biol Chem. 1959; 193: 265-275.
25. Lukanova LD. [Bioenergetic hypoxia: definition, mechanisms, and methods of correction]. Biull Eksp Biol Med. 1997; 124(9): 244-254. Russian.
26. Marchi M, Risso F, Viola C, Cavazzani P, Raiteri M. Direct evidence that release-stimulating alpha7 nicotinic cholinergic receptors are localized on human and rat brain glutamatergic axon terminals. J Neurochem. 200; 80(6): 1071-1078.
27. Metherate R. Nicotinic Acetylcholine Receptors in Sensory Cortex. Learning & Memory. 2004; 1(1): 50-59.
28. Morris RG; Frey U. Hippocampal synaptic plasticity: role in spatial learning or the automatic recording of attended experience? Philos Trans R Soc Lond B Biol Sci.1997; 352(1360): 1489-1503.
29. Mugnaini M, Tessari M, Tarter G, Merlo Pich E, Chiamulera C, Bunnemann B. Upregulation of [3H] methyllycaconitine binding of alpha7 or alpha6 subunit mRNA: an autoradiography and in situ hybridization study in rat brain. Eur J Neurosci. 2002; 16(9): 1633-1646.
30. Mukhin EI, Zakharova EI, Kikteva EA. Comparison of the cholinergic system in neocortical field Ep in cats with strong and weak cognitive abilities. Neurosci Behav Physiol. 2002; 32(4): 379-387.
31. Mumby DG. Perspectives on object-recognition memory following hippocampal damage: lessons from studies in rats. Behav Brain Res. 2001; 127(1-2):159-181.
32. Mumby DG, Pinel JP, Rhinal cortex lesions and object recognition in rats. Behav Neurosci. 1994; 108(1): 11-18.
33. Ni JW, Matsumoto K, Li HB, Murakami Y, Watanabe H. Neuronal damage and decrease of central acetylcholine level following permanent occlusion of bilateral common carotid arteries in rat. Brain Res.1995; 673(2): 290-296.
34. Nuutinen S, Ahtee L, Tuominen RK. Time and brain region specific up-regulation of low affinity nicotinic receptors during chronic nicotine administration in mice. Eur J Pharmacol. 2005; 515(1-3): 83-89.
35. Ogawa N, Haba K, Asanuma M, Mizukava K, Mori A. Super-delayed changes of muscarinic acetylcholine receptors in the gerbil hippocampus following transient ischemia. In: Kito S e.a., eds., Neuroreceptor Mechanisms in Brain, Plenum Press, New York, 1991: 343-347.

36. O'Neill M, Canney M, Earley B, Junien JL, Leonard BE. The novel sigma ligand JO 1994 protects against ischemia-induced behavioural changes, cell death and receptor dysfunction in the gerbil. Neurochem. Int. 1996; 28(2): 193-207.
37. Onodera H, Sato G, Kogure K. Quantitative autoradiographic analysis of muscarinic and adenosine A_1 binding sites after transient forefrain ischemia in the gerbil. Brain Res.1987; 415(2): 309-322.
38. Parnavelas JG, Kelly W, Franke E, Eckenstein F.P. Cholinergic neurons and fibers in rat visual cortex. J. Neurocytol. 1986; 15 (3): 329-336.
39. Paxinos G, Watson C. The rat brain in stereotaxic coordinates. Academic Press, New York, 1982: 152P.
40. Porter JT, Cauli B, Tsuzuki K, Lambolez B, Rossier J, Audinat E Selective excitation of subtypes of neocortical interneurons by nicotinic receptors.J.Neurosci.1999; 10(13): 5228-5235.
41. Pullman ME, Penevsky HS, Datta A, Racker E. Partial resolution of the enzymes catalyzing oxidative phosphorilation. J Biol. Chem. 1960; 235(11): 3322-3329.
42. Rossi S, Singer S, Shearman E, Sershen H, Lajtha A. The effects of cholinergic and dopamiergic antagonists on nicotine-induced cerebral neurotransmitter changes. Neurochem Res. 2005; 30(4): 541-558.
43. Rossor MN, Svendsen C, Hunt SP, Mountjoy CQ, Roth M, Iversen LL The substantia innominata in Alzheimer's disease: a histochemical and biochemical study of cholinergic marker enzymes. Neurosci Lett. 1982; 28(2): 217-222.
44. Rylett RJ. Synaptosomal "membrane-bound" choline acetyltransferaseis most sensitive to inhibition by choline mustard. J. Neurochem.1989; 61(5): 1774-1781.
45. Santos MD, Alkondon M, Pereira EF, Aracava Y, Eisenberg HM, Maelicke A, Albuquerque EX. The nicotinic allosteric potentiating ligand galantamine facilitates synaptic transmission in mammalian central nervous system. Mol Pharmacol. 2002; 61(5): 1222-1234.
46. Schmidt BM, Rylett RJ. Phosphorylation of rat brain choline acetyltransferase and its relationship to enzyme activity. J.Neurochem. 1993; 61(5): 1774-1781.
47. Schmidt BM, Rylett RJ. Basal synthesis of acetylcholine in hippocampal synaptosomes is not dependent upon membrane-bound choline acetyltransferase activity. Neuroscience. 1993; 54(3): 649-656.
48. Scremin OU, Jenden DJ. Cholinergic control of cerebral blood flow in stroke, trauma and aging. Life Sci. 1996; 58(22): 2011-2018.
49. Scremin OU, Torres C, Scremin AM, O'Neal M, Heuser D, Blisard KS. Role of nucleus basalis in cholinergic control of cortical blood flow. J Neurosci Res. 1991; 28(3): 382-390.
50. Sudweeks SN, Yakel JL. Functional and molecular characterization of neuronal nicotinic ACh receptors in rat CA1 hippocampal neurons. J Physiol. 2000; 527(Pt 3): 515-528.
51. Sun MK, Xu H, Alkon DL. Pharmacological protection of synaptic function, spatial learning, and memory from transient hypoxia in rats. J Pharmacol Exp Ter. 2002; 300(2): 408-416.
52. Tanaka K, Wada N, Hori K, Asanuma M, Nomura M, Ogawa N. J. Chronic cerebral hypoperfusion disrupts discriminative behavior in acquired-learning rats. Neurosci. Methods. 1998; 84(1-2): 63-68.
53. Watanabe H, Ni JW, Sakai Y, Matsumoto K, Murukami Y, Tohda M. Permanent occlusion of bilateral internal carotid arteries produces cognitive deficits in two learning behavior tasks. Nihon Shinkei Seishin Yakurigaku Zasshi. 1996; 16(1): 19-24.
54. Zakharova EI, Dudchenko AM, Svinov MM, Ivanov DS, Ignatiev IV. [Comparative characteristics of the cholinergic systems in the brain of rats with the low and high resistance to the oxygen deficient]. Neurokhimia (Mosk). 2001; 18(2): 119-131. Russian.

Adaptation Biology and Medicine (Volume 5: Health Potentials)
Editors: L. Lukyanova, N. Takeda and P.K. Singal

11. Adaptation to Stress in Physical and Mental Illness

Maryrita K. Mallet*

Private Practice in Psychiatry, Weatherford, Texas, USA

Abbreviations

ACTH: adrenocorticotropic hormone; AD: Alzheimer's dementia; AVP: arginine vasopressin; CRH: corticotropin releasing hormone; DA: dopamine; GR: glucocorticoid receptors; HPA: hypothalamus-pituitary-adrenal axis; HPG: hypothalamus-pituitary-gonadal axis; HPT: hypothalamus-pituitary-thyroid axis; IBS: irritable bowel syndrome; IL: interleukin; ILC: infralimbic cortex; MBSR: mindfulness based stress reduction; MI: myocardial infarction; nAChR: nicotinic acetylcholine receptor; PTSD: post-traumatic stress disorder; TNF-α: tumor necrosis factor-alpha; TSH: thyroid stimulating hormone

Stress, Resilience and Frailty of the Psychiatric Patient

The concept of 'stress' has myriad definitions and has considerably evolved from the original work of Hans Selye.[1] A modern clarification of the stress concept uses the term 'allostasis,' defined as "the adaptive processes used by an organism to maintain homeostasis."[2] 'Allostatic load' is the cost to the organism to continually adapt.[2,3] 'Allostatic overload,' which may result from an unpredictable stressor, is the equivalent of being "stressed out" such that the organism is no longer able to adapt or maintain homeostasis and, consequently, is injured by the stressor.[2] In my clinical work, I use the terms 'frail' and 'resilient' to describe two opposite extremes of the stress adaptation spectrum. Resilient individuals see a stressor as a challenge, an opportunity to learn and grow. These persons have an internal locus of control, seek social support, and use coping mechanisms that include taking an active role in their life and healthy interactions with others.[4] They also have altruism and a positive spirituality which helps them find meaning in suffering, enhanced self efficacy and enhanced quality of life when faced with illness or other difficulties. Optimism, a sense of humor, regular exercise, and pet ownership are also protective factors against mental stress.[5,6] I almost never see highly resilient individuals in my practice, although I do see quite a few moderately resilient patients.

On the other hand, I see far too many frail individuals. After describing the extremes of frail and resilient, I ask my frail patient to tell me where on that scale s/he falls, with 1 = frail and 10 = resilient. Interestingly, my patients usually pick a number higher than my own estimation of their function; they proceed to give me multiple examples of their frailty and few examples of resilience. "Yes, but"

*Address for Correspondence: Maryrita K. Mallet, M.D., P.O. Box 1976, Weatherford, TX 76086 USA. Phone: 817-613-1942, E-mail: mmallet@itexas.net

responses abound to my various recommendations of ways to modify behavior or other aspects of treatment. They tell me (directly or indirectly) they are awaiting changes in their circumstances, their medication, their family, or other external factors in order to get better. "I'm such a burden, they'll all be better off without me." "Why is everybody against me?" "Living with my parents makes me this way." "Nothing you do/give me helps." These patients tend to have severe mental illness with high levels of distress. In fact, they tell me that *life itself* is a stressor. These patients feel hopeless, helpless and worthless. They are pessimistic, have an external locus of control and low self esteem,.[5-7]

The perception of life as more stressful, along with a perception of an increased impact of daily stress and increased intensity of major stress is associated with poor sleep.[8] Decreased sleep efficiency is common among frail individuals, with increased latency to stage one and difficulties with persistent sleep[8] due to pre-sleep physiological hyperarousal.[8,9] Physical illnesses are more prevalent and severe and chronic physical pain is common in frail *vs.* resilient patients. Their coping style tends to be emotional, passive, inhibited, and overly sensitive, or they may use dissociation.[7,8,10-12] For example, frail individuals with poor coping can unconsciously generate both epileptiform[7,13] and non-epileptiform seizures.[12] These patients are far more likely to have suffered traumatic events,[14,15] especially early in life, such as abuse, neglect or early maternal separation, making them more susceptible to stress, depression, anxiety[14] and chronic pain.[15] They have little or no psychosocial support, or even antagonistic, rejecting caregivers.[16] These patients experience high levels of distress at work, have difficulty keeping a job and are often unemployed or on disability.[17,18]

Frail patients may have a negative spirituality and believe that God hates them, has abandoned them, and/or is punishing them, so their suffering is unbearable.[19] Some of these patients are actually somewhat protected from suicide because they believe it to be an unforgivable sin that will condemn them to eternity in hell. In my effort to keep these frail individuals alive, I do not attempt to dissuade them of this belief. The relative risk of suicide attempts in those with severe anxiety is nine fold in men and three fold in women.[20] My frail patients are far more likely to attempt suicide multiple times and some, unfortunately, have completed suicide. Such high-risk patients cannot be hospitalized indefinitely, so I document their chronic risk of completed suicide as well as notating discussions with the family about this risk. In fact, I will even discourage repeated hospitalizations of my frail patients; each admission brings a different psychiatrist who changes their medications, often leading to further destabilization.

Physical Illness in the Patient with Mental Illness

Persons with mental illness are at heightened risk of physical illnesses. Patients with high levels of psychological distress have an increased relative risk of premature mortality (respiratory 5.39, ischemic heart disease 1.90, and all causes 1.71).[21] In another study, subjects with high psychological distress had a 2.56 relative risk of having a fatal stroke after controlling for all confounders, but did not have a heightened risk of non-fatal stroke.[22] The high rate of mortality from respiratory illnesses may be a corollary of the belief that smoking and other forms of tobacco use calms feelings of mental stress, when in fact it eases the symptoms of nicotine withdrawal that *feel like* stress. Indeed, patients who are able to quit smoking, after an initial period of increased stress feelings, actually report that their stress subsides – because they are no longer experiencing the stress of nicotine withdrawal.[23] Smoking is also increased in persons with substance use (4.6), affective disorders (1.8), anxiety (1.6), and somatoform disorders (1.4).[24] Schizophrenic patients also have extremely high rates of smoking.[25]

Depression has been shown to have a cyclical relationship with cardiovascular disease, each one increasing the risk of the other.[26,27] The underlying mechanism is, at least in part, the dysregulation of the hypothalamus-pituitary-adrenal (HPA) axis, as well as an imbalance of the autonomic nervous system and immune system. In the healthy, resilient individual, stress activates the HPA axis, but homeostatic, negative feedback mechanisms return the axis to its basal activity. On the other hand, in the frail individual, who is far more likely to have mental illness, the negative feedback mechanism is attenuated, leading to chronic dysregulation of the HPA axis. Furthermore, such individuals have increased circulating concentrations of pro-inflammatory cytokines, enhanced sympathetic activity and diminished parasympathetic activity.[5,14,26]

Hostility and strong negative emotions also increase the risk of myocardial infarction (MI).[5,28] Fifteen months after MI, patients whose MI was influenced by negative emotions still had increased platelet aggregation and delayed recovery of cardiac output and systolic blood pressure after a mental stress task compared to a group of MI survivors whose event was not preceded by negative emotions.[28] Sudden emotional stress may also lead to myocardial stunning. In one study, a cohort of 19 patients (18 women) presented with rapidly reversible left ventricular dysfunction. Only one of them had clinical evidence of coronary artery disease as diagnosed by angiography. The patients with cardiac stunning had only a mild elevation of cardiac troponin isoform in systemic venous blood compared to the striking increase in the MI patients. However, the stunned group had an exaggerated catecholamine response with a fourfold increase in circulating epinephrine, a twofold increase in norepinephrine, and a 1.6-fold increase in dopamine. Myocardial biopsy revealed contraction-band necrosis and inflammation.[29]

Mental Stress and the Hypothalamic-pituitary-adrenal Axis

In a rat model of stress-induced depression, the animals had increased plasma concentrations of the pro-inflammatory cytokines tumor necrosis factor alpha (TNF-α) and interleukin (IL)-1-beta (IL-1β). TNF-α was also found in the hypothalamus and pituitary, and IL-1β was found in the hypothalamus. Plasma activities of renin, angiotensin and aldosterone were also elevated.[26] Dysregulation of these hormones contributes to hypertension. The enhanced sympathetic drive, along with increased concentrations of homocysteine, platelet activating factor and nitric oxide can cause vascular endothelial changes that damage the microcirculation of the heart, brain and kidney.[26,27] As these lesions accumulate in the brain, the patient develops dementia. Job stress and burnout increase fibrinogen and C-reactive protein, which are markers of microinflammation, especially in men.[30] Academic exam stress leads to enhanced secretion of immunoglobulins, complement and acute phase proteins, especially in frail students with a high perception of stress.[31]

Young firefighters are more likely than firefighters over age 45 to deny ubiquitous shortcomings and have a defensive coping pattern with decreased sensitivity to aversive stimuli, decreased sexual arousal and decreased response to pleasurable stimuli. Even when young firefighters have low anxiety, they still have elevated activities of the stress hormone cortisol, equivalent to those with high anxiety who do not have a defensive coping pattern. Firefighters over age 45 who have this coping pattern are nevertheless less likely to have cortisol elevations, suggesting enhanced adaptability with age and/or experience. The author's alternative explanation was that poorly coping firefighters adversely affected by the high levels of job stress had retired before the age of 45.[32]

Chronic elevations of cortisol or increased cortisol from a severe acute stressor can cause changes in memory in the amygdala (emotional), hippocampus (declarative, spatial and contextual) and prefrontal cortex (working), all of which have glucocorticoid receptors (GR).[3,14,33] Acute stress causes the hypothalamus and amygdala to release corticotropin releasing hormone (CRH) and activate the HPA axis. Glucocorticoids then reduce CRH activity in the hypothalamus. However, chronic stress attenuates negative feedback of glucocorticoids on the hypothalamus leading to persistent activation of the HPA axis, which further enhances CRH activity in the amygdala.[14,33] Quantitative magnetic resonance imaging studies in post-traumatic stress disorder (PTSD) demonstrate a hyperresponsive amygdala, which intensifies fear conditioning and fearful responses. The anterior cingulate cortex is involved in the affective evaluation of sensory stimuli, visceral perceptions and homeostatic responses.[34] Decreased activity in the anterior cingulate cortex diminishes negative feedback to the amygdala. Moreover, decreased function in the hippocampus diminishes habituation to aversive stimuli and increases the likelihood of stimulus generalization.[35] These mechanisms may be responsible for impaired memory in anxiety, depression, dissociation and PTSD. At the time of combat trauma military veterans have higher cortisol levels, but in chronic PTSD veterans do not have elevations in cortisol secondary to diminished activity of the HPA axis and hypersensitivity to steroid feedback.[14] Swaab *et al.*[14] suggest that patients with PTSD may have an increased sensitivity to cortisol or an upregulation of glucocorticoid receptors. I have many patients with PTSD in my practice. Their hyperarousal contributes to the symptoms of flashbacks, nightmares, avoidance of triggers, a marked startle response and difficulty with initial and middle insomnia. Some military veterans can only sleep in the daytime when others are up and around "to stand guard," an example of stimulus generalization. Triggers can be mild stressors which are perceived as repeated trauma and activate the amygdala as above. For example, some patients with a history of sexual abuse are triggered by healthy sexual intimacy, while others slip into promiscuity in an unconscious effort to master the trauma.

In a study on hypnosis, subjects were first guided to a suggested neutral stimulus and then an unpleasant stimulus. During the neutral stimulus, subjects had no activation of the autonomic nervous system and their electroencephalogram continued to have minimal γ waves and enhanced θ waves demonstrating complete relaxation. In contrast, the high emotional valence of the unpleasant stimulus overrode the hypnosis as demonstrated by a rise in autonomic activity and the return of γ, $\beta3$ and $\beta2$ waves.[36] Patients with animal phobias also revealed autonomic activation along with a rise in γ and a fall in $\alpha1$, $\theta1$ and $\alpha2$ activity. This effect was far more pronounced in subjects with low hypnotizability. Subjects with high hypnotizability who practice hypnosis are more protected from the deleterious effects of mental stress, including cardiovascular disease.[37]

Overactivation of the HPA axis may lead to structural alterations in the hippocampus. These alterations may include diminished neurogenesis in the dentate gyrus and dendritic remodeling in the CA3 area of the hippocampus.[3] The rat hippocampus shows neurotoxicity with chronic stress[3] that is lateralized to the left.[14] Some studies have shown similar lateralization of hippocampal volume in humans, though one study showed lateralization to the right, while the other study showed lateralization to the left. Cushing's syndrome, characterized by marked elevations of cortisol, is associated with reductions in cerebral, cerebellar and hippocampal volumes.[14] Studies in patients treated with glucocorticoids showed an alteration in hippocampal volume, attributed to a decrease in water volume, but these changes were reversible when glucocorticoid treatment was discontinued. Depressed patients have also been shown to have cortical atrophy, but not to the same extent seen in

Cushing's syndrome. Diminished volume in the hippocampus may also be secondary to a decrease in arborization of the dendrites. Post-mortem studies, however, did not show any histological changes in the hippocampus. However, most of these patients had been on antidepressant treatment which may reduce apoptosis and increase neurogenesis in the hippocampus. Some studies suggest that the decrease in hippocampal volume may be a pre-morbid risk factor for depression and PTSD,[14] but others found no decrease in hippocampal volume in PTSD.[35] While chronically stressed rodents show neurotoxicity of the hippocampus,[3,14,35] humans experiencing chronic stress show adaptive and reversible changes in the hippocampus and do not demonstrate hippocampal atrophy, massive cell loss or loss of plasticity, but may have changes in hippocampal water content instead.[14,35] On the other hand, hippocampal damage seen in Alzheimer's dementia is irreversible. Swaab *et al* suggest this damage may involve the HPA cascade as part of its etiology or, conversely, the increased activation of the HPA axis in Alzheimer's may be secondary to the hippocampal damage.[14]

Depression is associated with strong activation of the HPA axis, which is driven by chronic activation of the hypothalamic paraventricular nucleus. CRH-secreting neurons in the paraventricular nucleus can co-express arginine vasopressin (AVP) which potentiates the action of CRH. Increased production of AVP by the supraoptic nucleus has been associated with an increased risk of suicide.[14] Decreased amounts of AVP in the suprachiasmatic nucleus, the 'biological clock,' have been associated with diminished circadian amplitude that could be related to disruption of sleep in depression.[14] Increased production of oxytocin in the paraventricular nucleus may be associated with eating disorders.[14] Depression also inhibits the hypothalamus pituitary gonadal (HPG) axis, but estrogen exerts negative feedback inhibition of the HPA axis[14] and enhances synaptogenesis in the CA1 region of the hippocampus in rats, the area responsible for declarative, spatial and contextual memory.[3]

Depressed women have higher levels of androgens from activation of the HPA axis, and lower levels of estrogen due to suppression of the HPG axis, while depressed men have lower levels of testosterone due to suppression of the HPG axis.[14] The relative levels of the sex steroids are not as important as the fluctuations of their circulating concentrations. The increased prevalence of depression in women begins at menarche and subsides after menopause. The fluctuations found in pre-menstrual syndrome, antepartum and postpartum and in perimenopause may lead to increased dysphoria, irritability, tearfulness and vulnerability to stress.[14] Depressed women also have an increased risk of fractures including non-vertebral (odds ratio 2.5) and osteoporotic (3.1) fractures.[38] With the use of psychotropic medicines the respective odds ratios increase to 4.4 and 4.7.[38] This increased risk may involve dysregulation of the HPG axis.[39]

Depression and stress also suppress the hypothalamus-pituitary-thyroid (HPT) axis.[14,38] Thyroid stimulating hormone (TSH), and free T4 in the event of abnormal TSH, are variables we monitor in depression. Thyroid hormone may be used to augment the treatment of depression when TSH is at the upper end of the clinically normal range. Activation of the HPA axis and suppression of the HPT axis suppresses the HPG axis, which can lead to infertility, including anovulation. Hormone replacement therapy alone may result in ovulation, but unless the HPA and HPT axes are restored to normal function, the mother and fetus (and later the neonate) can have poor outcomes.[38] Many infertility clinics now offer psychological services to promote changes in behaviors and attitudes which facilitate a return to fertility via normalization of the HPA, HPT, and HPG axes.[38]

Psychological Stress and Exaggerated Pain Perception

The anterior cingulate cortex is involved in the affective evaluation of sensory stimuli, visceral perceptions and homeostatic responses which play a role in asthma and irritable bowel syndrome (IBS).[34,40]. Moreover, decreased function in the hippocampus diminishes habituation to aversive stimuli and increases the likelihood of stimulus generalization.[35] In asthma, emotional stress increases inflammation and airway obstruction which in turn influences the brain.[34] In a large epidemiological study of the general population,[41] those with some (odds ratio 1.7), much (2.6) or very much (3.6) emotional stress were more likely to have wheezed in the last 12 months, compared to respondents with little emotional stress (odds ratio 1.0). Likewise, an increase in waking with cough or shortness of breath (odds ratios of 1.5, 2.4, and 4.0 in the respective emotional stress groups) was also reported.[41]

The anterior cingulate cortex, limbic area and locus ceruleus mediate gut signals. IBS involves a dysregulation of the brain-gut axis, which has visceral hyperalgesia and abnormal cognitive processing of emotional and visceral stimuli. These patients are often alexithymic.[40] Patients with IBS, especially men, have a dysregulation of the autonomic response to a visceral stressor, resulting in an exaggerated sympathetic response and a diminished parasympathetic response.[42] The sympathetic stress hormone epinephrine has been shown to influence encoding of memories in the limbic structures, yet it cannot cross the blood-brain barrier. However, the vagus nerve has afferent connections to the limbic system. Epinephrine causes a dramatic sustained increase in vagal afferent firing, which is attenuated by beta-adrenergic blockers.[43] In a functional magnetic resonance imaging study of IBS,[44] subjects underwent balloon rectal distensions. Control subjects registered urge to defecate and pain in the right anterior insula and right anterior cingulate cortex. Subjects with IBS registered urge in the primary sensory cortex and pain in the medial thalamus and hippocampus. The IBS pain may be secondary to alterations in the hippocampus brought on by chronic emotional stress,[3] along with decreased habituation to aversive stimuli.[35] These alterations may include diminished neurogenesis in the dentate gyrus and dendritic remodeling in the CA3 area of the hippocampus that is responsible for contextual memory and may contribute to chronic physical pain.[3] Alterations in the HPA axis and pro-inflammatory cytokines are also seen in IBS. The marked increases in adrenocorticotropic hormone (ACTH) and cortisol following CRH stimulation correlate with the increase in IL-6.[45]

Gastro-esophageal reflux is not intensified in patients with psychological distress, despite elevated circulating cortisol concentrations. However, the *perception* of symptoms is enhanced, perhaps because of the higher levels of state anxiety. This altered perception contributes to a belief that clinical therapy for this condition is ineffective. Monitoring the esophageal pH for 24 hours establishes if the problem is ineffective perception or ineffective medication.[46] Another example of exaggerated perception is in somatoform illnesses. These patients have a tendency to somaticize, or experience in their body, high levels of psychological distress. One study demonstrated symptoms of dyspnea in highly somatic women exposed to either mental stress or 5% $CO_{2,}$ but not during mild physical exercise. These women also experienced more tension.[47]

Patients with chronic physical pain also have increased muscular tension. High levels of anxiety may result in hyperventilation. The subsequent loss of CO_2 leads to a respiratory alkalosis, provoking increases in muscle tension, spasm, ischemia and hypoxia .as well as an exaggerated response to catacholamines. Furthermore, thoracic breathing contributes to a biomechanical stress with pain in the muscles of the neck and upper thorax.[18] Changes in posture and movement also create a biomechanical stressor.[17] Pain may be associated with muscle cell activation or decreased energy reserves, as well as damage from increases in intracellular calcium or changes in buffering mechanisms. Interactions of blood vessels and nociceptive nerve fibers in the muscle may also produce pain. These fibers travel in the spinal medulla which further modulates the degree of transmission to the brain.[17] The greatest modulator of pain, however, is cognitive appraisal, which is effected by pain beliefs. Increases in the emotional valence of threat as well as the degree of attention to and perception of pain can amplify and maintain chronic pain.[17,48] Women with chronic pain, especially neck pain, had increases of cognitive dysfunction, catastrophizing, depression, PTSD, fatigue and perception of pain severity.[48] Changes in physiological mechanisms such as activation of the HPA axis may cause changes in the hippocampus that alter the contextual memory of pain,[3] along with decreased habituation to aversive stimuli.[35] Chronic pain is common in patients with temporomandibular disorder. In one study, 50% of such patients had a history of traumatic stressors which was associated with myogenous pain. This subset had greater psychological distress, higher psychometric scores, higher pain perception, more disability and more ineffective coping compared to the 50% of patients who did not have prior trauma. Patients with fibromyalgia also have chronic myogenous pain, which has been associated with multiple chronic hormonal changes possibly due to a change in setpoint of the hormonal axes along with changes in nociception.[49]

Prenatal and Childhood Factors Predisposing to Vulnerability to Mental Stress

Both animal and human studies demonstrate that autonomic and HPA functions are programmed in the fetus.[50] Prenatal administration of estrogen can increase the risk of later depression in women,[14] while prenatal administration of androgens can inhibit the HPA axis in rodents[14] which may explain the twofold greater prevalence of depression in women than in men.[50] Nutrition in pregnancy and for the neonate and child is also important. Maternal malnutrition during pregnancy substantially increases the risk of depression in the child, regardless of sex.[14] Dietary deficiency of polyunsaturated fatty acids combined with mental stress-induced increases in catecholamines, cortisol, serotonin and proinflammatory cytokines (TNFα, IL- 1, 2 and 6) can damage the ventromedial hypothalamus and insulin receptors in the brain during the critical period of rapid brain development and growth in the fetus and young child. This malnutrition-evoked hormonal cascade can also damage pancreatic islet cells, increasing the risk of metabolic syndrome and non-insulin dependent diabetes mellitus. Omega-3 polyunsaturated fatty acids can prevent this damage and have been shown to increase parasympathetic activity and anti-inflammatory cytokine secretion.[51]

Early environment has profound effects on later vulnerability to stress. Neonates and infants with small birth weights are more easily stressed than those with normal birth weights. Prenatal exposure to nicotine increases the risk of depression later in life.[14] Maternal stress increases the child's vulnerability to stress, emphasizing the importance of treating mood disorders in the postpartum period, and during the prenatal period, although the latter is controversial. One alternative to prenatal pharmacotherapy is light therapy.[14] Depression, anxiety and distress in both parents can affect a child's adaptation to and behavior in a variety of chronic illnesses such as childhood diabetes and juvenile rheumatoid arthritis.[10,11] The child's adaptation, in turn, influences disease adjustment and treatment compliance. For example, children can inhibit or generate their own epileptic seizures.[13] Children with an internal locus of control[54] fared better than those with an external locus of control accompanied by outbursts and avoidance coping.[11]

Early maternal separation, child abuse, and bereavement can also sensitize the HPA axis, leading to the development of depression and anxiety disorders.[14] Handled rats are much less vulnerable to stress than rats which were not handled postnatally.[52] This paradigm could be a model for the child who is nurtured and cared-for versus the child who is neglected, abandoned and/or abused. In this rat model, the mesocortical dopamine (DA) system sends afferents to the infralimbic cortex (ILC). Release of this DA inhibits the HPA axis which facilitates adaptation to stress. Injection of a DA antagonist into the ILC produced exaggerated ACTH responses to restraint stress in both handled and not-handled rats. However, handled rats only showed such elevations when the right ILC was injected. Neurochemical staining revealed higher DA content in the right ILC of handled *vs.* non-handled rats following repeated stress. This laterality to the right ILC is believed to be one of the reasons that handled rats are more resilient since stress modulation is generally found in the right prefrontal cortex.[52]

Nicotine and Mental Stress

The mesolimbic DA system is modulated by nicotinic acetylcholine receptors (nAChRs). These receptors are found in many other areas of the brain including the lateral septum, dorsal raphe and hippocampus. Activation of these receptors impacts many other systems including monoamine transmission and the release of neurotransmitters. Nicotine can both activate and desensitize the nAChRs, which explains how it can be both anxiogenic and anxiolytic depending on the route, timing and model.[53] Nicotine withdrawal results in anxiety, depression and other symptoms of mental stress.[23,53] Nicotine also plays a role in cognition, most likely by modulating DA and acetylcholine. Deprivation of nicotine leads to impaired cognition in smokers. Administration of nicotine to non-smokers and non-deprived smokers improved psychometric testing scores in the area of finger tapping, focus and sustained attention, recognition memory and reasoning.[54] One of the reasons patients with schizophrenia have impaired cognition is a marked loss of nAChRs in the brain (50%) and blood (90%), even in non-smokers. These patients may also be low in arachidonic acid and other fatty acids. Arachidonic acid is the precuser to anandamide, an endogenous cannabinoid that may require activation of nAChRs to be synthesized. Anandamide may also play a role in cognition. Improving cognition is important in schizophrenia, which may be why so many schizophrenics smoke.[55] Those with better cognition are better able to problem-solve and less likely to ignore stressors, including pleasurable ones. Treatment of positive symptoms, *e.g.* psychosis, lowers state anxiety and facilitates active coping. However, the persistence of negative symptoms, such as apathy, leaves the patient in a more frail state with passive, avoidant coping and increased trait anxiety.[56]

Enhancing Patient Resilience

In a study of PTSD, an increase in resilience was documented in patients treated with medication with or without cognitive behavioral therapy.[57] Measures of baseline resilience, disability and a sense of humor were predictive of the response to treatment. The greatest changes were in confidence, control, coping, knowing where to seek help, and adaptability.[57] A study in patients with bipolar disorder who had stayed well for two or more years identified multiple factors that increased resilience including understanding and acceptance of their diagnosis, identification of triggers and warnings of impending relapse along with intervention strategies, sufficient sleep, medication, learning stress management and accessing support networks.[58]

A study in melanoma patients also identified multiple factors promoting resilience. These include confronting the illness, processing the emotional impact, keeping perspective (the cancer is not their identification) as well as support from within themselves and from others. Identification of cherished reasons for living includes meaning, joy, relationships, values and spirituality. Knowledge that these reasons will not be obliterated, if they die from melanoma, comforts them as well.[59] Existential concerns in another oncology study do not necessarily lead to psychological distress. Factors that enhance resilience include early palliative care, increased family support, effective coping and religious beliefs.[60] Enhanced spirituality and spiritual well being are associated with improved functional well being, while increased meaning and peace was associated with physical well being. Spirituality was associated with all coping styles except the unhealthy one of avoidance in women with breast cancer.[61]

Several structured group programs have been shown to enhance resilience in multiple patient populations. Mindfulness based stress reduction (MBSR) seeks to increase awareness of the moment, enhance a more realistic perception, diminish negative affect and improve coping.[62] Mood disturbances and stress symptoms were still diminished six months after MBSR in cancer outpatients.[63] MBSR, along with gentle hatha yoga, in solid-organ transplant recipients continued to promote increased quality and duration of sleep six months later. This improved sleep was associated with positive mental health and well being.[64] Music therapy combined with progressive deep muscle relaxation in abused women in shelters also improved sleep quality and lowered anxiety.[65] Progressive muscle relaxation has also been shown to have immediate and cumulative effects on lowering blood pressure and pulse. Perception of stress was diminished, while perception of health was enhanced.[66]

Conclusion

My goal for my patients is not just response, but remission of their psychological distress, mental and physical illness by helping them adapt to the mental stress in their lives. Ongoing remission requires a plan of action. I use written questionnaires, mood charts and 'relapse warning signs' cards with my patients to both achieve remission and to prevent relapse. Patients who are moderately resilient have the best chance of achieving that goal. Using the bio-psycho-social-spiritual model will assist my patients to live happier, healthier lives.

Acknowledgements

The author thanks Ms. Terrie Bumpas for excellent secretarial support in typing the references and Dr. Robert Mallet for his expert editing of the manuscript.

References

1. Selye H. A syndrome produced by diverse nocuous agents. Nature 1936; 138: 32.
2. McEwen BS. Stressed or stressed out: what is the difference? J Psych Neurosci 2005; 30: 315-318.
3. McEwen BS. Plasticity of the hippocampus: adaptation to chronic stress and allostatic load. Ann NY Acad Sci 2001; 933; 265-277.
4. Meijer SA, *et al.* Coping styles and locus of control as predictors for psychological adjustment of adolescents with a chronic illness. Soc Sci Med 2002; 54: 1453-1461.
5. Das S, O'Keefe JH. Behavioral cardiology: recognizing and addressing the profound impact of psychosocial stress on cardiovascular health. Curr Atheroscler Rep 2006; 8: 111-118.

6. Robinson-Smith G. Prayer after stroke. Its relationship to quality of life. J Holist Nurs 2002; 20: 352-366.
7. Watten VP, Watten RG. Psychological profiles in patients with medically refractory epilepsy. Seizure 1999; 8: 304-309.
8. Morin CM, Rodrigue S, Ivers H. Role of stress, arousal, and coping skills in primary insomnia. Psychosom Med 2003; 65: 259-267.
9. Drake C, *et al.* Vulnerability to stress-related sleep disturbance and hyperarousal. Sleep 2004; 27: 285-291.
10. Frank RG, *et al.* Disease and family contributors to adaptation in juvenile rheumatoid arthritis and juvenile diabetes. Arthritis Care Res 1998; 11: 166-176.
11. Hanson CL, *et al.* Coping styles in youth with insulin-dependent diabetes mellitus. J Consult Clin Psychol 1989; 57: 644-651.
12. Goldstein LH, *et al.* Dissociation, hypnotizability, coping styles and health locus of control: characteristics of pseudoseizure patients. Seizure 2000; 9: 314-322.
13. Fenwick PB. The relationship between mind, brain, and seizures. Epilepsia 1992; 33(suppl 6): S1-S6.
14. Swaab DF, Bao A-M, Lucassen PJ. The stress system in the human brain in depression and neurodegeneration. Ageing Res Rev 2005; 4: 141-194.
15. De Leeuw R, Bertoli E, Schmidt JE, Carlson CR. Prevalence of traumatic stressors in patients with temporomandibular disorders. J Oral Maxillofac Surg 2005; 63: 42-50.
16. Rosenfarb IS, Bellack AS, Aziz N. A sociocultural stress, appraisal, and coping model of subjective burden and family attitudes toward patients with schizophrenia. J Abnorm Psychol 2006; 115: 157-165.
17. Knardahl S. Psychological and social factors at work: contribution to musculoskeletal disorders and disabilities. G Ital Med Lav Ergon 2005; 27: 65-73.
18. Schleifer LM, Ley R, Spalding TW. A hyperventilation theory of job stress and musculoskeletal disorders. Am J Ind Med 2002; 41: 420-432.
19. Hills J, Paice JA, Cameron JR, Shott S. Spirituality and distress in palliative care consultation. J Palliat Med. 2005; 8: 782-788.
20. Ringback Weitoft G, Rosen M. Is perceived nervousness and anxiety a predictor of premature mortality and severe morbidity? J Epidemiol Community Health 2005; 59: 794-798.
21. Robinson KL, McBeth J, Macfarlane GJ. Psychological distress and premature mortality in the general population: a prospective study. Ann Epidemiol 2004; 14: 467-472.
22. May M *et al.* Does psychological distress predict the risk of ischemic stroke and transient ischemic attack? The Caerphilly Study. Stroke 2002; 33: 7-12.
23. Parrott AC. Smoking cessation leads to reduced stress, but why? Int J Addict 1995; 30: 1509-1516.
24. John U, Meyer C, Rumpf HJ, Hapke U. Smoking, nicotine dependence and psychiatric comorbidity—a population-based study including smoking cessation after three years. Drug Alcohol Depend 2004; 763: 287-295.
25. Monterrubio S, Solowij N. Letter to the editor. Am J Psych 2006; 163: 553.
26. Grippo AJ, *et al.* Neuroendocrine and cytokine profile of chronic stress-induced anhedonia. Physiol Behav 2005; 84: 697-706.
27. Plante GE. Depression and cardiovascular disease: a reciprocal relationship. Metabolism 2005: 54(suppl 1): 45-48.
28. Strike PC, *et al.* Pathophysiological processes underlying triggering of acute cardiac events. Proc Natl Acad Sci USA 2006; 103: 4322-4327.
29. Wittstein IS, *et al.* Neurohumoral features of myocardial stunning due to sudden emotional stress. N Engl J Med 2005; 352: 539-548.
30. Toker S, *et al.* The association between burnout, depression, anxiety, and inflammation biomarkers: C-reactive protein and fibrinogen in men and women. J Occup Health Psychol 2005; l0: 344-362.

31. Maes M, *et al.* Effects of psychological stress on serum innunogloblin, complement and acute phase protein concentrations in normal volunteers. Psychoneuroendocrinology 1997; 22: 397-409.

32. Brody S, *et al.* Social desirability scores are associated with higher morning cortisol levels in firefighters. J Psychosomat Res 2000; 49: 227-228.

33. Makino S, Hashimoto K, Gold PW. Multiple feedback mechanisms activating corticotropin-releasing hormone system in the brain during stress. Pharmacol Biochem Behav 2002; 73: 147-158.

34. Rosenkranz MA, *et al.* Neural circuitry underlying the interaction between emotion and asthma symptom exacerbation. Proc Natl Acad Sci USA 2005; 102: 13319-13324.

35. Grossman R, Buchsbaum MS, Yehuda R. Neuroimaging studies in post-traumatic stress disorder. Psychiatr Clin N Am 2002; 25: 317-340.

36. Sebastiani L, *et al.* Human hypnosis: autonomic and electroencephalographic correlates of a guided multimode cognitive-emotional imagery. Neurosci Lett 2003; 338: 41-44.

37. Gemignani A, *et al.* Hypnotic trait and specific phobia: EEG and autonomic output during phobic stimulation. Brain Res Bull 2006; 69: 197-203.

38. Berga SL, Loucks TL. The diagnosis and treatment of stress-induced anovulation. Minerva Ginecol. 2005; 57: 45-54.

39. Sogaard AJ, *et al.* Long-term mental distress, bone mineral density and non-vertebral fractures. The Tromso Study. Osteoporos Int 2005; 16: 887-897.

40. Porcelli P. Psychological abnormalities in patients with irritable bowel syndrome. Indian J Gastroenterol 2004; 23: 63-69.

41. Oh YM, *et al.* Association between stress and asthma symptoms: a population-based study. Respirology 2004; 9: 363-368.

42. Tillisch K, *et al.* Sex specific alterations in autonomic function among patients with irritable bowel syndrome. Gut 2005; 54: 1396-1401.

43. Miyashita T, Williams CL. Epinephrine administration increases neural impulses propagated along the vagus nerve: Role of peripheral beta-adrenergic receptors. Neurobiol Learn Mem 2006; 85: 116-124.

44. Kwan CL, *et al.* Abnormal forebrain activity in functional bowel disorder patients with chronic pain. Neurology 2005; 65: 1268-1277.

45. Dinan TG, *et al.* Hypothalamic-pituitary-gut axis dysregulation in irritable bowel syndrome: plasma cytokines as a potential biomarker? Gastroenterology 2006; 130: 596-600.

46. Wright CE, *et al.* The effects of psychological stress on symptom severity and perception in patients with gastro-oesophageal reflux. J Psychosom Res 2005; 59: 415-424.

47. Houtveen JH, Rietveld S, de Geus EJ. Exaggerated perception of normal physiological responses to stress and hypercapnia in young women with numerous functional somatic symptoms. J Psychosom Res 2003; 55: 481-490.

48. Roth RS, Geisser MA, Theison-Goodvich M, Dixon PJ. Cognitive complaints are associated with depression, fatugue, female sex, and pain catastrophizing in patients with chronic pain. Arch Phys Med Rehabil 2005; 86: 1147-1154.

49. Neeck G, Crofford LJ. Neuroendocrine perturbations in fibromyalgia and chronic fatigue syndrome. Rheum Dis Clin North Am 2000; 26: 989-1002.

50. Phillips DI, Jones A. Fetal programming of autonomic and HPA function: do people who were small babies have enhanced stress responses? J Physiol 2006; 572: 45-50.

51. Singh RB, Pella D, Mechirova V, Otsuka K. Can brain dysfunction be a predisposing factor for metabolic syndrome? Biomed Pharmacother 2004; 58(suppl 1): S56-S68.

52. Sullivan RM, Dufresne MM. Mesocortical dopamine and HPA axis regulation: role of laterality and early environment. Brain Res 2006; 1076: 49-59.

53. Picciotto MR, Brunzill DH, Caldarone BJ. Effect of nicotine and nicotinic receptors on anxiety and depression. Neuroreport 2002; 13: 1097-1106.
54. Heishman SJ. Behavioral and cognitive effects of smoking: relationship to nicotine addiction. Nicotine Tob Res 1999; 1(suppl 2): S143-S147.
55. Monterrubio S, Solowij N. Letter to the Editor. Am J Psych 2006; 163: 553.
56. Lysaker PH, *et al.* Association of neurocognition, anxiety, positive and negative symptoms with coping preference in schizophrenia spectrum disorders. Schizophr Res 2005; 80: 163-171.
57. Davidson JR, *et al.* Trauma, resilience and saliostasis: effects of treatment in post-traumatic stress disorder. Int Clin Psychopharmacol 2005; 20: 43-48.
58. Russell SJ, Browne JL. Staying well with bipolar disorder. Aust N Z J Psychiatry 2005; 39: 187-193.
59. Kneier AW. Coping with melanoma—ten strategies that promote psychological adjustment. Surg Clin North Am 2003; 83: 417-430.
60. Blinderman CD, Cherny NI. Existential issues do not necessarily result in existential suffering: lessons from cancer patients in Israel. Palliat Med 2005; 19: 371-380.
61. Levine EG, Targ E. Spiritual correlates of functional well-being in women with breast cancer. Integr Cancer Ther 2002; 1: 166-174.
62. Grossman P, Niemann L, Schmidt S, Walach H. Mindfulness-based stress reduction and health benefits. A meta-analysis. J Psychosom Res 2004; 57: 35-43.
63. Carlson LE, *et al.* The effects of a mindfulness meditation-based stress reduction program on mood and symptoms of stress in cancer outpatients: 6-month follow-up. Support Care Cancer 2001; 9: 112-123.
64. Kreitzer MJ, *et al.* Longitudinal impact of mindfulness meditation on illness burden in solid-organ transplant recipients. Prog Transplant 2005; 15: 166-172.
65. Hernandez-Ruiz E. Effect of music therapy on the anxiety levels and sleep patterns of abused women in shelters. J Music Ther 2005; 42: 140-158.
66. Sheu S, Irvin BL, Lin HS, Mar CL. Effects of progressive muscle relaxation on blood pressure and psychosocial status for clients with essential hypertension in Taiwan. Holist Nurs Pract 2003; 17: 41-47.

Adaptation Biology and Medicine (Volume 5: Health Potentials)
Editors: L. Lukyanova, N. Takeda and P.K. Singal

12. Resistance to Stress and Neurodegenerative Damage in Rats of Different Genetic Strains: Protective Effects of Adaptation to Hypoxia

M.G. Pshennikova*, E.V. Popkova, D.A. Pokidyshev, I.P. Khomenko, O.M. Zelenina, S.V. Kruglov, E.B. Manukhina, M.V. Shimkovich and I. Yu. Malyshev

Institute of General Pathology and Pathophysiology, Moscow, Russia

Keywords: Antioxidants, neurodegenerative, hypoxia and magnocellular

Introduction

Pathogenesis, prevention and therapy of neurodegenerative damages of brain resulting in Alzheimer's disease is an urgent problem of pathophysiology and medicine today. These diseases of the central nervous system are characterized by progressive memory loss and other cognitive disorders and these diseases affect from 5 to 10% of people older than 65 years [5, 20]. The key point of this problem is a mechanism of resistance to these damages [5, 6]. Up to the present, however, this mechanism has not been studied.

It is known that neurodegenerative injuries of brain resulting in Alzheimer's disease are promoted by emotional stress [1, 11, 19]. It was shown that the neurodegenerative injures in Alzheimer's disease and stress damages have the same key pathogenetic factors including oxidative stress [3, 13], disturbed synthesis of NO [6, 7], increase in intracellular Ca^{2+} concentration [9], and altered mitochondrial function [11, 19]. Therefore we suggested that resistance to neurodegenerative damage of brain might be associated with resistance to emotional stress and the mechanism of resistance to neurodegenerative damage might resemble the mechanism of resistance to stress.

We have shown earlier that the resistance to stress damage in rats is predetermined by inherited peculiarities of regulatory stress-limiting factors which include antioxidants, dopamine, serotonin, adenosine, NO, GABA etc. Rats with higher inherited efficiency of stress-limiting factors (August rats) are characterized by higher resistance to acute stress, whereas rats with lower inherited efficiency of stress-limiting factors (Wistar rats) display lower resistance to stress injuries [10, 14, 15, 17].

*Address for Correspondence: Pshennikova M.G., PhD., DSc., Prof., Institute of General Pathology and Pathophysiology, Baltijskaya 8, Moscow 125315, Russia. Tel. +7(495)6012344; +7(495)6012248 Fax: +7(495)1511756 E-mail: v.kapelko@cardio.ru for M.G.Pshennikova

Therefore we suggested that the resistance to neurodegenerative damage of brain might be depended also on the efficiency of these protective factors.

It is known that adaptations to hypoxia, exercise and other environmental factors are good means for the enhancement of the organism's resistance to stress, ischemia and other detrimental impacts and adaptive defense is associated with increase in efficiency of stress-limiting factors [10, 16, 18]. We suggest that these adaptations can protect the neurodegenerative damage of brain.

The aim of the present work was to study the resistance to neurodegenerative damage in experimental Alzheimer's disease, developed by injection of a β-amyloid peptide fragment 25-35 (Aβ) into the brain [4], of two strains of rats, August and Wistar, which differ in their inherited resistance to emotional stress [15, 16]:

1. We compared the resistance of the brain to neurodegenerative damage, induced by Aβ with intensity of oxidative stress (detrimental factor) and with plasma level NO (protective, stress-limiting factor) in August and Wistar rats.
2. We studied the protective effects of prior adaptation to hypoxia on neurodegenerative damages of brain induced by Aâ in August and Wistar rats and the possible mechanisms of these effects.

Materials and Methods

Experiments were carried out on male August rats weighing 220-260 g, and on male Wistar rats weighing 270-330 g.

Alzheimer's disease was developed by bilateral injection of the β-amyloid peptide fragment 25-35 (Aβ) 2 μl 0.4×10^{-9} into the basal magnocellular nucleus [4]. Fig. 1 shows the location of Aβ injection into the rat brain.

Severity of brain damage was evaluated by the extent of cognitive disorders using the tests of passive avoidance conditioned reflex (PACR) and open field [4, 22]. The PACR test evaluates the retention of memory of an electric shock. For this purpose the rat was placed in an illuminated compartment of an apparatus and the latency time (L1) of entering a dark compartment was recorded for 3 minutes. Upon entry into the dark compartment the rat received a foot electric shock. The electric shock was delivered until the rat escaped the dark compartment. The test was repeated and the latency time (L2) was recorded again in 24 hours. The difference L2 – L1 = ΔL reflects the memory retention. The memory of a rat was better if greater was the difference between latency times of the pre-shock (L1) and post-shock (L2) tests, that is ΔL.

The 2-minute open field test evaluates horizontal and vertical motor activity which reflect orientation and exploratory activity of the rat [4]. For this purpose the rat was placed in the center of an open area in the box. Each test lasted for 3 min, during which (i) the degree of horizontal ambulation (the number of line-crossings) and (ii) the number of an upright positions of the rat (vertical activity) were recorded. PACR test was carried out in 14 days post-surgery (Aβ injection) and open field test was carried out in 14 days after PACR test in accordance with the method used [4].

Intensity of oxidative stress was measured by using thiobarbituric acid reactive substance (TBARS), malonaldehyde (MDA) in the hippocampus and fronto-parietal cortex participating in memory processing.

Activity of the NO system was evaluated by plasma level of nitrite and nitrate, stable NO metabolites using the well-known Griess spectrophotometric method.

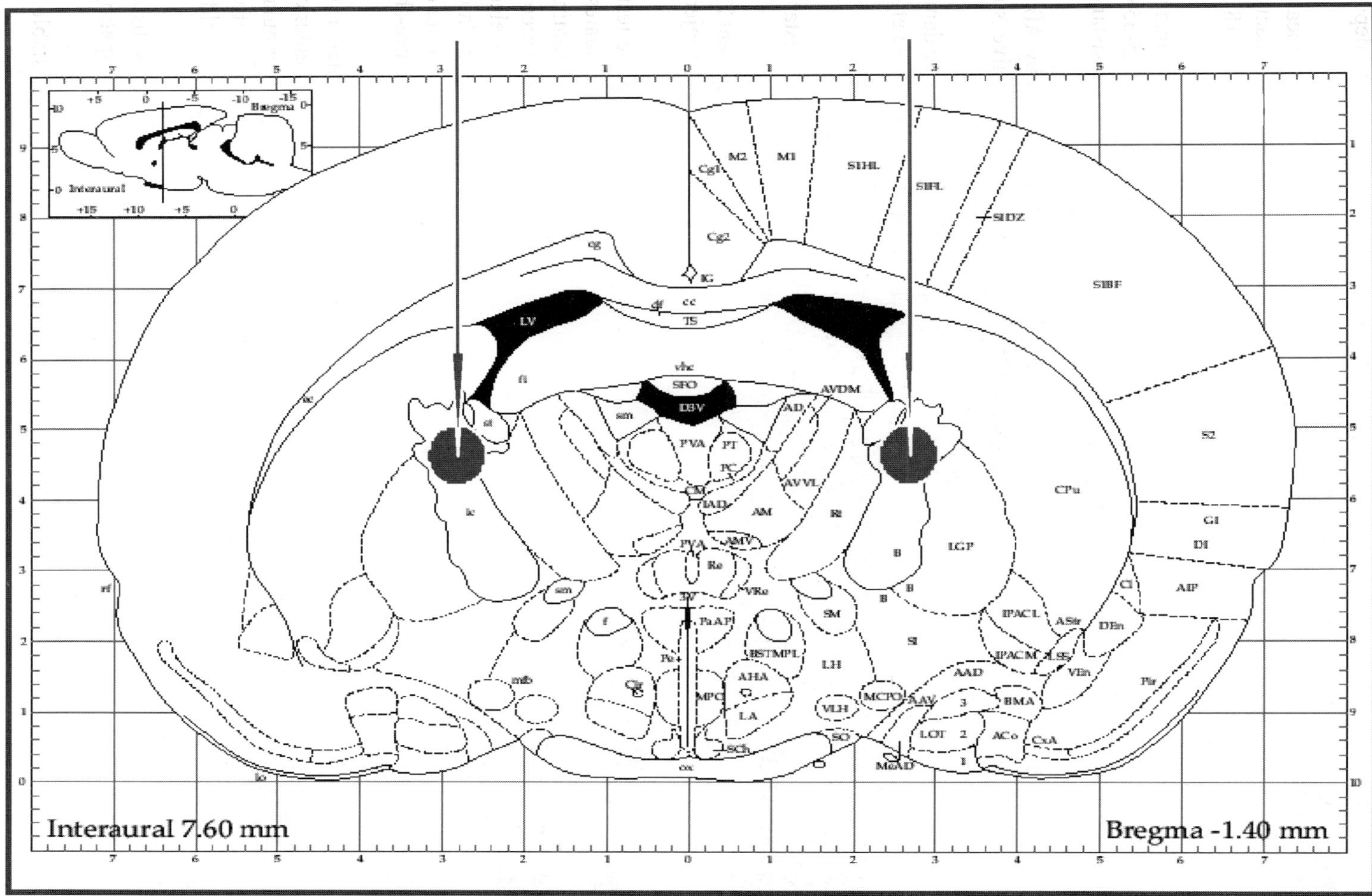

Fig. 1. **Location of β-amyloid peptide fragment 25-35 (Aβ) into the rat brain [12]. Arrows show basal magnocellular nucleus.**

Adaptation to hypoxia was produced in an altitude chamber for 14 days in the following regimen: day 1, 1 hour at a simulated altitude of 1,000 m; day 2, 1 hour at 2,000 m; day 3, 1.5 hours at 3,000 m; day 4, 2 hours at 4,000 m; day 5, 3 hours at 4,000 m; days 6 – 14, 4 hours at 4,000 m.

Eight series of experiments were conducted (4 series for each rat strain): (1) Control; (2) Aβ treatment; (3) Adaptation to hypoxia; (4) Aβ treatment following adaptation to hypoxia.

Statistical Analysis

Results on altered rat behavior, expressed as PACR test, were statistically evaluated with the non-parametric Mann-Whitney U-tests. Statistical significance for the degree of memory retention (ΔL) was evaluated using paired Wilcoxon's T-criterion: when $P_T < 0.05$, the memory retention occurs and when $P_T > 0.05$, the memory retention is absent. Changes in the rat behavior in open field test, changes in the plasma nitrite/nitrate level and in the brain MDA level were analyzed using paired Student's t-test; statistical significance was defined as $P < 0.05$.

Results

Table 1 shows the effect of Aβ treatment and prior adaptation to hypoxia on the memory retention (ΔL) in the PACR test in August and Wistar rats and on the horizontal and vertical motor activity of these rats in the open field. It was shown that the Aβ treatment decreased the memory of Wistar rats, which have "remembered" the electric shock. The number of August rats, which have remembered the electric shock, in contrast, did not lose memory. Adaptation to hypoxia alone did not affect the number of rats, which have remembered the electric shock. Prior adaptation protects from the disorders induced by Aβ in Wistar rats, while in August rats, on the contrary, the prior adaptation provoked Aβ induced disorders. None of the adapted August rats remembered the electric shock. The Aβ treatment also decreased ΔL (memory retention index) in Wistar rats by 43% as compared to the control. Prior adaptation to hypoxia protects the Wistar rats from these disorders induced by Aβ. In August rats Aβ treatment did not decrease memory retention. The adaptation alone decreased this parameter of cognitive reaction in August rats. Prior adaptation provoked Aβ induced injury and adapted August rats had no memory retention of the electric shock.

Table 1 also shows the effect of Aβ and adaptation to hypoxia on the horizontal and vertical motor activity of rats in the open field. It was shown that the Aβ treatment decreased the horizontal activity in Wistar rats, but not in August rats. Prior adaptation protected Wistar rats from these disorders. In August rats, on the contrary, adaptation alone decreased this motor activity by 40% as compared to the control. Prior adaptation provoked Aβ induced decrease in horizontal activity in these rats. The vertical motor activity in open field in August rats was higher than that in Wistar rats in all series of experiments. This result is in agreement with our previous data on comparison of behavior reactions in open field: in August rats with those in Wistar rats [15]. The Aβ treatment and adaptation alone had no effect on this parameter of cognitive reaction in August and Wistar rats.

Therefore Wistar rats, which were less resistant to stress as compared to August rats [15, 16], were also less resistant to neurodegenerative damage induced by Aβ. August rats, which were more resistant to stress as compared to Wistar rats [15, 16] were also more resistant to neurodegenerative damage induced by Aβ. Studies on the adaptive protection against the disorders induced by Aβ treatment in Wistar and August rats provided paradoxical result. Prior adaptation to hypoxia protected Wistar rats

Table 1. Effect of Aβ treatment and adaptation to hypoxia on the retention ability (in PACR test) and on the horizontal and vertical motor activity (in open field test) in August and Wistar rats

Series of experiments	Rat strains	PACR test		Open field test	
		ΔL, sec	The number of rats, which have remembered the electric shock, %	Horizontal activity (the number of line-crossings)	Vertical activity (the number of upright positions of the rat)
Control	August (28)	80.34 $P_T < 0.01$	89	32.15 ± 1.54	13.37 ± 0.82
	Wistar (45)	133.25* $P_T < 0.01$	89	47.75 ± 1.80 *	9.42 ± 0.56 *
Aβ treatment	August (27)	60.10 $P_T < 0.01$	82	33.80 ± 1.85	13.85 ± 0.83
	Wistar (48)	58.10$^+$ $P_T > 0.05$	75 $^{+}$ *	35.9 ± 2.5 $^+$	8,70 ± 0.56 *
Adaptation	August (10)	7.90$^+$ $P_T < 0.05$	88	19.90 ± 4.40 $^+$	10.28 ± 2.10
	Wistar (25)	130.29* $P_T < 0.01$	90	46.48 ± 1.69 *	8.39 ± 0.52
Aβ treatment following adaptation	August (10)	– 22.75$^{+\omega}$ $P_T > 0.05$	0$^{+\ \omega}$	27.51 ± 2.10 $^\omega$	15.28 ± 3.42
	Wistar (30)	86.97$^\omega$ $P_T < 0.05$	90 $^\omega$ *	42.88 ± 1.94 $^\omega$ *	9.38 ± 0.63 *

* P < 0.05 vs. August rats. $^+$ P < 0.05 vs. control. $^\omega$ P < 0.05 vs. series Aβ treatment. The number of rats in the series are shown in brackets. Values in open field test are means ± SE.

preventing the memory loss and behavioral disorders in the open field test, i.e., adaptation increased the resistance to injuries induced by Aβ. In August rats, in contrast, prior adaptation to hypoxia aggravated memory loss and behavioral disorders induced by Aβ, i.e., adaptation reduced the resistance to neurodegenerative injury.

These data are in agreement with our previous results which show that prior adaptation to repeated non-damaging emotional stress and prior adaptation to hypoxia increased resistance to acute stress (induced stomach lesions) in Wistar rats, while in August rats, in contrast, the adaptation considerably reduced the resistance to acute stress and area of stress induced stomach lesion was several times larger in adapted than in naive August rats. Furthermore it is important that adaptation itself induced stomach lesions in August rats but not in Wistar rats; these lesions were similar to those induced by the test acute stress exposure [14, 16, 18]. It was also shown that prior adaptation increased the resistance to heat shock in Wistar rats and decreased this resistance in August rats [8].

We studied possible mechanisms and obtained data from these experiments. For this purpose we compared the resistance to neurodegenerative damage and protective effects prior to adaptation to hypoxia with the intensity of oxidative stress (detrimental factor) and with the activity of NO system (protective, stress-limiting factor) in August and Wistar rats.

Table 2 shows the effect of Aβ treatment and adaptation to hypoxia on intensity of oxidative stress (MDA level) in hippocampus and fronto-parietal cortex of August and Wistar rats. It was shown that Aβ treatment increased the hippocampus MDA level in Wistar rats (by 14% as compared to the control). Prior adaptation protected from the MDA level increase was induced by Aβ in Wistar rats. Therefore in Wistar rats the Aβ-induced disorders of memory and motor activity were associated with

Table 2. Effects of Aβ treatment and adaptation to hypoxia on malonic dialdehyde (MDA) level, nmol/g, in fronto-parietal cortex and hippocampus in August and Wistar rats

Series of experiment	Rat strains	Brain structures	
		Fronto-parietal cortex	Hippocampus
Control	August (16)	65.90 ± 2.10	70.91 ± 2.40
	Wistar (19)	68.99 ± 2.90	61.20 ± 1.30
Aβ treatment	August (17)	64.70 ± 2.50	72.60 ± 2.60
	Wistar (19)	67.47 ± 2.05	70.18 ± 1.04 +
Adaptation	August (9)	65.30 ± 3.70	62.65 ± 3.35
	Wistar (9)	67.46 ± 5.20	58.64 ± 3.56
Aβ treatment following adaptation	August (10)	67.20 ± 4.20	67.90 ± 3.00
	Wistar (20)	66.82 ± 1.57	61.34 ± 1.16 *

+ $P < 0.05$ vs. control. * $P < 0.05$ vs.series Aβ treatment. The number of rats in the series are shown in brackets. Values are means ± SE.

increased oxidative stress in hippocampus. The adaptive protection against Aβ-induced oxidative stress in hippocampus is associated with activation of the antioxidant systems during the adaptation course [10, 14]. In August rats the Aβ treatment did not increase the MDA level in hippocampus. It agrees with the data which show that in August rats the antioxidant systems are more effective and stable than those in Wistar rats [2]. The Aβ treatment and the adaptation did not change the MDA level in cortex in Wistar and August rats.

Table 3 shows the effect of Aβ and adaptation to hypoxia on NO production (plasma nitrite/nitrate level) in Wistar and August rats. It was shown that in Wistar rats Aβ treatment decreased plasma NO level by 45% as compared to the control. Prior adaptation has protected from this decrease. In August rats Aβ treatment did not change plasma NO level. But prior adaptation provoked decrease in this parameter induced by Aβ in these rats. Therefore in Wistar rats, adaptive protection against the Aβ treatment-induced cognitive disorders is associated with the increased plasma level of NO, i.e., with

Table 3. Effect of Aβ treatment and adaptation to hypoxia on NO production (plasma NOx) in August and Wistar rats

Series of experiments	Rat strains	Concentration of nitrite/nitrate in plasma, μM
Control	August	38.8 ± 4.80
	Wistar	39.9 ± 2.79
Aβ treatment	August	42.65 ± 5.52
	Wistar	21.81 ± 2.85 + *
Adaptation	August	50.35 ± 10.80
	Wistar	38.33 ± 5.88
Aβ treatment following adaptation	August	18.30 ± 4.00 + ω
	Wistar	61.08 ± 4.17 + * ω

+ $P < 0.01$ vs. control. * $P < 0.01$ vs. August rats. ω $P < 0.01$ vs. series Aβ treatment. Values are means ± SE.

the activation of NO stress-limiting system, while in August rats, in contrast, prior adaptation to hypoxia provokes cognitive disorders under the action of Aβ. These disorders are associated with decreased plasma level of NO, i.e., with attenuated efficiency of the NO stress-limiting system. Experiments with NO synthesis inhibition and NO donors confirm these conclusions.

Table 4 shows effect of NO synthase inhibitor L-NNA and NO donor dinitrosyl iron complex [21] on the plasma level of NO and cognitive reaction in PACR test and the activity in the open field test in Wistar rats. It was shown that L-NNA treatment (20 mg/kg every other day for 14 days) decreased the plasma level of NO (NO production) by 39% as compared to the control. Simultaneously this L-NNA treatment decreased the memory retention (ΔL in PACR test) and the horizontal and vertical activities in the open field. So the memory retention was decreased by 14% as compared to the control. The horizontal activity was decreased by 31% and vertical activity by 48% as compared to the control.

Table 4. Effect of NO synthase inhibitor N^{ω}-nitro-L-arginine (L-NNA) and NO donor dinitrosyl iron complex (DNIC) on plasma level of NO and on cognitive reactions in PACR test and open field test in Wistar rats.

Indices		Series of experiments		
		Control	L-NNA	DNIC
Open field test	Horizontal activity	62.17 ± 7.35	43.30 ± 3.56 $^{++}$	61.60 ± 3.30
	Vertical activity	9.17 ± 0.90	4.74 ± 0.50 $^{++}$	14.25 ± 1.10 $^{++}$
PACR test	ΔL, sec	138.10 ± 3.5	120.00 ± 6.8 $^{+}$	150.30 ± 4.36
Concentration of nitrite/nitrate in plasma, ìM		52.44 ± 3.90	32.33 ± 9.00 $^{+}$	82.25 ± 16.40 $^{+}$

$^{+}$ $P < 0.05$ vs. control. $^{++}$ $P < 0.01$ vs. control. Values are means ± SE.

DNIC (200 μg daily for 14 days), in contrast, increases plasma NO level (by 54% as compared to the control) that testifies the increase in NO production. Simultaneously DNIC treatment increased the vertical activity (by 55% as compared to the control) and memory retention (ΔL in PACR test) by 8.2% as compared to the control. Therefore NO donor increased the motor activity to a great extent in rats in open field which reflects orientation and exploratory activity. In summary, these data confirm our conclusions that in Wistar rats adaptive protection against the Aβ-induced cognitive disorders is associated with the increased NO production, i.e., with the activation of the NO stress-limiting system. In preliminary adapted August rats cognitive disorders under the action of Aβ are associated with decreased NO production, i.e., with attenuated activity of the NO stress-limiting system in adaptation process. Therefore these data showed that prior adaptation to hypoxia resulted in efficiency of NO system and activation of this system response to Aβ treatment in Wistar rats and, vice versa, decreased this system responsiveness in August rats.

Conclusion

On the whole, based on the results obtained from this study together with other data, the effect of adaptation to hypoxia in neurodegenerative damage induced by Aβ can be presented in the following hypothetical diagrams (Fig. 2, 3).

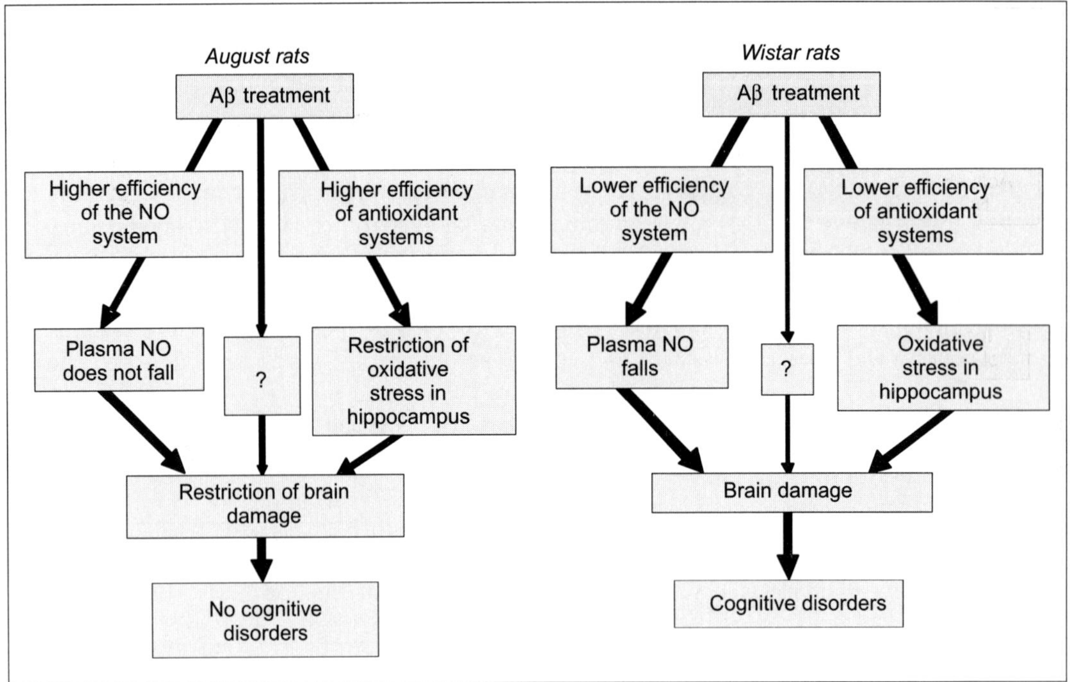

Fig. 2. Effect of Aβ treatment in naive (non-adapted) August and Wistar rats. The role of efficiency of the NO and antioxidant stress-limiting factors. Question mark means other stress-limiting factors.

Fig. 2 shows possible mechanism of the Aβ treatment effect in non-adapted (naive) August and Wistar rats. In non-adapted August rats the efficiency of NO and antioxidant stress-limiting systems is higher. As a result, Aβ treatment did not decrease NO production and did not provoke oxidative stress in hippocampus. Therefore, we see the restriction of brain damage and no cognitive disorders.

In non-adapted Wistar rats the efficiency of NO and antioxidant stress-limiting systems is lower. As a result, Aβ treatment decreased NO production and provoked oxidative stress in hippocampus. Therefore, we see brain damage and cognitive disorders in non-adapted Wistar rats. Question mark (Fig. 2 and 3) means other stress-limiting systems which might participate in the resistance to neurodegenerative damage of brain. Our previous studies showed that the dopaminergic, serotonin, opioidergic and other regulatory stress-limiting systems restrict the stress response and stress damage. It was shown that in August rats inherited efficiency of these systems is higher than that in Wistar rats [10, 14, 16]. Therefore we suggest that the resistance to neurodegenerative injures of brain might be dependent also on the efficiency of these protective systems.

Fig. 3 shows possible mechanism of adaptation effects on neurodegenerative injures induced by Aβ in August and Wistar rats. In Wistar rats the prior adaptation to hypoxia activated stress-limiting NO and antioxidant systems. As a result, we see an increase in plasma NO level and restriction of oxidative stress in hippocampus. There are restrictions of brain injury and cognitive disorders induced by Aβ treatment.

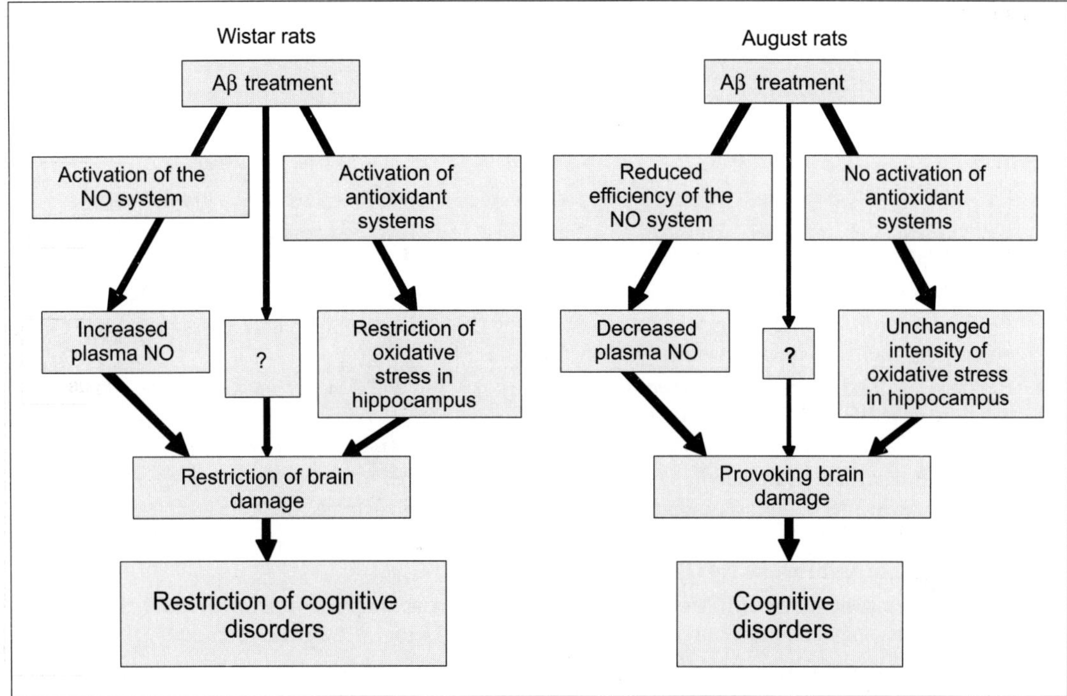

Fig. 3. Effect of Aβ treatment in preliminary adapted August and Wistar rats. The role of efficiency of the NO and antioxidant stress-limiting factors. Question mark means other stress-limiting factors.

In August rats the prior adaptation to hypoxia decreases the efficiency of the NO system and does not activate antioxidant stress-limiting systems. As a result, it was shown the decreased NO production and unchanged intensity of oxidative stress in hippocampus, brain damage and cognitive disorders. When prior adaptation reduces the efficiency of stress-limiting systems, the resistance of animals to neurodegenerative disorders induced by Aβ treatment also is reduced (as in August rats). When prior adaptation activates the efficiency of stress-limiting systems, the resistance to the neurodegenerative injures also is increased (as in Wistar rats).

Mechanism of resistance to neurodegenerative disorders and its changes in adaptation to hypoxia and other environmental factors require further investigations. However, the above notions should be undoubtedly considered in developing methods of adaptive prevention and therapy for these disorders since the notions indicate the necessity of evaluating the state of stress-limiting systems in patients for defining the advisability and regimens of adaptation.

Based on our studies, we conclude that

1) Resistance to neurodegenerative injury of brain is predetermined to a great extent by inherited resistance to emotional stress and efficiency of stress-limiting systems and 2) adaptation to hypoxia can be an efficient means for prevention and, possibly, correction of neurodegenerative processes. However, one should consider inherited peculiarities of a specific organism and its stress-limiting systems.

Acknowledgements

The study was supported by NOW (grant 047.011.2001.010) and the Russian Foundation for Basic Research (grant 99-04-48823).

References

1. Ball LJ, Birge SJ. Prevention of brain aging and dementia. Clin Geriatr Med. 2002; 18: 485-503.
2. Belkina LM, Lakomkin VL, Zhukova AG, Sazontova TG. Heart resistance to oxidative stress in rats of different genetic strains. Bull Eksp Biol Med. 2004; 138: 250-253.
3. Butterfield DA, Howard B, Yatin S, Koppal T., Drake J, Hensley K, Aksenov M, Aksenova M, Subramaniam R, Varadarajan S, Harris-White ME, Pedigo Jr, Carmey JM. Elevating oxidative stress on models of human brain aging and Alzheimer's disease. Life Sci. 1999; 65: 1883-1892.
4. Harkany T, O'Mahony S, Kelly JP, Soos K, Toro I, Penke B, Luiten PGM, Nyakas C, Gulya K, Leonard BE. â-Amyloid (Phe(SO_3H)^{24}) 25-35 in rat nucleus basalis induces behavioral dysfunctions, impairs learning and memory and disrupts cortical cholinergic innervation. Behav Brain Res. 1998; 90: 133-145.
5. Kawas CH, Katzman R. Epidemiology of dementia and Alzheimer's disease. In: Terry RD, Katzman R, Bick KL, eds., Alzheimer's Disease, Lippincot Williams Wilkins, Philadelphia, 1999; 95-116.
6. Law A, Gauther S, Quirion R. Say NO to Alzheimer's disease: the putative link between nitric oxide and dementia of the Alzheimer's type? Brain Res Rev. 2001; 53: 73-96.
7. Malyshev IYu, Manukhina EB.Stress, adaptation and nitric oxide. In: Pandolf BB, Takeda N, Singal PK, eds., Adaptation Biology and Medicine, Narosa Publishers, New Delhi, 1999: Vol. 2, 375-392.
8. Malysheva EV, Zamotrinskii AV, Malyshev IYu. The role of heat shock proteins in formation of stress resistance in different strains of animals. Bull Eksp Biol Med. 1994; 118: 11-13.
9. Mark RJ, Hensley DA, Butterfield DA, Mattson MP. Amyloid beta-peptide impairs ion-motive ATPase activities: evidence for a role in loss of neuronal Ca^{2+} homeostasis and cell death. J Neurosci. 1995; 15: 6239-6249.
10. Meerson FZ, Pshennikova MG, Malyshev IYu. Adaptive defense of the organism: architectural trace and cross protective effects of adaptation. Ann NY Acad Sci. 1996; 793: 371-385.
11. Nappi G, Genazzani AR, Martignoni E, eds., Stress and the aging brain. Integrative mechanisms. Raven Press, NY, 1990.
12. Paxinos G, Watson C. The rat brain in stereotaxic coordinates. Acad. Press. Inc., San-Diego, USA, 1998.
13. Pratiko D. Alzheimer's disease and oxygen radicals: new sights. Biochem Pharmacol. 2002; 63: 563-567.
14. Pshennikova MG. Hereditary efficiency of stress-limiting systems as a factor of the resistance to stress induced disorders. Uspekhi Fiziol Nauk. 2003; 34: 55-67 (in Russian).
15. Pshennikova MG, Bondarenko NA, Shimkovich MV, Bondarenko ON, Malyshev IYu. Differences in the behavior and resistance to stomach ulcer during stress in August and Wistar rats adapted and non-adapted to hypoxia. Bull Eksp Biol Med. 1999; 128: 638-641.
16. Pshennikova MG, Malyshev IYu, Manukhina EB, Meerson FZ. Distinction between stress resistance and protective effects of adaptation in rats of different genetic strains: role of regulatory systems. In: Hargens A, Takeda N, Singal PK, eds., Adaptation Biology and Medicine, Narosa Publishing House, New Delhi, 2005: Vol. 4, 29-40.
17. Pshennikova MG, Popkova EV, Bondarenko NA, Malyshev IYu, Shimkovich MV, Smirin BV, Manukhina EB. Catecholamines, nitric oxide and resistance to stress lesions: effect of adaptation to hypoxia. Ross Fiziol Zh im IM Sechenov. 2002; 88: 485-495.
18. Pshennikova MG, Popkova EV, Shimkovich MV. Adaptation to stress improves resistance to gastric damage during acute stress in Wistar rats and decreases resistance in August rats: role of serotonin. Bull Eksp Biol Med. 2002; 134: 383-386.

19. Riga S, Riga D. An antistress and antiaging neurometabolic therapy. Ann NY Acad Sci. 1995; 771: 535-550.

20. Sadik K, Wilcock G. The increasing burden of Alzheimer's disease. Alzheimer Dis Assoc Disord. 2003; 17: S75-S79.

21. Vanin AF. Dinitrosil iron complex and S-nitrosothiols are two possible forms for stabilization and transport of nitric oxide in biological systems. Biochemistry (M0scow). 1998; 63: 924-938.

22. Zee van der, EA, Douma BR, Bohus B, Luiten PGM. Passive avoidance training induces enhanced levels of immunoreactivity for muscarinic acetylcholine receptor and co-expressed PKC gamma and MAP-2 in rat cortical neurons. Cereb Cortex. 1994; 4: 376-390.

Adaptation Biology and Medicine (Volume 5: Health Potentials)
Editors: L. Lukyanova, N. Takeda and P.K. Singal

13. The Processes of Adaptation of Microglia in Brain Trauma and Metastasis

M. Noda[1*], T. Seike[1], K. Fujita[1], M. Kido[2], T. Tanaka[2] and H. Iguchi[3]

[1]Laboratory of Pathophysiology, Graduate School of Pharmaceutical Sciences
[2]Laboratory of Anatomy and Cell Biology, Graduate School of Dental Sciences, Kyushu University, Fukuoka, Japan
[3]Shikoku Cancer Center Research Institute, Ehime, Japan

Abstract

Following any type of brain injury such as lesion, stroke, and tumor/cancer invasion, microglia are rapidly activated and recruited to the site of injury. Microglia is the main immune effector cell population of the central nervous system and control immune cell recruitment. In the case of brain metastasis, the first event was the invasion of cancer cells with a breakdown of local blood brain barrier, followed by a progression of cancer cells in the brain. However, the molecular mechanism of brain metastasis and the role of glial cells are poorly understood. Therefore, we established an animal model for brain metastasis using human lung cancer-derived cells (HARA-B) in nude mice. Brain metastases were histologically proven 4-5 weeks after intracardiac inoculation of HARA-B cells in mice. Accumulation of activated microglia was noted around the metastatic lesions in the early stage of brain metastasis. However, its number rather decreased along with the progression of brain metastasis. On the other hand, the number of activated astrocytes increased dramatically around brain metastasis with tumor development. In vitro experiments showed that co-culture with microglia or conditioned medium of microglia inhibited the growth of HARA-B cells, while co-culture with astrocytes enhanced the growth of HARA-B cells. Though the responsible factors released from microglia and time-dependent changes in glial cells are still under investigation, these studies will contribute to understand the mechanism of glial cell-adaptation and may help finding a new tool to prevent brain metastasis.

Keywords: Microglia, metastasis, cytokines, astrocytes, and macrophage

Introduction

The probability of metastasis is dependent on the tumor cell phenotype including the invasive, adhesive and growth ability, and interaction between the tumor cell and the organ environment, such as the cellular components, cytokines, organ-derived growth factors (Nicolson, 1993). Brain metastases from lung cancer occur with higher frequency as well as from breast cancer. In pathological studies the occurrence of brain metastases is between 30 and 50% of total case of non-small lung cancer (NSCLC) (Galluzzi & Payne, 1956; Magilligan *et al.*, 1986; Knights, 1954). The reason for high frequency of

*Address for Correspondence: Mami Noda, Ph.D., Laboratory of Pathophysiology, Graduate School of Pharmaceutical Sciences, Kyushu University, 3-1-1 Maidashi, Higashi-ku, Fukuoka 812-8582, Japan. Tel/Fax: +81-92-642-6574, E-mail: noda@phar.kyushi-u.ac.jp

brain metastases may be that the brain microenvironment plays an important role in the progression of lung or breast cancer cells in the brain.

In the brain, microglial cells are considered as the pathologic response element (Kim and de Vellis, 2005; Kreutzberg, 1996; Perry *et al.*, 1993). Microglial cells are dispersed throughout the entire central nervous system (CNS) and exhibit a ramified morphology under normal conditions. As recently shown by Davalos *et al.*, (2005) and Nimmerjahn *et al.*, (2005), microglial processes are highly dynamic in the intact brain, suggesting that microglial cells scan the brain parenchyma with their processes and potentially shield it from injury. Under pathologic conditions such as a lesion, stroke, neurodegenerative disorders or tumor invasion, activated microglia migrate rapidly to the affected sites of the CNS. At the same time, microglial activation is accompanied by the release of immunocompetent molecules such as cytokines or chemokines, and other substances such as growth factors (Hanisch, 2002). On the other hand, astrocytes are the most abundant cell population and may play an important role in maintaining homeostasis by performing variety of functions (Miller *et al.*, 1989). Astrocytes can also produce cytokines or growth factors in response to extracellular stimulation, such as inflammatory reaction (Aloisi *et al.*, 1992; Hertz *et al.*, 1990; Lee *et al.*, 1993). Excess production of inflammatory agents such as TNF-α, IL-1β and IL-6 result in over-immune response and ultimately cause gliosis (Schobitz *et al.*, 1994).

Activated microglia and astrocytes contribute to a part of the innate immune response in the brain and produce a large variety of different mediators of inflammatory reaction (Aloisi *et al.*, 2000). Therefore, glial cells around the metastatic focus may induce inflammatory response to tumor and provide pro-inflammatory cytokines such as IL-1β, TNF-α to tumor cells, followed by promoting the progression of brain metastasis. To test this hypothesis, we investigated how glial cells around the metastatic focus behave in brain metastatic model. These results may help understanding the mechanism of brain metastases and may contribute to a new therapeutic strategy to prevent or attenuate brain metastases.

Materials and Methods

Animal Model for Brain Metastasis

Male 5-week-old BALB/c *nu*/*nu* mice (Kyudo, Kumamoto, Japan), kept in specific pathogen-free environment, were used. A single suspension of HARA-B cells (2×10^5 cells/ 0.2 ml PBS) was inoculated subcutaneously into the left ventricle of the nude mice. Then, mice were subjected for *in vitro* experiments 4 to 5 weeks after cell inoculation.

Immunohistochemistry

Nude mice were perfused transcardialy with 50 ml of 10 U/ml and 0.5% procaine in PBS and 4.0% paraformaldehyde in PBS. Then, the brain was removed and post-fixed for 3 hour, cryoprotected for 24 hour in PBS containing 20% sucrose. The brain was sliced into 20 μm in thickness by using a cryostat and the sections were placed on a slide glass pre-coated with gelatin, permeabilized with 0.3% TritonX-100 in PBS for 15 min and blocked in PBS containing 1% BSA and 5% normal donkey serum (Jackson Immuno Research) for 1 hour at room temperature. Then, brain sections were incubated over night at 4°C with primary antibodies, including FITC-conjugated anti-human cytokeratin antibody (CAM5.2) (Becton Dickinson, undiluted solution) and or rabbit anti-Iba1 antibody (Wako Chemicals,

1:500) or Cy3-conjugated anti-GFAP antibody (Sigma, 1:250). The secondary antibody for Iba-1 was Cy3-conjugated donkey anti-rabbit IgG (1:250). Every treatment was followed by washing three times for 5 min with PBS containing 0.3% TritonX-100. Sections were mounted in the Perma Fluor Aqueous Mounting Medium (Thermo, Shandon, Pittsburgh, PA) and were analyzed with a confocal microscope (LSM510 META, Carl Zeiss, Co. Ltd. Germany).

Cell Culture

The study was approved by the Animal Research Committee of Kyushu University and Max-Delbrück Center for Molecular Medicine, and carried out in accordance with the National Institutes of Health Guide for the Care and Use of Laboratory Animals.

Primary glial cell cultures were obtained from cerebral cortex of 1-day-old C57BL/6 mice (Kyudo, Kumamoto, Japan). The cortex was isolated under a dissecting microscope and carefully separated from the choroid plexus and meninges. The cerebral cortex was minced and treated with trypsin-EDTA solution (0.25% trypsin, 1 mM EDTA and 1500 U DNase in HBSS) at 37°C for 10 min. Cell suspensions were filtered through 70 μm pore size mesh (BD Falcon, USA) into Dulbecco's modified Eagle medium (DMEM; Nissui, Tokyo, Japan) containing 10% fetal calf serum (FCS; Hyclone, UT, USA), 2 mM L-glutamine, 100 U/ml penicillin, 100 μg/ml streptomycin, 0.37% $NaHCO_3$, 110 μg/ml pyruvic acid. After centrifugation, cells were filtered through 40 μm pore size mesh (BD Falcon, Bedford, MA, USA) and plated into 75 cm^2 cell culture flask at density of 10^7 cell /75 cm^2 and maintained at 37°C in 10% CO_2-90% air with change of medium twice per week.

Microglial cells were first obtained after 10-14 days of mixed glial cell cultures, and then every 1 week, from the surface of the mixed primary cultured cell layer by gently shaking the flasks for 20 min. The resulting cell suspension was seeded onto glass coverslips and allowed to adhere for 30 min at 37°C. Then, microglial cells were isolated as strongly adhering cells after unattached cells were removed. The purity of microglia as measured with the isolectin technique, ranged from 95-99%.

Astrocytes were obtained after the last isolation of microglia at 28 days of mixed glial cell cultures. Astroglial layer was removed from the flasks by treating with trypsin-EDTA solution (0.06% trypsin, 0.25 mM EDTA in serum free DMEM) at 37°C for 45 min. Suspended astrocytes were filtered through 40 μm pore size mesh and seeded. Purify of astrocyte was ranged from 90 to 95% as determined by immunostainig with anti-glial fibrillary acidic protein (GFAP) antibody (Sigma, St. Louis, MO, USA). Astrocytes were maintained in the medium mentioned above at 37°C in 10% CO_2-90% air.

HARA-B cells were maintained in the medium mentioned above at 37°C in 10% CO_2-90% air. Cells were grown as a monolayer culture in 25 cm^2 cell culture flask (Nalge Nunc International), and single-cell suspension of cells was obtained by trypsin treatment.

Cell Proliferation Assay

In co-culture experiment, HARA-B cells (0.5×10^3 cells/well) and astrocytes or microglia (2.5×10^3 or 5×10^3 cells/well) were seeded into 8-well cell culture slide (BD Falcon) with DMEM for 24 hour. Then, cells were rinsed twice with PBS and incubated in serum free DMEM. After 72 hour of co-culture, cells were fixed with 4% paraformaldehyde for 30 min at room temperature and permeabilized with 0.3% TritonX-100 in PBS for 15 min and blocked in PBS containing 1% BSA and 5% normal donkey serum (Jackson Immuno Research) for 1 hour at room temperature. Cells were incubated with primary antibody of monoclonal mouse anti-human cytokeratin antibody (AE1/AE3 pool of cytokeratin) (Dako, 1:100) over night at 4°C, followed by incubated for 5 hour at room temperature with secondary antibody of FITC-conjugated anti-mouse IgG (Sigma, 1:500) and then for 30 min at room temperature with 300 nM 4', 6'-diamidino-2-phenylindole hydrochloride (DAPI, Sigma). Numbers of all HARA-B cells in a well were counted by using a digital camera system (Axio Cam, Carl Zeiss) mounted on a light and fluorescent microscope (Axopscope2 plus, Carl Zeiss). The results were indicated as the percentage of control (single cell culture of HARA-B cells). The glial cell-conditioned medium was obtained by culturing mice primary cultured microglia or astrocytes at a density of 10^4 cells/well for 72 hours. The medium was removed, centrifuged at 2000 rpm for 10 min and then added to the well of HARA-B cell culture at the volume of 25% or 50%. The number of HARA-B cells were counted as mentioned above.

Statistical Analysis

One-way analysis of variance (ANOVA) and post-hoc Bonferroni/Dunn test were used to examine the statistical differences. Differences were considered significant at $p<0.05$.

Results

Glial Cells Around Metatatic Region

To investigate the interaction between cancer cells and glial cells, we initially established an experimental model for brain metastasis using HARA-B cells. Cancer cells (HARA-B cells) metastasized to whole region of the brain, especially cortex, 4-5 weeks after intracardiac inoculation. In brain slices obtained from normal and tumor-bearing mice, cancer cells and microglia/macrophage or astrocytes were stained with anti-cytokeratin antibody and anti-Iba1 or anti-GFAP antibody, respectively. In our study, the progression of metastases was defined in terms of the size of metastatic focus, *i.e.* its size less than 50 μm was considered as an early stage and that more than 100 μm was as an advanced stage in the metastatic tumor development. In cortex of normal mice, microglial cells were distributred evenly, while in cortex of metastatic mice microglial cells showed activated morphology showing large cell bodies and thick processes. Around small metastatic foci (an early stage), a large number of activated microglial cells were found, while the number of microglial cells was rather decreased around large metastatic foci (an advanced stage) (Fig. 1A). On the other hand, GFAP-positive astrocytes were few in normal mouse brain but accumulated increasingly around metastatic foci along with metastaitic tumor development, inducing gliosis as metastases were progressed (Fig. 1B).

Effects of Glial Cells on Proliferation of Cancer Cells

To investigate the effects of glial cells on the proliferation of cancer cells, we examined the interaction between HARA-B cells and glial cells *in vitro*. We cultured HARA-B cells, HARA-B cells with microglia, or HARA-B cells with astrocytes for 3 days and compared the proliferation of HARA-B cells. In the condition of co-culture with microglia with the ratio of 1:5 and 1:10 (cancer cells : microglia), the relative number of HARA-B cells was attenuated to 62.8±3.8% ($p<0.05$, n=4) and 41.7±2.3 % ($p<0.01$, n=4), respectively, compared to the one without microglia (control; 100%) (Fig. 2A). On the other hand, in the condition of co-culture with astrocytes with ratio of 1:5 and 1:10 (cancer

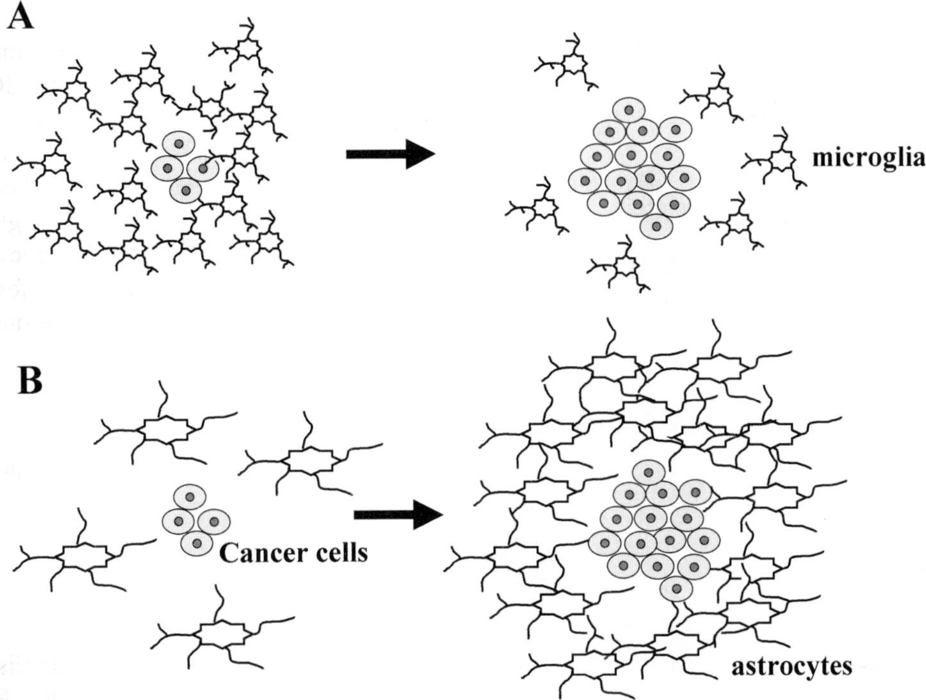

Fig. 1. Schematic relationship between glial cells and metastatic cancer cells in an animal model for brain metastasis. (A) Activated microglial cells gathered around small metatatic cancer cells (an early stage) but less number of microglial cells was observed around large metastatic cancer cells (an advanced stage). (B) More astrocytes gathered around metastatic cancer cells according to the size of metastatic cancer cells.

cells : astrocytes) in the same way, the relative number of HARA-B cells increased to 285±9.5% ($p<0.01$, n=6) and 441±11.2% ($p<0.01$, n=6), respectively, compared to the one without astrocytes (control; 100%) (Fig. 2B).

To elucidate the mechanism for such effects of glial cells on the proliferation of HARA-B cells, we further examined the effects of conditioned medium obtained from glial cell cultures on the proliferation of HARA-B cells. In the presence of 25% and 50% of microglial conditioned medium (MGCM) in the total volume of HARA-B cell culture-medium, the proliferation of HARA-B cells was attenuated to 65.9±6.1% ($p<0.05$, n=5) and 42.6±2.1% ($p<0.01$, n=5), respectively, compared to the one in the absence of MGCM (control; 100%) (Fig. 2C). These values were similar to those observed in the co-culture experiments. Astroglial conditioned medium (AGCM) also enhanced the proliferation of HARA-B cells, however, its effect was much smaller than that observed in the co-culture experiments (Fig. 2D).

Discussion

In the animal model of brain metastasis using human lung cancer-derived cells (HARA-B cells), large number of activated microglia, and also macrophages presumably invaded together with cancer cells,

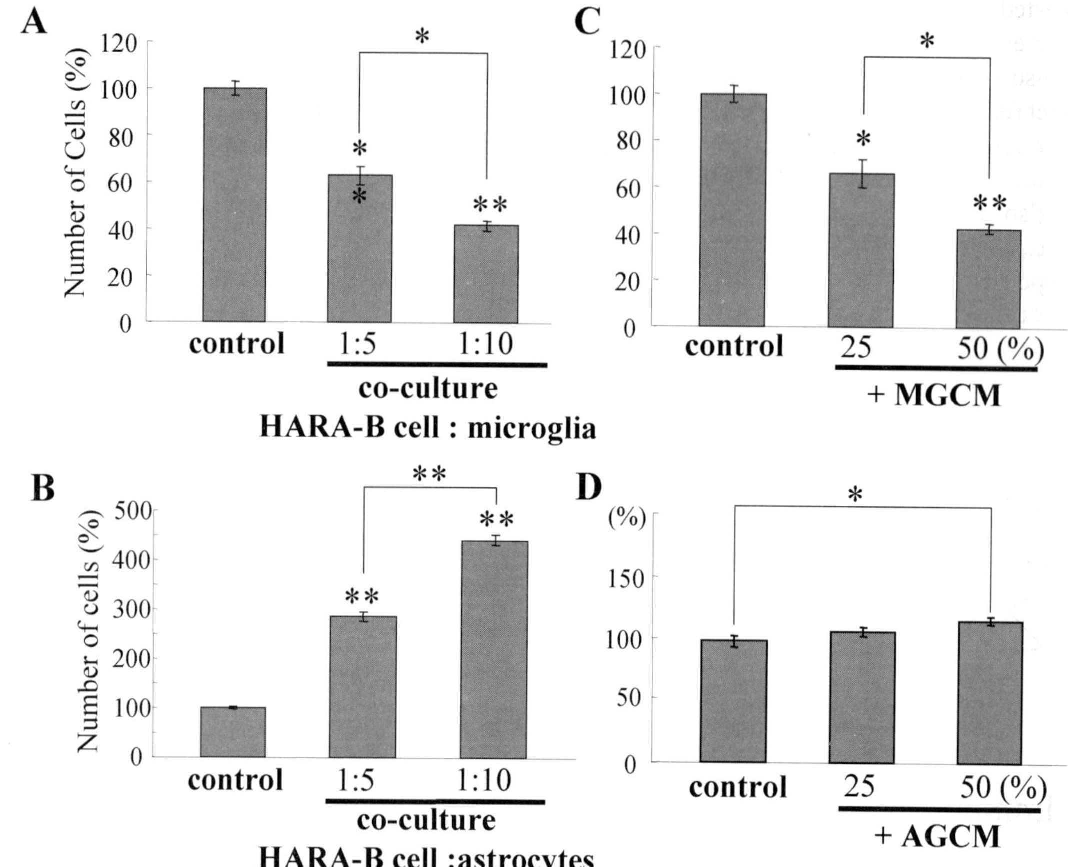

Fig. 2. Effects of co-culture or conditioned medium of glial cells on the proliferation of cancer cells. (A) Co-culture of cancer cells (HARA-B cells) and mouse primary cultured microglia inhibited the proliferation of cancer cells depending on the number of microglial cells. (B) Co-culture of HARA-B cells and mouse primary cultured astrocytes enhanced the proliferation of HARA-B cells depending on the number astrocytes. (C) Microglial conditioning medium (MGCM) also inhibited the proliferation of HARA-B cells in a volume-dependent manner. (D) Astroglial conditioning medium (AGCM) had little or much smaller effect on the proliferation of HARA-B cells. * $p < 0.05$, ** $p < 0.01$.

gathered to small metastatic regions (an early stage) but not around large metastatic regions (an advanced stage). Instead, large number of GFAP-positive astrocytes, namely activated astrocytes, was observed around large metastatic regions (an advanced stage). In the present study, microglial cells attenuated the proliferation of HARA-B cells when they were co-cultured. And the inhibitory effect on the proliferation of HARA-B cells was reproduced by microglial conditioned medium. These results suggest that microglial inhibitory effect on the proliferation of HARA-B cells could be conducted through soluble factors released from microglial cells, although such factors are not yet identified. On the other hand, astrocytes enhanced the proliferation of HARA-B cells when they were co-cultured. Astroglial conditioned medium (AGCM) also showed an enhancement of proliferation of HARA-B cells, however, this effect was much smaller than that observed in the co-culture. Therefore it is suggested that co-existence of astrocytes and HARA-B cells might produce an extra factor which further enhanced the proliferation of HARA-B cells. Such enhanced proliferation of HARA-B cells was also observed when astrocytes were cultured in an insert-culture well to avoid direct cell-cell contact (data not shown). From these findings, astrocytes seem to be activated by the interaction with cancer cells, and then presumably release soluble factors such as cytokines and growth factors. In addition to the development of metastatic cancer cells in the brain, astrocyte-derived soluble factors may also inhibit the activation of microglia. Astrocyte-induced cytokines, such as IL-1, IL-6, or TNF-α, could be good candidates for attenuation of microglial activation or proliferation of microglia in response to cancer cell invasion, resulting in an adaptation of microglia and immune evasion of cancer cells (Fig. 3).

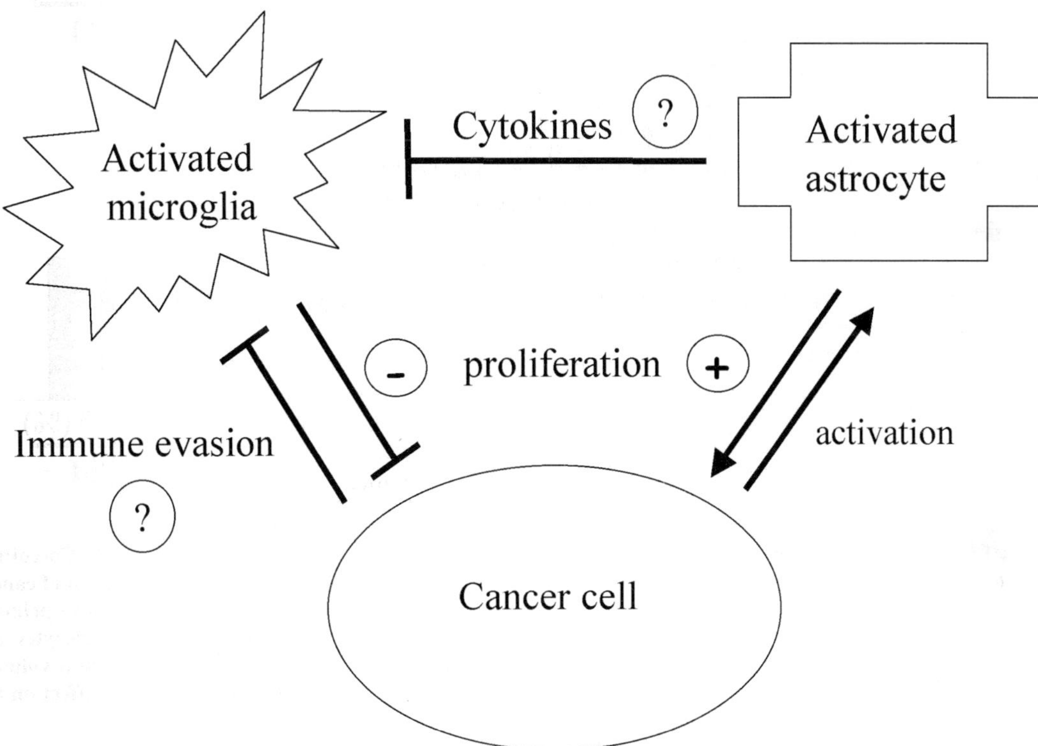

Fig. 3. Predicted mechanism of microglial adaptation in the progression of metastatic cancer cells. According to the invasion of cancer cells, microglial cells or macrophages are activated and migrate or proliferate at the site of metastasis and try to inhibit the proliferation of cancer cells. Astrocytes are also activated and release cytokines which enhance the proliferation of cancer cells. Astrocyte-induced cytokines and/or cancer cell-derived immune evasion may attenuate microglial activation, resulting in the apparent adaptation of microglia during the progression of cancer cells.

In conclusion, glial cells (microglia and astrocytes) are associated with tumor development of brain metastasis through the interaction between these two glial cells and tumor cells, in the microenvironment of brain metastasis. Especially microglia plays an important role in inflammatory diseases and traumatic injury in the CNS, a better understanding of microglial function is essential for the development of immune-based treatment strategies against trauma and metastases in the CNS, as well as in brain tumor (Watters *et al.*, 2005). The present studies may contribute to understand the mechanism of microglial adaptation and may help developing a new strategy for the treatment of brain trauma and metastasis.

References

1. Aloisi F, Care A, Borsellino G, Gallo P, Rosa S, Bassani A, Cabibbo A, Testa U, Levi G, Peschle C. Production of hemolymphopoietic cytokines (IL-6, IL-8, colony-stimulating factors) by normal human astrocytes in response to IL-1 beta and tumor necrosis factor-alpha. J immunol 1992; 149: 2358-2366.
2. Aloisi F, Ria F, Adorini L. Regulation of T-cell responses by CNS antigen-presenting cells: different roles for microglia and astrocytes. Immunol Today 2000; 21: 141-147.
3. Davalos D, Grutzendler J, Yang G, Kim JV, Zuo Y, Jung S, Littman DR, Dustin ML, Gan WB. ATP mediates rapid microglial response to local brain injury in vivo. Nat. Neurosci. 2005; 8: 752-758.
4. Galluzzi S, Payne PM. Brain metastases from primary bronchial carcinoma: a statistical study of 741 necropsies. Biol J Cancer 1956; 10: 408.
5. Hanisch U. K. Microglia as a source and target of cytokines. Glia 2002; 40: 140-155.
6. Hertz L, McFarlin DE, Waksman BH. Astrocytes: auxiliary cells for immune responses in the central nervous system? Immunol Today 1990; 11: 265-268.
7. Kim S. U. and de Vellis J. Microglia in health and disease. J. Neurosci. Res. 2005; 81: 302-313.
8. Knights EM. Metastatic tumors of the brain and their relation to primary and secondary pulmonary cancer. Cncer 1954; 7: 259.
9. Kreutzberg GW. Microglia: a sensor for pathological events in the CNS. Trends. Neurosci. 1996; 19: 312-318.
10. Lee SC, Liu W, Dickson DW, Brosnan CF, Berman JW. Cytokine production by human fetal microglia and astrocytes. Differential induction by lipopolysaccharide and IL-1 beta. J immunol 1993; 150: 2659-2667.
11. Magilligan DJ, Duvernoy C, Malik G, Lewis JW, Knighton R, Ausman JI. Surgical approach to lung cancer with solitary cerebral metastasis: twenty-five years' experience. Ann Thorac Surg 1986; 42: 360-364.
12. Miller RH, Ffrench-Constant C, Raff MC. The macroglial cells of the rat optic nerve. Annu Rev Neurosci 1989; 12: 517-534.
13. Nicolson GL. Cancer progression and growth: relationship of paracrine and autocrine growth mechanisms to organ preference of metastasis. Exp cell Res 1993; 204: 171-180.
14. Nimmerjahn A., Kirchhoff F. and Helmchen F. R. Resting microglial cells are highly dynamic surveillants of brain parenchyma in vivo. Science 2005; 308: 1314-1318.
15. Perry VH, Andersson PB Gordon S. Macrophages and inflammation in the central nervous system. Trends. Neurosci. 1993; 16: 268-273.
16. Schobitz B, De Kloet ER, Holsboer F. Gene expression and function of interleukin 1, interleukin 6 and tumor necrosis factor in the brain. Prog Neurobiol 1994; 44: 397-432.
17. Watters JJ, Schartner JM, Badie B. Microglia function in brain tumors. J Neurosci Res. 2005; 81: 447-455.

Adaptation Biology and Medicine (Volume 5: Health Potentials)
Editors: L. Lukyanova, N. Takeda and P.K. Singal

14. Genetic Peculiarities of Neurohumoral Regulation and Resistance to Stress Damages

Ludmila Belkina[1*], Valerii Kapelko[2], Tatyana Kirillina[1], Vladimir Lakomkin[2], Elena Popkova[1], Elena Smirnova[1] and Mariya Usacheva[1]

[1]Institute of General Pathology and Pathophysiology, Russian Academy of Medical Sciences, Moscow, Russia
[2]Russian Cardiological Research-and-Production Complex^, Ministry of Health, Moscow, Russia.

Summary

The interrelation between genetic features of neurohumoral regulation and resistance to stress damages has not been investigated sufficiently. In this respect rats of different genetic lines represent worthwhile models for our study. It was earlier established that, August rats, as compared to Wistar rats, have less resistance to long-term stress because of sharp fall of pressure; but these rats have more resistance to stomach ulcers, induced by stress, and lower mortality from acute myocardial infarction. In the given work in Wistar and August rats by spectral analysis of heart rate variability and catecholamine levels, it was established that in August rats sympathetic activity was increased and the activity of the parasympathetic system (PNS) was reduced in comparison with Wistar rats. The latter was documented first by a lowered baroreflex sensitivity. However, without reference to the rigidity of the autonomic regulation in August rats, the heart resistance to stress was higher than in Wistar rats. This resistance is connected at a systemic level to greater activity of NO system, at a cellular level to improved resistance to oxygen damage. In contrast to Wistar rats, the adaptation to short-term stresses in August rats stimulates the formation of stress ulcers and increased PNS tone. Thus, the disturbance of the initial autonomic balance as a result of adaptation decreases the resistance to ulcer damages.
Keywords: Stress, adaptation, genetic differences, heart rate variability, oxidative stress, stomach ulcer.

Introduction

Stress plays an important role in the pathogenesis of different illnesses including ischemic disease. Therefore an ability for effective adaptation to stress actions is an important factor of prevention of pathologic developments (16, 17). At the same time neither pharmacological nor adaptive protection are effective against such injuries because a large spectrum of the individual features of reaction depend on environment factors. In this connection, at the present time the study of individual (genetic) peculiarities of neurohumoral regulation, which determine resistance and effectiveness of adaptation to damaging factors, becomes more urgent. An important aspect of this problem is the influence of

*Address for Correspondence: Ludmila Belkina, Institute of General Pathology and Pathophysiology, Russian Academy of Medical Sciences, Moscow, Russia. E-mail: belkinal@yandex.ru

environment factors on adaptation, inherent patterns of neurohumoral regulation and resistance to damages. Resistance to injuries depends on balance between systems which realize and limit stress-reaction (2, 17, 18).

It is clear that non-invasive methods are very important for studying individual peculiarities of regulation. In recent years, the variability of hemodynamic parameters has been used for studying the character of neurohumoral regulation, because it reflects the activity of many regulatory systems. Clinical researches show that different pathologic processes are accompanied by decrease of heart rate (HR) variability and therefore this parameter is used for diagnostics and prognosis of disease development (1, 29). At the same time, the connection between variability of hemodynamic parameters and resistance to stress has not been investigated. On this point, animals of different genetic lines are of interest.

Recently we have showed that rats of August strain have less power spectral density of HR and more arterial pressure (AP), than rats of Wistar population (7, 1, 29). However, comparative researches of the resistance to stress in these rats show ambiguous result. In comparison of resistance to stress in these rats on classical criterion, on stress ulcer, it was found that August rats have less ulcers than Wistar rats (27). On the other hand August rats have less resistance to long immobilization stress (30 hours), because of sharp arterial pressure (AP) fall. Based on this, a view has appeared that these rats are less resistant to stress damages, than Wistar rats (32). However, in this work there is no comparative information about the resistance of their heart to stress. At the same time, the August rats with acute myocardial infarction have less mortality and less pronounced disturbances of heart function, than Wistar rats (6). Also in August rats the frequency and duration of malignant arrhythmia, induced by short-term myocardial ischemia and reperfusion are smaller as compared to Wistar rats (3).

The ability of August rats to effectively adapt to stress remains poorly investigated. It is known that adaptation to environmental factors increases resistance of organism to stress and ischemic injures (14, 16, 17, 18). But it was found that adaptation to short-term stress actions in August rats, unlike Wistar rats, formation of ulcers in stomach mucous coat (25). It is known that sympathetic nervous system (SNS) activation plays an important role in pathogenesis of stress and ischemic injures. The depression of parasympathetic nervous system (PNS) activity is associated with increased vulnerability to stress and ischemic arrhythmias (34), accompanied with reduced survival in patients with heart failure (22). In this connection it is impotent to note, that August rats are different from Wistar rats in their low adrenoreactivity (6) and higher NO system activity (14, 20), which plays a significant role not only in regulation of AP, but also has an ability to limit SNS activity (30). However features of neurohumoral regulation in August rats as compared to Wistar rats is investigated insufficiently. Therefore, for the purpose of analysis of different resistance mechanisms to injury factors in August and Wistar rats following were investigated: (1) Autonomic regulation of cardiovascular system by spectral analysis of HR and AP variability at rest and in condition of NO synthesis blockade; Baroreflex sensitivity and catecholamine levels in adrenal glands and blood, were compared with heart resistance to stress; (2) the effect of adaptation to repeated short-term stresses on the sympathovagal balance and resistance to ulcer injuries of stomach, induced by hard stress.

Materials and Methods

Animals. The study was performed in male Wistar and August rats of 4-6 month's age. Weight of Wistar rats in different series of experiences was from 372 g up to 420 g, and weight of August rats, for which smaller weight of a body, from 238 g up to 250 g is characteristic. Animals were housed under a 12 h (08.00-20.00 h) light-dark cycle and had free access to food and water.

Variability of HR and MAP. Arterial pressure (AP) was recorded in awake freely moving animals. A catheter was introduced into the femoral artery under sodium thiopental anesthesia (40 mg/kg, ip) 2 days before experiment. Tensometric gauge Statham 8200 P23AA (USA), 16-bit precision analog-to-digital converter (L-Card) and IBM PC (500 Hz) were used. The measurements were performed in a dark quiet room. Mean AP (MAP) and pulse interval (PI) in each cardiac cycle were calculated using original software. Variability of MAP and PI was estimated by standard deviation of cardiac cycles recorded over 1 h. Changes in MAP and PI were studied by spectral analysis. Values differing by 0.1 sec were determined for each parameters by the method of linear interpolation. A 1-h interval was divided into non-overlapping 51.2 sec segments (512 points). Linear drift was corrected, and each segment was brought to a mean of zero. Power spectra were calculated using Fourier rapid transform and averaged. The spectrum of MAP was analyzed in very-low-(VLF, 0,0195-0,2500 Hz), low- (LF, 0,25-0,75 Hz), and high frequency ranges (HF, 0,75-2,00 Hz), and also was summarized the total power (TP) spectral density of MAP and PI. The variability of hemodynamic parameters was being monitored for 1 h at rest and during 1 h after injection NO synthesis blocker N^{ω}-nitro-L-arginin (L-NNA, «IGN» company, USA,10 mg/kg).

Adaptation to short-term stress. During 11 days the rats were being adapted to stress actions by means of plunging into cold water (21°C) in dwelling plastic cage ($50 \times 30 \times 20$ sm^2) covered with metal net. The cage was being filled with water in a way that water was up to animal's neck (the animal stood). As usual the rats didn't swim at that, they hung, catching on the cage with arms. The scheme of adaptation is the following: during the first 3 days the duration of staying in water was increased from 5 to 15 minutes and then one-day break followed, after that in the next 3 days this course was repeated and after 2-days break the rats were kept in water 1 time a day for 15 minutes during 3 days. After several hours, when the last sense had been over, the rats were implanted a catheter (under anesthetic) for AP recording, and the next day after finishing adaptation course the variability of hemodynamic parameters was being analyzed. After study of hemodynamic parameters the animals were exposed to stress by immersing into cold water (21°C). 30 min and 1 h after, the stomach ulcers were estimated.

The baroreflex sensitivity was evaluated in sodium-thiopental-anaesthetized rats on changing the PI duration in response to the increase of AP, induced by phenylephrine hydrochloride (5 mkg/kg, i/v). AP was registered in femoral artery with the help of "Mingograf-34" ("Siemens", Sweden). The baroreflex sensitivity was calculated as $\Delta PI/\Delta AD$ where Δ is relative change of proper index on the peak of response development in comparison with background value (as 10 second interval immediately before injection of the drug).

Stress models for the heart study. For revealing differences between Wistar and August rats in resistance of heart to stress injuries 2 kinds of stress have been used: (1) short-term, (2-hour immobilization), by attachment to a board in position on their back; (2) long, (30-hour immobilization) in a close cage 4, 5 cm width. The experiments were conducted on the isolated hearts, and on the whole animals. As it is known, activation of free radicals production plays an important role in pathogenesis of stress injuries. In this connection the stress influence on isolated heart resistance to oxygen damages (which were caused by H_2O_2) was evaluated. At realization of experiment on the whole animals, we proceeded from the data about lowered resistance of August rats to long (30 h) immobilization stress

(32). In these experiments, the resistance to stress was evaluated by heart response to maximal isometric load which was caused by aorta clamping.

The isolated heart. In 2 hours after stress, the hearts were taken under urethane anesthesia and were perfused through an aorta by means of the peristaltic pump at 37°C modified Krebs-Henzeleit solution (pH 7,4) oxygenized with gas mixture (98% O_2+2% CO_2) containing glucose (11 mM). The Initial speed of the flow was established so that perfusion pressure in an aorta reached the average of 70 mmHg. A catheter with latex balloon for measurement of pressure was inserted in a left ventricle cavity. The diastolic pressure in left ventricle was set at a level of 10-15 mm Hg. The pressure in the balloon was registered by means of the gauge Gould Statham P23Db on polygraph Gould 2400S (USA). The scheme of the experiment included a assessment of heart function in an initial condition (30 min of perfusion after extraction of the heart), at gradual increase of perfusion speed from 10 to 22 ml/min during 10 min and at introduction hydrogen peroxide (H_2O_2), carried out by means of infusion pump (Sage) with the speed provided constant concentration of 100 mcM. The following parameters were evaluated of heart function: HR; systolic, diastolic and developed pressure; velocities of the pressure development and drop, an index of heart work or double product was calculated – product of HR on developed pressure.

The heart function on the whole animal. The experiments were performed under urethane anesthesia (1,5 g/kg) in conditions of the opened chest and artificial pulmonary ventilation by atmospheric air. The pressure recorded in a left ventricle cavity (LV) through the catheter, inserted in the apex of LV on "Mingograf-34" ("Siemens", Sweden) before and during the maximal isometric loading created by clamping ascending arch of an aorta for 30 sec. The same parameters of heart contractile function were estimated, as on the isolated hearts.

Catecholamine content in adrenal glands and in blood were assessed in an initial condition by high-performance liquid chromatography method with electrochemical detection (11). The left adrenal gland and blood from jugular vein were taken in rats under urethane anesthesia.

In animals, exposed to stress, the area of the stomach ulcers in a mucous membrane was measured by a method of planimetry. An hour, after stopping stress, under urethane anesthesia stomachs were taken away. Were cut on the big curvature, washed with cold water and stretched on a board. The total area of ulcers was evaluated. Statistical analysis of results was made by estimating standard deviation and evaluating differences by Student's t-test and U-criterion.

Results

In conscious August rats HR and AP is always higher, than in Wistar rats. In these experiments in August rats (n = 9) HR and MAP were 464 ± 11 beat/min and 127 ± 1,9 mmHg accordingly, and in Wistar rats (n = 10) 439 ± 12 beats/min and 114 ± 1,6 mmHg ($p < 0,01$). August rats have a little less statistical variability of PI, that is standard deviation, than Wistar rats (8,5 ± 0,8 ms vs 10,0 ± 0,9 ms, $p < 0,05$ on U-criterion), and there was no differences in MAP variability between groups (7,4 ± 0,2 mmHg vs 6,7 ± 0,4 mmHg). At the same time August rats considerably differed from Wistar rats on the spectral characteristic of PI in all frequency ranges (Fig. 1): so, in August rats power of PI oscillations in VLF-, LF- and HF -ranges made 24%, 20% and 59%, accordingly from Wistar rats' PI spectrums and the total power spectral density was less in 4 times ($p < 0,01$). In August rats, density of MAP fluctuations, on the contrary, is more, than in Wistar rats in all frequency ranges, especially – HF-band

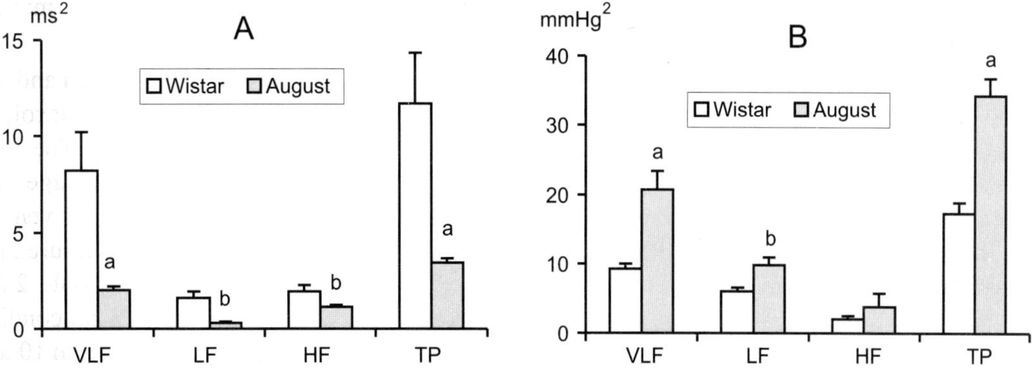

Fig. 1. Power spectral density of HR (A) and (MAP) in very-low- (VLF), low- (LF), and high- (HF) frequency bands in Wistar and August rats. TP – total power. Significantly different between these rats: [a]P < 0,01, [b]P < 0,05.

(in 2 times) and the MF-band (in 1,5 times). As a result total power of MAP fluctuations has been increased twice.

It has been shown earlier, that August rats at rest have higher activity of sympathetic-adrenal system (SAS), than in Wistar rats – the raised CA level in blood is the evidence of it (10, 24). We have obtained similar data (Fig. 2). August rats had higher norepinephrine (NE) content, adrenaline and dopamine (DA) in adrenal glands at rest, than Wistar rats, on 140%, 44% and 45% accordingly, and the relation DA/NA, which reflects the activity of catecholamine (CA) re-synthesis, is less on 40% that points to more active metabolism of CA in August rats' adrenal glands. Thus in August rats the NE level in blood was increased by 95%, ($p < 0,05$), and adrenaline by 127% ($p < 0,02$). The negative correlation between power spectral density in LF-and MF-ranges of PI fluctuations and NE level in rats blood (which we found out) is correlated with the data received on the human being (33).

While defining the sensitivity of chronotropic component of baroreflex, it turned out, that August rats had not only the low tone of PNS, but also the reduced ability of this system to activation. In this series, the weight of August rats (n = 9) and Wistar (n = 10) made 210 ± 11 and 315 ± 16 Γ accordingly. Under anesthesia the initial systolic AP in August and Wistar rats was 92 ± 7,0 mmHg and

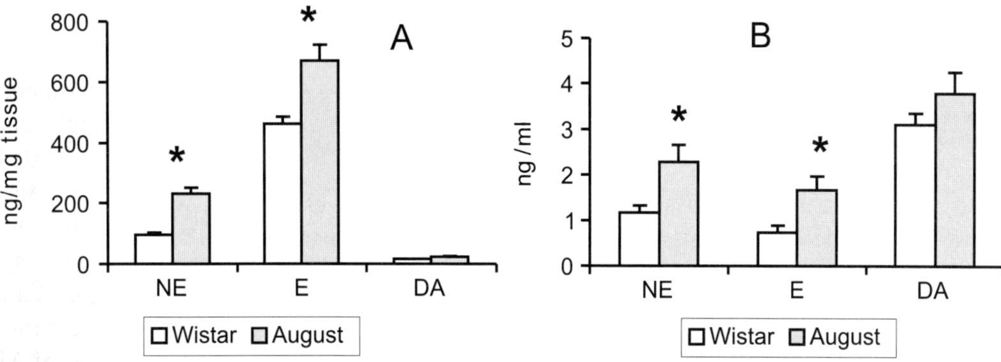

Fig. 2. Catecholamine level in adrenal glands (A) and blood in Wistar and August rats. Significantly different between these rats:*P < 0,05.

111 ± 4,6 mmHg, and HR - 325 ± 9 beat/min and 370 ± 23 beat/min accordingly. Phenylephrine HCL increased AP in the August rats by 49%, and in the Wistar rats – by 38%. Thus the bradycardia at August rats was much less, than at Wistar rats: PI of the first rats increased from 186 ± 4,9 ms up to 197 ± 4,1 ms, and of the second rats - from 165 ± 9,4 ms up to 204 ± 5,0 ms (p < 0,001), and as a result the parameter of baroreflex sensitivity in August rats turned to be almost 4 times less, than in Wistar rats (0,27 ± 0,05 vs 1,06 ± 0,16, p < 0,001). Thus, the shift of sympathovagal balance in August rats in prevalence of tone SNS direction is connected with the lowered sensitivity of baroreceptors and owing to this there was the suppression of PNS activity that specifies on certain rigidity of cardiovascular system regulation in these animals.

While studying the NO contribution in MAP variability, it was found that the blockade of NO synthesis with the help of L-NNA caused in August rats rather less expressed, than in Wistar rats, the pressor reaction: in the first rats MAP increased from 124 ± 1,2 mmHg up to 143 + 1,6 mmHg, and in the second rats - from 116 ± 3,9 mmHg up to 150 ± 4,2 mmHg. Despite of it the statistical variability of MAP in August rats (Table 1), that is a standard deviation, increased considerably in a greater degree, than in Wistar rats (83% vs 47%, p < 0,05). And, this parameter in Wistar rats was completely restored within 60 min after injection of blocker and in August rats remained reliably increased by 32% (p < 0,05). Thus, in August rats NO system in a greater extent, than in Wistar rats, limits the AP fluctuations, that is the contribution of this system in regulation of cardiovascular system function at first is more, than at second. Increased activity of NO system in August rats can partly compensate insufficiency of PNS, namely to limit activity of SNS. Really, previously by us was shown, that August rats have less expressed activation of SAS (that is caused by stress), than Wistar rats have (5). These facts along with reduced adrenoreactivity (6) give the basis to assume, that in August rats the heart resistance to stress injuries should be improved, than in Wistar rats. The results of the following series of experiments confirm this assumption. Also, as well as in the previous series, in August rats (n = 12) spectral power of HR fluctuations in all ranges was much less, and the power of MAP spectrums is more, than Wistar rats (n = 9) have.

Table 1. Effect of L-NNA on statistical variability of MAP (mmHg) in Wistar and August rats

Series	Before injection	After injection	
		30 min	60 min
Wistar (n=6)	6,36 ± 0,28	9,36 ± 0,91^	6,22 ± 0,74
August (n=7)	7,1 ± 0,27*	13,0 ± 1,43** ^^	9,38 ± 0,36** ^

Values are means ± SD. n-number of animals. significantly different from the initial level: ^P < 0,05, ^^P < 0,01. Significantly different between Wistar and August rats: *P < 0,05, **P < 0,01.

The experiments on the isolated hearts have shown, that in August rats HR was raised 8% in the initial condition, and developed pressure was lowered 10% in comparison with Wistar rats, therefore double production (DP) was identical (Table 2). After the immobilization stress, depression of heart function was observed in both groups, it was expressed in DP reduction (in August rats on 15% and in Âèñòàð rats on 24 %), due to developed pressure at poorly changed HR. In August rats velocities of myocardium contraction and relaxation decreased after stress to a lesser degree, than in Wistar rats. The load caused by the increase of coronary perfusion speed activated the heart function (Fig. 3A), in

Table 2. Contractile function of isolated heart in control and after 2-h immobilization stress in Wistar and August rats

Indices	WISTAR		AUGUST	
	Control (n = 12)	Stress (n = 16)	Control (n = 11)	Stress (n = 18)
HR (beats/min)	244 ± 6	261 ± 5	266 ± 6*	275 ± 6
Developed pressure (mmHg)	159 ± 8,5	113 ± 9,4^^	144 ± 4,7	120 ± 5,9^^
Double product (mmHg ŏ beats/min)	37888 ± 2049	28895 ± 2210^	38354 ± 1405	32730 ± 1545^
Velocity of myocardium contraction (mmHg/sec)	3660 ± 204	2576±244^^	3427±152	2862 ± 151^
Velocity of myocardium relaxation (mmHg/sec)	2758 ± 110	2093±141^	2755±152	2363 ± 135^

Values are means ± SD. n-number of animals. Significantly different from control: ^P < 0,05, ^^P < 0,01; Significantly different between Wistar and August rats: *P < 0,05.

control groups, approximately, to the same extent and after the stress the maximal magnitude DP in August group moderately differed from the control level while in Wistar rats this parameter decreased on 20% ($p < 0,05$). The described difference was substantially provided due to August rats' higher HR (Fig. 3B).

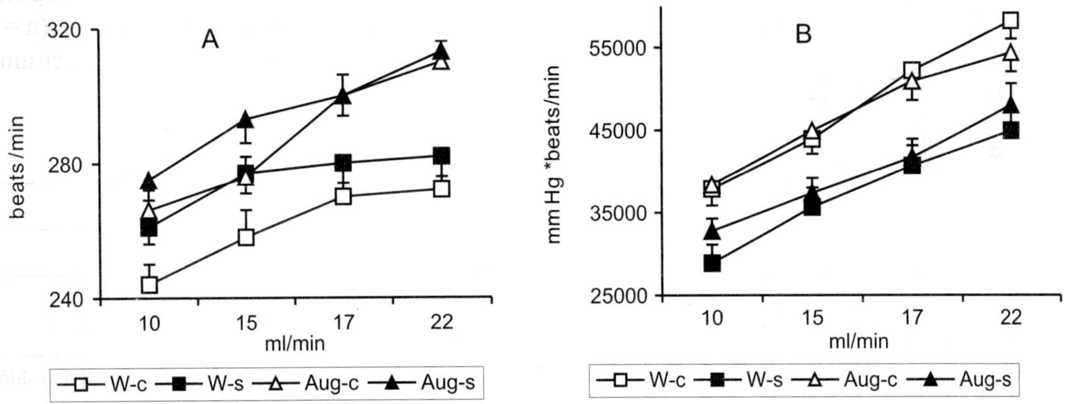

Fig. 3. Effect of the coronary flow load on the heart rate (A) and double production (B) in Wistar and August rats. From here on "W-c"-Wistar-control, "W-s"-Wistar-stress, "Aug-c"-August-control, "Aug-s"-August-stress.

In experiences with H_2O_2 was supported the earlier established fact, that in August rats raised resistance of heart to oxidizing stress (4). After stress H_2O_2 caused deeper, than in the control, depression of heart function (Fig. 4) in both groups, but in August rats hearts kept higher level of function in comparison with Wistar rats. It is significant that within the first 15 minutes of H_2O_2 action

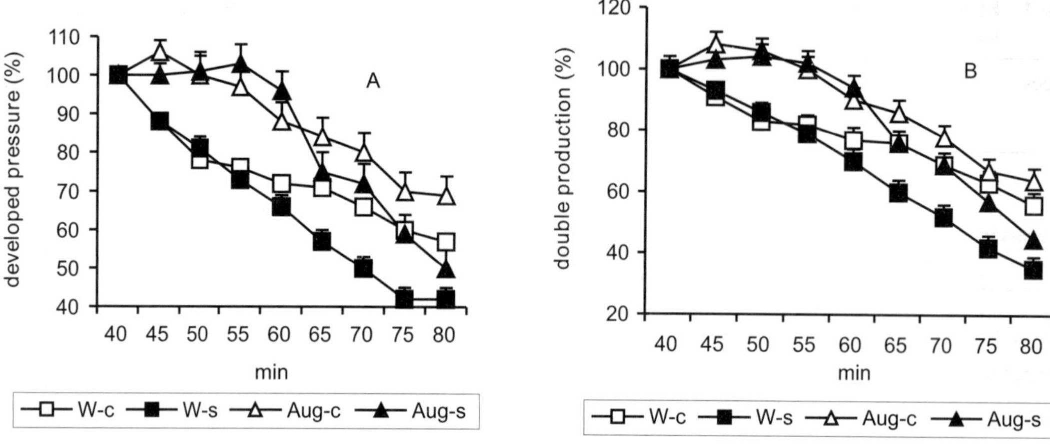

Fig. 4. Effect of hydrogen peroxide on the heart developed pressure (A) and double production (B) in Wistar and August rats, exposed to 2-h immobilization stress. Initial parameters were quantified as 100%.

in August rats the heart function depression were not observed and DP was kept on steady level, while in Wistar group this parameter decreased more, than on 23% ($p < 0,05$). After 20 mines of H_2O_2 action the depression of heart function increased by both groups of the rats after stress, but the difference between groups was kept. Thus, August rats' hearts, in contrast to Wistar rats, kept the raised stability of heart function to oxidizing stress after immobilization stress. It is necessary to note, that in August rats' ulcer injuries in a stomach are less expressed as usual, than in Wistar rats: the total area of ulcers made 7,8 and 26 mm^2 accordingly.

At study of influence of 30 h immobilization stress on heart function here was no case of death in one group, as it was informed in above mentioned work (32). The results of the these experiments show, that in conditions of the whole organism stress injuries of heart is not revealed in compared groups of animals even after long heavy stress. However and in the given conditions of experiment in August rats the efficiency of heart work always is more, than in Wistar rat. It is significant that, that after the termination of stress in Wistar ($n = 11$) rats HR before the thoracotomy has appeared considerably increased as compared to control (458 ± 7 beats/min vs 393 ± 13 beats/min, $p < 0,001$) and was kept on more high level after opening chest. At the same time at rats August ($n = 9$), on the contrary, in the post-stress period HR decreased in comparison with the control (367 ± 15 beats/min vs 400 ± 12 beats/min, $p < 0,05$). These facts speak that at Wistar rats after stress there was an activation of SNS, and at rats August - activation PNS. It is clear from Table 3, that under urethane narcosis in August rats HR in control is lower, and developed pressure is a little higher, than in Wistar rats. As a result, DP in compared groups was identical. At the same time it has appeared, that at the maximal isometric load in August rats the heart function depression was pronounced considerably to a lesser degree, than in Wistar rats: in the first rats DP decreased in 22%, and in the second rats - in 3 times. This difference has was caused by two factors: first, August rats have the less expressed bradycardia, typical for the given load. It is known, that this effect is connected to PNS activation and thus the stability of heart rate in given conditions in August rats is quite probably explained by the reduced activity of PNS, characteristic for these animals; secondly, with a greater rise of developed pressure.

Table 3. Effect of long-term immobilization stress on the heart contractile function in rest and in maximal isometric load in Wistar and August rats

Indices	Series	Relative physiologic rest	Clamping aorta	
			5 sec	30 sec
HR (beats/min)	W-control (8)	395 ± 16	300 ± 15	98 ± 10
	Aug-control (9)	337 ± 8##	343 ± 8#	206 ± 14
	W-stress (10)	422 ± 10*	389 ± 13**	128 ± 5*
	Aug-stress (8)	340 ± 13##	337 ± 18#	230 ± 6##
Developed Pressure (mmHg)	W-control	92 ± 6,1	109 ± 4,1	125 ± 3,4
	Aug-control	104 ± 2,2#	126 ± 4,3#	132 ± 2,4
	W-stress	118 ± 7,6*	127 ± 3,5*	134 ± 5,9
	Aug-stress	138 ± 5,1**#	142 ± 3,8*#	145 ± 8,5
Double product (mmHg × beats/min)	W-control	36143 ± 2500	32625 ± 2661	12188 ± 1204
	Aug-control	35015 ± 1371	43347 ± 1847##	27184 ± 2385##
	W-stress	49880 ± 3560**	49259 ± 2434**	17071 ± 1905*
	Aug-stress	47033 ± 2752**	47694 ± 3250	33350 ± 2197*##

Values are means ±SD. Figures in brackets - number of animals. Abbreviations W and Aug – Wistar and August rats. Significantly different from control: *P < 0,05, **P < 0,01; Significantly different between Wistar and August rats: #P < 0,05, ##P < 0,01.

The stress did not cause abnormalities of heart function of in both groups, but on the contrary, caused significant and similar activation of heart in compared groups: as a result in Wistar rats DP increased by 38%, and in August rats - by 34%. During 30 sec the actions of the maximal isometric load in August rats, exposed to stress, the DP was also considerably higher, than in Wistar rats, that is between animals the initial difference on a level of heart functioning was kept. The described distinctions also were connected with better HR stability in August rats, in comparison with Wistar rats, that substantially provides higher effectiveness of heart work in these animals. This peculiarity of August rats was shown in the previous works (6). It is also necessary to note, that the rats' ulcer injuries in the stomachs under long immobilization action were poorly expressed and made in August and Wistar rats in total - 12,1 mm^2 and 17,5 mm^2 accordingly. Thus, the damaging action of stress is revealed on a stomach, but not on heart.

At comparing of adaptation influence to short-term stress on HR variability and vulnerability to stress ulcers it was found out, that in Wistar rats (n = 8) adaptation did not have an influence on MAP and HR (368 ± 8,6 beat/min) in comparison with the control (367 ± 2,7 beat/min). In August rats (n = 7) adaptation also did not have an influence on MAP, but certainly reduced HR (from 420 ± 3,5 beat/min up to 398 ± 8,1 beat/min, p < 0,03). The power spectral density of PI in VLF band and LF band in Wistar rats under the influence of adaptation tended to reduction (Fig. 5) while in August rats, on the contrary, the power of these fluctuations small increased. More, than double increase in spectral power of HF band in August rats attracts attention. As a result TP was reliably increased by 65%. The reduced HR and increased the power in the HF band of PI fluctuations testify, that the adaptation to short-term stress influences led to the substantial growth of PNS tone in August rats. This appearance was accompanied by the increase of vulnerability to stress ulcer in August rats. Figure 6 shows, that the stress in non-adapted August rats, in comparison with Wistar rats as always, causes much less ulcers in stomach while the heavy stress in adapted August rats stimulates formation of more ulcer injuries, than

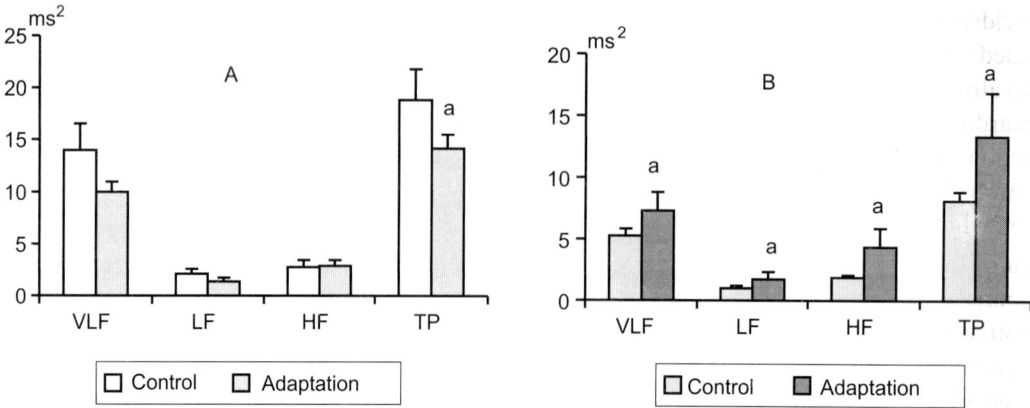

Fig. 5. Effect of adaptation to repeated short-term stress on power spectral density of HR in Wistar (A) and August rats (B). Significantly different from control: [a]P < 0,05.

in Wistar rats. As a result of adaptation in Wistar rats the size of ulcer has decreased in 2 times, and in August rats it has increased in 4 times (p < 0,05). Nevertheless, it is necessary to note, that on absolute size in August rats area of stress ulcers is less after adaptation than in adapted Wistar rats.

Thus, the disturbance of the initial balance of neurovegetative regulation as a result of adaptation, that is a shift of inherent balance of August rats towards increase in PNS activity, lowered resistance to the stomachs ulcer damages, caused by stress.

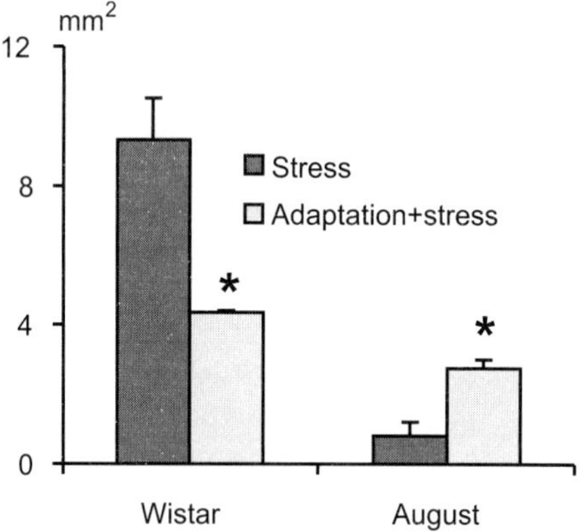

Fig. 6. Effect of preliminary adaptation to short-term stress actions on size of the stomach ulcer, induced by 30-h immobilization, in Wistar and August rats. Significantly different from "Stress":*P < 0,05.

Discussion

We studied interrelation between genetically determined peculiarities of neurohumoral regulation and resistance to stress on the rats of different genetic lines. It was found that in August rats the heart resistance to stress is higher than in Wistar rats, in spite of the lower liability of vegetative nervous system. The reduced HR variability which is connected with the higher SAS activity in August rats is the evidence of the latter. We showed that this phenomenon in August rats is connected with the lessened PNS activity, and the lowered sensitivity of baroreflex chronotropic component verifies it. As it was already marked, rigidity of vegetative regulation itself is a risk factor for the ischemic injury of myocardium (34). Also under stress action, the rise of vulnerability to arrhythmias is accompanied by a reduction of PNS activity (12). However, reduced PNS activity in August rats is not accompanied by the lowered resistance of heart to injuries. Analyzing this phenomenon, it is necessary to take some factors which provide more effective stress protection of myocardium into consideration.

On of the most important factors which are realized on the systematic level is obviously the less pronounced SAS activation in stress situations, and the decreased CA level in August rats in blood in post-stress period in comparison with Wistar rats. It was established that this effect meshes with nearly equal activation of CA synthesis in the adrenal glands (5). As the adrenalin concentration in blood of both groups can increase 8-16 times more (10, 24) after stopping the stress, it is assumed that the diminished SAS activation in August rats is provided by reverse capture in chromaffine cells of adrenal glands and sympathetic terminals in the first hours after the stress. Probably the increased SNS activity in August rats can be compensated by the better functional power of the reverse capture system and the decay of CA.

It was found that the August rats' level of corticosterone in blood under stress increases to a lesser degree that Wistar rats' (28). On the other hand, the results of spectral analysis of NO contribution to AP variability and increased NO level in August rats' organisms are the evidence of the higher SNS activity and the insufficiency of PNS can be compensated by the increased activity of NO system, which can limit the coming out of NE at the level of sympathetic terminals (30). It is known that NO system has cardio-protective effect at stress and ischemic injuries (14, 26). Therefore the better resistance of August rats to stress and ischemic injuries (in comparison to Wistar rats) can be substantially connected with the higher activity of this system. Besides there are the facts that August rats have enhanced activity of opioidergic system, which is known to limits stress-reaction as evidenced by fact that in August rats exposed to stress, the blood level of beta-endorphin increases to a greater extent than in Wistar rats (31).

At the same time, the experiments showed that decreased PNS activity in August rats can be the factor which protects mucous coat of stomach from stress ulcers. It is known that PNS induces the secretion of gastric acid and as a result it plays an important role in the pathogenesis of ulcer formation in conditions of stomach's ischemia (9). A close correlation between susceptibility to ulcer formation and the higher secretion of gastric acid was shown on people. As this takes place, the increased level of acid in a stomach correlated with the enhanced PNS activity, and the higher power of HR oscillation in HF band (13). It is also known that the increase of PNS tone, which is caused by stress, stimulates ulcer formation (9), and adaptation to stress increases resistance to the given injury to a great extent (17). It is ascertained that in August rats, in contrast to Wistar rats, the adaptation to short-term stress actions doesn't result in increase of stomach's resistance to ulcer formation, induced by hard stress (25). On the contrary, after the adaptation course, hard stress stimulates the formation of more ulcers in August rats, than in Wistar rats. It was established that the adaptation to short-term stress influence increases in August rats' PNS tone, as evidenced by the substantial increase of power in the HF band of HR. This phenomenon was accompanied by the increase of stomach's vulnerability to stress ulcer formation, that is evidently connected with the shift of sympathovagal balance towards the increase of PNS activity. Our experiments showed that after long immobilization the August and Wistar rats have the similar activation of the heart function in conditions of the relative physiological rest. At the same time at maximum isometric load, August rats have less expressed abnormalities of the heart function (both under control and after the stress) than Wistar rats. This August rats' advantage to a greater extent is connected with the better HR stability, than Wistar rats have.

One of the mechanisms that provides higher resistance of heart at the cellular level of August rats is reduced myocardium adrenoreactivity (in comparison with Wistar rats) that was ascertained earlier (6). From the large share of probability higher resistance of heart to oxygen injuries in August rats can be connected with the increased CA level in blood. It is known that auto oxidation of CA results in formation of reactive oxygen radicals and thus it makes major contribution on activation of free-radical processes. The elevated adrenergic tone in August rats obviously promotes the formation of adaptation to the higher level of oxygen radicals in the organism, that to a greater extend along with the lowered adrenoreactivity causes the higher resistance of heart to stress injuries. This inherently reduced myocardium adrenoreactivity and the higher resistance to oxygen injuries are the mechanisms of genetically determined resistance to stress and ischemic injuries.

From the submitted data follows, that the increased CA level in August rats' organisms is compensated by the less expressed SAS activity under stress and ischemic injuries on the one hand, and by lower adrenoreactivity and higher resistance to oxygen injuries on the other hand. Thus the lack of neurovegetative regulation liability in August rats is effectively compensated by protective mechanisms under stress influence both at system and cellular levels. It is typical that the change of inherent patterns of regulation caused by adaptation can result not in an increase but in a decrease of resistance to stress – in this case August rats' stress ulcer formation. However, it is evident that the adaptation to environmental factors, as well as the stress, have selective effects on the function of organs, i.e., it can exert protective action on one organ and simultaneously damaging action on the other. The facts described above show that pathogenetic value of HR and AP parameters variability and also another indices of sympathovagal balance for prognosis of resistance to injuring factors is highly limited. In spite of negative prognosis indexes of neurohumoral regulation, animals can have better resistance to stress and ischemic injuries as shown in August rats.

References

1. Baevskii PM. Analysis of heart rate variability in space medicine. Phyziologia of Man. 2002; 28: 70-82.
2. Baevskii PM, Kirilov OI, Klezkin S. Z. Mathematic analysis of heart rate changes in stress. Moscow. Nauka. 1984.
3. Belkina LM, Kirillina TN, Pshennikova MG, Arkhipenko YuV. Rats of August strain have more resistance to arrhythmias, induced by myocardial ischemia and reperfusion, compared to rats of Wistar population. Biull Eksp Biol Med. 2002; 133: 625-628.
4. Belkina LM, Lakomkin VL, Kirillina TN, Zhukova AG, Sazontova TG, Kapelko VI. The heart resistance to oxygen stress in rats of different genetic lines. Biull Eksp Biol Med. 2004; 138: 250-253.
5. Belkina LM, Popkova EV, Lakomkin VL, Kirillina TN, Zhukova AG, Sazontova TG, Usacheva MA, Kapelko VI. Variability of hemodynamic parameters and resistance to stress damages in rats of different lines. Rus J Physiol Im. I. M .Sechenov. 2006; 92: 221-231.
6. Belkina LM, Saltykova VA, Pshennikova MG. Genetic determined differences in resistance to myocardial infarction in Wistar rats and rats of August strain. Biull Eksp Biol Med. 2001; 131: 624-628.
7. Belkina LM, Tarasova OS, Kirillina TN, Borovik AS, Popkova EV. Effect of stress on variability of systemic hemodynamics in rats of various genetic strains. Biull Eksp Biol Med. 2003; 136: 269-273.

8. Glick G, Braunwald E. Relative role of the sympathetic and parasympathetic nervous systems in the reflex control of the heart. Circul Res. 1965; 16: 363-375.
9. Desiderato O, Mac Kinnon JR, Hisson H. Development of gastric ulcers in rats following termination. J Comp Physiol. 1974; 87: 208-214.
10. Kvetnanskii R, Belova TI, Oprshalova Z, Ponez I, Iindra A, Dushkin VA. Blood catecholamine in rats of August strain and Wistar hats in emotional stress. Fiziol Zh SSSR. Im. I.M. Sechenova. 1981; 67: 516-523.
11. Larskii EG, Barchatova VP, Demina EG. Analysis of blood catecholamine by method of high-performance liquid chromatography with electrochemical detection. Laboratornoe delo. 1985; No 9: 519-522.
12. Lown B, Verrier RL. Neural activity and ventricular fibrillation. New Engl J Med. 1976; 294: 1165-1170.
13. Lucini D, Cerchiello M, Basilisco G, Cainelli M, Bianchi PA, Fiorelli G, Malliani A, Pagani M. Autonomic control of heart period in duodenal ulcer patients insight from spectral analysis oh heart rate variability. Auton neurosci. 2000; 84: 122-129.
14. Malyshev IYu, Manuchina EB. Stress, adaptation and nitric oxide. Biochimia. 1998; 63: 992-1006.
15. Mancia G, Parati G, Pomidossi G, Casadei R, Di Rienzo M, Zanchetti A. Arterial baroreflexes and blood pressure and heart rate variabilities in humans. Hypertension. 1986; 8: 147-53.
16. Meerson PhZ. Disturbances of metabolism and cardiac function under the action of emotional painful stress and their prophilaxis. Basic Res Cardiol. 1980; 75: 479-500.
17. Meerson PhZ. Adaptation, stress and prophylaxis. Springer-Verlag. Berlin. 1984.
18. Meerson PhZ, Pshennikova MG. Adaptation to stress situations and to exercises. Moscow. Medicine. 1988.
19. Meerson PhZ. Adaptive medicine: mechanisms and protective effects of adaptation. Moscow. Hypoxia Medical LTD. 1993.
20. Mikoyan VD, Kubrina LN, Manuchina EB, Malyshev IYu, Vanin APh. Differences in NO synthesis stimulation in rats of different population, exposed to heat shock. Biull Eksp Biol Med. 1996; 121: 634-637.
21. Murphy CA, Sloan RP, Myers MM. Pharmacologic responses and spectral analysis of spontaneous fluctuation in heart and blood pressure in SHR rats. J Auton Nerv Syst. 1991; 36: 237-250.
22. Osterziel K, Hanlein JD, Willenbrock R, Eichorn C, Luft F, Dietz R. Baroreflex sensitivity and cardiovascular mortality in patients with mild to moderate heart failure. Brit Heart J. 1995; 73: 517-522.
23. Perlini S, Giangregorio F, Coco M, Radaelli A, Solda PL, Bernardi L, Ferrari AU. Autonomic and ventilatory components of heart rate and blood pressure variability in freely behaving rats. Amer J Physiol. 1995; 269: H1729-H1734.
24. Perzov SS, Koplik EV, Krauzer V, Michael H, Emme P, Sudakov KV. Catecholamine in adrenal glands in August and Wistar rats in acute emotional stress. Biull Eksp Biol Med. 1997; 123: 645-648.
25. Pshennikova MG. Hereditary efficiency of stress-limiting systems as a factor of the resistance to stress-induced disorders. Uspechi Phiziol Nauk. 2003; 34: 55-67.
26. Pshennikova MG, Belkina LM, Bachtina LYu, Baida LA, Smirin BV, Malyshev I. Yu. Stress-protein hsp70, nitric oxide and resistance to the myocardial infarction in rats of August line and Wistar population. Biull Eksp Biol Med. 2001; 132: 153-155.
27. Pshennikova MG, Bondarenko NA, Shimkovich MV, Bondarenko ON, Malyshev IYu. Differences in behavior and resistance to stomach ulcer formation, induced by stress in August and Wistar rats. Biull Eksp Biol Med. 1999; 128: 638-641.
28. Pshennikova MG, Golubeva LYu, Kuznezova BA, Shimkovich MV, Malysheva EV, Malyshev IYu. Differences in stress-reaction and formation of adaptation to stress in August and Wistar rats. Biull Eksp Biol Med. 1996; 122: 156-159.
29. Ryabykina GV, Sobolev AV. The heart rate variability. Moscow. Overlei. 2001.
30. Schwarz P, Diem R, Dun NJ, Fostermann U. Endogenous and exogenous nitric oxide inhibits norepinephrine release from rat heart sympathetic nerves. Circ Res. 1995; 774: 841-848.

31. Sudakov KV, Coghlan JP, Kotov AV. Delta-sleep-inducing peptide sequels in the mechanisns of resistance to emotional stress. Ann N Y Acad Sci. 1995; 771: 240-251.
32. Sudakov KV, Yumatov KV, Ulyaninskii LS. Hemodynamic in emotional reaction and emotional stress. Fiziol Zh SSSR. Im. I.M.Sechenova. 1988; 4: 1535-1545.
33. Tulen JH, Boomsma F, Man in't Veld AJ. Cardiovascular control and plasma catecholamines during rest and mental stress: effects of posture. Clin Sci (Lond). 1999; 96: 567-576.
34. Vanoli E, Schwarz PJ. Parasympathetic interaction and sudden death. Basic Res Cardiol. 1990; 85. Suppl 1. 305-321.

Adaptation Biology and Medicine (Volume 5: Health Potentials)
Editors: L. Lukyanova, N. Takeda and P.K. Singal

15. Hypothalamic Chemical Mechanisms of Resistance to Emotional Stress

Konstantin V. Sudakov*

Anokhin Institute of Normal Physiology, Russian Academy of Medical Sciences, Moscow, Russia

Abstract

The developed by us methods have permitted to prognostically investigate individual mechanisms of resistance and stress-proneness to emotional stress in rats.

The experiments showed that noradrenalin, epinephrine, β-endophrin, substance P, delta sleep inducing peptide (DSIP) contents in hypothalamus were higher in animals prognostically resistant to emotional stress in comparison with stress-prone animals. In prognostically stress-prone rats more pronounced C-fos gene expression in hypothalamus was revealed. Additional injections of substance P and DSIP to prognostically stress-prone rats increased their resistance to acute stress loads. Intraperitoneal DSIP injection decreased C-fos gene expression in hypothalamus of stress-prone and resistant to stress rats.

Microiontophoretic DSIP injection to single neurons of paraventricular and ventromedial hypothalamic nuclei along with inhibition of their spike activity produced in most cases a reaction of activation. At the same time microiontophoretic DSIP application to single hypothalamic neurons blocked their activating reaction to glutamate microiontophoretic injection.

Intracerebroventrical (ICV) application of cycloxemide – an inhibitor of protein synthesis – to brain neurons blocked the suppression of C-fos gene expression in paraventricular nuclei of hypothalamus of stress-prone animals.

Keywords: hypothalamus, emotional stress, individual resistance, noradrenalin, epinephrine, β-endorphin, substance P, delta sleep inducing peptide (DSIP)

Emotional stress is shaped in so called conflict situations when a subject motivated by a biological or social need, is for long or short period limited in the possibility of its satisfaction. Just in these cases when there is no possibility of achieving a useful adaptive result, negative emotions deepen, sum up and turn into "stagnant" fixed excitation in the brain.

Unlike Hans Selye who considered as a universal reaction activation of the axis-hypothalamus-hypophysis-adrenal cortex on stressors's action, the concept of emotional stress made it possible to regard mechanism of emotions as a primary reaction on a conflict situation [1]. Long term investigations of Paul MacLean showed that morpho-functional substratum of emotions was presented by limbic-reticular structures of brain [2].

*Address for Correspondence: Konstantin V. Sudakov, Anokhin Institute of Normal Physiology, Russian Academy of Medical Sciences, Moscow, Russia. Email: ksudakov@mail.ru

All that gives grounds to think that hypothalamo-limbico-reticular structures in which, as known, excitations circulate protractedly along special closed circles [3], can be regarded as a morpho-functional basis of "stagnant emotional excitation" that under emotional stress determines the disorders in the most genetically weak autonomic mechanisms.

Our experiments has shown that emotional excitation springing up in a conflict situation at first extends on limbic-reticular structures of brain, in which hypothalamic structures play a central role, and only after that like an integrated excitation embraces effector autonomic centers [4]. So the changes in limbic-reticular brain structures present a primary reaction on emotional stress. The adaptive syndrome under stress, discovered by Hans Selye, reflects already secondary reactions of the organism on conflict situations.

Immobilization is most widely used as a model of psychoemotional stress in animal experimentation. Our investigation has shown that under these conditions the main stress-inducing factor is the forced confinement of an animal in restraining situations. More severe stress was induced in immobilized rats by alternating electrical stimulation of ventromedial hypothalamus defense centers and electrical skin stimulation.

In H. Selye's works and the ones of his followers in the analysis of different stress manifestations as a rule averaged physiological and biochemical parameters were used [5].

However, as early as in 1978 we paid attention to the fact that in similar experimental conflict situations inducing emotional stress, appeared animals, resistant and predisposed to disorders in this or that physiological functions.

The methods developed by us have permitted to prognostically investigate individual mechanisms of stress-resistance and stress-proneness to emotional stress in rats.

Individual resistance to stress could be determined in rats using tail-flick test, measuring the general motor activity (e.g. in Varimex device), analyzing the ECG, impedance rheogram, blood pressure and respiration in response to an orthostatic challenge.

Open field test was also used for prognostical characteristic. Our experiment showed that rats prognostically predisposed to emotional stress are characterized by significant differences in latency to first movement and entry into center of the open field, and peripheral and central ambulation [6].

In our experiments prognostically resistant and predisposed to emotional stress rats were used.

The degree of resistance to emotional stress differed in rats of distinct strains. Thus, Vistar rats appeared more resistant than August rats.

Stress-resistant and predisposed rats showed different catecholamine contents in brain structures. Catecholamine levels were measured using radioenzyme tests in different brain nuclei in Vistar and August rats. Vistar rats had higher levels of norepinephrine in hypothalamus (Fig. 1).

When exposed to immobilization stress, the rats differing in their blood pressure (BP) response, also had a different time course of catecholamine contents in n.arcuatus hypothalami (group A_{12}). The norepinephrine content did not differ from the control values in rats with stable BP, and was significantly decreased in rats predisposed to BP changes.

Under conflict immobilization inducing emotional stress both lines of rats showed a decrease of norepinephrine in hypothalamus. At that in rats resistant to emotional stress unlike those predisposed to emotional stress the decrease in norepinephrine level in hypothalamus was less pronounced. Our

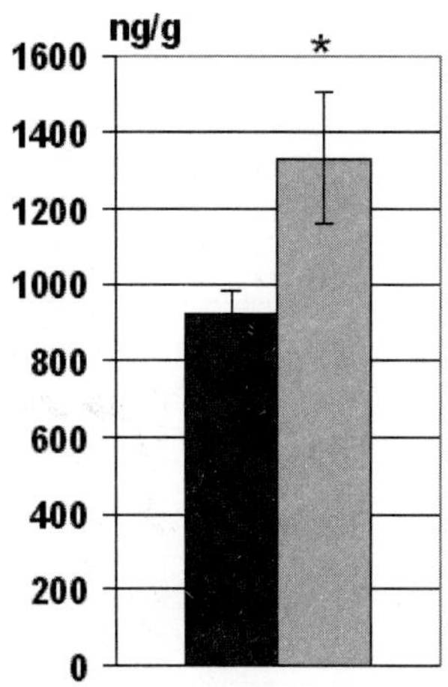

Fig. 1. Concentration of norepinephrine (ng/g of brain tissue) in hypothalamus in August (n = 12) – black columns and Wistar (n = 12) rats – light columns. * – $p < 0.05$.

experiments showed that in rats adapted to chronic emotional stress could be seen normalization of the decreased under the influence of acute emotional stress of norepinephrine level in hypothalamus. In animals adapted to emotional stress an additional single conflict immobilization did not decrease norepinephrine content in hypothalamus. So in rats adapted to chronic emotional stress a change of catecholamine metabolism in hypothalamus took place which could determine an increase of animal's resistance to emotional stress.

A line of our experiments showed that in conflict situations inducing emotional stress in response to microiontophoretic application of neuromediators: noradrenaline, acetilholine and serotonin pronounced changes in chemical neuron properties of hypothalamus took place.

When the skin was electrically stimulated in immobilized animals of 189 neurons studied in the ventromedial hypothalamic nucleus, 74 (39.2%) changed their sensitivity to microiontophoretically applied neurotransmitters; in 41.9% of these the change was related to acetylcholine, in 40.5% to norepinephrine and in 17.6% to serotonin. In that series of experiments 61 neurons were subsequently exposed to all three neurotransmitters. Twenty five neurons altered sensitivity to only one transmitter, 13 – to two and 4 to all three.

Hence, after brief negative emotional stress some of hypothalamic neurons alter their sensitivity to the most prevalent neurotransmitters – acetylcholine, serotonin and to norepinephrine.

Our special experiments have shown that the animal's resistance to emotional stress is determined by the content of protein-like endogenous substances in the brain. The content of oligopeptides in the hypothalamus of animals differing in their resistance to stress was analyzed by immunoenzyme technique. Hypothalamic levels of immunoreactive β-endorphin-like substance were higher in stress-resistant Vistar rats as compared to predisposed animals. That level increased throughout 3 h of acute emotional stress induced by immobilization combined with aperiodic electrical ventromedial hypothalamus stimulation alternating with electrocutaneous stimulation (Fig. 2).

The content of immunoreactive delta-sleep inducing peptide (DSIP) was also measured in the hypothalamus of animals with different resistance to emotional stress. Stress-resistant Vistar rats had

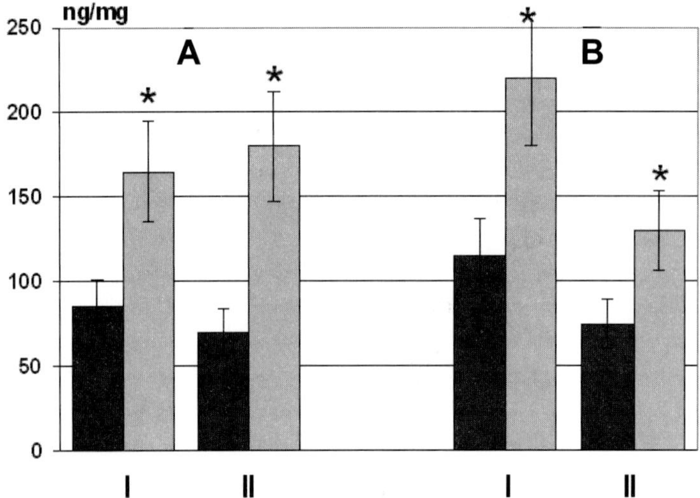

Fig. 2. **Content of immunoreactive β-endorphin (A) and DSIP (B) in hypothalamus of Vistar and August rats. On ordinate axes – content of investigated material (in ng/mg). I – rats resistant to emotional stress: black columns – August (n = 9) rats, light columns – Vistar (n = 13) rats; II –animals predisposed to emotional stress: black – August (n = 16) rats.**

higher levels of the peptide in hypothalamus than their stress-prone counterparts. By the 3rd h of stress stimulation the hypothalamic DSIP content continued to increase (Fig. 2).

We have also investigated the contents of substance P (SP) in the hypothalamus of rats with different resistance to emotional stress using radioimmunoassay. Stress-resistant rats had significantly higher levels of SP in the hypothalamus than predisposed animals (Fig. 3).

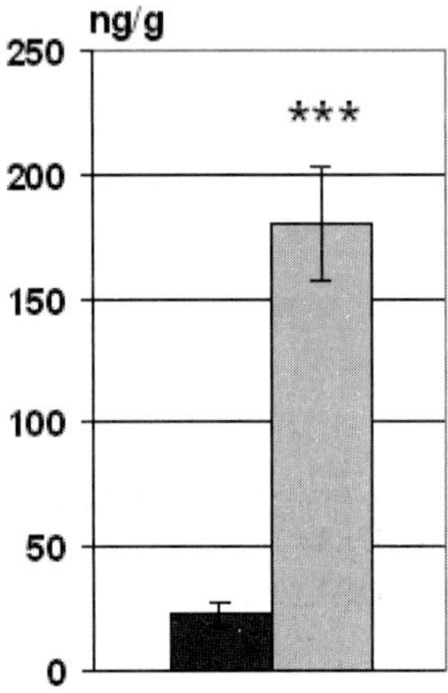

Fig. 3. **Content of Substance P (in ng/g) in hypothalamus of Vistar rats, resistant to emotional stress (light column) and August rats, predisposed to emotional stress (black column).**

SP content was studied in the hypothalamus of emotional stress-prone and emotional stress-resistant intact August (n = 51) and Vistar (n = 27) rats with its content substantially higher (8-fold; $\rho < 0.001$) in stress-resistant Vistar rats.

Our investigations revealed that under emotional stress sensitivity of hypothalamic neurons to microiontophoretically applied oligopeptides changed. It was shown that substance P significantly increased sensitivity of ventromedial hypothalamic neurons to microiontophoretic application of norepinephrine.

The changes in brain neurons sensitivity to biologically active substances have a complex integrative character. Some neurons at that increase sensitivity to neuropeptides and neuromediators others, on the contrary, inhibit it.

A single i.p. SP injection elevated significantly ($\rho < 0.01$) the hypothalamic SP content in all rats. That was consistent 24 h as well as 2 weeks after the injection. The duration of that effect (no less than 2 weeks) was equal to that of the stress-protective action after SP injection.

The contents of norepinephrine and dopamine in the hypothalamus in intact animals and those under emotional stress were studied in 83 August rats. After the i.p. SP injection the catecholamine content in the hypothalamus of control rats was found to rise sharply ($p < 0.05$). SP injected just before a 2-week immobilization stress prevented the stress-induced decrease in NE content in the hypothalamus. That is, norepinephrine content in the studied structures was kept at normal level. This effect was distinctly observable even 2 weeks after the SP injection. SP was also shown to increase the dopamine content in the hypothalamus of rats subjected to emotional stress. This suggests a modulating effect of SP on catecholamine brain metabolism under emotional stress. The drug evoked long-lasting changes of norepinephrine and dopamine content in the brain structures studied, incident to the duration of stress-protective effect after the SP injection (Fig. 4).

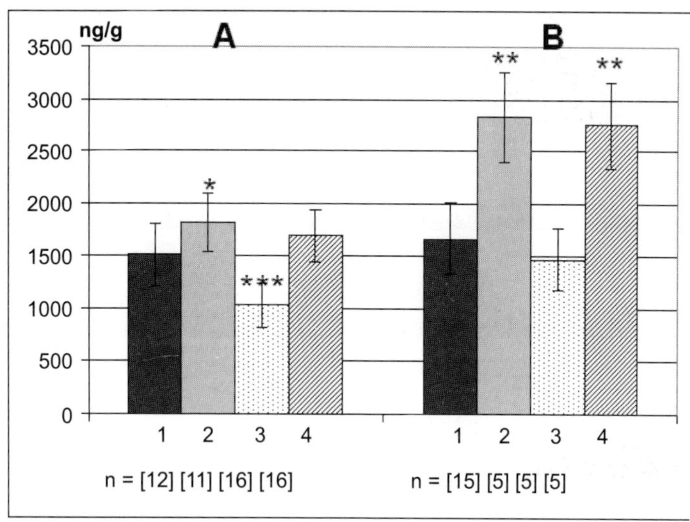

Fig. 4. Norepinephrine (A) and dopamine (B) contents (ng/g) in the hypothalamus of rats under different experimental conditions: 1 – control; 2 – SP injection (125 µg/kg); 3 – 1–2 weeks of 24-h immobilization stress alternating with 24-h rest periods; 4 – immobilization stress after the same SP dose. * – $p < 0.05$, * – $p < 0.001$ as related to control.**

Microiontophoretic experiments showed that 52 (57.1%) of the 91 neurons in hypothalamus were chemically sensitive to SP and responded to 111 of 221 SP applications. SP microiontophoresis revealed qualitative changes in neuronal responses under emotional stress. Thus, in 20 (34.5%) of the 58 neurons studied these changes were found, which was proved by the effect elicited in 29 of 183 SP applications. The changes in neuron chemosensitivity led to the appearance or disappearance of responses as well as to alterations in their pattern.

Before studying the effects of SP on brain neuron sensitivity to norepinephrine, we investigated their responses to norepinephrine itself. Two hundred and thirty four norepinephrine applications were performed on 92 neurons. Of these 54 (58.6%) were shown to be norepinephrine-sensitive, i.e. activating or inhibiting their firing on norepinephrine iontophoresis. Sixty-three neurons were tested at least twice and in 28 of these (44.4%) the sensitivity to norepinephrine changed upon repetitive applications.

SP effects on the chemosensitivity of the medial hypothalamic neurons were studied by comparing the responses of each to microiontophoretically applied norepinephrine before and after SP application. The changes in the neuron norepinephrine-sensitivity were detected as alterations of the response pattern after SP administration. Such alterations were observed in 26 (41.3%) of the 63 tested neurons. Of 114 SP applications 40 (35.1%) changed patterns of response to norepinephrine (Fig. 5).

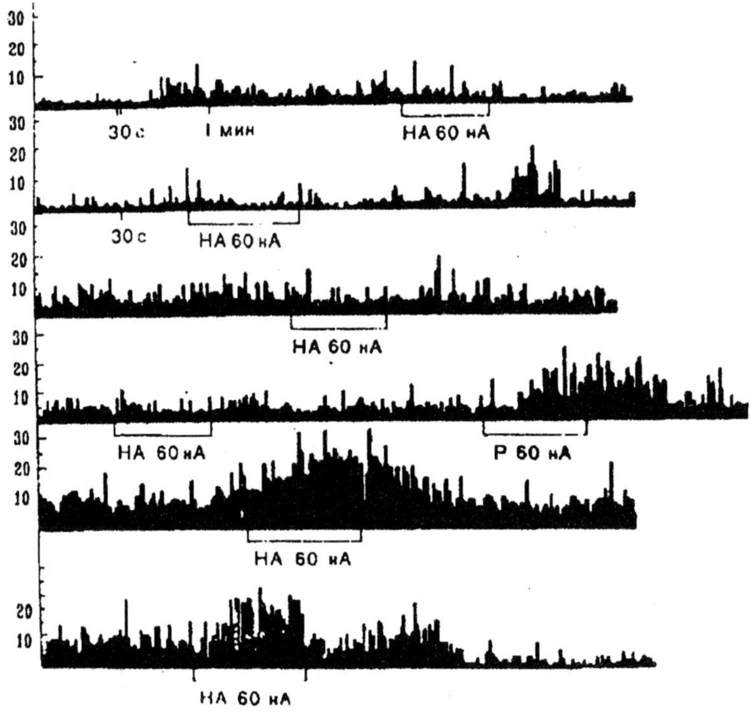

Fig. 5. Chemosensitivity of neurons in the medial hypothalamus to microiontophoretic applications of SP and NE. Ordinate: pulse rate (pps); abscissa: time (s). Microiontophoresis time and current values are indicated under each track.

This showed that SP can change the neuron sensitivity to norepinephrine, even when the peptide itself does not affect neuronal activity. Since the sensitivity to both SP and norepinephrine was shown to be subject to changes, the next question was: whether these changes were intercorrelated. Accordingly we performed 61 series of sequential microiontophoretic applications of norepinephrine and SP, designed so as to study whether the changes in neuron sensitivity to NE were related to changes in sensitivity to SP, and *vice* versa. The changes were judged to be correlated when the second testing showed simultaneous changes in responses to norepinephrine and SP, and uncorrelated when we observed a changing pattern of response to only one substance.

Our analysis revealed that in 7 of 61 tests the responses to SP and norepinephrine were correlated, i.e. one drug changed the responses to another.

We may thus infer that despite the changes in chemical sensitivity to norepinephrine observed in a part of neurons after SP application, the general neurochemical properties of the neuron population studied remained constant. This might be due to intercorrelation and quantitative equivalence of changes in chemical properties of individual neurons.

Our experiments showed that under basal conditions the number of Fos-IR cells in the paraventricular nuclei of hypothalamus in control animals potentially predisposed to emotional stress was greater ($p < 0.05$) than in potentially resistant control rats. In stress-resistant rats given DSIP, but not stressed, the number of Fos-positive cells was greater ($p < 0.001$) than after vehicle injection.

Additional application of oligopeptides to the rats with decreased content of oligopeptides increased their resistance to emotional stress.

A single intra-abdominal injection of SP significantly raised the survival of predisposed animals under chronic emotional stress and reduced manifestations of classic stress such as hypertrophy of adrenal glands, thymus involution, and stomach ulcer.

DSIP injection also increased rats' resistance to acute stress. The percentage of the survivals under acute emotional stress significantly increased. DSIP not only heightened the resistance to emotional stress but as well changed the level of catecholamines in brain hypothalamo-reticular structures. DSIP also increased the decreased under emotional stress SP content in hypothalamus and alternately changed β-endorphine and corticosteron content in hypothalamus and blood [7]. After SP and DSIP injections cardiovascular functions that were changed under emotional stress became normal.

In a series of experiments the Vistar rats subjected to immobilization received i.a. injections of DSIP. Animals tested under acute emotional stress could be divided into the following groups: (a) resistant; (b) adapted with increasing or decreasing mean BP; (c) predisposed which died at different times in the acute conflict situation.

DSIP increased the percentage of animals surviving the 3 h conflict situation. Under the effect of the peptide the percentage of stress-prone animals reduced from 65% to 25% and that of resistant rats increased from 14% to 60%.

Additional injections of Substance P and DSIP to prognostically stress-prone rats increased their resistance to acute stress loads.

Intraperitoneal DSIP injection decreased C-fos gene expression in hypothalamus in stress-prone and resistant to stress rats (Fig. 6).

DSIP SUPRESSES C-FOS EXPRESSION IN PARVENTRICULAR NUCLEUS OF HYPOTHALAMUS

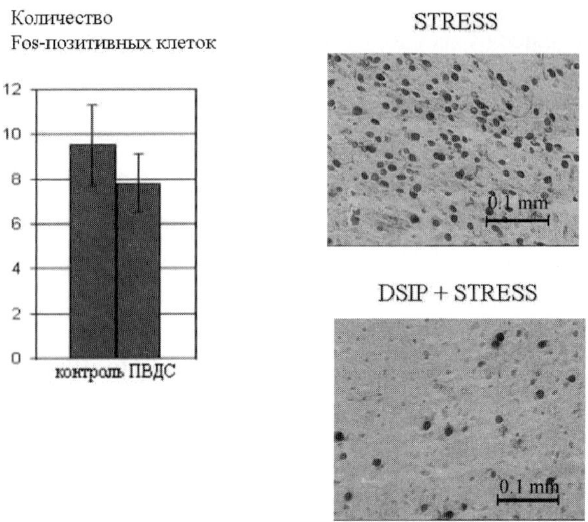

Fig. 6. DSIP suppresses c-fos expression in paraventricular nucleus of hypothalamus.

In stress-resistant and in stress-predisposed rats, again without stress, an injection of the ACTH analogue ($ACTH_{4-10}$)-Pro-Gly-Pro increased the number of Fos-IR cells in the paraventricular nuclei of hypothalamus compared with the control rats ($p < 0.002$ and $p < 0.02$ respectively).

In stressed rats the number of Fos-IR cells was greater ($p < 0.001$) than in the control group of rats. Rats predisposed to emotional stress that received injection of DSIP had a lower ($p < 0.001$) level of Fos expression in the paraventricular nuclei of hypothalamus after stress compared to predisposed stressed rats given vehicle. Similarly, in rats predisposed to emotional stress and injected with ($ACTH_{4-7}$)-Pro-Gly-Pro, the number of Fos-IR cells after stress was lower ($p < 0.0005$) than in predisposed stressed animals given vehicle [8].

Thus, either DSIP or $ACTH_{4-7}$ injection alone increased Fos-IR cell number in the paraventricular nuclei of hypothalamus in rats resistant to stress; ($ACTH_{4-7}$)-Pro-Gly-Pro also increased Fos-IR cell number in predisposed rats. DSIP decreased the number of Fos-IR cells in the paraventricular nuclei of hypothalamus of both resistant and predisposed rats subjected to restraint/electric shock stress. $ACTH_{4-7}$ decreased the stimulation of Fos expression in only predisposed stressed rats.

Microiontophoretic DSIP injection to single neurons of paraventricular and ventromedial hypothalamic nuclei along with inhibition of their spike activity produced in most cases a reaction of activation. At the same time microiontophoretic DSIP application to single hypothalamic neurons blocked their activating reaction to glutamate microiontophoretic application.

Intracerebroventricular application of cycloheximide – an inhibitor of protein synthesis – to brain neurons blocked the suppression of C-fos gene expression in paraventricular nuclei of hypothalamus in stress-prone animals.

So the experiments point to the distinction of brain chemical mechanisms in animals with different resistance to stressors and open new ways of elevation of resistance to emotional stress in stress-prone individuals on the bases of scientifically substantiated knowledge.

References

1. Sudakov K.V. New accents of a classical concept of stress. Bull. Exp. Biol. and Med. 123(2): 124–130. 1997. (In Russian).
2. MacLean P. The Triune Brain in Evolution. Role in Paleocerebral Functions. N-Y-Lond. Plenum Press. 1989. 672 p.
3. Papez J.W. A proposed mechanism of emotion. Arch. Neural and Psychiatry. 38: 725–743. 1937.
4. Sudakov K.V. Brain Mechanisms of Emotional stress. The Human Mind. Proceed. of the Fifth Int. Conf. 1990. 97–100.
5. Selye H. A syndrome produced by divers notions agents. Nature. 138: 32. 1936.
6. Koplik E.V., Gorbunova A.V., Salieva R.M. Open field test as a prognostic criterion of resistance to emotional stress in Vistar rats. Pavlov. J. High. Nerv. Activ. 4: 775–781. 1995. (In Russian).
7. Sudakov K.V., Cohlan J.P., Kotov A.V. et. al. Delta-sleep inducing peptide sequels in mechanisms of resistance to emotional stress. Ann. N-Y Acad. Sci. 771: 240–252. 1995.
8. Sudakov K.V., Umriukhin P.E., Koplik E.V., Anokhin K.V. Delta-Sleep Inducing Peptide (DSIP) and $ACTH_{4-10}$ Analoque Influence Fos-Induction in the Limbic Structures of the Rat Brain Under Emotional Stress. J. Stress. 4(2): 143–153. 2001.

Adaptation Biology and Medicine (Volume 5: Health Potentials)
Editors: L. Lukyanova, N. Takeda and P.K. Singal

16. Physiological Significance of Endogenous Satiety Substances During Food Intake

Yutaka Oomura[1*], Shuji Aou[1], Nobuaki Hori[1] and Kazuo Sasaki[2]

[1]Department of Integrative Physiology, Faculty of Medicine, Kyushu University, Fukuoka, 812-0054,
[2]Department of Life Science, Faculty of Engineering, Toyama University, Toyama, 930-8555, Japan

Abstract

During food intake, endogenous satiety proteins, one is acidic fibroblast growth factor (aFGF), released from the ependymal cells responding to the increase in CSF glucose and the other is leptin released from the adipocytes are responding to the increase in blood glucose. Both first reach the hypothalamus and suppress food intake and activate sympathetic nerve activity and facilitate learning and memory. We want to discuss these physiological mechanisms.

Keywords: Fibroblast growth factor, ependymal cells, adipocytes, mitogens and hypothalamus

Physiological Roles of Acidic Fibroblast Growth Factor (aFGF)

It has been well known that aFGF is one of the potent mitogens and these receptors are present throughout the nervous system. A postprandial increase in aFGF from 11.0 ± pg/ml (0.73 ± 0.2 pmol) to 10 ng/ml (0.7 nmol) was detected in rat CSF 2 h after feeding (Fig. 1). Released aFGF diffused into the brain parenchyma and was taken by neurons through their receptors in the hypothalamus and hippocampus. Food intake was suppressed by aFGF reached the lateral hypothalamic area, the feeding center (1). The minimum concentration of aFGF applied to the III cerebral ventricle for suppression of food intake was 3.3 pmol. Antibody against aFGF applied to the lateral hypothalamic area facilitated food intake. Neurons in the lateral hypothalamic area, especially the glucose-sensitive neurons were significantly suppressed by aFGF applied electrophoretically through the multibarreled electrode (1). Neurons in the ventromedial nucleus, the satiety center, were not affected by aFGF since the aFGF receptors could not be found there. Anyway, aFGF is one of the strongest anorectic substances at present. Histochemical staining of the brain with anti-aFGF antibody revealed aFGF in ependymal cells in the III cerebral ventricle walls (2, 3). However 2 h after meal or 300 mg/kg glucose injection intraperitoneally, the ependymal cells were clear and aFGF was observed in neurons in the lateral hypothalamic area, dorsomedial nucleus, septum, hippocampus, amygdala and other central regions, but not be in the ventromedial hypothalamic nucleus (3). When cDNA of the 30-100 base from the

*Address for Correspondence: Yutaka Oomura, Kawamo 6-14-17, Takarazuka, 665-0842, Japan, Tel. 81-797-85-0964, Fax: 81-797-85-0964, E-mail: y_oomura25@hotmail.com

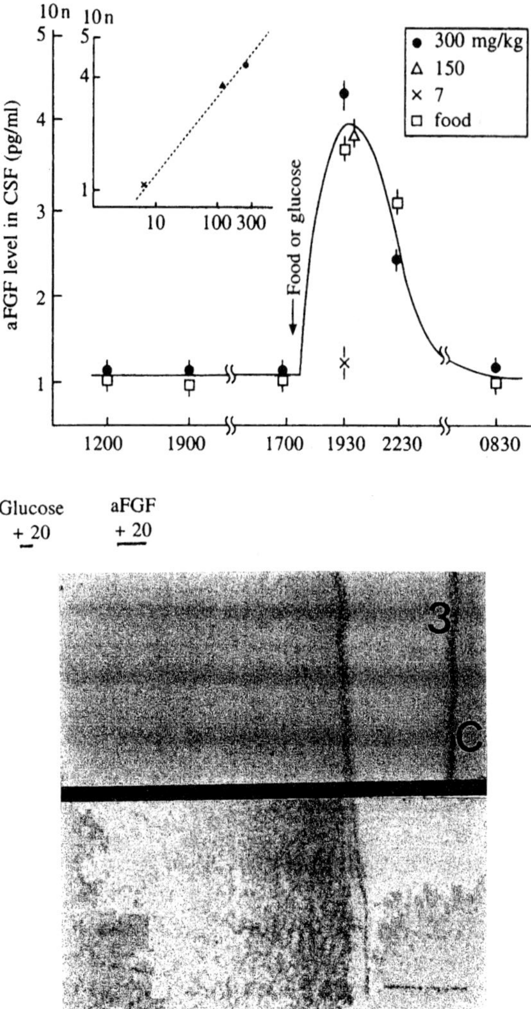

Fig. 1. Top. aFGF change in CSF before and after feeding or glucose injection. □, food intake. •, Δ, X, glucose IP injection. Mean ± SD. n = 7. Inset: Dose dependent effect of glucose injection; symbols, same as above; mean values only.
Bottom: Immunohistochemical staining by anti-aFGF antibody. Rat brain. Upper, before glucose application (300 mg/kg, ip). Dense staining of ependymal cells, but not neurons. Lower, 2 h after the glucose application. Cleaning of ependymal cells, but stain diffused from ependymal cells and moved to brain parenchyma. 3, third cerebral ventricle. C, cerebral parenchyma. Bars, 25 μm (l, 2).

middle part of aFGF were synthesized and hybridized, only ependymal cells were stained (2, 3). This revealed mRNA in the ependymal cell and hence aFGF is produced in those cells.

Facilitation of Learning and Memory and Synaptic Plasticity

The performances of emotional (passive avoidance task) and spatial learning and memory (water maze task) were significantly more reliable after taking aFGF in the hippocampus (4). On the emotional learning and memory task, 4 groups of mice were injected intraperitoneally with 300 mg/kg glucose 1, 2, 3 or 5 h before giving electric shock when they entered a dark box (acquisition). After 24 h, the same trials were repeated (retention, affective memory). As shown in Fig. 2 upper, behavior of the 1, 3 and 5

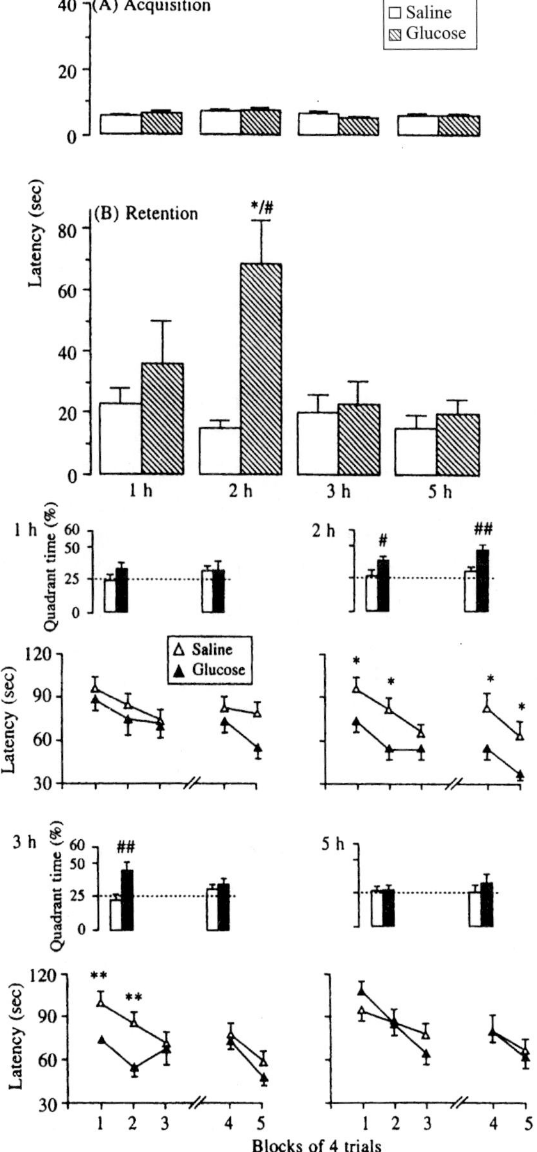

Fig. 2. **Effect of 300 mg/kg glucose intraperitoneal injection on learning and memory task by mice. Glucose or control saline administered at various time before the task. Upper (A and B): Passive avoidance task. Retention trial, carried out 24 h after acquisition trial. Ordinates: Latency (mean ± SEM) for acquisition trial (A) and retention trial (B). n = 20 for each group. $^{*}P < 0.002$, $^{\#}P < 0.002$ (vs 5h glucose group) (B). Lower: Water maze task. n = 8 for each group. Latency to escape onto the platform, shown for each block (mean ± SEM). Mean time spent in platform quadrant in the 100 s after the last trial of block 2 and 4 (probe test, shown as a percentage) is shown in the insets for glucose group (filled bar) and saline group. $^{*,\#}P < 0.05$, $^{**,\#\#}P < 0.01$ (4).**

h groups was not significantly different from that of the controls. Mice injected with Krebs Ringer solution were the controls. The latency of the 2 h group to enter the box (retention) was significantly longer than that of the controls ($p < 0.002$, $n = 20$) and that of the 1, 3 and 5 groups. In similar trials, anti-aFGF antibody was administrated into III cerebral ventricle and glucose was injected 30 min later. Based on the first trials as controls, all comparisons were made after 2 h. In these trials the latency tended to be longer than the acquisition time, but was significantly shorter than that of the glucose alone group. Infusion of the same amount of pre-immune IgG did not affect the retention.

The mean latency of 8 groups ($n = 10$ per group) to climb on the platform in the pool in the Morris water maze task was measured (4). Each mouse received one block of 4 trials in one day with inter-trial interval of 10 min. Each point (one block) is the mean of 4 times. Generally the mice showed a progressive decrease in latency as they underwent blocks 1, 2 and 3. However, only those mice that received their glucose injection 2 or 3 h before the trial showed a more rapid enhancement of performance than the corresponding saline controls. First block performance was significantly more rapid after intraperitoneal glucose administration 2 and 3 h before each first trial, than after saline administration at the same intervals (Student's t test). On the other hand mice that received glucose injections 1 and 5 h before task had latencies similar to those of the saline control group. These results indicate that the glucose application on 2 and 3 h groups facilitated spatial learning and memory with time dependent induction. The probe test, carried out 10 min after the last trail in block 2 with the platform removed, revealed that the 2 h and 3 h glucose injected mice spent longer time than the saline controls in the quadrant in which the platform had been situated, thus indicating a better spatial awareness than the corresponding saline controls as shown in Fig. 2 upper insets.

At the cellular level to study the learning and memory process, as shown in Fig. 3, long-term potentiation (LTP) to hippocampal CA1 neurons was measured. The amplitudes of population excitatory postsynaptic potentials (pEPSP) were recorded from the apical dendrites layer of the CA1 pyramidal cells by mean of saline filled glass electrodes by stimulation of Schaeffer collateral/ commissural afferents. LTP was induced by the application of a brief high-frequency stimulation (7 pulses at 100 Hz). As shown in the figure, LTP was dose-dependently facilitated by aFGF (5).

Effect of aFGF on Hypothalamo-Pituitary-Adrenal Axis and Neuroimmune Function

Significantly marked dose dependent increase in plasma ACTH, corticosterone as well as norepinephrine was detected after intracerebroventricular (0.7-7 pmol/rat) or intravenous (0.7-7 nmol/ rat) administration of aFGF. Concomitant increases in dose- dependency occurred in the activity in the efferent sympathetic nerve as well as the body temperature. These results were due to the facilitation of the neuronal activity of parvocellular neuron groups in the paraventricular nucleus by aFGF. These findings suggest that aFGF has activated the hypothalamo-pituitary-adrenal axis and sympathetic outflow (6).

Phagocytosis in peritoneal macrophages in vitro was significantly enhanced by aFGF with a dose-dependent manner from 10^{-9} to 10^{-8} M (7).

These results indicate that aFGF is not only the most potent substance yet found for the suppression of feeding, but also extremely effective substance as memory facilitation, sympathetic efferent outflow as well as biodefence activation.

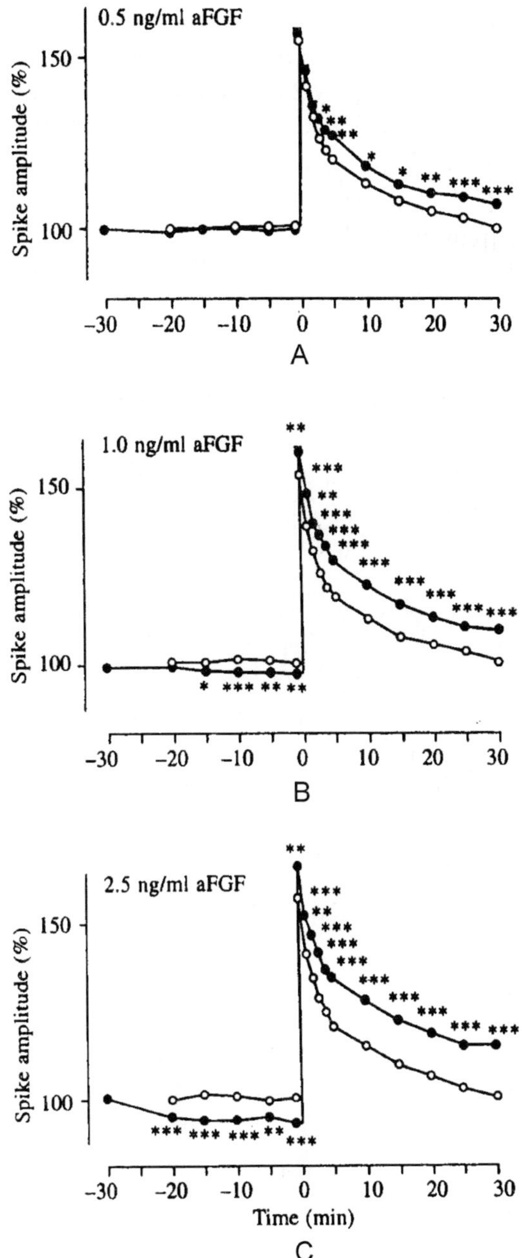

Fig. 3. Effect of aFGF on LTP on CA1 neurons in hippocampal slices induced by brief tetanic stimulation (7 pulses, at 100 Hz). Ordinate: Amplitudes of population excitatory postsynaptic potentials (pEPSP), expressed % change. Mean basal pEPSP before tetanic stimulation defined as 100 %. After the control potentiation (open circles) induced by the first tetanic stimulation was observed for 30 min in Krebs Ringer solution, aFGF was perfused for 30 min. The second tetanic stimulation was applied 30 min after beginning of aFGF perfusion (A, B and C). Effect of 0.5 (n = 5).1.0 (n = 9) and 2.5 (n = 9) ng/ml aFGF on potentiation (filled circles) induced by the second titanic stimulation, respectively. $^{*}P < 0.05$, $^{**}P < 0.01$, $^{***}P < 0.001$ (5).

II. Physiological Role of Leptin

Leptin, a protein encoded by an obesity gene and expressed in adipocytes, is increased in rat plasma during eating, the rise being from approximately 2.3 ng/ml to 3.8 ng/ml (i.e. to 2.4×10^{-10} M). In the mouse, approximately 1/5000 of peritoneally administered ^{125}I-leptin reaches the central nervous system (8). Leptin entered the brain, approximately at 10^{-13} to $^{-12}$M concentrations and binds to widely distributed long-form leptin receptors, especially in the hypothalamus and affects the feeding behavior through alternating neuronal activity in feeding related hypothalamic neurons (9). Food intake was suppressed by 300 nM leptin, administered in the peritoneal cavity. The glucose-sensitive neurons in the lateral hypothalamic area and the glucoreceptor neurons in the ventromedial hypothalamic nucleus were significantly suppressed and facilitated respectively.

Facilitation of Learning and Memory and Synaptic Plasticity

The effects of leptin, applied intravenously on emotional learning and memory task, are measured (10). The data are presented in Fig. 4. On day 1, all groups performed similarly. On day 2 - 4, retention latency increased day by day in all five groups. The performance of rats in 0.5 and 500 µg/kg leptin groups was not significantly better than the control (vehicle) group on days 2 through 4. On day 2 and 3, a significant dose-dependent increase in retention latency was seen during learning in the 5.0 and 50 µg /kg groups. Leptin, at 5 and 50 µg/kg, therefore improved learning and retention of passive shock-avoidance responses. These results indicate that leptin shows an inverted U dose-related function in terms of its effects on emotional learning and memory.

As shown in Fig. 5 A and B, there were no differences in swimming time or distance between the five groups on day 1. On day 2 through 4, swimming time and distance became short, day by day, in all groups including the vehicle group. The only value significantly shorter than that for the control group

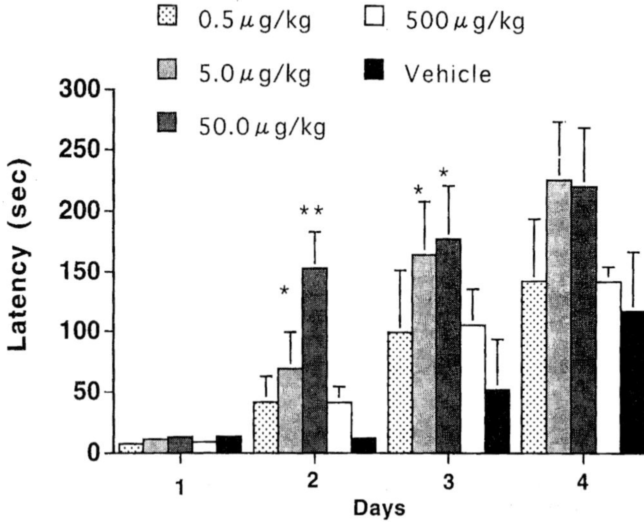

Fig. 4. Enhancement of passive avoidance learning by two concentrations of leptin. Leptin 0.5 µg/kg group, n = 16; other groups, n = 20. By comparison with the control group on day 2 and 3, significantly longer retention latencies occurred with 5 and 50 µg/kg leptin, but not with 500 mg/kg. $^{*}P < 0.05$, $^{**}P < 0.01$ (10).

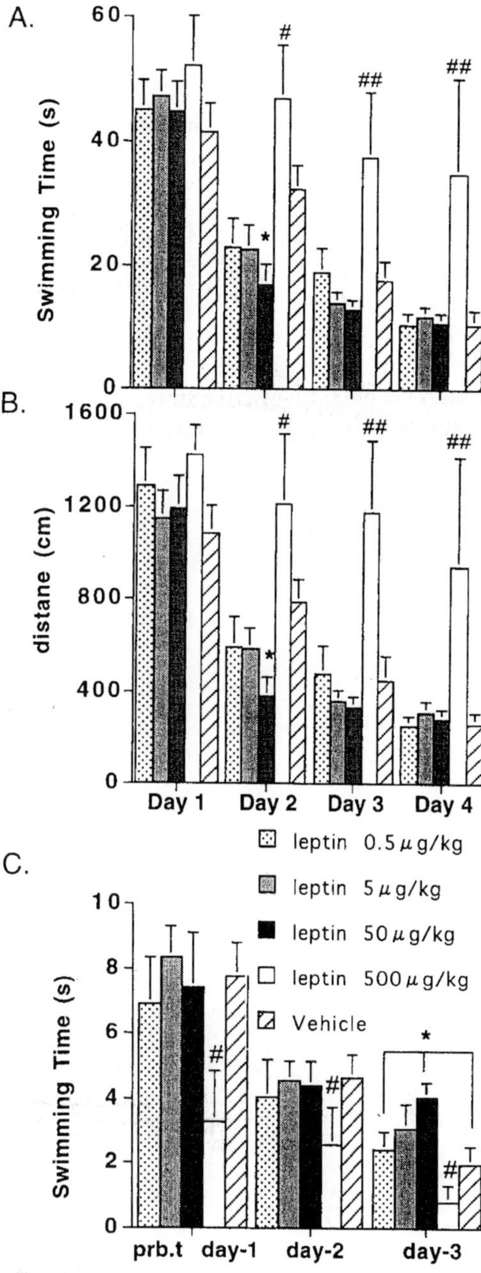

Fig. 5. Enhancement of water-maze performance by the low concentration of leptin and impairment by the high concentration of leptin. Leptin 0.5 μg/kg group, n = 16, other group, n = 20. The shortest swimming time (A) and the shortest swimming distance (B) occurred on day 2 in the 50 μg/kg leptin group. Swimming time and distance were each significantly greater at 500 μg/kg than in controls on day 2 to 4. (C), probe test after withdrawing the platform. The 50 μg/kg leptin group spent a longer time in the goal area than the vehicle group on probe test (prb.t) day 3 ($p < 0.05$), whereas the 500 μg /kg leptin group spent less time than the vehicle group on probe test days 1, 2, and 3 ($p < 0.05$). $^{*}P < 0.05$, $^{**}P < 0.01$ (10).

was obtained on day 2 for the 50 μg/kg group. Swimming time and distance were significantly longer for the 500 μg/kg group than for the control group on day 2 through 4. These data obtained in the Morris water maze task indicate that leptin has a U-shaped dose-related function in terms of its effects on spatial learning and memory. In probe test on day 3, as shown in Fig 5C, the 50 μg/kg leptin group spent significantly more time swimming in the goal-area than the vehicle group. As a result of the poor learning and memory in the Morris water maze, the 500 μg/kg group spent significantly shorter time in the goal-area on probe test days 1 to 3 than the vehicle control. These results obtained in the probe trials suggest that 50 μg/kg leptin consolidates memory process so that it can last at least for 2 days after the Morris water maze test.

Figure 6 provides the effects of various concentrations of leptin on the LTP on CA1 neurons. The percentage of the increase in pEPSP amplitude in the 10^{-12} M leptin-treated hippocampal slice was approximately 250%, whereas that in the normal Krebs Ringer treated slice was approximately 150 %. At the high concentration of 10^{-10} M leptin LTP was suppressed.

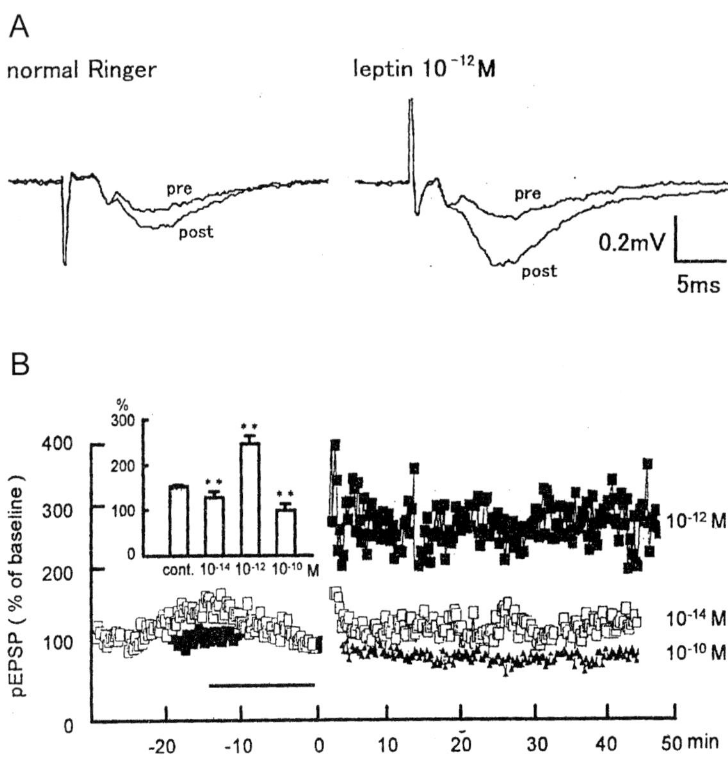

Fig. 6. Leptin effects on LTP in CA1 neurons. A: Example records of pEPSPs 20 min before (pre) and after (post) brief tetanic stimulation. Left, normal Krebs Ringer; right, 10^{-12} M leptin. Leptin applied for 15 min just before tetanic stimulation. B: Percent change of pEPSP amplitudes. Tetanic stimulation took place at 0 time (see abscissa). Solid horizontal line, leptin application for 15 min. Ordinate, size of pEPSP as percentage of baseline (100%). Representative data from individual slices, each from a different animal. Inset, mean size of pEPSPs (as percentage of baseline) following leptin administration; measurements were made over a 10-min period (n=3, 10^{-14}, 10^{-12}, and 10^{-10}M leptin; n = 6 control). $^{*}P < 0.05$ (10).

Effect of Leptin on Postsynaptic Responses in CA1 Neurons

Postsynaptic potentials recorded intracellularly from CA1 neurons using a glass pipette electrode of 0.1 μm tip diameter were evoked by microelectrophoretic applications of quisqualic acid (Quis), y-aminobutyric acid (GABA) and N-methyl D aspartic acid (NMDA) through a multibarrel array to the distal dendrites of the same neuron (10). The postsynaptic responses evoked by Quis, GABA and NMDA ejected on to the apical dendrites of CA1 neurons are obtained in normal Krebs Ringer perfusate. The measurements were repeated in the same neuron over the next 15 min with 10^{-12} M leptin in the perfusate, and then repeated in normal Krebs Ringer solution. As shown in Fig. 7 left, postsynaptic responses to Quis, GABA and NMDA tended to enhancement during the application of 10^{-2} M leptin. However, as shown in Fig. 7 right with 10^{-10} M leptin, an attenuation of postsynaptic responses was observed for all three transmitters and usually endured for approximately 20 min after the cessation of leptin application.

Effects of Leptin on Intracellular Ca^{2+} in Single Isolated CA1 Neurons

As mentioned above, the different effects caused by the different leptin concentrations could be related to the $[Ca^{2+}]_i$ in the postsynaptic CA1 neurons. Therefore we measured the in $[Ca^{2+}]_i$ single CA1

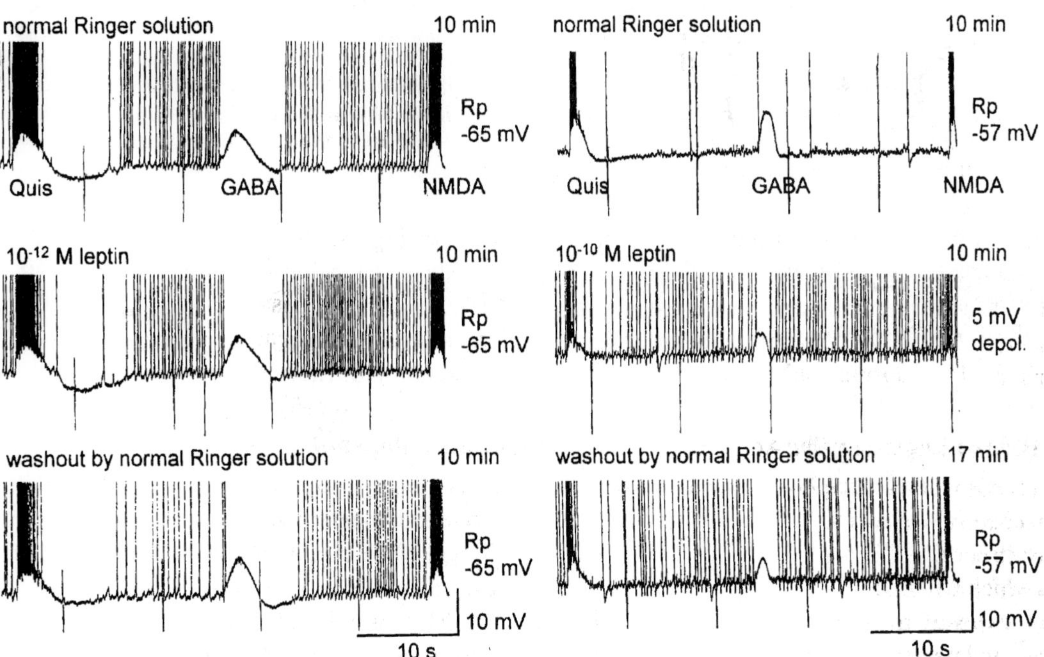

Fig. 7. Effects of leptin on postsynaptic responses induced in a CA1 neuron by electrophoretic applications (25 nA, for 1 s) of Quis, GABA and NMDA.
Upper left and upper right traces: Normal Krebs Ringer solution, 10 min after recording had stabilized. Middle left trace: 10 min after 10^{-12} M leptin application. Lower left trace: 10 min after washout. No observable changes in postsynaptic responses following 10^{-12} M leptin application. Middle right trace: 10 min after 10^{-10} M leptin application showing suppression of postsynaptic responses to all three transmitters and a 5 mV depolarization (depol) in the membrane potential. Lower right trace: 17 min after washout (10).

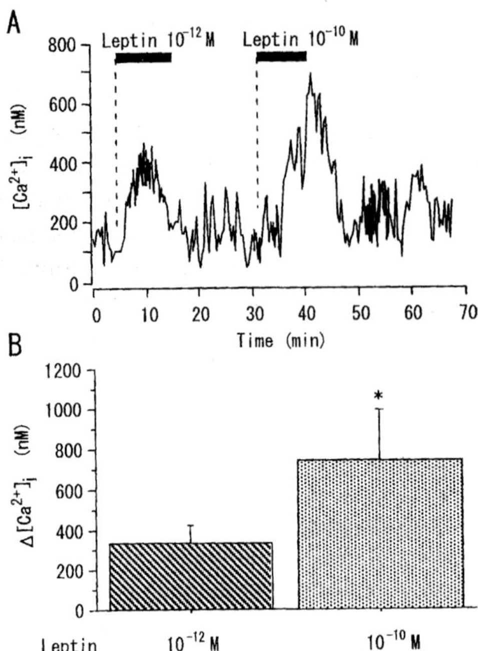

Fig. 8. Increased $[Ca^{2+}]_i$ in CA 1 neurons following leptin application. A: $[Ca^{2+}]_i$ in a single isolated CA 1 neuron. Upper horizontal lines, two applications of leptin, each for 10 min. Broken line, beginning of exposure to leptin. B: Increase in mean $[Ca^{2+}]_i$ inducedby 10^{-12} M and 10^{-10} M leptin (n = 6, each group). $^*P < 0.05$ between the two leptin concentrations (10).

neurons at different leptin concentrations. As shown in Fig. 8, the basal level of $[Ca^{2+}]_i$ during superfusion with Krebs Ringer solution was 115 ± 44 nM (n = 6). Administration of leptin at 10^{-12} and 10^{-10} M induced marked increase in $[Ca^{2+}]_i$ was 329 ± 97 nM (n = 6) for 10^{-12} M leptin and 739 ± 250 nM (n = 6) for 10^{-10} M leptin, the latter being significantly larger than the former, and returned to its original level during washing within 10 min after the cessation of the leptin application (10).

Effect of Leptin on the Activity of Ca^{2+} /Calmodulin -dependent Protein Kinase II

Activation and autophosphorylation of CaMK II is one of the molecular signal transduction mechanisms known to LTP in the hippocampus and to learning and memory (11). The purpose was to determine the effects of leptin, at 10^{-12}M and 10^{-10} M, on CaMK II activity in the CA1 regions of slices in which LTP had been induced as described above. Control slices were stimulated and potentiated, but not exposed to leptin. As shown in Table 1, 10^{-12}M leptin induced a significant increase in Ca^{2+}-independent CaMK II activity without a significant increase in either total CaMK II activity or "Ca^{2+}-independence" (Ca^{2+}-independent activity as a percentage of total activity). On the other hand, the Ca^{2+}-independent activity and "Ca^{2+}-independence" were both significantly lower at 10^{-10} M than at 10^{-12} M leptin (10). These results indicate that a specific dose-dependent increase in CaMK II activity at 10^{-12} M leptin is associated with the induction of LTP seen with the concentration of leptin.

Table 1. Effects of leptin on CaMK II activity.

	(A) "Ca^{2+}–independent Activity (nmol/mg/min)	(B) Total Activity (nmol/mg/min)	"Ca^{2+}–independence" (A/B × 100%)
Control	25.3 ± 6.87	232.5 ± 15.46	15.3 ± 3.38
n = 8			
10^{-12}M leptin	42.3 ± 5.15†	239.1 ± 18.65	18.0 ± 3.25
n = 10			
10^{-10M} leptin	31.3 ± 3.74†	221.6 ± 21.02	14.2 ± 2.55*

Effects of leptin on absolute level of Ca^{2+} independent activity (nmol/mg/min) of CaMK II, together with a measure of "Ca^{2+}-independence" (the former expressed as a percentage of the latter), in rat CA 1 hippocampal tissue 20 min after brief tetanic stimulation. Absolute Ca^{2+}-independet activity was significantly greater at 10^{-12} M leptin than in control ($^{\dagger}p < 0.05$). Both measures of Ca^{2+}-dependent activity (absolute value and as percentage of total) were significantly smaller at 10^{-10} M leptin than at 10^{-12} M leptin ($^{\dagger}P < 0.05$) (10).

Physiological Mechanism of Leptin Effects

In the present in vivo experiments, emotional and spatial learning and memory tasks were clearly enhanced by 50 μg/kg leptin but impaired by 500 μg/kg leptin. In the in vitro slice experiments, 10^{-12} M leptin enhanced LTP and 10^{-10} M suppressed it. Thus, the present in vivo and in vitro experiments demonstrate that leptin shows an inverted U dose-related function in terms of its effects on learning and memory and LTP. The close association between enhanced hippocampal LTP and facilitated learning and memory has been reported (12). If we assume that a 200-250 g rat has a plasma volume of approximately 10 ml, our 50 μg/kg doses of leptin would be equivalent in plasma to about 10^{-8} M and 500 μg/kg dose to 10^{-7} M. As mentioned before, less than 1/5000 of ^{125}I-leptin administered peritonealy reaches the central nervous system (8), since 50 μg/kg leptin is the physiological concentration in the brain whereas 500 μg/kg is the pharmacological concentration. The enhancement of LTP at 10^{-12} M leptin was partly contributed to the increase in the amount of presynaptic transmitter release (10) and the suppression of LTP at 10^{-10} M leptin was contributed to the decrease in the postsynaptic sensitivity to the transmitters (Fig. 7).

It is well known that in the NMDA receptor-dependent LTP in the hippocampus, intracellular Ca^{2+}, which increases transiently via NMDA receptors following tetanic stimulation, binds to calmodulin, and Ca^{2+}/calmodulin activated CaMK II through autophosphorylation process at the threonine 286/287 residues (11). Simultaneously, Ca^{2+}/calmodulin-activated cyclic-protein kinase A phosphorylates the inhibitor 1, and thereby inhibitor 1 cannot activate PP 1. PP 1 is a strong phosphatase and acts as a factor of the dephosphorylation of CaMK II. Therefore the activation of protein kinase A inhibits the CaMK II-dephosphorylation. Thus persistent potentiation of CaMK II might be elicited in the LTP. Indeed, it has been reported that persistent activation of the Ca^{2+} independent activity of CaMK II as ell as of its "Ca^{2+}-independence" is closely associated with LTP production in the hippocampus (11, 16). These studies indicate that CaMK II, especially its Ca^{2+}-independent activity, plays a key role in the induction of NMDA receptor-dependent LTP in the hippocampus. Further, the facilitation and inhibition of CaMK II activity might be explained by different modes of $[Ca^{2+}]_i$ increase. It is known that a moderate increase of $[Ca^{2+}]_i$ in CA1 neurons activates the Ca^{2+}-independent activity of CaMK II, whereas a large and prolong increase of $[Ca^{2+}]_i$ inhibits it, possibly through the activation of calcineurin (13). Calcineurin is a protein phosphatase that activates PP1 through the dephosphorylation of inhibitor 1, and activated PP 1 dephosphorylates CaMK II, thereby deactivating CaMK II. As shown in Fig. 8, an increase more than 500 nM $[Ca^{2+}]_i$ at the 10^{-10} M leptin condition in CA1 neurons for 10 min preferentially activates calcineurin and partially deactivates CaMK II (14). Thus high $[Ca^{2+}]_i$ at 10^{-10}M leptin, instead of activation of CaMK II may activate the calcineurin.

The same as aFGF, the phagocytosis in peritoneal macrophages was enhanced by leptin in a dose dependent manner from 10^{-9} M to 10^{-8} M.

Our previous study revealed that Zucker *fa/fa* rats and *db/db* mice that have abnormal leptin receptors do not produce both the increase of Ca^{2+}-independent activity of CaMK II and LTP following tetanic stimulation, and consequently show impaired spatial learning and memory performance (17).

Conclusion

An increase in CSF glucose during food intake, makes release aFGF from the ependymal cells. Similarly an increase in plasma glucose during food intake makes release leptin from the adipocytes and this enters the brain. Both substances could act not only as endogenous satiety proteins but also as prime hippocampal synapses for an enhancement of LTP and thereby facilitate learning and memory function. Both proteins facilitate neuronal activity of the parvocellular part of the paraventricular nucleus and thus the activity of the efferent sympathetic nerve is facilitated. These findings suggest that aFGF and leptin activate the hypothalamic-pituitary-adrenal axis and sympathetic outflow. These evidences indicate that food intake maintains not only the body energy homeostasis but also the prime brain function as well.

References

1. Hanai K, Oomura Y, Kai Y, Nishikawa K, Morita H, Plata-Salaman CR. Central action of acidic fibroblast growth factor in feeding regulation. Am J Physiol. 1989; 256: R217-R223.
2. Oomura Y, Sasaki K, Hanai K. Chemical and neuronal regulation of food intake. In: Oomura Y, Tarui S, Baba S, Inoue S, eds., Progress in Obesity Research. John Libby, London, 1991: 3-12.
3. Tooyama I, Hara Y, Yoshihara O, Oomura Y, Sasaki K, Muto T, Suzuki K, Hanai K, Kimura H. Production of antisera to acidic fibroblast growth factor and their application to immunohistochemical study in the rat brain. Neuroscience 1991; 40: 769-779.
4. Li AJ, Oomura Y, Sasaki K, Suzuki K, Tooyama I, Hanai K, Kimura H, Hori T. A single pre-training glucose injection induces memory facilitation in rodent performing various tasks: contribution of acidic fibroblast growth factor. Neuroscience 1998; 85: 785- 794.
5. Sasaki K, Oomura Y, Figrov A, Yagi H. Acidic fibroblast growth factor facilitates generation of long-term potentiation in rat hippocampal slices. Brain Res Bull. 1994; 33: 505-511.
6. Matsumoto I, Oomura Y, Niijima A, Sasaki K, Aikawa T. Acidic fibroblast growth factor activates hypothalamo- pituitary- adrenocortical axis in rats. Am J Physiol. 1998; 274: R503-R509.
7. Ichinose M, Sawada M, Sasaki K, Oomura Y. Enhancement of phagocytosis in mouse peritoneal macrophages by fragments of acidic fibroblast growth factor (aFGF). Int Immunopharmcol. 1998; 20: 193-204.
8. van Heek M, Mullins DE, Wirth MA, Graziano MP, Fawzi AB, Compton DS, France CF, Hoos LM, Casale RL, Sybertz EJ, Strader CD, Davis HR. The relationship of tissue localization, distribution and turnover to feeding after intraperitoneal ^{125}I-leptin administration to ob/ob and db/db mice. Horm Metab Res, 1996; 28: 653-658.

9. Shiraishi T, Oomura Y, Sasaki K, Wayner MJ. Effects of leptin and orexin A on food intake and feeding related hypothalamic neurons. Physiol Behav 2000; 71: 251-261
10. Oomura Y, Hori N, Shiraishi T, Fukunaga K, Ishibashi M, Takeda H, Tsuji M, Matsumiya T, Aou S, Li XL, Kono D, Uramura K, Sougawa H, Yada T, Wayner MJ, Sasaki K. Leptin facilitates learning and memory performance and enhances hippocampal CA 1 long-term potentiation and CaMK II phosphorylation in rats. Peptides 2006; 27: 2738-2749.
11. Fukunaga K, Stoppini I, Miyamoto E, Muller D. Long-term potentiation is associated with an increased activity of Ca^{2+}/calmodulin protein kinase II. Biol Chem. 1993; 268: 7863-7867.
12. Manabe T, Noda Y, Mamiya T, Katagiri H, Houtani T, Nishi M, Noda T, Takahashi T, Sugimoto T, Nabeshima T, Takeshima H. Facilitation of long-term potentiation and memory in mice lacking nociceptin receptors. Nature, 1998; 394: 577-581.
13. Lisman J. The CaM kinase II hypothsis for the storage of synaptic memory. Trends Neurosci. 1994;17: 406-412.
14. Morioka M, Fukunaga K, Nagahiro S, Kurino M, Ushio Y, Miyamoto E. Glutamate-induced loss of Ca^{2+}/ calmodulin dependent protein kinase II activity in cultured rat hippocampal neurons. J Neurochem, 1995; 84: 2132-2139.
15. McKernan MG, Shinick-Gallagher P. Fear conditioning induces a lasting potentiation of synaptic currents in vitro. Nature 1997; 390: 607-611.
16. Makhinson M, Chotiner JK, Watson JB, O'Dell, TJ. Adenyl cyclase activation modulates activity-dependent changes in synaptic strength and Ca^{2+}/calmodulin -dependent kinase II autophosphorylation. J Neurosci. 1999; 19: 2500-2510.
17. Li XL, Aou S, Oomura Y, Hori N, Fukunaga K, Hori T. Impairment of long-term potentiation and spatial memory in leptin receptor-deficient rodents. Neuroscience, 2002; 113: 607-615.

Adaptation Biology and Medicine (Volume 5: Health Potentials)
Editors: L. Lukyanova, N. Takeda and P.K. Singal

17. Clinicopathological Aspects and Risk Factors of Breast Cancer: Adaptation of Biologically Aggressive Carcinomas in Young Arab Women

D.T. Lincoln[1*], P.K. Singal[2], M. Al-Khaledi[3], F. Al-Yatama[3] and F.M. Mohammed[3]

[1]Entity Systems, Independent Research Foundation, 53 Ashburton Street, Chapel Hill QLD 4069, Australia
[2]Institute of Cardiovascular Sciences, St. Boniface General Hospital Research Center, University of Manitoba, Winnipeg, Canada
[3]Faculty of Allied Health Sciences, Kuwait University, Kuwait.

Keywords: Heterogeneous, mammography, menarche, adenocarcinoma and nuclear pleomorphism

Introduction

Risk Factors and Prognostic Markers in Breast Cancer

Breast cancer is an extremely heterogeneous disease. It is characterized by a long history and marked heterogeneity in growth rates and clinical manifestations. It is estimated that the pre-clinical course of breast cancer is between five and ten years for most patients and that the interval from the point when the cancer can be diagnosed by Mammography but not by physical examination or symptoms to the time when the cancer is 1 cm. in diameter is in the range of several years for most patients.The clinical course ranges from 5 to 40 years, and it is not unusual for patients to secure more than 5 years after the initial diagnosis and treatment (Feigelson *et al.*, 1997). Substantial progress has been made in the diagnosis and treatment of breast cancer and to detect "risk factors" that will help to identify those women likely to develop breast cancer apart from the genetic factors that contribute to these risks (Kerlikowska *et al.*, 1997).

Certain risk factors have been identified that may produce insight into the genesis of breast cancer (Kerlikowska *et al.*, 1997).

Established risk factors include inherited genetic defects, menarche before the age of 12, menopause after 55 years of age, having either no pregnancy or late child bearing, no lactation, exposure to radiation, prolonged use of hormone replacement therapy, increased breast density,

*Address for Correspondence: Dr. David Lincoln, Entity Systems, Independent Research Foundation, 53 Ashburton Street, Chapel Hill, QLD 4069, Australia. E-mail: dtlhabibi@yahoo.com.au

higher socioeconomic status and obesity. Most of these factors can be linked with increased life time exposure to oestrogen and other hormones. Risk factors can be distinguished into the following classes (Figure 1):

Risk factors that directly cause breast cancer:

These include ionizing radiation and inherited genetic damage. Both can be said to cause breast cancer in that they produce alterations in cell growth, repair and metabolism which are known to be linked with the development of breast cancer.

1. Risk factors that extend the time period of vulnerability of breast cancer by prolonging breast development and consequent susceptibility to damage:

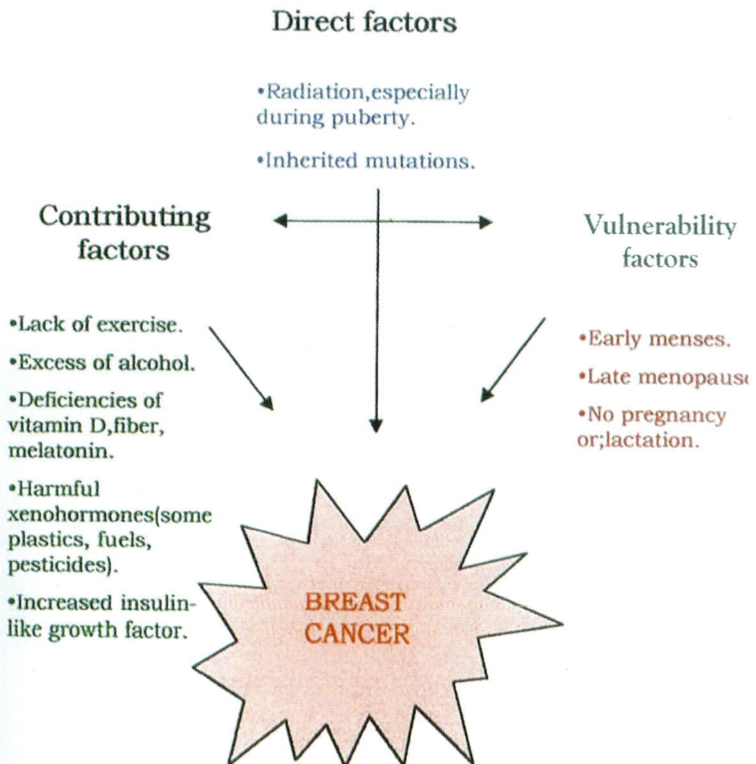

Fig. 1. Risk factors can be distinguished in terms of their ability to cause breast cancer directly, through genetic damage or by altering hormonal metabolism. Vulnerability factors prolong the duration of breast cell growth, while contributing factors can distort hormone levels. (Adapted from Davis *et al.*, 1998).

Vulnerability factors for breast cancer development are those that extend the time during which the breast is growing and the rate of growth. Early menses, no lactation and late menopause prolong the time period during which the developing breast is potentially exposed to astraddle and deleterious agents.

2. Risk factors that contribute to altered levels of cancer causing endogenous or xenohormones: Contributing factors can affect the overall exposure to circulating hormones and can interact with vulnerability factors. Some hormones are potent stimulators of cell growth and some hormonal metabolites can bind to DNA and trigger aberrant cell growth. Alcohol, lack of exercise, diets low in fibre and vitamin D, and exposure to some pesticides, solvents, fuels and pharmaceuticals have the ability to increase the total amount of bio-available astraddle

Pathology of Breast Cancer

More than 95 percent of breast malignancies arise from the breast epithelial elements and are classified as carcinomas (specifically adenocarcinoma). Breast carcinoma encompasses a diverse group of lesions that differ in microscopic appearance and biologic behaviour (Bässler, 1995). It can be divided into major groups: Non invasive (*in situ*) cancers in which the tumour cells remain confined to the ducts or lobules and show no microscopic evidence of invasion into the surrounding stroma, and invasive (infiltrating) cancers characterized by tumour cells that invade the breast stroma (Bässler, 1995). The non invasive lesions are of two main types, ductal carcinoma *in situ* (intraductal carcinoma) and lobular carcinoma *in situ*. These two types of cancers differ in their clinical features, morphology and biological behavior (Silverstein, 1995). The microscopic appearance of lobular carcinoma *in situ* is histologically heterogeneous.

Most tumours of the breast are invasive adenocarcinomas, which arise from the terminal ducts and lobular units, forming invasive ductal and lobular carcinomas. The invasive breast carcinomas also comprise a heterogeneous group of lesions (Bässler, 1995). There are six main types of invasive breast carcinoma, which are named according to their histological pattern and occur with varying frequency. Invasive ductal carcinoma is the most common type of breast cancer. It may occur as a pure form or appear mixed with another pattern of carcinoma, most commonly lobular carcinoma. In contrast to in situ neoplasia, there is invasion into local tissue, with the possibility of metastatic spread. These tumours are usually hard, grey, gritty masses that invade the surrounding tissue. On microscopic examination, infiltrating ductal carcinomas are divided into three grades, well differentiated (Grade I), moderately differentiated (Grade II) and poorly differentiated (Grade III), based on a combination of architectural and cytologic features (Bässler, 1995). The prognostic significance of this grading system has been repeatedly documented in clinical studies (Page, 1991).

Infiltrating lobular carcinomas are the second-most common type of invasive breast cancer. Importantly, lobular invasive carcinoma is often multifocal within the breast and these tumours have a higher frequency of bilaterality and multicentricity of breast involvement than infiltrating duct carcinoma. Other variants of invasive breast carcinomas all have a better prognosis than infiltrating ductal carcinomas (Bässler, 1995). Tubular carcinoma, mucoid carcinoma and medullary carcinoma are less common and they are associated with a better prognosis. A number of histological features of breast carcinoma are associated with an unfavourable prognosis (Page, 1991). These include lymphatic vessel and blood vessel invasion, tumour necrosis and a prominent lymphoplasmacytic infiltrate. An extensive intraductal component within an infiltrating ductal carcinoma is associated with an increased risk of tumour recurrence in the breast after conservative surgery and radiation therapy.

Tubular carcinoma accounts for approximately 5 percent of all breast cancers. It has a favourable prognosis, with a five-year survival rate of 95 to 100%. Tubular carcinoma usually presents as a small mass, 1 cm in diameter or smaller, and usually is detected by mammography. The lesions appear firm, white and stellate. Microscopically these tumours are composed of well differentiated cells that form regular tubular structures resembling small glands or tubules that have round, oval or angular shapes. The glands are uniformly distributed in centrally dense fibrous stroma with prominent elastosis.

Mucinous carcinoma accounts for approximately 2 to 3% of all breast cancers, commonly occurring in post-menopausal women. Typical mucinous carcinomas have a very good prognosis with a 10 year survival rate in the range of 85 to 90%. Tumours present as soft palpable masses that produce vague, nonspecific findings on mammography. On cross section they appear circumscribed, glistening, tan, soft and gelatinous. Histologically, the tumours are composed of small islands of malignant cells in a contiguous pool of extracellular mucinous material. Cytological and nuclear features are those of low to intermediate-grade malignant tumour. The mucinous part must constitute 90 percent of the total lesion if the tumour is to be classified as mucinous.

Prognostic Markers in Breast Cancer

Prognostic markers in breast cancer are many, but the most important ones are:

1. **Histopathological clinical markers are:** Axillary lymph nodes- nodal status, tumour size, histopathological type and tumour grade, and tumour staging.
2. **Biochemical/cellular and molecular markers are:** Hormone receptors (OR, PR), proliferation markers (Ki-67; PCNA), cathepsin D, P53 (Tumour suppressor gene), HER 2/ neu oncogene, C-cerb 2 protein, epidermal growth factor (EGF) and its receptor (EGFR), growth hormone (GH) and its receptor (GHR), Thioredoxin (TRX), Thioredoxin reductase (TR), and angiogenesis markers (Factor VIII antigen, CD-31).
3. **Serological markers are**: CA-15-3, mucin-like cancer antigen, tissue polypeptide antigen (TPA), lactic dehydrogenase (LDH), G-glutamyl transaminase (GGT) and alkaline phosphatase (ALP)

Axillary Lymph Nodes - Nodal Status

Lymph node status, which reflects the ability of a tumour to spread, is one of the strongest prognostic indicators and used as an indication for adjuvant systemic therapy. The prognosis is poorer for most patients if the tumour has spread to regional lymph nodes. Five year survival for node-positive patients ranges from 63% for one node to 42% for patients with 4 to 12 positive nodes (Elias, 1993). In node-negative patients the estimated recurrence rate is as high as 30%, and whether adjuvant systematic therapy should be offered to all these patients remains a major challenge to oncologist. The greater the number of positive lymph nodes, the more likely the patient will die within 5 years (Diel ,1997). With the advent of better diagnosis and early detection (screening) fewer women are presenting with metastatic disease. In developed countries, only 30-35% of patients have metastatic disease to the axillary lymph nodes at diagnosis. In contrast, axillary lymph node involvement is more than 70% in Arab women.

Tumour Size

Primary tumour size is a factor which reflects tumour growth and is strongly related to the degree of metastatic disease, the longer a tumour has been growing the greater the likelihood is of lymph-node metastasis. In patients with a tumour diameter of 3.5 cm, metastatic spread is over two times greater than tumours of 1.5 cm (Elias, 1993). Within the subset of patients with node negative breast cancer, tumour size is the most powerful and consistent predictor of breast cancer recurrence (Mc Guire and Clark, 1992).

Histopathological Type and Tumor Grade

More than two-thirds (67%) of breast cancers are of a single histological type, infiltrating duct carcinomas, with patients having a 5 year relative survival of 79 per cent (Berg and Hunter, 1995). Of the remainder 'special-type cancers' are those which offer an improved prognosis especially when present in pure form (Simpson and Page, 1994). Histological grading of invasive mammary cancers yields important prognostic information with many studies demonstrating a significant association between tumour grade and both relapse-free and overall survival (Elston and Wilis, 1991). Determination of histological grade using the Scarff, Bloom and Richardson (SBR) method involves a subjective semiquantitive assessment of tubule formation, nuclear pleomorphism and mitotic index to arrive at a combined histological grade (Page *et al*., 1991). The modified SBR grading system uses the two nuclear elements of SBR grade to give a highly independent factor with increased prognostic capability, especially for node negative patients. Tumor grade is a standard component of the pathology report and it is a powerful predictor of the course of a breast cancer tumour.

Serological Markers

Serological markers for breast cancer include mucin-like cancer antigen (MCA), CA-15-3, carcinoembryonic antigen (CEA), tissue polypeptide antigen (TPA) and the enzymes alkaline phosphatase (ALP), g-glutamyl transaminase (GGT), and lactic dehydrogenase (LDH). However, the high number of false positives, particularly when combinations of these markers are used, together with the general disagreement of the therapeutic outcome based on elevation of serological markers, has resulted in the general consensus that these markers are only of limited value for the individual patient (Sell and Garrett, 1995).

DNA Ploidy and S-phase Fraction

DNA flow cytometry can be performed on fresh tissue specimens, frozen biopsy samples, needle aspirates taken directly from the tumor, or paraffin embedded tissue (Van Derluden *et al*., 1989). This technique produces a DNA histogram from which a measure of DNA content (DNA ploidy) and cell cycle components can be estimated. A strong correlation exists between high S-phase fraction and other prognostic factors, including poor histological or cytological grade (Wenger *et al*., 1993). Generally, patients with DNA diploid tumours with low numbers of proliferating cells have a longer disease-free survival and better prognosis than patients with DNA aneuploid tumours of high proliferative indices (Taylor *et al*., 1993). Aneuploid tumours are also associated with negative oestrogen receptor status.

Tumour Staging

The size of a primary tumor, the degree to which it has locally invaded and the extent to which it has spread, ultimately determine the chances of survival once a neoplasm has been diagnosed. There are several schemes for standardized staging of individual tumours. The TNM (Figure 2) system is based upon the extend of local tumor spread, regional lymph node involvement and the presence of distant metastasis. It can be applied to many different types of tumours, although the criteria are different for

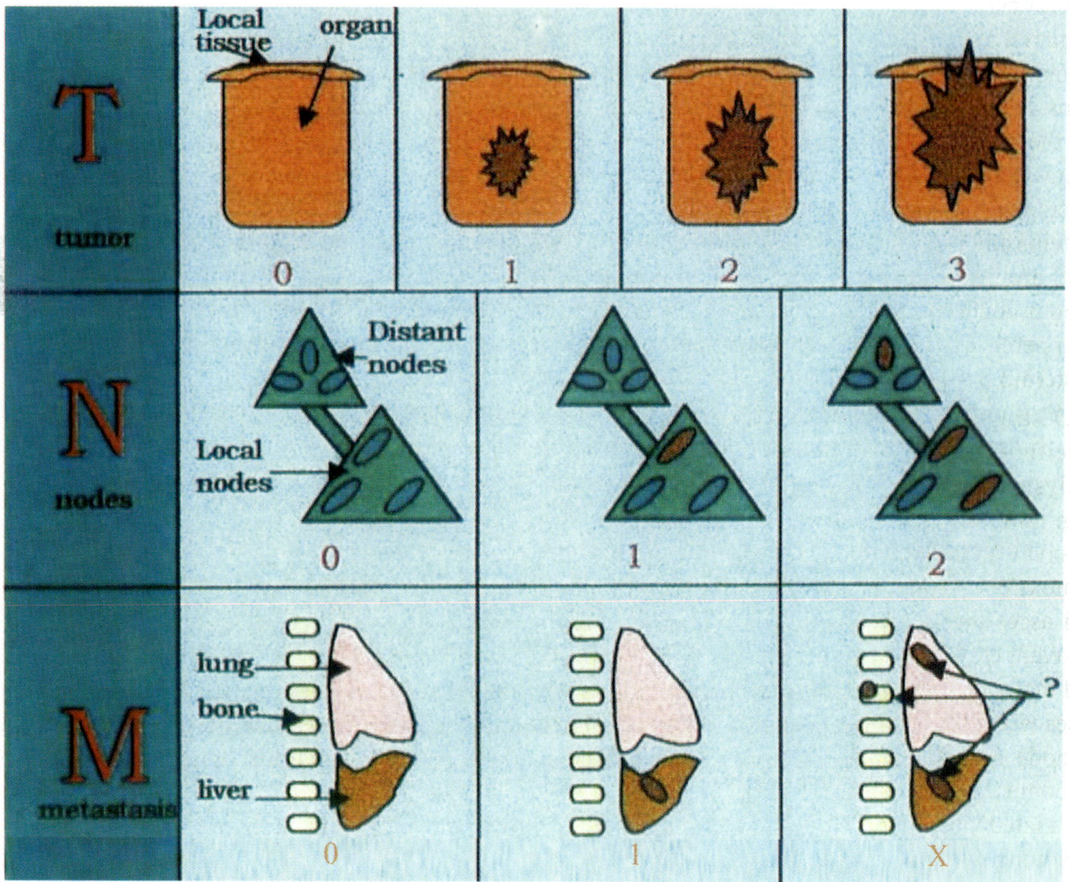

Fig. 2. **Staging of carcinoma by the TNM system. The general principle of TNM staging is as follows:**
T refers to primary tumor. The accompanying number denotes the size of the tumor and its local extent. The number varies according to the size.
N refers to lymph node involvement and a high number denotes increasing extent of involvement.
M refers to the extent of distant metastasis.
As an example, the TNM system for staging malignant neoplasm of the breast is:

T0 = breast free of tumor
T1 = lesion <2 cm in size
T2 = lesion 2-5 cm
T3 = skin and/or chest wall involved by invasion

N0 = no axillary nodes involved
N1 = mobile nodes involved
N2 = fixed nodes involved
M0 = no metastasis
M1 = demonstrable metastasis
MX = suspected metastasis

each tumor site. The stage of a tumor is generally the most important indicator of likely prognosis and appropriate therapy. Tumours at advanced stage may require aggressive treatment, while early stage tumours may be treatable by relatively conservative measures.

Hormonal Effects

Normal breast epithelial cells and their malignant counterparts are very sensitive to the effects of hormones, including oestrogen, progesterone and androgens (Clark, 1988). These effects are mediated through respective hormone receptors (Ojasoo, 1996). Oestrogen primarily stimulates hormonal ductal growth, whereas progesterone is responsible for lobulo-alveolar development. Steroid hormones diffuse easily through the cell membrane and bind to cytoplasmic receptors, which are then transported to the nucleus where they interact with DNA (Thorpe, 1987). The resulting interaction then stimulates appropriate genetic response that dictates normal cellular growth and functions. A certain fraction of breast cancers express oestrogen (OR) and progesterone receptor (PR) proteins and cell growth and viability appear to be hormone dependent. Tumours with high OR or PR level are better differentiated and patients with these tumours have a better prognosis. In addition, tumour concentration of OR and PR proteins is strongly predictive of response to endocrine therapy (Hupperets *et al.*, 1997). Manipulations of the hormonal environment influence division and viability of steroid hormone-dependant breast cancer cells.

Oestrogen receptors (OR) with Progesterone receptors (PR) are intracellular proteins with high specificity and affinity that bind oestrogens and progesterone respectively. Oestrogen receptor analysis is used for prognosis and in the prediction of hormonal response in breast tumours. Patients with tumours showing positivity for both OR and PR tend to have longer disease-free and overall survival periods than patients with receptor negative tumours. Anti-oestrogen therapy (e.g. Tamoxifen) improves the survival rates of patients with receptor positive tumours (Osborne, 1991) and assessment of hormone receptor status will identify patients who are unlikely to respond to adjuvant endocrine therapy. Progesterone receptor expression is an oestrogen dependent phenomenon and may be a better indicator of hormone sensitivity in breast cancer. Prognosis is less favourable in cases with only positive PR than those with only OR positivity (Elias, 1993). In multivariate analyses of node negative patients, either OR or PR, but seldom both, is a significant predictor of clinical outcome. Often, OR is a stronger predictor of disease free survival, whereas PR is more closely associated with overall survival, perhaps since it is a better indicator of response to endocrine therapy following disease recurrence. Both OR and PR levels are strongly inversely correlated with measures of proliferation (Wenger *et al.*,1993) and they are directly related to histological grade. Oestrogen concentrations increase almost linearly with increasing age, whereas PR levels are more dependent on menopausal status.

Growth Hormone Receptors

Growth hormone (GH) is secreted by the pituitary gland into the bloodstream and is the major hormone acting endocrinally. It plays a crucial role in controlling metabolism and acts throughout the body to promote balanced growth and differentiated cell expression. The processes responsible for this modulation are not fully understood. The original somatomedin hypothesis suggested that the mitogenic actions of growth hormone were mediated by circulating insulin-like growth factors

(IGFs), derived from the liver in response to growth hormone synthesized in the pituitary gland (Salmon and Daughaday, 1957). However, evidence from animals and developing humans shows that insulin-like growth factor I (IGF-I) is also widely synthesized locally in supporting tissues (Hoyt *et al.*, 1988) and the concept that it may be directly regulated by local autocrine/paracrine action of GH, to promote both stem cell differentiation and proliferation *in vitro* (Stracke *et al.*, 1984) and *in vivo* (Isgaard *et al.*, 1988), has become firmly established. A local response to GH requires the presence and subsequent binding to appropriately located receptors.

The biological actions of GH are mediated through specific trans-membrane receptors belonging to cytokine receptor superfamily (Kelly *et al.*, 1991). The receptors for GH modulate specific gene expression leading to target-tissue proliferation and/or differentiation. This is achieved by activating the Janus protein tyrosine kinases (JAK) with tyrosine phosphorylate the cytoplasmic STAT (signal transducer and activators of transcription) transcription factors (Schindler and Darnell, 1995). Several reports have shown that GH controls the differentiation, cellular growth and functions of the normal mammary gland (Feldman *et al.*, 1993; 1998; Plaut *et al.*, 1993) and the presence of its receptors has been characterized and localized in different species. Thus, GH-receptor mRNA and protein have been identified in mammary epithelial components in rodents (Lincoln *et al.*, 1990; 1995), bovines (Sinowatz *et al.*, 2000) and human (Sobrier *et al.*, 1993; Mertani *et al.*, 1995). Because of its proliferative effect, an important role for circulating GH has been ascribed to mammary carcinogenesis, particularly in rodents (Nandi and Guzman, 1995), and in humans (Bonneterre *et al.*, 1987). GH-receptor gene expression has been evidenced in human breast cancer by RT-PCR (Decouvelaere *et al.*, 1995). The involvement of GH in human breast carcinogenesis is supported by the demonstration of endogenous synthesis of GH (Mol *et al.*, 1995) by the mammary gland suggesting the existence of a possible autocrine/paracrine loop. Nuclear STAT activation levels are increased in human breast carcinoma (Watson and Miller, 1995), indicating altered expression of GH-receptor coupled to local or endocrine GH action.

Proliferative Indices

Immunohistochemical analysis of nuclear antigen-associated cell proliferation has successfully been used to demonstrate levels of Ki-67 and Proliferating Cell Nuclear Antigen (PCNA) in fresh and fixed tissues (Leong *et al.*, 1993). Immunohistochemical detection of Ki-67, which labels a nuclear protein expressed in cycling cells, correlates well with other proliferative markers such as thymine labelling, mitotic activity and S-phase fraction, and when present in high levels, is an independent prognostic indicator (Simpson and Page; 1994). Ki-67 monoclonal antibodies are specific for a nuclear antigen that is expressed only in proliferating cells in late G1, S, M and G2 phases of the cell cycle (Iatropoulos and Williams, 1996). Ki-67 staining correlates directly with tumor size histological grade, vascular invasion and axillary lymph node status (Brown *et al.*, 1990; 1993). In univariate analysis of survival data, Haerslev *et al.* (1996) showed that Ki-67 was a parameter of a poor overall survival in both lymph node-positive and -negative patients. In multivariate analysis using a Cox model stratified by nodal status, Ki-67 failed to be of prognostic significance, whereas classical histopathological parameters such as tumor size and histopathological parameters such as tumor size and histological grade, are of independent prognostic significance in both lymph node-positive and -negative patients.

Proliferating Cell Nuclear Antigen (PCNA), detected by immunohsitochemistry, is highly representative of proliferative activity and is associated with poor prognostic factors such as high histological grade, numerous mitosis and presence of metastases. PCNA is a 36 kD nuclear protein which is primarily associated with the synthesis phase of the cell cycle. A high PCNA index was significantly related to risk of relapse in univariate but not multivariate analysis (Russo *et al.*,1994). PCNA is a nuclear protein associated with DNA polymerase, which is present throughout the cell cycle in proliferating cells. Whereas some studies found weak correlation with other prognostic factors and poor association with clinical outcomes, especially in node negative patients (Haarslev *et al.*, 1996), others found that PCNA expression was the only independent prognostic factor in lymph node negative breast cancer patients (Schonborn *et al.*, 1995). Thus, in individual cases the PCNA proliferative fraction can help the clinician to decide on the therapy.

erbB-2 Family of Growth Factor Receptors

The erbB-2 family of growth factor receptors and their ligands is one of the most successful areas of breast cancer research. Included in this group are the epidermal growth factor receptors (EGFR) as well as HER-2/*neu*/*erb*B-2, HER-3//erbB-3 and recently HER-4/erbB-4. These receptors are structurally similar and are frequently implicated in the pathogenesis and biological behaviour of breast tumours with many cancers expressing more than one receptor of this family (Bacus *et al.*, 1994). Human breast carcinomas with a high level of EGF receptor show a significant correlation with high histological grade, an inverse relationship with oestrogen receptors and are associated with a more aggressive disease course ((Elston and Wilis, 1991). The c-erb-B2 gene is located on chromosome 17 q21 and is transcribed into a 4.5 kb mRNA. Amplification of the c-erbB-2 and proto-oncogene is reflected in overexpression of the c-erbB-2 oncoprotein, a 185 kD transmembrane glycoprotein which can be detected immunocytologically in various human adenocarcinomas. The c-erb-B2 protein is expressed at low levels in the epithelial and myoepithelial cells of normal breast tissue (Schonborn *et al.*, 1995). Overexpression of c-erbB-2 occurs in poor prognostic factors such as lymph node involvement and negative oestrogen and progesterone receptor status. Such tumours are often less responsive to adjuvant therapy and appear to correlate with a worse disease-free and overall survival (Bacus *et al.*, 1994). Corkill and Katz (1994) found a significant correlation between positive c-erbB-2 staining on cytology smears and high nuclear grade tumours, but not with other prognostic indicators. Using image analysis, Bacus *et al.*, (1994; 1998) quantified c-erbB-2 protein levels to show that breast cancers overexpressing c-erbB-2, possessed polyploid DNA amounts. Furthermore, membrane staining of c-erbB-2 was associated with a higher proliferating index than tumour cells exhibiting only cytoplasmic staining, and may be useful in differentiating fast growing tumours from slow growing tumours.

Tumour Suppressor Gene p53

The p53 tumour suppressor gene is located on chromosome 17 p13 and encodes a 53 kDa nuclear phosphoprotein. Alterations in this gene are the most frequent genetic changes found in many malignant diseases, including breast cancer (Hallstein *et al.*, 1991). If several sensitive regions of the gene are mutated, the protein conformation changes and stability increases, so that it finally functions as an oncogene product (Engels *et al.*, 1997). p53 gene alterations occur early in breast carcinoma genesis, as mutations were detected not only *in situ* carcinomas but also in atypical ductal hyperplasia

(Lisboa *et al.*, 1997). The p53 tumor suppressor gene acts as a negative regulator of cell growth (Spandidos *et al.*, 1992) and alterations in the gene lead to loss of the negative growth regulation and to a more rapid cell proliferation. Expression of p53 is relatively independent of axillary lymph node and menopausal status. It is weakly related to tumor size, but is strongly associated with DNA ploidy and measures of proliferation, steroid receptors and nuclear grade (Elledge *et a*l,, 1994). The majority of multivariate analyses suggest that p53 is an independent prognostic factor and a predictor of shortened survival (Dowell and Hall, 1994). In general, p53 positivity is associated with highly proliferative and aneuploid tumours, particularly breast carcinomas, often with negative oestrogen and progesterone receptor status. No significant association was found with age, lymph node status, tumour size or invasion (Dowell and Hall, 1994).

Cathepsin D Measure of Invasion and Metastasis

Cathepsin D is a widely expressed lysosomal aspartyl protease which plays an important role in tumor invasion and metastasis. It is initially synthesized as an inactive precursor (pro-cathepsin D) before being converted by proteolytic processing via an intermediate to its mature forms (Engel *at al.*, 1997). Cathepsin D is a comparatively strong mitogenic factor for breast carcinoma cell lines (Vignon *et al.*, 1996) and degrades basement membranes (Briozzo *et al.*, 1988). It also facilitates the release of the potent angiogenic factor bFGF (Briozzo, 1988) and thus stimulates neovascularization (Engels *et al.*, 1997). In breast carcinomas, high cathepsin D expression is regarded as a predictor of a shorter recurrence-free survival (Briozzo *et al.*, 1988), metastasis-free survival (Spyratos *et al.*, 1989; 1992) and overall survival (Engels *et al.*, 1997).

Angiogenesis

There is increasing interest in tumour angiogenesis as a prognostic marker, as neovascularization is an essential component of tumour growth and metastasis. Microvessels can be identified by immunocytochemical staining for factor VIII-related antigen on formalin-fixed tissue sections. Angiogenesis can also be evaluated by immunostaining with monoclonal antibody against CD-31, which labels endothelial cells, and quantified by image analysis. The CD-31 molecule may be involved with tumour spread and extracellular matrix interactions (Charpin *et al.*, 1995).

Characteristics of Breast Cancer in Arab Women

The majority of breast cancer studies have focused on Western societies and reports are limited on populations of women in the Arabian Gulf region. The state of Kuwait is located at the northeast extremity of the Arabian Peninsula with a population of 2.4 million in 2004 and with 10% of the world known oil reserves has one of the highest per capita incomes. Kuwaitis account for 41.6% , most of the population is young with 50% below the age of 15 years. Because of its extreme wealth and low population, it attracts many other Arab nationals from Palestine, Jordan, Irak, Egypt, Oman, Yemen, Saudi Arabia and North Africa. The Arab population in Kuwait is genetically homogeneous because of the high rate of consanguinity (Radovanovic *et al.*, 1999). Hence the Kuwaiti population is ideal for studying breast cancer of Arab women.

Breast cancer is on top of the list in incidence in Kuwaiti and non Kuwaiti resident females (Table 1). It accounts for 32.8% according to age per 100,000 of all cases (Al-Jarallah, Kuwait Cancer

Table 1. Most prevalent cancer cases in Kuwait females 1998 - 2003.

Type of Cancer	Percentage of cases	Percentage according to age per 100,000
Breast cancer	27.6%	32.8%
Cancer of the Thyroid gland	6.4%	6.1%
Lymphoma (other than Hodgkin's disease)	5.7%	7.1%
Cancer of the Cervix	5.6%	7.6%
Leukemia	5.6%	4.6%
Cancer of the Brain and CNS	3.6%	2.9%
Colon Cancer	3.3%	4.8%
Ovarian Cancer	3.3%	4.7%
Stomach cancer	3.1%	4.8%
Liver cancer	2.1%	3.5%

(Source: Ministry of Health, Kuwait)

Control Centre Registry). There are between 100-120 new diagnosed cases every year, 30% of them are Kuwaiti and the rest are mostly from other Arab countries. The mean age of the breast cancer patients is 48.7 years and the peak occurrence is at the age of 45-50 years with a significant percentage, 26% occurring below the age of 35 years and only 14.1% above the age of 60 years. Similar data have been reported in some of the Arab countries in North Africa (Franco Arab Cancer Association Symposium, Casablanca 1994; Cairo 1995). In contrast, breast cancer in Europe and USA occurs at an older age, the peak is between the age of 60-65 years, and only 14% occur below the age of 40 years (Feigelson *et al.*, 1997).

The delay in presentation ranges from one month to two years with an average of three to 4 months. Hence, most of the cases present late with advanced locoregional disease. Only 15% are eligible for conservative breast surgery. The average size of the tumor is 5.6 cm, this is similar to all reports from other Arab countries. The majority of cases (76.2%) are invasive ductal carcinoma of which only 6.8% are well differentiated tumours (Grade I), 41% are moderately differentiated (Grade II) and 22% are Grade III. The grade was unknown in the remaining 160 cases (30%). Data from 2,143 patients shows that the highest percentage of cases was in the age groups 31-40 and 41-50 years old, being 28% and 33% respectively and a very high proportion ((9%)of patients were in the young 20 to 30 years age group. The percentage of patients in the age groups studied was as follows:

Age Group	%
20-30	9
31-40	28
41-50	33
51-60	19
61-70	8
71-80	2
81-90	1

Parity

Kuwaiti females, similar to women in other Arab countries, are married at a relatively young age and the majority have their first child within two years of marriage. The protective effect of parity indicates that women, having their first baby birth under the age of 20 years, have about half the risk of developing breast cancer, compared to nulliparous women. Subsequent births have no influence on the risk of developing breast cancer (Mahon *et al.*, 1970). Therefore, it could be expected that, with its conservative kind of life style, habits and tradition, the incidence of breast cancer in Kuwait is low and that more cases would be seen in single or nulliparous women. However, this is not the case. Pregnancy has two contradictory effects on breast cancer risk that are particularly notable in the first pregnancy. This apparent paradox has a physiological explanation based on oestrogen secretions and metabolism during pregnancy. During the first trimester of pregnancy, the level of free oestradiol rises rapidly, an effect that is more apparent in the first than in subsequent pregnancies. The net effect of this early part of pregnancy in terms of oestrogen exposure to the breast is equivalent to several ovulatory cycles over a relatively short time (Hoel 1983).

Lactation

A prolonged lactation period has repeatedly been reported to protect against development of breast cancer. In a case control study in China, a population in which long duration nursing is the norm, a progressive reduction in breast cancer risk was observed with an increasing number of years of nursing experience (Freudenheim *et al.*, 1997). Most Arab women breast-feed their children for periods ranging between 4 to 18 months (mean 9 months).

Contraceptives

Oral contraceptives have been widely used in Kuwait since the 1960, as it is world wide. A substantial body of literature exists on the relationship between contraceptive pills and risk of breast cancer, mostly in perimenopausal women with sufficient exposure, to examine the effects of duration of oral contraceptive use. Women over the age of 45 years at diagnosis, showed a positive correlation (La Vecchia *et al.*, 1996). The combined results of studies show a satisfactory significant 3.1% increase in breast cancer risk per year of contraceptive pill users, indicating that young women who had taken contraceptive pills for 10 years, had 36% higher risk of breast cancer than women who had never used the pills (WHO collaborative study, 1991).

Diet

An association between dietary calorie consumption and breast cancer (Caroll, 1994), is supported by experimental demonstration of reduced incidence of cancer in animals on energy restricted diet. Willet (1989) also found a positive association between body height and the risk of cancer in several countries that had experienced famine. This may indicate that dietary contribution to breast cancer risk is due to amounts of calorie consumed and utilized, especially during critical periods such as adolescence, and not the intake of specific nutrients. Energy-rich diet in later life contributes to the occurrence of obesity and development of breast cancer . Cancer cells have higher and increased rates of glucose metabolism, compared to normal cells (Brown and Wahl, 1993). Insulin, which regulates glucose metabolism at the cellular level, has to compete with insulin like growth factors IGF-1 and

IGF-2 for receptor binding sites. However, insulin binding to receptor is only one hundredths that of IGF-1 (Peyrat, 1992), and IGF-2 is over-expressed and its binding is increased in breast cancer (Yballe *et al.*, 1996). Since insulin receptor itself is over-expressed in breast cancer, but insulin binding is low in subjects with diabetes mellitus, glucose concentration in the breast cancer and other tissues is expected to be high in patients with both disorders. It is therefore necessary to examine the relationships that might exist between diabetes mellitus, especially the non insulin dependant type (NIDDM), the form of the disease commonly associated with obesity, and breast cancer.

Materials and Methods

Obesity, Non-insulin Dependent Diabetes (NIDDM)

The case files of all breast cancer patients seen at the Kuwait Cancer Control Center (KCCC) over the 10 year period from January 1993 to December 2002. were reviewed. In addition, the demographic data of each case, specific information was recorded on the following:

- Height of patient on admission (in centimeters)
- Weight of patient on admission (in kilograms)
- Presence of diabetes mellitus

  These were either known diabetics or whose diabetes was ascertained from testing plasma glucose level estimated at the time of initial diagnosis of breast cancer, KCCC has a policy of complete laboratory investigation including testing blood glucose on admission.
- Duration of diabetes mellitus, if available from the files
- Method of treatment of diabetes mellitus, if stated in the files
- Presence and severity of hypertension

  Mild hypertension was defined as 140-160 mm Hg systolic and 90-100 mm Hg diastolic and severe hypertension > 180 mm Hg systolic and > 120 mm Hg diastolic.
- Type of breast cancer according to histological classification

  Stage of breast cancer based on the TNM classification into stages O, I, IIA, IIB, IIIA, IIIB, IV.

Any relationships among the parameters studied and their significance were examined using simple statistical methods such as chi-square test.

Breast Cancer in Young Age Group

Tissues

The case files of 122 cases of breast cancer women < 30 years old, diagnosed and treated at the Kuwait Cancer Control Centre during the period from 1993 to 2003, were reviewed. The samples included normal breast (n=5); benign breast (n=1); fibrocystic disease (n=9); sclerosingadenosis (n=2); gynecomastia (n=3); intraductal papilloma (n=1); ductal carcinoma (n=28); intraductal carcinoma (n=12); infiltrating ductal carcinoma (n=21); lobular carcinoma (n=14); infiltrating lobular carcinoma (n=9); ductal/lobular carcinoma (n=1); signet ring carcinoma (n=1); tubular carcinoma (n=1); medullary carcinoma (n=1); mucin producing carcinoma (n=1); metastatic carcinoma (n=12). The tumours were staged pathologically according to the TNM system of UICC. Light microscopical diagnosis was performed on formaldehyde fixed, paraffin-embedded tissue sections stained with haematoxylin and eosin. All tumours were histopathologically classified according to the WHO and the degree of differentiation was assessed according to the guidelines of Bloom and Richardson (1957).

Immunohistochemistry

Sections of 5 μm thickness were cut from formalin-fixed paraffin-embedded specimens, mounted on poly-L-lysine-coated slides and dried for one hour at 60°C. Sections were then treated with three changes of xylene and two changes of alcohol for five minutes, followed by treatment with 0.3% H_2O_2 in methanol for 20 minutes to block endogenous peroxidase and rinsed in distilled H_2O. To improve the staining pattern, the sections were treated with antigen retrieval solution at 90°C for 20 minutes, allowed to cool at room temperature in the same solution for 20 minutes and washed in distilled H_2O, followed by washing in three changes of Tris buffer. Sections were then treated with blocking reagent (PBS containing carrier protein) for 20 minutes at room temperature to reduce non-specific staining with the primary antibody, by inhibiting the non-specific binding. The sections were then incubated overnight at 4°C in a humidity chamber with the primary antibody at a concentration as specified by the manufacturer. Following overnight incubation with the primary antibody, sections were washed with Tris buffer, followed by incubation with biotinylated link antibody (biotinylated antibody to mouse immunoglobulin) for 30 minutes at room temperature. The sections were then washed with three changes of Tris buffer and incubated with streptavidin (4 ml of streptavidin buffer with 1 drop of streptavidin concentrate peroxidase conjugated) for 30 minutes ar room temperature, followed by washing in three changes of Tris buffer. Visualisation was by incubating slides at room temperature with 1 mM 3-amino-9-ethylcarbazole (AEC) containing 0.015% hydrogen peroxide in 0.1 M acetate buffer, pH 5.2 as chromagen for 6 minutes at room temperature. In some cases, the peroxidase reaction was performed using 3.3' diaminobenzidine tetrahydrochloride-hydrogen peroxide as a chromogen in Tris-buffer (pH 7.6) for 5-10 minutes. Sections were then washed with distilled H_2O, dehydrated in ascending series of alcohol, cleared in 3 changes of xylene and mounted with Depex.

Immunohistochemical methods were performed for the status of oestrogen receptor (OR), progesterone receptor (PR), growth hormone receptor (GHR), proliferating cell nuclear antigen (PCNA), measure of proliferation (Ki 67), tumor suppressor gene (p53), measure of growth factor (C-erb-B2), vascular density (CD-31) and measure of invasion and metastasis (Cathepsin D). The p53 specimens were subjected to five 2-minute pulses in a microwave oven (650-700 W) with a one-minute rest between each pulse. The Ki-67 specimens received two 5-minute pulses. After cooling for 20 minutes they underwent immunohistochemical staining. The primary antibodies were polyclonal in the case of Ki-67 (DAKO, Denmark) and monoclonal for p53 (Immunotech, France) at dilutions of 1:100 and 1:40, respectively. The specimens were recognized as positive if at least 5% of the tumor cells stained specifically for p53 protein content. For immunohistochemical estimation of the OR and PR status, two commercial kits (ERICA and PRICA, Abbot Lab, USA) were used following the kit instructions without modification.

Quantitative Analysis

The quantitative analysis of Ki-67, PCNA, ER and PR receptors on paraffin sections was done by calculating the percentage of stained cells in 10 low power magnification fields or 16 high power fields using a Zeiss micoscope connected to a computer using image analysis software. In cases of the other markers, the stained slides were qualitatively interpreted as being either positive or negative. Material from the same paraffin blocks was used for flow cytometric analysis of DNA and apoptosis. This technique produces a DNA histogram from which a measure of DNA content (DNA ploidy) and cell cycle components. Differences between the results of immunostaining for p53 and tumour size, degree of differentiation, lymph node metastasis, ploidy, Ki-67 and c-erbB-2 expression were analysed statistically by the Chi-square test (cut-off level for significance was $p<0.05$).

Results

Risk Factors

Obesity in Breast Cancer Patients

Calculation of the BMI shows that virtually all Kuwait breast cancer patients are overweight. Approximately 50% of Kuwaiti breast cancer patients having a high BMI are classified as diabetics. This prevalent rate of non-insulin dependent diabetes mellitus (NIDDM) in breast cancer patient of 50% can be considered as very high. A plasma glucose level above 7-8 mmol/L is diagnostic for diabetes mellitus. On this basis, 261 (50%) of the breast cancer patients are classified as diabetic. A total of 520 breast cancer patients were seen during the ten year period, with an age range between 26 and 88 years and a mean of 48.7 years. The body height of the patients ranged between 136 and 182cm with mean of 154.3cm. Table 3 summarizes the body height distribution of 520 patients. The body weights ranged between 40 and 165 kg (mean 81.3 kg).

Body Mass Index (BMI)

Mean body index (BMI) of all breast cancer patients was calculated, using the formula: body weight divided by the body height in meters squared (Thomas, 1977). A BMI below 20 is regarded as being under weight while a BMI of 20-25 is normal. BMI values above 25 are regarded as over weight (Abdella *et al.*, 1996). Using the above calculation, the mean BMI of the breast cancer patients, with a mean body height of 153.07 cm and a mean body weight of 79.15 kg, was 33.8 and therefore indicating that the breast cancer patients in this study were generally over weight.

Plasma Glucose Levels and Diabetes Mellitus in Breast Cancer Patients

In addition to history of diabetes mellitus, the fasting plasma glucose was determined for each patient on admission. Although the WHO classification and diagnostic criteria for diabetes mellitus recommends plasma glucose levels of over 7-8 mmol/litre for the diagnosis of diabetes mellitus, the KCCC uses a level of 8 mmol/litre as a cut off point. On this basis, 261 (49.9%) of the breast cancer patients were classified as diabetic.

Presence and Severity of Hypertension

Hypertension was present in all breast cancer patients. Over half of the patients (53.1%) were classified as having mild hypertension, while a significant proportion (14.8%) had severe hypertension. There was a significant correlation between the presence of hypertension and diabetes mellitus ($p < 0.05$).

Breast Cancer in Youg Age (<30 Years)

A total of 122 breast patients <30 years of age over the period from January 1993 to December 2002 were investigated. This particular group of patients accounted for 9% of all cases of breast cancer registered in the same period. The majority of cases (73%) underwent mastectomy and axillary clearance, 9% had lumpectomy and axillary clearance, while 17% had biopsy only. This latter group were given neoadjuvant combination chemotherapy. Table 2 shows the histopathological types of the 122 cases of breast cancer in the young age group.

The staging was pathological in 110 cases and clinical in 12 cases, the latter group received neoadjuvant combination chemotherapy. All patients underwent complete oncological workup. There were no cases of To N+Mo and none of the patients presented with immunohistochemical nodal involvement (N3). Table 3 shows stage grouping of 122 cases of women with breast cancer < 30 years old.

Table 4 shows the per cent tumour grade distinction of 122 cases of women with breast cancer < 30 years old.

Table 5 shows the distribution of breast cancer by tumour size in young women <30 years of age, according to largest diameter as seen by mammography, ultra sound or as reported in the pathology report.

Tumour Markers of Breast Cancer in Young Age (<30 Years)

The expression of different prognostic tumour markers was investigated in 122 primary breast carcinomas from young Arab women (age <30 years). The markers included tumour size, axillary node status, tumour grade, oestrogen and progesterone receptor status, DNA ploidy, HER-2/neu (c-erB-2) expression, growth hormone receptor expression, vascular density and the proliferation marker Ki 67 and PCNA. Positive immunostaining for oestrogen receptors was observed within the nuclei of tumour

Table 2. Histology of breast cancers in young patients.

Histological type	Number of cases
Infiltrating duct cancer	82
Infiltrating lobular cancer	6
Mucinous carcinoma	4
Medullary carcinoma	11
Tubular carcinoma	3
Duct carcinoma in situ	10
Lobular carcinoma in situ	6
Total	122

Table 3. Stage of breast cancers in young patients.

TNM	Stage	No. of Cases
T1 No Mo	I	3
T1 N1 Mo	II A	16
T2 No Mo	II B	4
T2 N1 Mo	II B	23
T3 No. Mo	-	-
T1 No Mo	III A	18
T3 No Mo	-	-
T1 N2 Mo	III A	8
T2 N2 Mo	-	-
T3 N1 Mo	III A	5
T3 N2 Mo	III A	24
T4 No N1 N2 Mo	III B	21
Total		122

Table 4. Tumour grading in young breast cancer patients.

Grade I	Grade II	Grade III
4%	25%	71%

Table 5. Distribution of breast cancer cases according to tumour size.

< 2 cm (T1)		2-5 cm (T2)		> 5 cm (T3)	
Number	per cent	Number	per cent	Number	percent
22	11	67	65	33	24

cells in 23 out of 122 (18.9%) breast cancer patients (Table 6, Figure 4). The percentage of OR-positive nuclei in these carcinomas was in the range from 17 to 99%. The cytoplasm of all cancer cells was negative. All other tumour tissue components, including stroma, adipose tissue, vasculature, nerve fibres were also OR negative. A distinct immunoreactivity for progesterone receptor was present within the nuclei (Figure 5) in 16 out of 122 (13.1%) breast carcinomas (Table 6). Quantitation of the positive cases showed that there was a considerable variation in the number of positive tumour nuclei, ranging from 12 to 99%. Stromal cells, adipocytes, nerves and blood vessels were consistantly negative for progesterone receptor.

Ki-67 immunoreactivity of was present to varying degree in all breast cancer specimens. Immunoreactivity was mostly confined to the nuclei of the cancer cells. Cells in mitosis at times also displayed weak staining in their cytoplasm. The number of cells with immunopositive nuclei varied between 9% and 89% of all malignant cells in a tumour specimen. There was no Ki-67 immunoreactivity in the nuclei of stromal, vascular and nerve cells present in the tumour tissue. PCNA immunoreactivity was present in all mammary carcinomas, the number of immunopositive malignant cells varied from 2% to 100%. The range of positive cells in the majority of carcinomas was

Table 6. Number and percent of cases showing the grade, lymph node status, hormonal receptors, ploidy, p53, Ki-67, HER-2/neu in breast cancers of young women <30 years of age.

Differentiation Grade	Number of cases	Percentage
I	8	6.5
II	31	25.4
III	83	68.1
Lymph node status		
Positive	70	57.4
Negative	52	42.6
OR-status		
Positive	23	18.9
Negative	99	81.1
PR-status		
Positive	16	13.1
Negative	106	86.9
p53		
Positive	70	57.4
Negative	52	42.6
Ki-67		
(≤20%)	88	72.1
(<20%)	34	27.9
HER-2/neu		
Positive	80	65.6
Negative	42	34.4
Ploidy		
Diploid	51	41.8
Aneuploid	71	58.4

between 20 to 35% of the tumour cells. A distinct to strongly positive staining intensity was present in the nuclei of these cells. The nuclei of other tissue cells, including fibroblasts, vascular endothelial cells, adipocytes smooth muscle cells, lymphocytes, neutrophills, mast cells and plasma cells were mostly negative.

According to the modified grading system of Bloom Richardson (1957), 6.5% of the tumours were classified as Grade 1, while 25.4% were of Grade II and 68.1% were of Grade III. 14.6% of all carcinomas were <2 cm in size, 53.3% were between 2.1 and 5.0 cm and 28% were larger than 5 cm in size. 42.6% of all cases were lymph node negative, while 15.9% had 1 to 3 and 41.4% had 4 or more lymph node metastases. 41.8% of all carcinomas were diploid and 58.4% were aneuplpoid. Grading correlated significantly with ploidy (p=0.008). All of Grade I tumours , 47.4% of Grade II and 61.6% of Grade III tumours were aneuploid. 81.1% of all carcinomas were OR-negative and 86.9% were PR-negative. 65.6% of all mammary carcinomas displayed positive immunostaining for HER-2/neu (Tables 6, 8). A distinct to strongly pronounced staining for immunoreactive p53 was present in 70 (57.4%) breast carcinomas of the young age group of 122 specimens.

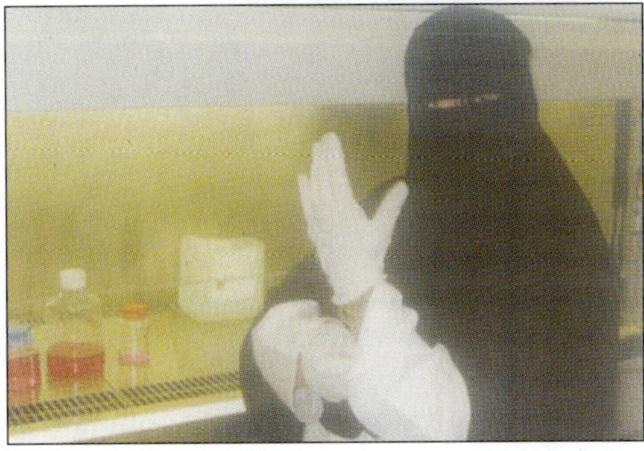

Fig. 3. Young Arab women's conservative dress expresses traditional values.

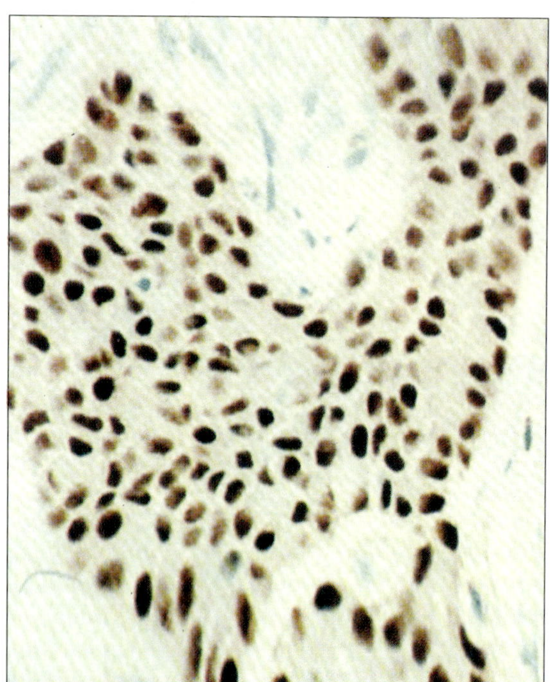

Fig. 4. Strong nuclear oestrogen receptor expression in infiltrating ductal carcinoma.

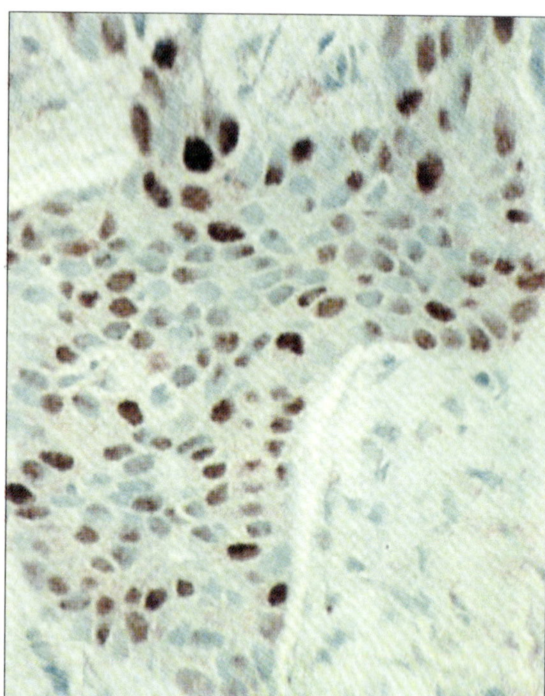

Fig. 5. Distinct progesterone receptor staining in nuclei of nfiltrating ductal carcinoma.

Expression of p53 was confined to tumour cells only, being totally absent in the stroma or in the vasculature (Figure 6). 63.7% of tumours with p53 overexpression were aneuploid and 93.6% of these were OR negative and 95.7% were PR negative. 64.8% of the p53 positive tumours were also node positive. 93.5% of the p53 immunopositive carcinomas were OR-negative and in 95.7% of this subclass of patients no PR could be detected. The vast majority of p53 positive carcinomas were Grade III (76.6%), 21.3% were Grade II and 2.1% were Grade I, but neither tumour grade (p=0.066) nor tumour size (p=0.53) showed a correlation with p53 expression. 60.9% of p53 stained breast carcinomas had positive lymph nodes. Of all the parameters studied, only (negative) OR and PR content significantly (p=<0.001) correlated with p53 staining.

Of the 122 breast carcinomas from patients of young age group *<30* years old investigated, 48 cases (39%) of the cases did display cathepsin D immunoreactivity. Staining was present in all grades of malignancy, including low grade I tumours. Generally only a small number (maximum 10%) of the carcinoma cells displayed a distinct immunoreactivity which was confined to the cytoplasm only. Cells of the stromal, adipose and vascular tissue were always negative, as were cells of the immune system, including plasma cells, lymphocytes, eosinophils, mast cells and neutrophil granulocytes.

Growth hormone receptor (GH-R) immunostaining was observed both in plasma membrane-cytoplasm and nuclei. The metabolically less active vascular endothelial cells, fibroblast cells and adipocytes were either negative or only weakly immunopositive. No immunoreactivity was found in the controls performed to validate the specificity of the immunoreaction. In the medullary carcinoma

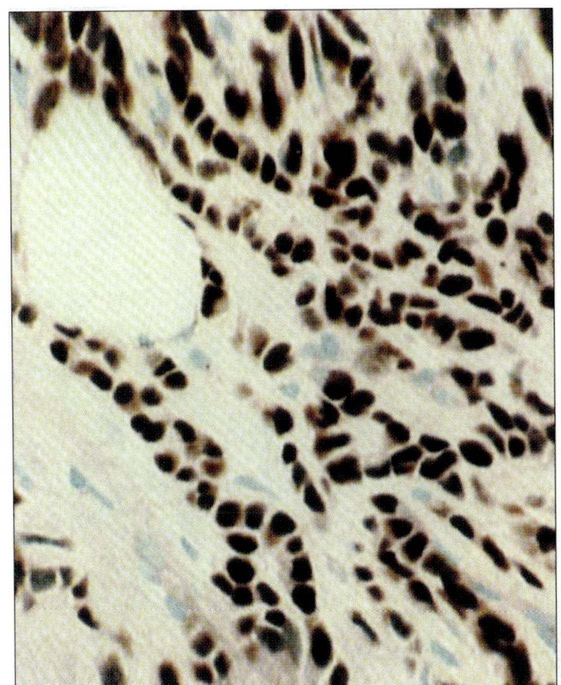

Fig. 6. Intense nuclear expression immunostaining of p53, infiltrating ductal carcinoma.

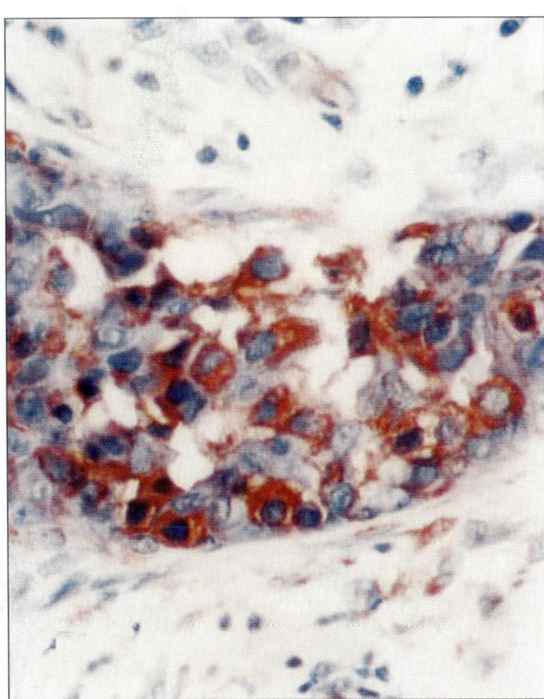

Fig. 7. Growth hormone receptor expression in infiltrating ductal carcinoma.

the proportion of positive cytoplasmic and nuclear immunoreactivity was 16-30%. In signet ring carcinoma the proportion of immunoreactivity was in excess of 31% in the membrane-cytoplasmic component and 16-30% of the nuclei (Table 7). The proportion of membrane-cytoplasmic and nuclear immunoreactivity in the intraductal carcinoma was greater than 31% and 5-15% respectively (Table 7). All cases of infiltrating ductal carcinomas investigated were GH-receptor immunopositive with a high (>31)% immunopositive nuclear component in this tumour etiology. The majority of ductal carcinomas were immunopositive (Figure 7), with proportion of membrane-cytoplasmic and nuclear immunoreactivity ranging from 5% to greater than 31%. In lobular carcinomas the proportion of positive cells showing membrane-cytoplasmic immunorstaining ranging from 16 to greater than 31%, while the nuclear component less than 5%. Intense immunostaining was present in the cytoplasmic component of majority of tumour cells in the infiltrating lobular carcinomas (Table 7). Immunoreactivity was primarily in the cytoplasm of malignant cells, but nuclei also showed immunostaining (16-30%) for GH-receptors. All of the 12 cases of metastatic lymph node cases investigated were immunopositive, the proportion cytoplasmic immunoreactivity was 16-30% with a lesser nuclear component.

Strong expression of HER-2/neu (also known as c-erbB-2) was present in 80 (65.6%) of breast carcinomas from patients of young age group <30 years old. (Table 8). Immunoreactivity was concentrated to the cell membrane of the tumour cells (Figure 8). At times, the cytoplasm displayed a weaker staining for HER-2/neu. No immunoreactivity was present in stromal cells, adipocytes,

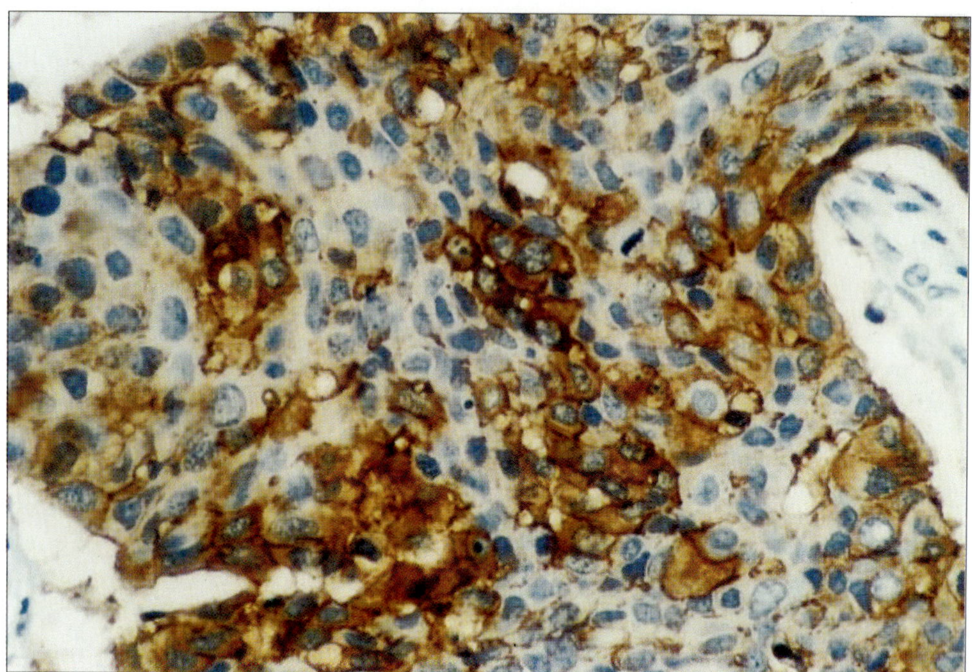

Fig. 8. Strong expression of HER-2/neu is concentrated in the cell membrane and weaker cytoplasmic staining in a lymph node positive infiltrating ductal carcinoma.

vascular endothelial cells and cells of the immune system. Among the 14 cases of HER-2/neu positive infiltrating ductal carcinomas, 10 were strongly positive and associated with lymph node involvement, while only 4 cases were moderately positive. The strong positive cases of the intraductal carcinomas were all lymph node positive, indicating a relationship between strong HER-2/neu expression and the lymph node status. In the lobular carcinoma, there were 8 positive cases showing intensity ranging from 5 (+), 1 (++), 2 (+++). The strongly positive cases of both lobular and infiltrating lobular carcinoma have a relationship with lymph node involvement. The signet ring carcinoma was strongly HER-2/neu positive with lymph node involvement. The medullary carcinoma and the mucin producing adenocarcinoma stained positively, the former being weak (+) and the mucin producing adenocarcinoma moderately HER-2/neu positive (++). Finally, all the metastatic carcinoma cases were positive at different levels of staining intensity, ranging from weak to strong.

Immunoreactivity of the vascular density CD-31 was present in the vascular endothelial cells of the larger blood vessels (arteries and veins) and in the endothelium of the capillaries (Figure 9). All other components of the breast carcinoma tissue, including tumour cells, stomal cells, adipocytes, nerve cells) were negative for CD-31 antigen.

Discussion

Risk Factors

Breast cancer is the most common neoplasm amongst women worldwide. In Kuwait, with respect to population size, it is a significant and growing disease. It has certain unusual features in Arab women,

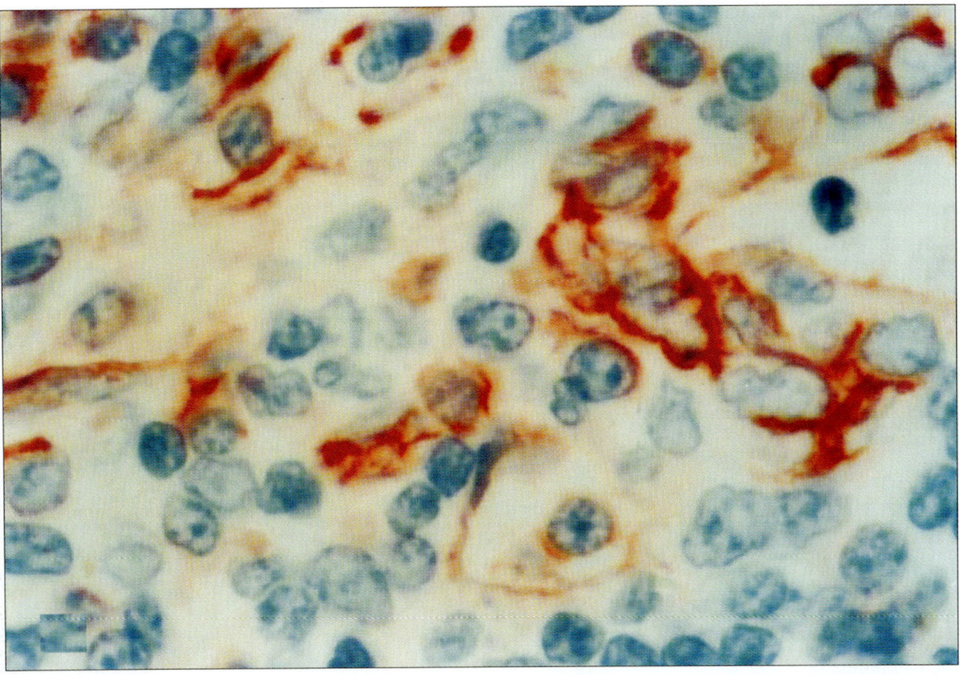

Fig. 9. Strong CD-31 immunoreactivity in the vascular endothelial cells of a lymph node positive ductal carcinoma.

such as early onset and high incidence, despite early age of childbirth, and high parity, all traditionally indicators for lower risk. Statistics from the Kuwait Ministry of Health (Al-Jarallah, 1998) show that 53% of breast cancer cases were aged between 13-44 years. Similarly, statistics from the same source, also suggest a much greater proportion of younger Kuwaiti women developing breast cancer than was observed for their counterparts in western Europe. Kuwaiti and other Middle East Arabs share a common lifestyle and cultural and religious heritage that includes traditional conservative factors such as diet, clothing, exercise and family planning, ultimately increasing the risk of breast cancer. An earlier publication on 315 Kuwaiti patients reported that 78% of cases were aged 21-50 years (Paszko *et al*, 1993). The mean age of breast cancer patients seen in this study is 49 years, being the peak occurrence in the 45-50 years age group, with a significant proportion (26%) below the age of 40 years. Only 14% of all patients investigated were aged over 60 years. This is in marked contrast to reports from developed Western countries, where the peak incidence falls in the 60- 65 years age group and only 14% of breast cancer patients are less than 40 years of age (Haagensen 1971). However, the mean age and peak age of occurrence of breast carcinoma in this study of Arab women in Kuwait is similar to that reported from Saudi Arabia (Anim, 1990, 1994) and other developing countries (Chiedozi, 1987).

Kuwaitis and other Arabs have similar height and weight distribution. Calculation of the body mass index (BMI) gives a better understanding to the demographic data and shows that virtually all Kuwaiti and other Arab women with breast cancer seen in this study are overweight. Abdella *et al.*, (1996) have reported a high BMI in a group of known Kuwaiti diabetics and the prevalent rate for NIDDM in the Kuwait population is 12%. In this study, 50% of all the patients were classified as diabetics.

Table 7. Expression of growth hormone-receptor in the membrane/cytoplasm and nucleus of normal and neoplastic breast tissue cells.

Breast tissue type	Number of cases		Proportion of receptors	
	Total	Positive	Cytoplasm	Nuclei
Normalbreast	6	4	+	–/+
Fibrocystic disease	9	7	+	–/+
Sclerosingadenosis	2	2	+/++	–/+
Gynaecomastia	3	3	+/++	–/+
Intraductal papilloma	1	1	+/++	+
Signet ring carcinoma	1	1	+++	++
Medullary carcinoma	1	1	++	++
Mucin producing adenocarcinoma	1	1	++	+/++
Tubular carcinoma	1	1	+++	–/+
Intraductal carcinoma	12	10	+++	+
Ductal carcinoma	28	21	++/+++	+/+++
Infiltrating ductal carcinoma	9	9	+++	+++
Ductal/Lobular carcinoma	1	1	++	+
Lobular carcinoma	14	9	++/+++	–/+
Infiltrating lobular carcinoma	9	9	+++	++
Metastatic carcinoma, L.N.	12	12	++	–/+

Proportion of positive cells and nuclei immunoreactivity:
-/+ = < 5%; + = 5-15%; ++ = 16-30%; +++ = > 31%

Although our study looks at a highly selected population, the prevalence rate of non insulin-dependant diabetes mellitus (NIDDM) of 50% in breast cancer patient mirrors the high prevalence rate of NIDDM in the Kuwaiti population and is evidence of an association between the two conditions, which may be linked together through a common predisposing or risk factor. Obesity and high calorie intake have been associated with breast cancer and NIDDM can also be associated. Although glucose metabolism is known to be increased in breast cancer cells, glucose utilization in normal tissue is impaired in diabetes mellitus. Thus, when the two conditions co-exist, excess blood glucose due to the diabetic condition, could theoretically be used by breast cancer cells in order to meet their increased metabolic needs. Therefore, diabetes mellitus can accelerate tumour cell growth in breast cancer patients. We have found a significant correlation between the presence of diabetes and advanced breast cancer in the patients investigated (<0.0001). While it is unlikely that NIDDM is implicated in the genesis of breast cancer, we suggest that it may be involved in the progression of the disease and that patients with NIDDM are more likely to present with advanced breast cancer. On the other hand, the presence of breast cancer may unmask diabetes mellitus in susceptible subjects or worsen an already established disease. This may be supported by the fact that 27% of breast cancer patients with NIDDM in this study had high plasma glucose levels of over 16 mMol/L, i.e more than twice the level for classifying subjects as diabetic. Advanced breast cancer implies an increased number of cancer cells utilizing more glucose (Brown, 1993). Thus, the association between NIDDM obesity and breast cancer may be through peripheral glucose utilization. Interactions between insulin receptor and receptors for the insulin like growth factors IGF-1 and IGF-2 may play a part. Moreover, IGF-2 released from activated

Table 8. HER-2/neu oncoprotein expression in normal breast and in neoplasm in young Arab women <30 years of age.

Breast tissue type	Reactivity		Intensity			
	Tested	Positive	–	+	++	+++
Normal breast	5	1	4	1		
Benign breast	1		1			
Fibrocystic disease	9	1	8	1		
Sclerosingadenosis	2		2			
Gynecomastia	3		3			
Intraductal papilloma	1		1			
Ductal carcinoma	28	16	12	11	3	2*
Intraductal carcinoma	12	7	5	4	1	2**
Infiltrating ductal carcinoma	21	14	7		4	10*
Lobular carcinoma	14	8	6	5	1	2*
Infiltrating lobular carcinoma	9	7	2	3	3	1*
Ductal/Lobular carcinoma	1	1				
Signet ring carcinoma	1					1*
Tubular carcinoma	1		1			
Medullary carcinoma	1	1		1		
Mucin producing adenocarc.	1	1			1	
Metastatic carcinoma, L.N.	12	12		1	10	1

Intensity: - = negative; + = weak; ++ = moderate; +++ = strong
* = Lymph node positive; ** = Grade II or III.

endothelium in tumours has a paracrine effect on neoplastic cells and enhances tumour progression (Fox, 1997).

Tumour Markers in Breast Cancer

The application of new biomarkers for breast cancer has led to an improved prognostic index or has influenced the treatment selection (Chapman *et al.*, 1996). This study evaluates the expression of OR, PR, HER -2/neu, p53, GHR, Ki-67, PCNA, cathepsin D and vascular density in breast carcinomas of young Arab women (<30 years). Of the 122 cases investigated, 91.5% had invasive ductal carcinomas, of which 68% were Grade III and in 57% of the patients' axillary lymph node metastases were detected. Expression of Ki-67, PCNA, p53, and c-erbB-2 correlated significantly with each other and with the histological grade. Many of the carcinomas showed a high Ki-67 and PCNA proliferative activity, indicative of poor prognosis (Bacus *et al.*, 1989; Sirvent *et al.*, 1994). A positive correlation was demonstrated between Ki-67 staining and histological and nuclear grade (Bacus *et al.*, 1998), mitotic index and overexpression of c-erbB-2 (Vanlemmens *et al.*, 1998). On the other hand, a negative correlation has been shown to exist with oestrogen and progesterone receptor status (Kristen *et al.*, 1997), which we could also show in our present study. However, no statistical difference in proliferative activity among the histological subtypes was observed and no strong correlation was found between Ki-67, PCNA, p53 and HER-2/neu on one side and cathepsin D on the other. The latter could be demonstrated in only 39% of patients. In the cathepsin D-positive tumours (34%) no involvement of the lymph nodes was found. Therefore, no correlation between the invasiveness of the breast cancer and immunoreactivity for cathepsin D can be postulated from our results. This finding is supported by a study done by Glikman *et al* (1997), which demonstrated that cathepsin D, a lysosomal protease, did not show any association with known prognostic factors or with lymph nodes. Cathepsin D gene expression in oestrogen receptor (OR) positive breast cancer cell lines is known to be induced by oestrogens (Cavailles *et al.*, 1991). Cathepsin D gene expression is also induced by growth factors, including epidermal growth factor, insulin, insulin-like growth factor 1, and basic fibroblast growth factor (Cavailles *et al.*, 1989). Briozzo *et al.* (1988) have shown that through autoactivation in acid pH, the 52 kd cathepsin D in vitro degrades extracellular matrix, suggesting a role of the enzyme in facilitating metastasis. Studies have shown a shorter disease free or overall survival period in cases with high cathepsin D levels (Thorpe *et al.*, 1993). In contrast, immunohistochemical studies clearly indicate that positive cathepsin D staining in breast cancer is associated with a longer survival in ER positive or lymph node positive patients (Henry *et al.*, 1990).

DNA aneuploidy has been found more commonly in high grade tumours, as was also the case in this study. Of 48 aneuploid breast carcinomas, 39 were grade 3 carcinomas. The correlation of ploidy with other clinicopathological variables is less convincing. No significant corelation was found between the frequency of DNA aneuploidy and tumour size, nodal status or steroid receptor status. The results of other groups on a correlation between ploidy and hormone receptor status are also controversial. Merkel and McGuire (1992) found that in studies involving 2,553 patients, DNA aneuploidy occurred more commonly in oestrogen receptor negative tumours, whereas in six other studies involving 752 patients, no relationship or a non-significant trend was observed.

The genetic changes most commonly found in breast cancer are alterations in p53 tumour suppressor gene, with an incidence ranging from 15 to 50% (Sirvent *et al.*, 1994). In our study, nuclear p53 positivity was even higher, 57.3% of all carcinomas.displayed a distinct staining for p53. Walker *et al* (1997) found a high incidence of p53 positivity in breast carcinomas among women aged under 35 years. These tumours showed a more aggressive nature compared with the breast carcinomas of a control group with was 50-67 years old. They concluded that this may indicate genetic instability and could account for the observed poor prognosis of tumours in young patients. The accumulation of p53 in breast cancer in our study was associated with negative oestrogen and progesterone receptor content, high histological grade, aneuploidy and positive c-erbB-2 status. Therefore, p53 mutation seems to play an important role in the development of breast cancer at a young age. The high incidence of p53 mutations might be explained by the fact that most of the young patients had Grade III tumours in which p53 mutations were frequent. Alterations of p53 are correlated with negative oestrogen and progesterone receptor status (Rosanelli *et al.*, 1995) and this may be explained by the findings of Lin *et al.* (2000) who demonstrated that oestrogen receptor alpha protects p53 from being deactivated. Cancer cells with p53 mutations are clinically more aggressive and patients with p53 mutations have a worse prognosis. Zellars *et al.* (2000) found that expression of p53 is an independent prognostic marker on survival curves and disease-free interval and Alfred *et al.* (1993) stated that p53 is an important indicator of early relapse of the disease, followed by tumour size and high percentage of cells in the S-phase. In our study, no complete data on disease-free interval or survival curves were available, therefore no final conclusions of the importance of p53 as a prognostic factor for breast cancer in young patients can be drawn. However, this study gives strong evidence that the association of high tumour grade, negative hormone receptor status, high cell proliferation rate (Ki-67) and positivity for p53 and c-erbB-2 points to greater tumour aggressiveness.

The erbB-2 family of growth factor receptors and their ligands represent a promising area of breast cancer research. Included in this group are the epidermal growth factor receptors (EGFR) as well as HER-2/*neu*/*erb*B-2, HER-3//erbB-3 and recently HER-4/erbB-4. These receptors are structurally similar and are frequently implicated in the pathogenesis and biological behaviour of breast cancers with many tumours expressing more than one receptor of this family. Human breast carcinomas with a high level of EGF receptor show a significant correlation with high histological grade, an inverse relationship with oestrogen receptors and are associated with a more aggressive disease course (Elston and Willis, 1991). Membrane staining of breast carcinoma cells obtained by immunohistochemistry using the antiserum to c-erbB-2 protein is a reliable marker of HER-2/neu gene amplification. Amplification of the c-erbB-2 and proto-oncogene is reflected in overexpression of the c-erbB-2 oncoprotein, a 185 kD transmembrane glycoprotein. Overexpression of HER-2/neu occurs in association with poor prognostic factors such as lymph node involvement and negative oestrogen and progesterone receptor status. Such tumours are often less responsive to adjuvant therapy and appear to correlate with a worse disease-free and overall survival (Jorda *et al.*, 1994; Bacus *et al.*, 1994).

In our study, 70 cases (57.3%) showed positive staining with different degrees according to the intensity of gene expression, while 52 cases (42.6%) were HER-2/neu protein negative. Based on this data, results suggest that the HER-2/neu protein is mostly overexpressed in ductal carcinoma (16 cases), intraductal carcinoma (7 cases), infiltrating ductal carcinoma (14 cases), lobular carcinoma (8 cases), infiltrating lobular carcinoma (7 cases), and metastatic carcinoma (12 cases). From a total of 70 positive cases, 18 showed strongly positive staining in patients with positive lymph node involvement. Two cases for each category of ductal carcinoma, intraductal carcinoma and lobular carcinoma showed strong positive lymph node involvement and 10 were strongly immuno-positive for infiltrating ductal carcinoma. In infiltrating lobular and signet ring carcinomas, positive staining stained was associated with lymph node involvement. A number of factors useful for the prediction of the course of breast cancer have been identified. These include axillary lymph node involvement, tumour size, hormone receptor and nuclear grade. The present results show a significant association between positive staining and c-erbB-2 protein expression. There is also a correlation between strong positive staining (c-erbB-2 protein expression) and axillary lymph node involvement.

Anti-HER2 is a tumour specific therapy in which efficacy depends on HER2 overexpression by carcinoma cells. The rate of HER2 overexpression in breast carcinoma, which is related to the clinical presentation of the disease, is 15% in small breast tumours and increases to 30% in inflammatory and early onset tumours. In patients with metastatic disease that is resistant to chemotherapy (CT), a 50% response rate has been obtained using anti-HER2 in association with CT. Trastuzumab (Herceptin; Roche, Switzerland), a humanised anti-HER2/neu monoclonal antibody, has been developed as a drug that binds to the HER2/neu protein and inhibits the action of HER2/neu-enhanced tumour cell growth. This drug has been introduced for the treatment of metastatic breast carcinoma that overexpresses HER2/neu. An important factor in the choice of HER2/neu as immunotherapeutic target is the fact that HER2/neu is not merely a selective tumour marker, but actually plays an important role in the pathogenesis of malignant cells through its effects on mitogenic cell signalling pathways (Berg and Hunter, 1995). Furthermore, membrane staining of HER-2/neu was associated with a higher proliferating index than tumour cells exhibiting only cytoplasmic staining, and may be useful in differentiating fast growing tumours from slow growing tumours.

Evidence shows that polypeptide growth factors and hormones, particularly somatomedins (IGF-1, IGF-2), and growth hormone (GH) are involved in breast cancer cell proliferation via paracrine and autocrine loops. An IGF-related polypeptide is secreted by human breast cancer cell lines examined, and both IGF-1 and IGF-2 are mitogenic for breast cancer cells. IGF-1 related polypeptides can mediate autocrine growth of fibroblasts and smooth muscle cells and those produced by malignant breast epithelium may also act in a paracrine fashion to mediate stromal cell chemotaxis and proliferation. The results from this study demonstrate the expression of receptors for growth hormone, an important mammotrophic hormone, in breast cancer tissue. Cellular immunoreactivity of GH-receptors was present in either both the plasma membrane-cytoplasm and nucleus, or exclusively in the plasma membrane-cytoplasm. This is in agreement with the localization pattern previously described in rat and human tissues (Lincoln *et al.*, 1990, 1992, 1994, 1995, 1997, 1998, 1999, 2000) and the demonstration of GH induced nuclear translocation of its receptor (Lobie *et al.*, 1994). GH-receptor is expressed predominantly in epithelial components of the malignant breast, but some stromal cellular expression was also evident. Growth hormone exerts a direct action on the development and differentiation of rodent mammary gland in vivo (Feldman *et al.*, 1993) and in vitro (Plaut *et al.*, 1993). Because of its proliferative actions, circulating levels of mammotrophic GH was linked to mammary gland carcinogenesis in rats and mice (Nandi and Guzman, 1995). Some of the GH effects are mediated through the production of IGF-I and the high expression of IGF-I and its receptor in mammary gland carcinoma is linked to cellular proliferation, ultimately leading to tumour formation (Cullen *et al.*, 1990). It has been proposed that GH enhances cellular proliferation by acting on sensitive ductular stem cells, rendering them targets for the proliferative effects of IGF-I, resulting in propagation of genetic errors (Nandi and Guzman, 1995).

Growth hormone also facilitate cell proliferation through paracrine and/or autocrine mechanisms. Administration of GH significantly increases the expression of c-myc proto-oncogene (Murphy *et al.*, 1987). The GH gene is also expressed in normal and tumour mammary gland of cat and dog (Mol *et al.*, 1995) and promotes mammary carcinogenesis in a progesterone-dependent manner (Van Garderen *et al.*, 1997). To interpret the presence of mammary GH ligands in a functional sense, Liu *et al.* (1997) have shown that endogenous production of human GH by transient transfection of the human GH gene in mammary carcinoma cells increases STAT mediated transcription. GH-receptor gene expression remained constant throughout progression of breast lesions Liu *et al.* (1997), suggesting that the local concentration of ligand is the contributing factor to neoplastic change. In addition to their role in the proliferative response through a paracrine and/or autocrine mechanism, a GH intracrine role is possible involving nuclear receptors for GH. In humans the link between circulating mammotrophic GH and the genesis of breast cancer is controversial (Bonneterre *et al.*, 1987). In patients with advanced breast cancers, intra-muscular injection of recombinant human GH for 2 days before chemotherapy induced a 2-fold increase in proliferative activity of tumour cells in 50% of cases, apparently mediated by local IGF-I (Baldini *et al.*, 1994). The presence of GH-receptors in the nucleus means that growth hormone internalised to the nucleus bind to chromatin growth hormone binding protein (GHBP) and modulates transcription of GH-specific mRNA (Lincoln *et al.*, 1999). Growth hormone is transported to the nucleus to exert its action via nuclear receptors. The localisation in the nucleus of the receptor intracellular domain is a requirement for growth hormone to have a nuclear function. The use of anti-growth hormone receptor antibodies to block tumour progression is an interesting possibility.

The initiation, growth and metastatic spread of cancer depends to a large extend on the responses of the normal host tissue. Important interactions with the non-neoplastic vasculature must develop to facilitate metastases. Tumor angiogenesis (TA) is the growth of new vessel toward and within the tumour and is mediated by factors secreted by the tumor cells and/or tumor-associated inflammatory cells. Micro-vessel density in the area of the most intense neovascularisation in and around tumours in invasive breast carcinoma has been reported to correlate with the presence of metastases and survival and the quantination is an independant prognostic factor (Weidner *et al.*, 1991; 1992). The established method of enumeration of micro-vessel density is to count the vessels using an ocular raster. It is critical to carefully select fields for measurement. The areas showing the highest estimated density of vessels were chosen for the assay. In this study we have relied on the sensitive vascular marker CD-31 and immunostained 122 invasive breast carcinomas for the presence of CD31 and quantitated vascular surface area with ten 20x fields. Vascular surface area was prognostically significant in univariate analysis of disease free and overall survival, as were stage of the disease, tumor size, nodal status and combined histological grade. High grade carcinomas contained greater vascular surface area than low grade carcinomas (p=0.0001).

The well documented number of breast cancer cases observed in Kuwait represents unique features both clinically and pathologically. Breast cancer affects a significant proportion of young Arab women. Patients <30 years of age have a worse prognosis than older patients. The applications of new cytochemical markers are a valuable tool for therapeutic assessment and for clinical prognosis. The combination of obesity, hypertension and diabetes mellitus seems to be in close correlation with onset of the disease. Furthermore, the tumour size and the number of positive lymph nodes reflect the status of the cancer. Most early recurrences and aggressive growth were present in high grade tumours with more than 3 cm in size. Age, stage and nodal status were the most important prognostic factors. All these findings have direct impact on therapeutic strategies to be applied. Adjuvant chemotherapy was used in node-negative patients in the young age group with large tumours of high grade vascular lymphatic invasion and anapleudy and oestrogen negative status. There are highly significant trends for the prevalence of poor prognostic features such as Grade 3 histology, extensive intraduct component and vascular and lymphatic invasion. Most of the observed cases were ductal carcinomas with a high proportion of oestrogen- and progesterone-receptor negative tumours. The association of negative oestrogen receptor and progesterone receptor hormone status and positivity for p53 oncoprotein points to a greater tumour aggressiveness. This may reflect changes in steroid metabolism and usually excludes tamoxifen therapy. Ki-67 and PCNA immunostaining specifically demonstrate proliferating tumour cells and good correlation with the grade of the carcinoma. Immunostaining for p53, HER-2/neu, GH-receptor resulted in the staining of a varying number of tumour cells, which could not be correlated with tumour grade. High numbers of apoptotic cells observed in flow cytometry certainly represent high cellular turnover in a proliferating tissue. The direct role for growth hormone and their receptors in breast cancer proliferation and in the epithelial cell reaction during carcinogenesis suggests caution in the use of recombinant growth hormone in the treatment of short statue prepubertal girls. This applies to those without evidence of classic growth hormone deficiency, particularly in the presence of a family history of breast cancer.

References

1. Abdella N. Khogali M, Al-Alis G. Known type 2 diabetes mellitus among the Kuwaiti population a prevalence study Acta Diabetal. 1996, 3: 145-149.
2. Al-Jaralla, M.. Health and Vital Statistics Division, Department of Statistics and Medical Records,Kuwait Ministry of Health publication, 1998.
3. Alfred D, Clark G, Elledge R, Fuqua S, Brown R, Chamness G, Osborne C. Association of p53 expression in node-negative breast cancer. J. Natl. Cancer Inst. 1993; 85: 200-206.
4. Anim JT, Al-Sohaiban Mo, Grant CS, Tamimi D. Breast Cancer as seen in King Fahed University Hospital, Al-Khobar 1983-1986, J. Coll. Surg. Edinburgh 1990; 1983-1986.
5. Anim JT, El-Hassan Ay, Al Mullim F, Al-Dossary N, Al-Sohaiibani M. Breast cancer in a teaching hospital in Saudi Arabia J. Kuwaiti Med. Ass. 1994; 26: 40-42.
6. Bacus SS, Zelnick CR, Plowman G, Yarden Y. Expression of the erbB-2 family of growth factor receptors and their ligands in breast cancers. Am. J. Clin. Patho.l 1994; 102: S13-S24.
7. Bacus S, Goldschmidt R, Chin D, Moran G, Weinberg D.. Biological grading of breast cancer using antibodies to proliferating cells and other markers. Am. J Pathol. 1998; 135: 783-792.
8. Baldini E, Gianessi P, Gardin G, Alama A, Minuto F, Barreca A, Conte P. In vivo cytokinetic effects of recombinant human growth hormone (rGH) in patients with advanced breast carcinoma. J. Biol. Regul. Homeostatic Agents 1994; 8: 113-116.
9. Bassler R. Histopathologische Kriterien und aktuelle Klassifikation des Mammakarzinoms. Onkologe 1995; 1: 180-189.
10. Berg J, Hunter R. Breast cancer. Cancer 1995; 75: 257-269
11. Bloom HJ, Richardson W. Histological grading and prognosis in breast cancer. Br.J. Cancer 1957; 11: 359-377.
12. Bonneterre J, Peyrat JP, Beuscart R, Lefebvre J. Prognostic significance of prolactin receptors in human breast cancer. Cancer Res. 1987; 47: 4724-4728.
13. Briozzo P, Morisset M, Capony F, Rougot C. In vitro degredation of extracellular matrix with Mr 52, 3000 cathepsin D secreted by breast cancer cells. Cancer Res. 1988; 48: 3688-3692.
14. Brown R W, Allered DC, Clark Gmetal prognostic significance and clinical pathological correlations of cell cycle Kuwaities measured by KI 67 immuno cytochemistry in axillary node negative carcinoma of the breast. Breast Cancer Res. Treat. 1990; 16: 191.
15. Brown RS, Wahl R. Overexpression of Glut-1 glucose transporter in human cancer. Cancer 1993; 72:2979-2985.
16. Carroll K K. Influence of diet on breast cancer nutrition. Cancer 1984; 2:232.
17. Cavailles V, Augereau P. Cathepsin D gene of human MCF-7 cells contains estrogen responsive sequences in its 5' proximal flanking region. Biochem. Biophys. Res. Commun. 1991; 174:816-824.
18. Chapman JW, Murray D, McCready D, Hanna W, Kahn H, Lickley H, Trudeau M, Mobbs B, Sawka C, Fish E, Pritchard K. An improved statistical approach: can it clarify the role of the new prognostic factors for breast cancer? Eur. J. Cancer 1996; 32: 1949-1956.
19. Charpin C, Andrac L, Vacheret H. Multiparametric evaluation of growth fraction (Monoclonal Ki 67) in breast carcinonoma tissue sections. Cancer Res.1988;48:4368.
20. Chiedozi L C. The patient history: A source of prognostic data in breast cancer. J. Royal College Edinburgh. 1987; 142-144.
21. Clark GM, McGuire WL. Steroid receptors and otehr prognostic factors in primary breast cancer. Semin. Oncol. 1988; 15: 20-25.
22. Corkill ME, Katz R. Immunocytochemical staining of C-erbB-2 oncogene in fine needle aspirates of breast carcinoma: A comparison with tissue sections and other breast cancer prognostic factors. Diag. Cytopathol. 1994; 11: 262-265

23. Cullen KJ, Yee D, Sly WS, Perdue J, Hampton B, Lippman ME, Rosen N. Insulin-like growth factor receptor expression and function in human breast cancer. Cancer Res. 1990; 50: 48-53.
24. Davis AR. Breast cancer: the search for new prognostic markers. Br. J. Biomed. Sc.1996; 53: 157-161.
25. Decouvelaere C, Peyrat JP, Bonneterre J, Dijane J. Presence of the two growth hormone receptor messenger RNA isoforms in human breast cancer. Cell Growth. Differ. 1995; 6: 477-483.
26. Diel EC 1997. Lymph node as prognostic factor in breast cancer. Early breast cancer trialists collaborative group. Systemic treatment of early breast cancer by hormonal immunetherapy. Lancet 1992; 339; I, 71.
27. Dowell S, Hal P. The clinical relevance of the p53 tumour supressor gene. Cytopathol. 1994; 5: 133-145.
28. Elledge R, Alfred D. The p53 tumor suppressor gene in breast cancer. Breast. Cancer. Res. Treat. 1994; 32: 39-47.
29. Elledge R, Clark G, Fuqua S, Alfred D. p53 protein accumulation detected by five different antibodies. Relationship to prognosis and heat shock protein 70 in breast cancer. Cancer Res. 1994; 54: 3752-3757.
30. Elias J. The price of everything, the value of nothing. J. Histotech. 1993; 16: 105-106.
31. Elston C, Wilis I. Pathological factors in breast cancer. I. The value of histological grade in breast cancer: experience from a large study with long term follow up. Histopathology 1991; 19: 403-410.
32. Engel K, Fox S, Harris A. Angiogenesis as a biologic and prognostic indicator in human breast carcinoma. In: Regulation of Angiogenesis, Birkhauser Verlag, Basel 1997; pp 114 156.
33. Feldman M, Ruan W, Cunningham B, Wells J. Evidence that growth hormone receptors mediates differentiation and development of the mammary gland. Endocrinol. 1993; 133: 1602-1608.
34. Feigelson HS, Henderson B, Pike M. Recent trends in U.S. breast cancer incidence, survival and mortality rates. J. Natl. Cancer Inst. 1997; 89: 1810-1812.
35. Fox SB, Tumour angiogenesis and prognosis. Histopathol 1997; 30:294-301.
36. Freudenheim H. Lactation history and breast cancer risk. Am. J. Epidemiol. 1997; 146: 932-938.
37. Glickman P, Rogozinski A, Mosto J, Pollina A, Garbovesky C, Levy C. Relationship between cathepsin D and other prognostic factors in human breast cancer. Tumori 1997; 83: 685-688.
38. Haagensen CD Diseases of the breast. 2nd. Edition, WB Sounders, Philadelphia 1971:363.
39. Hallstein M, Sidransky D, Vogel B. P53 Mutations in human cancers. Science 1991; 253: 49.
40. Haarslev T, Jacobson G, Zedeler K. Correlation of growth fraction by Ki-67 and proliferating nuclear antigen (PCNA) immunohistochemistry with histpathological parameters and prognosis in primary breast cancer. Breast Cancer Res. Treat 1996; 37: 101-113.
41. Henry JA, McCarthy A, Angus B. Prognostic significance of the estrogen regulated protein, cathepsin D, in Breast cancer. An immunohistochemical study. Cancer Res. 1990; 65: 265-271.
42. Hoyt EC, Van Wyk JJ, Lund P K: Tissue and development specific regulation of a complex family of rat IGF-I messenger ribonucleic acids. Mol. Endocrinol. 1988; 2: 1077-1086.
43. Hupperets P, Volovica L, Schouten L, Jager J, Schouten H. The prognostic significance of steroid receptor activity in tumour tissues of patients with primary breast cancer. Am. J. Clin. Oncol. 1997; 20: 546-551.
44. Iatropoulos M, William G. Proliferation markers. Exp. Toxicol. Pathol. 1996; 48: 175-181.
45. Isgaard J, Moller C, Isaksson OG, Nilson A, Mathews LS. Regulation of IGF-I messenger ribonucleic acid in rat growth plate by growth hormone. Endocrinol. 1988; 122: 1515-1520.
46. Jorda M, Ganjej P, Nadji M. Retrospective C-erbB-2 immunostaining in aspiration cytology of breast cancer. Diag. Cytopathol. 1994; 11: 262-265.
47. Kelly P A, Djiane J, Postel-Vinay MC, Edery M. The prolactin/growth hormone receptor family. Endicrinol. Rev. 1991; 12: 235-251.
48. Kerlikowska K, Barclay J, Grady D, Sickles E, Ernster V. Comparison of risk factors for ductal carcinoma, in situ and invasive breast cancer. J. Natl. Cancer Inst. 1997; 89: 76-82.
49. Kristen P, Müller J, Caffier H. Prognostic relevance of p53 in node negative breast cancer. Anticancer Res. 1997; 17: 2809-2872.

50. La Vecchia C, Tavani A, Franceschi S, Parazzia F. Oral contraceptives and cancer. A review of the evidence. Drug Safety 1996; 14: 260-272.
51. Leong A, Milios J, Tang S. Is immuno-localisation of proliferating cell nuclear antigen (PCNA) in paraffin sections a valid index of cell proliferation? Appl. Immunohistochem 1993; 1: 127-135.
52. Lincoln DT, Waters MJ, Breipohl W, Sinowatz F, Lobie P. Growth hormone receptor expression in the proliferating rat mammary gland. Acta Histochem. 1990; 40: 47-49.
53. Lincoln DT, Breipohl W, Wegener A. Immunohistochemical demonstration of GH receptors in stromal cells from long-term human bone marrow cultures. Histochem. J. 1992; 24: 626-627.
54. Lincoln DT, El-Hifnawi E, Sinowatz F, Waters, M. Immunohistochemical localization of growth hormone receptor binding protein in the mammalian cerebellum. Annals of Anatomy 1994; 176: 419-427.
55. Lincoln DT, Sinowatz F, Hughes RL, Waters MJ. Evidence of a direct role for growth hormone in mammary gland proliferation and lactation. Anat. Histol. Embryol. 1995; 24:107-115.
56. Lincoln DT, Sinowatz F, Gabius H, Gabius S, Temmim L, Baker H, Mathew C. Subpopulations of stromal cells from long-term human bone marrow cultures: ontogeny of progenitor cells and expression of growth hormone receptors. Anat. Hist. Embryol. 1997; 26: 11-28.
57. Lincoln DT, Sinowatz F, Baker H, Kölle S, Waters MJ. Growth hormone receptor expression in the nucleus and cytoplasm of normal and neoplastic cells. Histochem. Cell Biol. 1998; 109: 141-159.
58. Lincoln DT, Sinowatz F, Kölle S, Takahashi H, Parsons P, Waters MJ: Up-regulation of growth hormone receptor immunoreactivity in human melanoma. Anticancer Res. 1999; 19:1919-1932.
59. Lincoln DT, Kaiser H, Waters M J. Growth hormone and colorectal carcinoma: Localization of receptors. In Vivo 2000; 14: 41-50.
60. Lisboa B, Vogtlander S, Gilster T, Riethdorf L, Milde-Langosch K, Loning T. Molecular and immunohistochemical analysis of p53 mutations in scrapings and tissue from preinvasive and invasive breast cancer. Virchows Arch. 1997; 431 :375381.
61. Liu G, Schwartz J, Brooks S. Estrogen receptor protects p53 from deactivation by human double minute-2. Cancer Res. 2000; 60: 1810-1814.
62. Liu N, Mertani HC, Norstedt G, Tornell J, Lobie PE. Mode of the autocrine/paracrine mechanism of growth hormone action. Exp. Cell Res. 1997; 237: 196-206.
63. Lobie PE, Wood TJ, Chen CM, Waters MJ, Norstedt G. Nuclear translocation and anchorage of the growth hormone receptor. J. Biol. Chem. 1994; 269: 31735-31746
64. Mc Guire WL, Clark GM. Prognostic factors and treatment decision in axillary node negative breast cancer. New Engl. J. Med.1992:326:1756.
65. Mc Mahon B, Cole P, Lin T. Age at first birth and cancer of the breast. A summary of an international study. Bull. WHO 1970:43:209.
66. Merkel DE, McGuire WL. Ploidy, proliferative activity and prognosis. DNA flow cytometry of solid tumours. Cancer 1990; 65: 1194-1205.
67. Mertani H, Pechoux C, Garcia T, Waters M., Morel G. Cellular localization of the growth hormone receptor/binding protein in the human anterior pituitary. J. Clin. Endocrinol. Metab. 1995; 80: 3361-3367.
68. Mol JA, Henzen-Logman SC, Hageman P, Misdorp W. Expression of the gene encoding growth hormone in the human mammary gland. J. Clin. Endocrinol. Metab. 1995; 80: 3094-3096
69. Murphy LJ, Bell GI, Friesen HG: Growth hormone stimulates sequential induction of c-myc and insulin-like growth factor expression *in vivo*. Endocrinol. 1987; 120:1806-1812.
70. Nandi S, Guzman R. Hormones and mammary carcinogenesis in mice, rats and human: A unifying hypothesis. Proc. Nat. Acad. Sci. (Wash) 1995; 92: 3650-3657.
71. Nixon AJ, Neuber D, Hayes DF. Relation ship of patient age to pathologic features of the tumour and prognosis for patient with stage I or II breast cancer. J. Clin. Oncol 1994; 12:888

72. Osborne CK. Receptors In: Harris Jr, Hellman S, Henderson IC, Kinne DW, Eds. Breast Diseases. Philadelphia: JB Lippincott, 1991; 301-325.
73. Page DL, Anderson TJ, Elston CW. Grading of invasive carcinoma of the breast. In: Page DL, Le Doussal V, Tubian-Hulin M, Friedman S, Hacene K. Prognostic value of histologic grade nuclear components of Scarff-Bloom-Richardson (SBR). Cancer 1989; 63: 1914-1921.
74. Paszko, Z, Omar, YT, Nasralla, M.Y, Jazzaf, H, Bouzubar.. Estrogen and progesterone receptor status in breast cancer in Kuwait female population. Neoplasma, 1993; 40: 127-132.
75. Peyrat J, Bonneterre J. Type I IGF receptor in human breast disease. Breast Cancer Res. Treat. 1992; 22: 59-67.
76. Plaut K, Ikeda M, Vonderhaar BK: Role of growth hormone and insulin-like growth factor-I in mammary development. Endocrino. 1993; 133: 1843-1848.
77. Radovanovic A, Shah N, Behbehani J: Prevalence and social correlates of consanguinity in Kuwait. Ann.. Saudi Med. 1999, 19:206-210.
78. Rosanelli G, Steindorfer P, Wirnsperger G, Klimpfinger M, Ratschek M, Puerstner P, Auner H. Mutant p53 expression and DNA analysis in human breast cancer. Comparison with conventional clinicopathological parameters. Anticancer Res. 1995; 15: 581-586.
79. Russo A, Bazan V, Morello V, Tralongo V, Nagar C, Nuara R. Vinmentin expression, proliferating nuclear cell antigen and flow cytometric factors. Analyt. Quant. Cytol. Histo.l 1994; 16: 365-374.
80. Salmon WH, Daughaday WH. A hormonally controlled serum factor which stimulates sulfate incorporation in cartilage in vitro. J. Lab. Clin. Med. 49: 825-836, 1957.
81. Scarff RW, Torioni H. Histological typing of breast tumours. WHO, Geneva 1968.
82. Schindler C, Darnell JE: Transcriptional responses to polypeptide ligands: the JAK-STAT pathway. Ann. Rev. Biochem. 1995; 64: 621-651.
83. Schonborn I, Minguillon C, Lichtenegger W, Zschiesche W, Spitzer E. The prognostic significance of recent tumor markers for lymph node negative breast cancer. Geburtshilfe Frauenheilkunde 1995; 55: 566-571.
84. Sell S, Garrett C. Clinical application of genetic, oncogenic and differentiation markers of cancer. In: Cellular cancer markers. Humana Press, Totowa, New Jersey 1995; pp 1-16.
85. Silverstein MJ, Poller D, waisman J, Colburn W, Barth A, Gierson, E, Lewinsky B, Ganagami P, Slamon D. Prognostic classification of breast ductal carcinoma in-situ. Lancet 1995; 345: 1145-1157.
86. Silvestrini R, Benini E, Daidone M, Veneroni S, Boracchi P. p53 as an independent prognostic marker in lymph-node negative breast cancer patients. J. Natl. Cancer Inst. 1993; 85: 965-970.
87. Simpson JF, Page DL. Status of breast cancer prognostication based on histopathologic data. Am. J. Clin. Pathol. 1994; 102: S3-S8.
88. Sinowatz F, Schams D, Koelle S, Lincoln DT, Waters M. Cellular localisation of GH receptor in the bovine mammary gland during mammogenesis, lactation and involution. J. Endocrinol. 2000; 166: 503-510.
89. Sirvent JJ, Santage M, Salvado M, Alvaro T, Raventos A, Palacios J. Hormonal receptors, cell proliferation fraction (Ki-67) and c-erbB-2 amplification in breast cancer, relationship between differentiation, degree and axillary lymph node metastases. Histol. Histopathol. 1994; 9: 563-570.
90. Sobrier ML, Duquesnoy P, Durietz B, Amselem S, Goossens M. Expression and binding properties of 2 isoforms of the human growth hormone receptor. FEBS Letters 1993; 319: 16-20.
91. Spandidos DA, Karajossifidi H, Malliri A, Linardopoulos S. Expression of ras RB1 and P53 proteins in breast cancer related to prognostic factors. Anticancer Res. 1992; 12: 81-89.
92. Spyratos F, Brouillet J-P, Defrenne A, *et al.* Cathepsin D: An independent prognostic factor for metastasis of breast cancer. Lancet 1989; Nov: 1115-1118.
93. Stracke H, Schulz A, Moeller D. Effect of growth hormone on osteoblasts and demonstration of somatomedin-C/IGF-I in bone organ culture. Acta Endocrinol. 107: 16-24, 1984.

94. Taylor SR, Zachariah S, Chakrabotry S, Overstreet J, Ramzy I, Mody D. Ploidy studies by image analysis on fine needle aspirates of the breast. Acta Cytol. 1993; 37: 923-928.
95. Tandon AK, Clark GM, Chamness GC, Chirgwin JM, McGuire WL. Cathepsin D and prognosis in breast cancer. New Eng.l J. Med. 1990; 322: 297-302.
96. Thorpe SM, Rochefort H, Garcia M, *et al.* Association between high concentrations or M_r 52,000 cathepsin D and poor prognosis in primary breast carcinoma. Int. J. Cancer 19931 55L 371-374.
97. Van der Tuden JC, Lindman J. The multivariate prognostic index and nuclear DNA contents are independent prognostic factors in primary breast cancer patients. Cytometry 1989; 10: 56.
98. Van Garderen E, De Witt M, Voorhout W, Rutteman G, Mol J, Nedersbragt H, Misdorp W. Expression of growth hormone in canine mammary tissue and mammary tumours. Evidence for a potential autocrine/paracrine stimulatory loop. Amer. J. Pathol. 1997; 150: 1037-1047.
99. Vignon F, Capony F, chambon M, fress G, Garcia M, Rochefort H. Autocrine growth stimulation of the MCF7 breast cancer cells by the oestrogen regulated 52 K protein. Endocrinology 1986; 118: 1537-1545.
100. Vanlemmens L, Hebbar M, Peyrat J, Bonneterre J. Age as a prognostic factor in breast cancer Anticancer Res. 1998; 18: 1891-1896.
101. Walker RA, Lees E, Webb MB, Dearing s. Breast carcinoma occurring in young women (<35 years) are different. Br. J. Cancer 1996; 74: 1796-1800.
102. Wallgren A, Siefevsward C, Hulthorn A. Carcinoma of the breast in women under 30 years of age, a clinical and histopathological study of all cases reported as carcinoma to the Swedish Cancer Registry. Cancer 1977; 40: 916.
103. Watson CJ, Miller WR. Elevated levels of members of the STAT family of transcription factors in breast carcinoma nuclear extracts. Brit. J. Cancer 1995; 71: 840-844.
104. Weidner N, Semple JP, Welch WR, Folkman J. Tumor angiogenecies and metastasis-correlation in incasive breast carcinoma. New Engl. J. Med. 1991; 324: 1-8.
105. Wenger CR, Bearadslee S., Owens MA. DNA ploidy, S phase and steroid receptors in more than 127127 breast cancer patients. Breast Cancer Res. Treat. 1993; 28 :9.
106. Willet WJ. The search for the cause of breast and colon cancer. Nature 1989; 338: 389-393.
107. World Health Organization: Histological typing of breast tumours. WHO, Geneva, 1981.
108. Yballe EM, Hoffman A. Imprinting and expression of insulin-like growth factor II and H 19 in normal breast tissue and breast tumour. J Clin. Endocrinol. Metab 1996; 81: 1607-1612.
109. Zellars RC, Hilsenbeck S, Clark G, Allred D, Herrman T, Chamness G. Prognostic value of p53 for local failure in mastectomy-treated breast cancer patients. J. Clin. Oncol. 2000; 18: 906-913.

Adaptation Biology and Medicine (Volume 5: Health Potentials)
Editors: L. Lukyanova, N. Takeda and P.K. Singal

18. Mitochondrial Signaling in Adaptation to Hypoxia

L.D. Lukyanova, A.V. Dudchenko, T.A. Tzybina, E.L. Germanova and E.N. Tkatchuk[1]

Summary

The mitochondrial respiratory chain participates in performance of the signal transforming system which activates realization of metabolic compensatory processes and the coupled functional response to both single and repeated, long-term exposure to acute hypoxia. The suppressed function of mitochondrial enzyme complexes I (MEC 1) and the alternative activation of mitochondrial enzyme complexes II (MEC II) are a signaling regulatory mechanism participating in formation of tissue-specific and general resistance of the organism in different types of hypoxia. This process is associated with recovery of the electron transport function in the major respiratory chain and changed in kinetic properties of respiratory chain enzymes. Degree of MEC I inhibition and MEC II activation in rat brain in the first 2 hours after a hypoxic exposure correlates with pronouncement of adaptation to hypoxia. Triggering the mechanism is necessary for stabilization and accumulation of HIF-1α. The mechanisms of regulatory roles of mitochondrial enzymes and the tricarbonic acid cycle substrates succinate and α-ketoglutarate in the modulation of HIF-1α activity are discussed.

Keywords: Oxygen homeostasis, hypoxia, mitochondrial enzyme complexes, signaling function of succinate and α-ketoglutarate, HIF-1α, adaptation.

Introduction

In recent years a new step of studies in the field of bioenergetics has become clear. It was proven that mitochondria critically contributed not only to producing most of intracellular energy but also performed a signaling function at both systemic and cell levels [1-7]. It appeared that such processes as development, aging, response to temperature, apoptosis, insulin secretion by β-cells, formation of adaptive responses, biogenesis, genetic functions of mitochondrial DNA, and diseases with mitochondrial dysfunction are related with the signaling function of mitochondria [8-16]. This is possible, in turn, only in the presence of close regulatory interactions between mitochondria and different intracellular energy-dependent functional and metabolic systems.

All this suggests a **new era in mitochondrial research** and the emergence of a new field in biology, **"mitochondrial physiology"**. By the way, already 4 international conferences have born this name. Nevertheless, the major function of mitochondria is, as before, their ***regulatory role in oxygen***

Address for Correspondence: Institute of General Pathology and Pathophysiology, RAMS, Moscow, Russia, [1]"Hypoxia Medical Academy", Clin. Res, Lab., Moscow, Russia

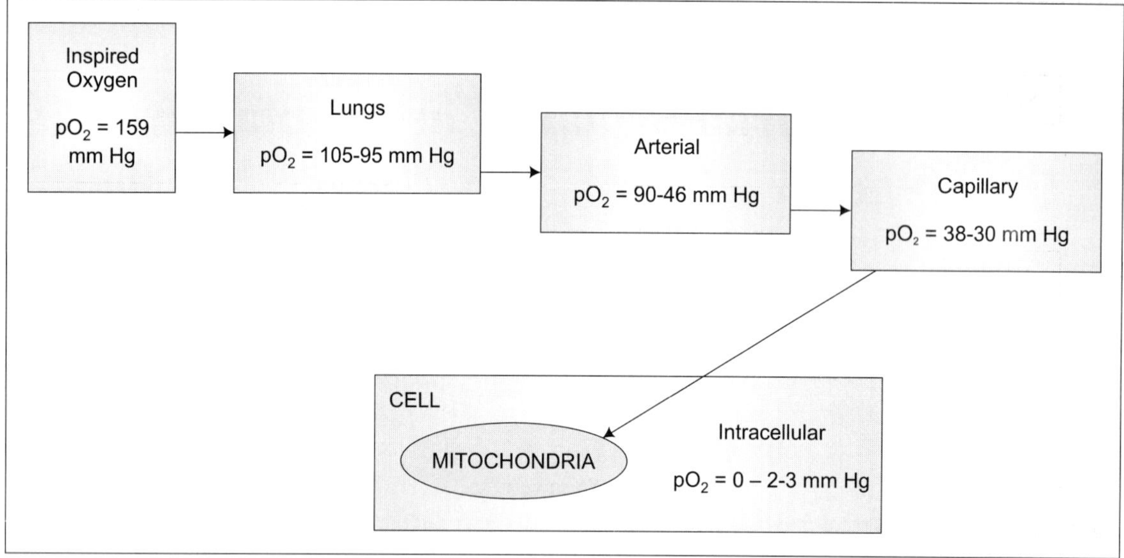

Fig. 1. The pathway of oxygen transport from ambient air to cell.

homeostasis, which is evident at both systemic and cellular levels. Mitochondria are the terminal biochemical step in production of reduction equivalents, which interact with molecular oxygen in the region of terminal oxidases in the respiratory chain.

This is why just mitochondria determine **at the systemic level** the concentration gradient of oxygen arriving into the cell from the environment (Fig. 1).

At the cell level the mitochondrial respiratory chain performs as a signal-transforming metabolic system, which activates the functional response and realization of the physiological response of the organism to hypoxia.

We have been studying this issue over recent twenty-five years. As long ago as in 1990, at the first constituent congress of ISAM in Freiburg, we provided evidence on the role of the first mitochondrial enzyme complex in the formation of individual organism resistance to hypoxia and in adaptation to hypoxia. Development of those ideas since 1984 has been published in detail in periodic literature [5, 17-22], presented at congresses on adaptive medicine, and reflected in chapters of 4 volums of Adaptation Biology and Medicine [5, 19-22].

To summarize briefly, the essence of this concept is that ***the mitochondrial respiratory chain is a sophisticated regulatory system, which responds to any stress, first of all to hypoxia, with successive changes in activities of mitochondrial enzymes.*** These changes determine the phase nature of cell, organ and organism hypoxic responses and correlate with changes in many physiological parameters. By convention we separate out three steps of the process [5, 18, 22].

Mitochondrial Physiology

The first step is activation of the major, NAD-dependent oxidation pathway, which provides stimulation of ATP synthesis in the conditions of initial hypoxia. This activation reflects the primary

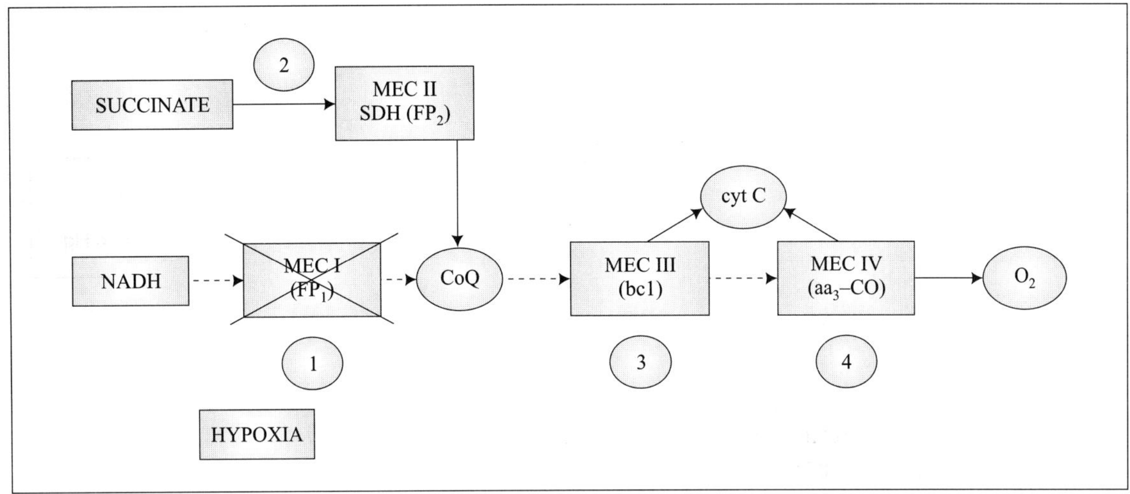

Fig. 2. Sequence of disturbances in the respiratory chain during hypoxia.
The first step of bioenergetic hypoxia: (1) inactivation of the mitochondrial enzymic complex I (MEC I) and (2) compensatory activation of the succinate oxidase pathway (MEC II) at the earlier stage of hypoxia. (3-4) inactivation of the mitochondrial enzyme complexes III-IV (the late steps of bioenergetics hypoxia:).

mechanism for mobilization of basal energy resources in response to relatively moderate external impacts including a relatively mild decrease in extracellular oxygen. This step is characterized by potentiated reactions of ATP dephosphorylation, with ATP synthesis under the control of the ATP/ ADP ratio during that period (Fig. 2, *pathway 1*).

The second step is suppression of mitochondrial enzyme complex I (MEC I) and compensatory activation of succinate oxidase oxidation, that is, activation of MEC II (Fig. 2, *pathway 2*).

Thirty years ago a special role of succinate oxidase oxidation in cell vital function was established as a compensatory oxidation pathway [23-29]. At early steps of hypoxia, this pathway has thermodynamic advantages over other substrates of the tricarbonic acid cycle and supplies reduced equivalents to the cytochrome region of the respiratory chain in the conditions of the first complex inactivation, thereby providing energy synthesis at two coupling points. This pathway has been extensively studied in Russia and formed the basis for a new trend related with the antihypoxic defense of the organism [24, 26, 28-33].

The third step (decompensation) occurs under the conditions of acute hypoxia and it is associated with disorders of, first, the electron transport function of the respiratory chain in the region of mitochondrial enzyme complex III (MEC III) (cytochromes *b-c*) and then of mitochondrial enzyme complex IV (MEC IV) (cytochrome oxidases) (Fig. 2, *pathway 3-4*). Therefore in the conditions of reduced oxygen delivery to cells, the mitochondrial respiratory chain is involved in the process as a single functional and metabolic unit which performs thereby as a regulator and modulator of oxygen consumption and rate of delivery from the extracellular medium to mitochondria. We have termed **bioenergetic hypoxia** just this specific process that can play a role of intracellular trigger system signaling changes in extracellular oxygen and initiating a cascade of functional and metabolic intracellular responses to oxygen shortage at the systemic level. We have earlier obtained a rich variety of evidence in support of the described sequence of changes in the performance of respiratory chain under the conditions of altitude chamber training as well as in high mountains. Some of the evidence are the increased reduction of pyridine nucleotides and flavins; changed sensitivity of mitochondrial respiration to inhibitors of the first and second mitochondrial complexes; changed intensity of oxidative phosphorylation and parameters of adenylate pool and their ratios as indices of the cell energy status [5, 19, 20].

Taking into account the broad recent use of interval hypoxic training in practical medicine, we have conducted a study on the effect of such training on bioenergetic parameters of the most hypoxia-sensitive tissue, the brain cortex from rats with different inherited sensitivity to acute hypoxia.

Kinetic parameters (K*m* and V*max*) of mitochondrial complex enzymes were measured in the rat brain cortex: rothenon-sensitive NADH-cytochrome *c* reductase (complexes I+III); succinate-cytochrome *c* reductase (complexes II+III); and cytochrome oxidase (complex IV). The effects of two types of hypoxia were studied: intermittent normobaric hypoxia interspersed with breathing air-O_{20}% (INH/O_{20}) and intermittent normobaric hypoxia interspersed with breathing 30% of oxygen (INH/O_{30}). Exposure lasted for one hour.

As early as 2 hours after the first training cycle of INH/O_{20}, we observed an increase in general resistance in low-resistance (LR) to hypoxia rats, which not only persisted over 24 hours (preconditioning effect) but also incrementally increased during the training reaching the maximum value after 8 to 15 training sessions. When the number of training sessions was increased to 8-15 LR animals displayed a decreased resistance (Fig. 3). As distinct from this effect, INH/O_{30} resulted in a

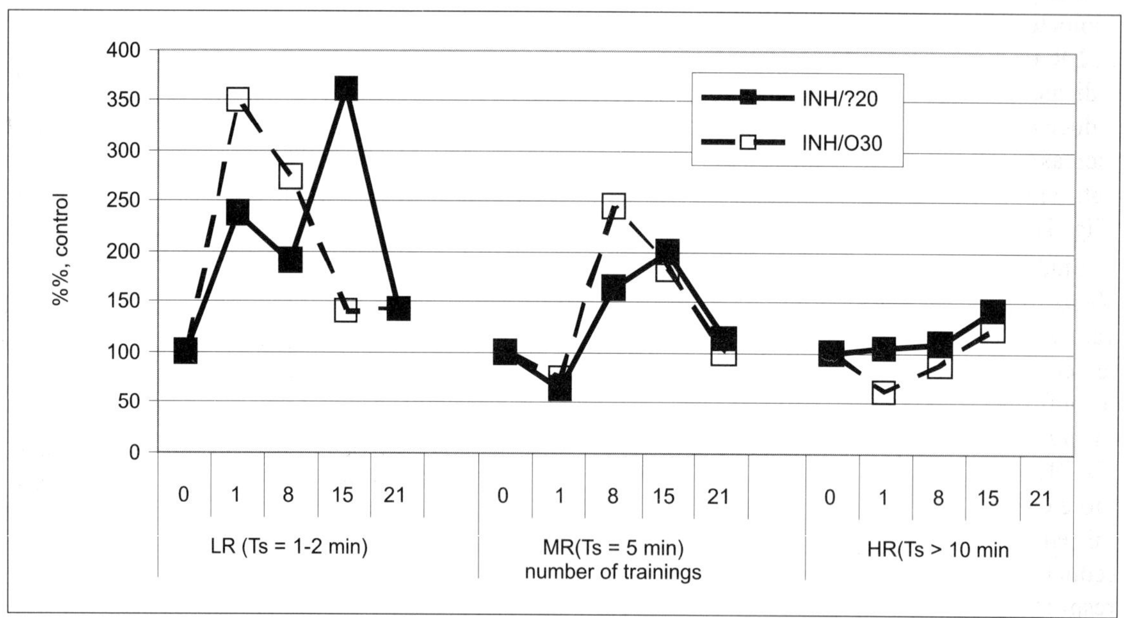

Fig. 3. Effects of hypoxia training on rat body resistance
INH/O_{20} - intermittent normobaric hypoxia interspersed with breathing air-O_{20}%
INH/O_{30} - intermittent normobaric hypoxia interspersed with breathing 30% oxygen
LR – low-resistance rats; MR – medium-resistance rats; HR – high-resistance rats
Ts – time of survival at the high altitude

maximum increase of LR rat resistance already 24 hours after the first training cycle; subsequently this parameter gradually decreased despite the continued training course.

The picture was absolutely different in training of high-resistance (HR) rats. In this instance, beneficial changes in animal resistance were slight or absent with either INH/O_{20} or INH/O_{30} training. In some cases animals even showed increased vulnerability, i.e., impaired resistance to acute hypobaric hypoxia (Fig. 3). This result is completely consistent with the fact we have previously observed that HR rats cannot adapt to hypoxia [4, 18-20]. The group of medium-resistance (MR) rats occupied an intermediate position between HR and LR rats. The MR rats lacked the preconditioning effect after the first training cycle with either INH/O_{20} or INH/O_{30}, and their resistance was reduced in 24 hours. Subsequently, however, positive changes in the resistance did occur although they were less pronounced than in LR rats. Maximum increase in resistance of the MR rat group developed after 8-15 training sessions, like in LR rats, and after that the resistance started to reduce (Fig. 3).

Therefore, the use of different versions of INH course provides formation of resistance to hypoxia in rats. However the extent of resistance depends on the animal genotype: LR>MR>HR. The latter group either develops a weak resistance or no resistance at all. Oxygenation degree of the gas mixture inhaled between hypoxic exposures influences the dynamics of resistance development in only LR rats. Higher than ambient oxygenation of the gas mixture inhaled during interhypoxic intervals (30% O_2) potentiated the preconditioning effect in such animals during the first day, as compared to INH/O_{20} but negatively influenced the development of resistance in the next 2 weeks.

Studying the effect of a single training cycle with different INH types on the weight of stress-competent organs revealed considerable differenced in responses of LR and HR rats to this exposure. In LR rats, INH/O_{20} primarily increased the weight of all studied organs in the first day, which indicated activation of the sympathetic-adrenal system in these conditions. However INH/O_{30} did not induce any significant changes in animals of that group. Contrastingly, HR rats showed a significant increase in thymus weight, a decrease in adrenal weight and, to a lesser extent, in spleen weight in both cases. Therefore HR rats showed clear signs of stress characteristic of the exhaustion stage (Fig. 4).

Studying kinetic parameters of mitochondrial enzymes demonstrated that baseline values of K*m* and V*max* for succunate cytochrome ***c*** reductase did not differ in brain cortex of LR and HR rats. However values of K*m* and V*max* for rothenon-sensitive NADH-cytochrome ***c***-reductase were significantly lower in the brain of LR than HR rats. This finding is in agreement with our earlier data [4, 18, 20] and it confirms a hypothesis put forward ibid. According to this hypothesis NADH cytochrome ***c***-reductase of brain mitochondria from HR rats possesses a more effective stechiometrical regulation at high NADH levels than the same enzyme in LR rats. Due to this fact the enzyme of HR rat brain can perform more effectively in a broader range of NADH concentrations, that is, at the concentrations saturating this enzyme complex in LR rat brain. Taking into account that cells accumulate excessive amounts of reduction equivalents in hypoxic conditions, more effective performance of the mitochondrial respiratory chain in the HR rat brain may underlie the higher resistance of HR rats to oxygen deficiency as compared to LR animals.

In the process of INH training, responses and dynamics of mitochondrial enzyme kinetic parameters were essentially different in the brain cortex of LR and HR rats. For example, the following were observed in the LR rat brain as early as 2 hours after a single exposure to either type of INH (Fig. 5):

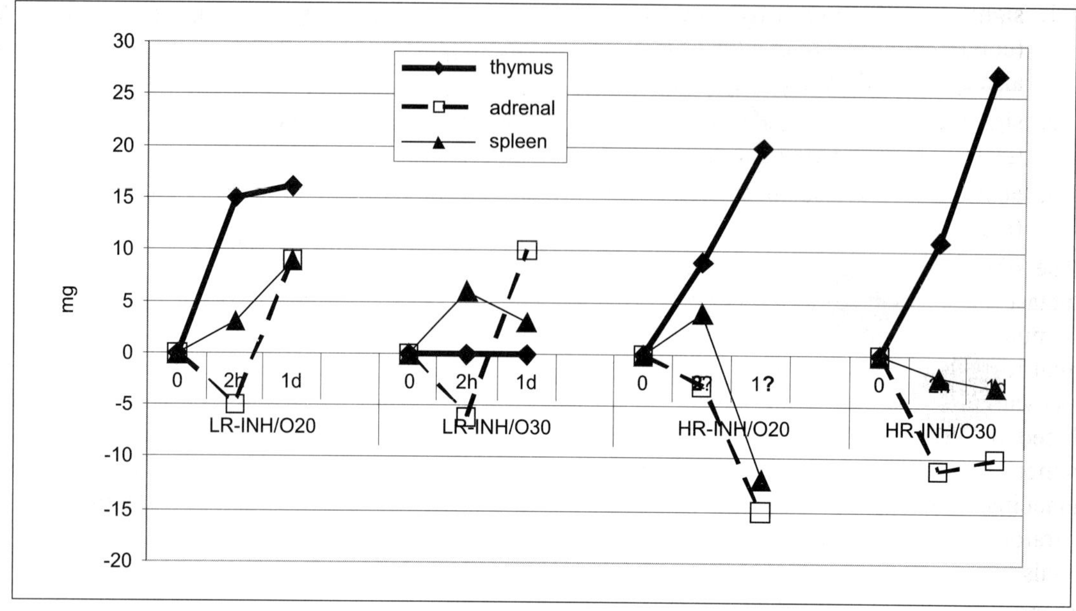

Fig. 4. Effect of a single training cycle with INH/O_2 and INH/O_{30} on weight of stress-competent organs from LR and HR rats.

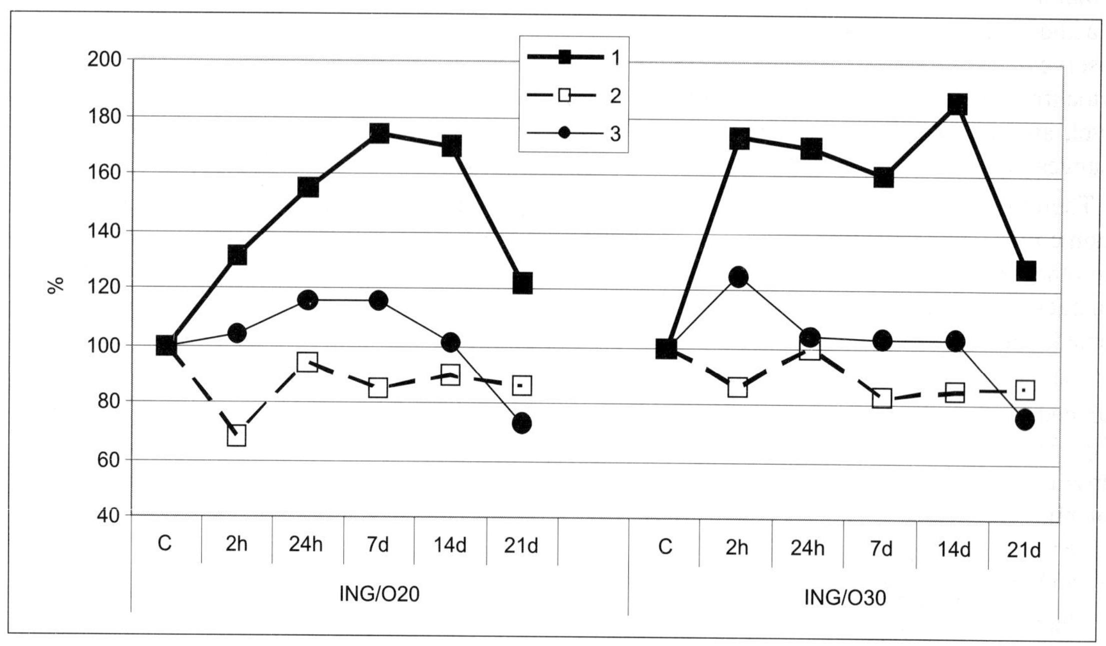

Fig. 5. Effects of hypoxia training on the dynamics of Km for mitochondrial enzymes in LR rat brain cortex
1 - rothenon-sensitive NADH-cytochrome *c* reductase (complexes I+III);
2- succinate-cytochrome *c* reductase (complexes II+III;
3 - cytochrome oxidase (complex IV).

1. significant increase in K*m* and V*max* for rothenon-sensitive NADH-cytochrome *c* reductase (complexes I+III) to the values approaching those in control HR rats. These changes persisted also 24 hours after the exposure;
2. significant decrease in K*m* for succinate-cytochrome *c* reductase (complexes II+III) indicating the increased substrate (succinate) affinity of the enzyme;
3. absence of significant changes in K*m* and V*max* for cytochrome oxidase in that period (complex IV).

The observed simultaneous increase in K*m* and V*max* for rothenon-sensitive cytochrome *c* reductase in the brain of LR rats not only persisted for a long time but also progressed reaching maximum values after 8-15 training cycles (Fig. 5), that is, at the same time as maximum changes in animal resistance (view fig. 3). When the number of training sessions was increased to 21, K*m* and V*max* values for NADH-cytochrome *c* reductase decreased to baseline values and correlated with the reduced resistance of animals to hypoxia. Therefore, INH led to very fast changes in kinetic parameters of mitochondrial complex I-III in the LR rat brain, which were aimed at optimizing (enhancing effectiveness) of respiratory chain performance in the conditions of increased reduction of respiratory carriers. This process had a time course similar to that of changes in general resistance of animals.

Opposite changes in K*m* and V*max* for succinate cytochrome *c* reductase (the decreases indicating an increased substrate (succinate) affinity of the enzyme, which were maximum 2 hours after the first INH training cycle, subsequently somewhat decreased. Nevertheless, these changes allow the enzyme to maintain high activity of the succinate oxidase oxidation pathway at early hypoxia, in the conditions of a known initial decrease in intracellular succinate due to the high oxidation rate [24, 25, 28, 29]. The observed changes in kinetic properties of respiratory chain substrate sites are directed at providing its cytochrome site with electrons, maintaining the cytochrome oxidase activity at a relatively steady-state level, and preserving the energy producing function of mitochondria. The absence of significant changes in cytochrome oxidase kinetic parameters in that period supports this suggestion.

Therefore, INH induced reciprocal changes in activity of the respiratory chain substrate site in the brain cortex of LR rat: increases in values for one enzyme complex were associated with decreases of the other. Taking into account that such regularity is characteristic of both INH/O_{20} and INH/O_{30}, changes in the extent of oxygenation during interhypoxic intervals obviously did not exert a considerable influence on this relationship.

It should be noted however, that cytochrome oxidase K*m* in the brain of LR rats was significantly reduced by the end of INH training, which indicated an increased substrate affinity of the enzyme (Fig. 5). Considering that mitochondria lose a part of cytochrome *c* in hypoxia, the physiological sense of such increased affinity may be in that the enzyme of LR rats adapted to hypoxia acquires a potential ability to function at low levels of reduced cytochrome *c*. Therefore in these conditions, cytochrome oxidase gains new kinetic properties which may provide more effective performance of the enzyme in impaired oxygen supply to the cell.

Changes in enzyme kinetic characteristics might be related with formation of isoforms with new properties. It is known, for example, that enzymic complex I is a flavin-adenine-mononucleotide-containing protein consisting of few dozens of subunits. For a part of them, information is translated from the mitochondrial genome and for the other part – from the nuclear genome. Such multipeptide nature of the enzyme may underlie emergence of new isoforms with new kinetic characteristics in long-term hypoxia. However, a question remains open on whether a two-hour period is sufficient for expression of new enzyme forms or changes in the kinetic characteristics are due to performance of other regulatory mechanisms.

Absolutely different K*m* dynamics were observed in the HR rat brain, which virtually could not adapt to INH as discussed above. In this instance, K*m* values of rothenon non-sensitive NADH-cytochrome *c* reductase not only did not significantly increase but even reduced by more than 40% 4 hours after a single training cycle with subsequent recovery to the baseline value. Changes in kinetic parameters of succinate-cytochrome *c* reductase were absent. These findings are in full agreement with our previous data on regulatory differences in properties of mitochondrial complexes I and II of LR and HR rat brain in hypoxia and ischemia [5, 18-20]. According to those data inhibition of complex I and activation of complex II are the initial responses to hypoxia in the LR rat brain. In contrast, activation of complex I without concomitant changes in complex II activity was observed in the HR rat brain.

These data document that INH is based on the same bioenergetic mechanisms as other types of hypoxia (ischemia, hypobaric hypoxia), and directly support our concept on involvement of mitochondrial complexes in responses to different types of hypoxia and tight relation of the complexes with development of organism resistance. Therefore it is obvious that:

1. INH induces an increase in general resistance of animals, which is most pronounced after 8-15 training cycles in LR rats. The effect declines in rats with higher baseline resistance and it is virtually absent in HR rats, in which INH results in strain of the sympathetic-adrenal system.
2. INH induces differently directed changes in kinetic parameters of mitochondrial complexes in the respiratory chain substrate region of LR and HR rat brain cortex as early as within the first 2 hours. In the former instance, rothenon-sensitive NADH cytochrome *c* reductase and succinate-cytochrome *c* reductase functioned more effectively in the conditions of increased reduction of the cell, which correlated with an increase in general resistance of animals. In the latter instance, significant changes in mitochondrial enzyme kinetic parameters were absent as well as signs of formation of organism resistance.
3. Reciprocal interrelations between activities of mitochondrial enzyme complexes at the respiratory chain substrate region perform as a signaling regulatory mechanism which participates in formation of tissue-specific and general resistance of rats in INH and other kinds of hypoxia.
4. Effects of INH/O_{20} and INH/O_{30} did not significantly differ by any of the studied parameters.

In this connection we should underline again the exclusive importance of activated succinate oxidase oxidation pathway. This is an alternative pathway, which takes up the function of providing electrons to the cytochrome region of respiratory chain in these conditions to maintain ATP production at the expense of MEC III and MEC IV. Therefore activation of this pathway in the conditions of oxygen shortage is an urgent compensatory mechanism used by the cell for ATP production. It becomes obvious that there a special role of succinate oxidase oxidation pathway as a signaling mechanism, on participation of which in the performance of respiratory chain formation of adaptive processes depends. Antihypoxic agents focused on activation of the succinate oxidase oxidation pathway early in hypoxia have been successfully used in Russia. All such agents are highly protective against various types of hypoxia [24, 28-33].

In long-term adaptation to hypoxia, maintenance of oxygen homeostasis at body and cell levels is closely related with the transcriptional expression of hypoxia-induced genes, which participate in regulation of many cell and systemic functions and are necessary for formation of adaptive features.

It is known that the process of adaptation to low oxygen level is controlled mostly by a ***specific transcription protein factor*** induced in all tissues under hypoxia, the hypoxia-inducible factor or HIF-1. This factor, which was discovered in early 90s [34, 35], performs as the major regulator of oxygen homeostasis. It is a mechanism, by which the organism exposed to tissue hypoxia controls expression of proteins responsible for oxygen delivery to cells, that is, it mediates cell adaptation to changes in tissue oxygenation [36-40].

At present more than 60 direct target genes for HIF-1 are identified. All of them provide improvement of oxygen delivery (erythropoiesis, angiogenesis), metabolic adaptation (glucose transport, enhanced glycolytic production of ATP, ion transport) and cell proliferation [36-40]. Products of HIF-1 regulated genes act at different functional levels. Such activation finally results in the increased oxygen delivery to cells. Hypoxia inducible factor-1 is heterodimer composed of the α and β subunits. The regulation of HIF-1 activity concerns mostly the α subunit and occurs at multiple level such as protein stabilization, posttranslational modifications, nuclear translocation, dimerization, transcriptional activation, and interaction with other proteins.

In the **normoxic** conditions, when HIF-1α and HIF-1β are constitutively transcribed and translated, the abrogation of HIF-1 activity results mainly from constitutive HIF-1α degradation by proteasomes. The rate of HIF-1α synthesis is low and its content is minimal [36-41]. A molecular basis for this regulation is the oxygen-dependent hydroxylation of two proline residues (from P402 and P564) incorporated in the HIF-1α structure. The proline residues are hydroxylated by one of the three enzymes known under a common name of "prolylhydroxylase domain proteins [42, 43], α-ketoglutarate, and iron are also obligatory components of the process. In parallel, the asparagine residue (Asn-803) in the C-terminal *trans*-activation domain (C-TAD) is hydroxylated, which results in suppression of the HIF-1 transcriptional activity. After hydroxylation of proline residues in Oxygen Dependent Degradation Domain (ODDD) and asparagine residue in C-NAD, HIF-1α binds to the Von Hippel Lindau (VHL) protein, which makes this subunit available for proteasomal degradation [42, 43].

In the conditions of severe oxygen deficiency, the oxygen-dependent process of prolyl residue hydroxylation typical for normoxia is suppressed. In this process, the bond of HIF-1α with VHL protein breaks to suppress the proteasomal degradation of HIF-1α. This makes possible HIF-1α accumulation, activation, translocation to the nucleus and dimerization with HIF-1α. The dimerization results in conformational modifications, formation of a transcriptionally active complex HRE, which triggers activation of numerous HIF-1-dependent target genes and synthesis of protective, adaptive proteins in response to hypoxia [36-43]. Therefore the HIF-1α activity is regulated by hypoxic signal transduction ***which is based on a unique oxygen-dependent posttranslational modification.***

Other signaling pathways for oxygen-independent HIF-1α synthesis via activation of the phosphatidyl-inositol- 3-kinase (PI3K) and MARK (MEK) – mitogen-activated protein kinase pathway are also available. However they play a minor role in the processes as a whole [16, 44, 45].

In normoxic conditions, HIF-1α is identified in nucleus of brain, kidneys, liver and heart. The presence of HIF-1α in normoxic tissues is apparently necessary for induction of basal gene level needed for energy production [45].

It is shown that *in vivo* the response to the acute hypoxia was immediate in all tissues with an induction of HIF-1α protein detected within 1 h of exposure. But the kinetics of HIF-1α expression varies in among different organs. In liver and kidney, HIF-1α reaches maximal level after 1 h and gradually decreases to baseline levels after 4 h of continuous hypoxia. In the brain, however, HIF-1α is maximally expressed after 5 h of hypoxia (increase in 9-10 times), and declined to basal level by 12 h. It expression increased when the inspired oxygen was reduced to 18% and reached maximum intensity at 6%. The level of HIF-1α is rapidly reduced on reoxygenation to undetectable level by 3-4 h [45].

HIF-1α protein accumulates significantly in the rat brain during adaptation to chronic hypobaric hypoxia (10% O_2). Its level remained significantly elevated for at least 14 days and returned to near baseline by 21 days. Acute exposure to 8% O_2, but not 10% O_2 normobaric oxygen of animals adapted to hypobaric hypoxia for 3 wk resulted in renewed accumulation of HIF-1α that is evidence of its synthesis was undisturbed [46].

Formation of adaptive features in this instance was demonstrated also by the transcription of HIF-1α-dependent target genes, (VEGF and Glut-1), which reached maximum between Days 12 and 14. After that it gradually decreased to the normal level at 21 days. Thus HIF-1α accumulation during hypoxia leads to transcriptional activation of HIF-1 target genes [46].

Recently evidence has appeared on the trigger role of mitochondria in intracellular signal transduction under hypoxia [46-50]. According to recent literature, stabilization of the hypoxia protein factor, HIF-1α is directly related to the activation of succinate oxidase oxidation. For instance, it was shown that even a partial (20%) suppression of MEC I activity almost completely inhibited the hypoxic induction of HIF-1α. However it recovered when MEC II was activated by succinate, which provided recovery of the electron flow at the cytochrome region of respiratory chain. (from MEC II to MEC III) [48-49].

The signaling role of succinate in maintaining the HIF-1α activity is confirmed also by the fact that succinate dehydrogenase (SDH) shortage in cells or inhibition of its activity associated with accumulation of succinate and HIF-1α or the reduced propylhydroxylase activity result in encephalomyopathy [51] or neoplasia [52]. It was shown, for instance, that succinate dehydrogenase mutations led to renal and stomach carcinoma and thyroid gland cancer.

In our cooperative study with Dr. Sazontova, which we highly appreciate, we observed that an increase in the HIF-1α protein started in the brain cortex of non-resistant rats as early as within 2 hours of a single training session in the conditions of INH/O_{20} and associated with MEC I inhibition and MEC II activation. This increase in HIF-1α confirmed both triggering of adaptive processes and the increase in HSP70 content (Fig. 6). The high content of HSP70 may reflect enhancement of repair processes.

At the same time, HIF-1α was not detected where the increased substrate affinity of the MEC II was absent (the brain of HR rats, 2 h after INH/O_{20}); the brain of LR rats, 2 h after INH/O_{30}) (Fig. 6). Therefore regulatory roles of mitochondrial enzymes and the tricarbonic acid cycle substrate succinate in the modulation of HIF-1α activity are obvious.

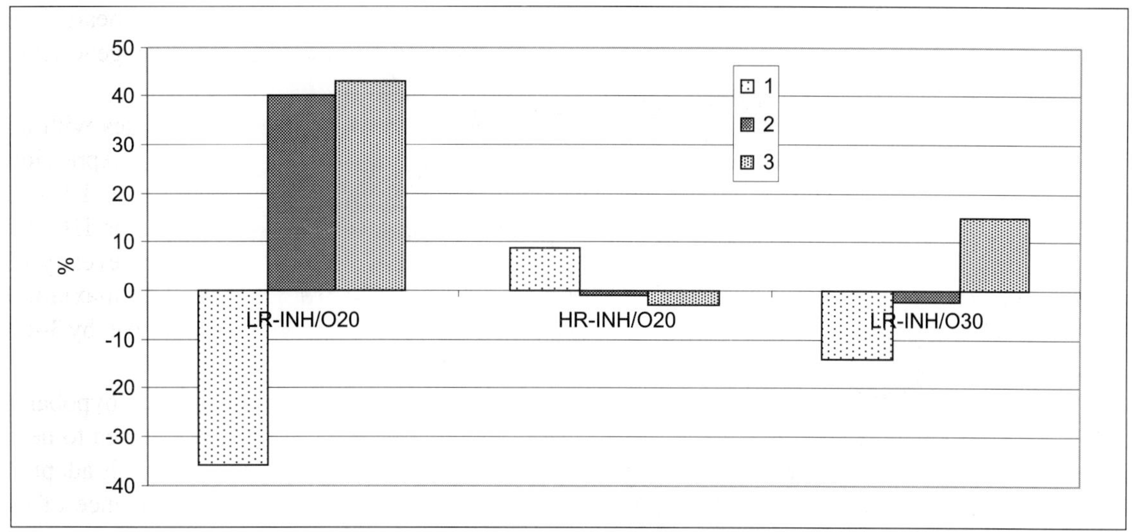

Fig. 6. Effects of INH/O_{20} and INH/O_{30} on the K*m* of MEC II in LR rat brain cortex (1), HIF-1α (2) and HSP (3) levels (2 h after the first training).

We believe that the mechanism of this phenomenon is as follows.

In the **normoxic** conditions, the regulatory oxygen-dependent process of HIF-1α prolyl hydroxylation and proteasomal degradation, which takes place in cell cytosol, depends on oxygen, iron, ascorbate and the substrates of these reactions, α-ketoglutarate and aspartate. This results in the formation of succinate, which is an allosteric inhibitor of the process. (Fig. 7A). The tricarbonic acid cycle substrate α-ketoglutarate participates in transamination of the aspartate-malate bypass, which results in the formation of aspartate. Both substrates, α-ketoglutarate and aspartate are delivered to cytosol due to the work of the bypass, and they can be used in propyl hydroxylase reactions.

In hypoxia, when MEC I is suppressed, the work of malate-aspartate bypass is inhibited to create the shortage of α-ketoglutarate in cytosol. This may be one of the reasons (along with oxygen and iron shortage) for inactivation of propylhydroxylase reactions and stabilization of HIF-1α.

At the same time, α-ketoglutarate in hypoxic brain can be metabolized in the GABA bypass (or Robets' cycle), which activation provides an additional source of succinate (fig. 7B). Succinate has thermodynamic advantages in hypoxia for oxidation in the respiratory chain [27, 29, 53, 54]. The respiratory chain actively uses succinate as a metabolic substrate and antioxidant to reduce metabolic acidosis and maintain the activity of HIF-1α (see Figs. 2, 3) [28-29].

Succinate moves freely between the mitochondria and the cytosol via the dicarboxylic acid translocator in mitochondrial inner membrane and the voltage-dependent anion channel (VDAC/porin) in the mitochondrial outer membrane. Thus, succinate can act as an intracellular messenger, liking TCA cycle function to HIF regulation.

Therefore ***in the conditions of hypoxia, functioning of the mitochondrial respiratory chain is coupled to transcriptional activity of HIF-1α,*** and due to this coupling, ***substrate of the tricarbonic acid cycle can perform the signaling function and regulate the activity of prolyl hydroxylase reactions.***

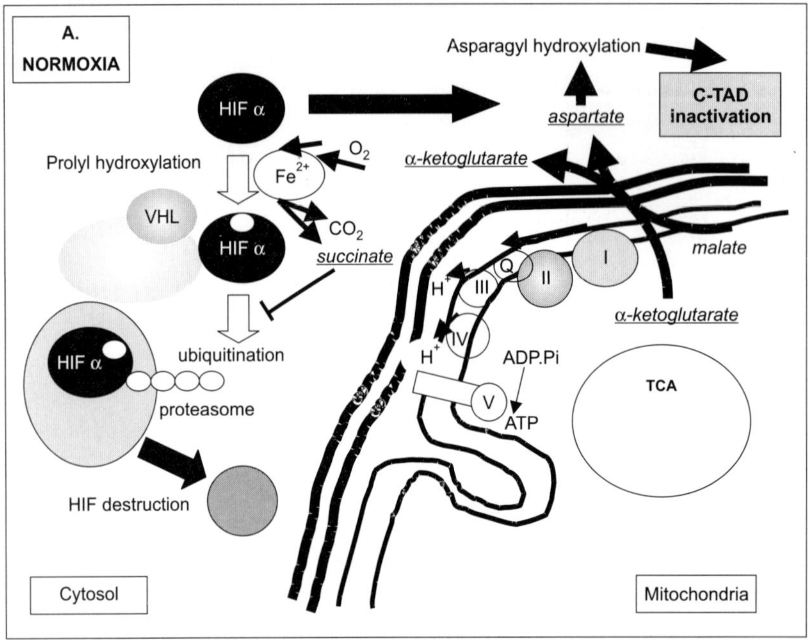

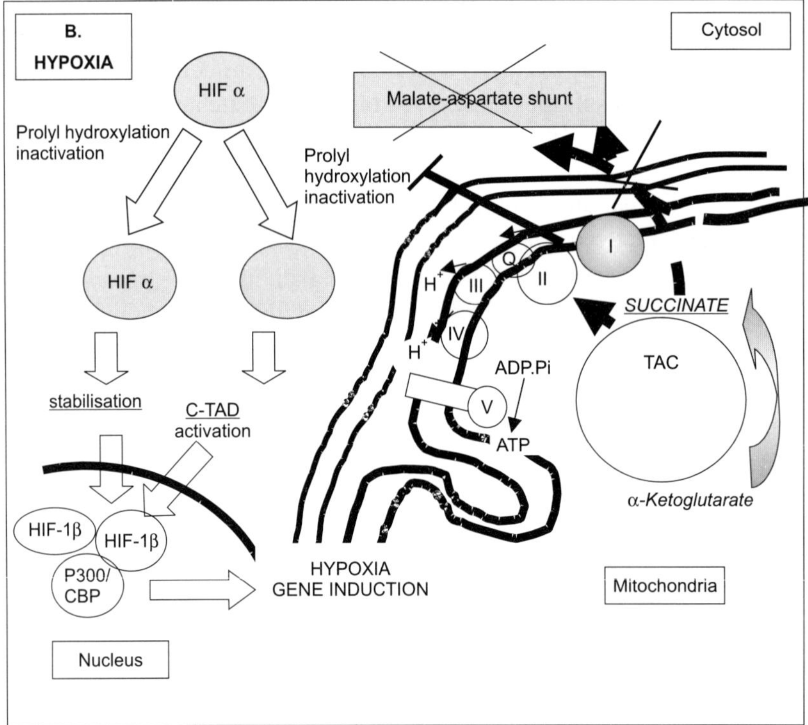

Fig. 7. Regulatory roles of the tricarbonic acid cycle substrate succinate in the modulation of HIF-1α activity: 7A - normoxia, 7B - hypoxia.

At this time the signaling ***function*** of ***the tricarbonic acid*** **cycle** ***substrate*** is being actively studied. For instance, data are available that GPR91 known as orphan G-protein-coupled receptors (GPCR) function as a receptor **for the citric acid cycle intermediate succinate.** It was also shown that GPR99, a close relative of GPR91, responds to α-ketoglutarate, another intermediate in the citric acid cycle [55]. Thus, by acting as ligand for GPCRs succinate and α-ketoglutarate are found **to have unexpected signaling functions beyond their traditional roles.** This suggests that succinate can also be released from cells under some physiological/pathological circumstances.

Conclusion

1. In the conditions of reduced oxygen delivery to cells, mitochondria become involved in the regulation of oxygen homeostasis. In this process they function as modulators of oxygen consumption, rate of oxygen delivery from the extracellular environment, energy synthesis, and activators of hypoxia-specific transcription factors.
2. The suppressed function of mitochondrial enzyme complex I and the alternative activation of mitochondrial enzyme complex II are evolutionarily developed signaling regulatory mechanisms participating in formation of tissue-specific and general resistance of the organism to different types of hypoxia. Triggering the mechanism and switching to succinate oxidation are necessary for stabilization of HIF-1α and expression of HIF-1α-dependent target genes.
3. The degree of mitochondrial enzyme complex I inhibition and mitochondrial enzyme complex II activation in rat brain in the first 2 hours after a hypoxic exposure correlates with pronouncement of adaptation to hypoxia.
4. The substrate of the tricarbonic acid cycle can perform the signaling function and regulate the transcriptional activity of HIF-1α.

References

1. Briere J-J., Chretien D., Benit P., Rustin P. Respiratory chain defects: what do we know for sure about their consequences *in vivo*. BBA. 2004; 1659: 172-177.
2. Chandel N.S., Schumacker P.T. Cell depleted of mitochondrial DNA (ρ°) yield insight into physiological mechanisms. FEBS Letters. 1999. **454**: 173-176.
3. Chandel N.S., Schumacker P.T. Cellular oxygen sensing by mitochondria: old questions, new insight. J.Am Physiol. 2000; 88:1880-1889.
4. Di Lisa F., Ziegler M. Pathophysiological revelance of mitochondria in NAD+ metabolism. FEBS Letters 2001; 492: 4-8.
5. Lukyanova L.D. Novel approaches to the understanding of molecular mechanisms of adaptation. In: Adaptation Biology and Medicine. Hargens A., Takeda N., Singal P.K. (eds). 2004; **4**: 11- 22.
6. Peers Ch., Kemp P.J. Acute oxygen sensing: diverse but convergent mechanisms in airway and arterial chemoreceptors. Respiratory Reseach. 2001; 2(3): 145-149.
7. Porwol T., Eheleben W., Brand V., Acker H. Tissue oxygen sensor function of NADPH oxidase isoforms, an unusual cytochrome aa_3 and reactive oxygen species. Respiration Physiology, 2001; 128, Issue 3, p. 331-348.
8. Brunk UT, Terman A. The mitochondrial-lysosomal axis theory of aging. Eur.J.Biochem. 2002. **269**, 1996-2002.
9. Butow RA, Avadhani NG. Mitochondria signaling. The retrograde response. Molecular Cell. 2004. **14**:1, 1-15.
10. Felty Q., Roy D. Estrogen, mitochondria, and growth of cancer non cancer cells. J.of Carcinogenesis. 2005; **4**:1, 1-34.

11. Kusnetsov AV, Schneeberger S, Seiler R., Brandacher G., Mark W, Steurer W, Sacs V, Usson Y, Margreiter R, Gnaiger E. Mitochondria: defects and heterogeneous cytochrome *c* release after cardiac cold ischemia and reperfusion. Am J Heart Circ Physiol. 2004; **286**: H1633-H1641.
12. McDonald MJ. Perspective: emerging evidence for signaling roles of mitochondrial anaplerotic products in insulin secretion. BBA. 2002; **1619:** 77-88.
13. Michiels K. Physiological and pathological responses to hypoxia. Am.J. Pathology. 2004; 164: 1875-1882.
14. Nishimura G, Proske RJm Doyama H, Higuchi M. Regulation of apoptosis by respiratory substrates. FEBS Letters. 2001. **505**:3, 399-404.
15. Voos W, Rotgers K. Molecular chaperones as essential mediators of mitochondrial biogenesis. BBA. 2002; **1592:** 51-62.
16. Zhu H., Bunn F. Oxygen sensing and signaling: impact on regulation of physiologically important genes. Respir Physiol . 1999; **2:** 239-247.
17. Lukyanova LD. Limiting steps of energy metabolism in brain in hypoxia. Neurochem. Intern. 1988, **13**:.I, 146-147.
18. Lukyanova L. Bioenergetic hypoxia: Definition, Mechanisms, and Methods of Correction. Bull. Exp. Biol. Med., 1997, 124: 9: 244-254.
19. Lukyanova L.D. Bioenergetic mechanisms of the cells adaptation to hypoxia and individual resistance of the organism. Proc. Int. Soc. Pathophysiol., Moscow. 1991. 249-250.
20. Lukyanova LD. Molecular, metabolic and functional mechanisms of individual resistance to hypoxia. In: Adaptation Biology and Medicine, Sharma B.K., Takeda N., et al. (eds), Narosa Publishing House New Delhi, India. 1997, **I**, 261-272
21. Lukyanova L.D., Dudchenko A.M. Regulatory role of the adenylate pool in the formation of hepatocyte resistance to hypoxia. In: Adaptation Biology and Medicine, K.B. Pandolf, N. Takeda, P.K. Singal (eds.). Narosa Publishing House, New Delhi, India. 1999. **2**, 139-150.
22. Lukyanova L.D. Cellular mechanism responsible for beneficial effects of hypoxic therapy. In: Adaptation Biology and Medicine, Moraveè, et al. (eds), Narosa Publishing House New Delhi, India. 2002. **3,** 290-303.
23. Sushila I., Rumasarma T. Modulation of succinate dehydrogenase in response to environmental stress conditions of hyporbaria and hypoxia. 1973. BBA; 321(2):423-436.
24. Kondrashova M.N. Therapeutic effects of succinate. Puschino. 1976.283 p. (russ)
25. Kondrashova M.N. The interaction of transamination and oxidation of carboxylic acids in different functional states of tissues. Biokhimiya 1991. **56,** 243-310.
26. Kondrashova M.N. The formation and utilization of succinate in mitochondria as a control mechanism of energization and energy state of tissue. In: Chance B. (Ed.). Biological and Biochemical Oscillators, Academic Press. New York, NY, 1993, 373-397.
27. Kondrashova M.N. The interaction of transamination and oxidation of carboxylic acids in different functional states of tissues. Biokhimiya 1991, 56, 243-310.
28. Kondrashova M.N. Succinate in medicine. Ed. Kondrashova M.N. Pushchino, Nauka, 1996, 300 P.
29. Maevsky E.I., Rosenfeld A.C., Grishina E.V., Kondrashova M.N. – Correction of metabolic acidosis by maintaining the mitochondrial function. – Pushchino. 2001. 155 P. (in Russian).
30. Lukyanova L.D. Bioenergetic mechanisms of hypoxic states and principles of pharmacological correction. In: Pharmacological Correction of Hypoxic States. Lukyanova L.D. (ed.), Moscow, 1989, 11- 44.
31. Lukyanova L.D. Bioenergetic mechanisms of hypoxic state development and strategies for its pharmacological correction. In: Antihypoxants. Lukyanova L.D. (ed)., Moscow, 1991, 5-35.
32. Lukyanova L.D. Actuàl problems and perspectives of hypoxic states pharmacological correction. Current problems and prospects in pharmacological correction of hypoxic states.- In: Pharmacotherapy in Neurology and Psychiatry. Lectures for Practitioners at the8th Congress "Man and Drug", Moscow, 2002, 22-34.

33. Lukyanova L.D. Actuàl problems and perspectives of hypoxic states pharmacological correction. Current problems and prospects in pharmacological correction of hypoxic states.- In: Pharmacotherapy in Neurology and Psychiatry. Lectures for Practitioners at the 8th Congress "Man and Drug", Moscow, 2002, p. 22-34.
34. Wang G., Semenza GL. Characterization of hypoxia-inducible factor 1 and regulation of DNA binding activity by hypoxia. J. Biol Chem 1993. **268**, 21513-21518.
35. Wang G.L., Jiang B.H., Rue E.A., Semenza G.L. Hypoxia-inducible factor 1 is a basic-helix-loop-helia-PAS heterodimer regulated by cellular O_2 tension. Proc.Natl.Acad.Sci.USA. 1995; 92: 5510-5514.
36. Semenza G. L. Expression of hypoxia- inducible factor 1: mechanisms and consequences. Bioch. Pharmacol. 2000; 59: 47-53.
37. Semenza G. L. Surviving ischemia: adaptive responses mediated by hypoxia-inducible factor 1. J.Clin.Investig. 2000; 106:(7): 809-812.
38. Semenza G. L. Signal transduction to hypoxia-inducible factor 1. Genes & Development.- 2000;14: 1983-1991.
39. Semenza G. L. Signal transduction to hypoxia-inducible factor 1. Bioch. Pharmacol. 2002; 64: 993-998.
40. Semenza G. L. Cancer. Targeting HIF-1α cancer therapy. Nature Reviews Cancer. 2003; 3: 721-732.
41. Stroka D.M., Burkhardt T., Desballerts I., *et al.* HIF-1 is expressed in normoxia tissue and displays an organ-specific regulation under systemic hypoxia. The FASEB J. 2001; 15:2445-2453.
42. Jaakkola P., Mole D.R., Tian Y.M. *et al.* Targeting of HIF-α to the von Hippel-Lindau ubiquitylation complex by O_2-regulated prolyl hydroxylation. Science. 2001; 292: 468-472.
43. Bruick R.K., McKnight S.L. A conserved family of prolyl-4-hydroxilases that modify HIF. Science. 2001; 294:1337-1340.
44. Chan-Hyung Kim, Young-Suk Cho, Yang-Sook Chun *et al.* Early expression of myocardial HIF-1α in response to mechanical stress. Circ. Res. 2002; 90:625-637.
45. Conrad PW, Freeman TL, Beitner-Johnson D, Millhorn DE. EPAS1 trans-activating during hypoxia requires p42/p44 MAPK. J.Biol. Chem. 1999; 274:33709-33713.
46. Chavez J.C., Pichiule P, Boero J., Arregui A. Reduced mitochondrial respiration in mouse cerebral cortex during chronic hypoxia. Neurosci Lett. 1995;193: 169-172.
47. Chavez J.C., Agani F., Pichule P, LaManna J.C. Expression of hypoxia-inducible factor - 1α in the brain of rats during chronic hypoxia. J Appl Physiol. 2000; 89(5):1937-1942.
48. Agani F.H. , Pichiule P., Chavez J.C., La Manna J.C. The role of Mitochondria in the regulation of Hypoxia-inducible Factor 1 Expression during Hypoxia – JBC. 2000; 275(46):35863-67.
49. Agani F.H, Puchowicz M, Chavez J-C, Pichiule P., LaManna J. Inhibitors of mitochondrial complex I attenuate the accumulation of hypoxia-inducible factor-1 during hypoxia in Hep3B cells. Compar. Biochemistry and Physiol. 2002; 132, Issue 1: 107-109.
50. Schroedl C., McClintock D.S., Budinger S., Chandel N.S. Hypoxic but not anoxic stabilization of HIF-1α requires mitochondrial reactive oxygen species. Am.J.Physiol.Lung Cell Mol Physiol 2002; 283:L922-L931.
51. Baysal B.E. On the association of succinate dehydrogenese mutations with hereditary paraganglion. Trends Endocrinil. Metab. 2003; 14: 453-459.
52. Selak M.A., Armour S.M., McKenzie E.D., *et al.* Succinate links TCA cycle dysfunction to oncogenesis by inhibiting HIF-prolyl hydroxylase. Cancer Cell. 2005; 7: 77-85.
53. Weinberg J.M., Venkatachalm M.A., Nancy F. *et al.* Anaerobic and aerobic pathways for salvage of ptoximal tubules from hypoxia-induced mitochondrial injury. Am.J. Renal Physiol. 200; 279: F927-F943.
54. Weinberg J.M., Venkatachalm M.A., Nancy F. *et al.* Mitochondrial disfunction during hypoxia/reoxigenation and its correction by anaerobic metabolism of citric acid cycle intermediates. PNAS. 2000; 97(3):2826-2831.
55. He W., Miao F.J., Lin D.C. *et al.* Citric acid cycle intermediates as ligands for orphan-G-protein-coupled receptors. Nature. 2004; 429 (6988): 188-193.

Adaptation Biology and Medicine (Volume 5: Health Potentials)
Editors: L. Lukyanova, N. Takeda and P.K. Singal

19. A Virtual Multicellular Organism with Homeostatic and Adaptive Properties

Rafik D. Grygoryan[1*] and Alan R. Hargens[2]

[1]Institute of Software Systems, National Academy of Sciences, 40, Prosp. Acad. Glushkov, 03187, Kiev, Ukraine

[2]Department of Orthopaedic Surgery, University of California, San Diego, California 92103-8894

Abstract

Two problems of adaptation physiology are analyzed by means of computer modeling. In the first one, a mechanism is proposed to understand the driving forces by which cells within a multicellular organism change their initial state in response to altered environmental factors. This new understanding helps identify internal mechanisms of systemic reactions empirically observed in different multicellular organisms and their general adaptive trends in unstable external environments. According to the new explanation, all mobilizing influences, including those in the brain, which cause a balanced deficit between current rates of energy synthesis and energy losses compel the "attacked" cell to activate its genetically-predetermined mechanism of biosynthesis. This mechanism is considered as the only reactive mechanism independently functioning in every cell. Under environmental challenges this reactive mechanism redistributes substrate inflow, thus modifying activity for practically all of cells of a multicellular organism. Therefore, the multicellular organism reacts to environmental challenges as an integrated system.

The mechanism by which blood is shifted towards the cranium during early adaptation to microgravity in humans is the second problem critically analyzed using a two-dimensional vascular net model. Computer simulation suggests that the loss of blood hydrostatic pressures during microgravity may be insufficient to maintain such an asymmetric redistribution of blood volume. Another proposed mechanism to obstruct venous return from the cranium during microgravity is pleural pressure elevation, caused by mechanical shifting of the diaphragm into the thorax. This mechanical compression results from loss of abdomen organ weight loss during microgravity and probably alters the mechanical forces acting on each side of the diaphragm muscle. Additional factors that may amplify this hemodynamic effect include: (1) simultaneous decrease of abdominal extravascular pressure, (2) active expiration, and (3) activation of vasoconstrictor reflexes from the aortic arch and lung arterial mechanoreceptor zones.

Keywords: multicellular balanced deficit, reactivity, adaptation, microgravity, computer models

Introduction

Physiological adaptation of different multicellular organisms to stable shifts of environmental factors is an intensively-studied problem area [1, 2, 4-6, 8, 12, 16, 17, 19]. After critically analyzing original

Address for Correspondence: Rafik D. Grygoryan, Ph.D., D.B.S., Chief of Bio.Med. Modeling & Software Department, Institute of Software Systems, National Academy of Sciences, 40, Prosp. Acad. Glushkov, 03187, Kiev, Ukraine. Phone:+38 044 526 5169, Fax: +38 044 526 6263, E-mail: graf@isofts.kiev.ua

investigations [1,2, 6,16,18], reviews [5,7,9], and the main generalizing publications [8, 15, 18], it is apparent that there exists no comprehensive theory to explain driving forces, mechanisms and general rules of adaptation within multicellular organisms.

Certainly, the extremely-high complexity of human organism, as well as organisms of many other species, used under empirical observations and experiments is a serious argument to mainly explain the absence of generalizing conceptual conclusions. This argument may explain why adaptation regularities on a sub-cellular biochemical organization level are better understood, than they are on organization levels of organ systems or on a scale of the whole multicellular organism [14, 18]. Currently there is a large gap between the large number of empirical observations and their theoretical interpretations [4,12,18]. Attempts for theoretical interpretations meet difficulties connected with an absence of knowledge necessary to understand the nature of forces compelling both the concrete structural units and the whole organism to change their current state. To our opinion, this lack of understanding is mostly a result of research methodology that remains almost exclusively empirical.

One of the main methodological problems is that the adaptation forces (generally speaking, every force) can not be directly measured. Conclusions concerning acting forces of every movement (mechanical, electrical, others) observed directly always have theoretical - indirect character based on measurements of several accompanied characteristics of movement as a process. It is clear that every dynamic process (including changes of physiological state of an organism in its multi-parametric state dimension) is possible until the resultant vector representing the vector sum of all acting forces is different from zero. The variable value and direction of this resultant vector force are the only providers of the movement. So, every steady-state physiological state of a multicellular organism or its structural-functional units in a dynamic environment must be considered as a temporary balanced situation. Such balance is possible on different organization levels (single cell, population of parallel or consequential acting cells, organism level). Taking into account the vast number of cells in complex multicellular organisms (for example, mammalians), and the fact that each cell produces chemicals that also are able to violate a temporary steady-state physiological state in multicellular organisms, it is reasonable to assert that we can never simultaneously monitor all of these forces. To aid our understanding of the intensity and direction of a multicellular organism's state change in reaction to environmental challenges, we must know the sources of such dynamic forces as well as the mechanism and theoretical principles of the origin of the forces. Without such knowledge, it is impossible to predict the consequence of altered physiological states. This knowledge relates directly to medical diagnostics and treatment which continue to be unreliable as long as a theory doesn't explain the general framework whereby an organism reacts to well-defined environmental challenges. To widen our research abilities in these directions, we must provide theoretical methods of computer modeling. Our objective in this chapter is to demonstrate that computer simulations provide new scientific understandings of two important problems of adaptation physiology.

This publication has two main goals. The first one concerns the theory of physiological adaptation and is aimed to show that every adaptation of multicellular organisms to external or internal (including mental) challenges is a process having driving forces from environmental factors. These forces disappear when internal structural changes on a subcellular level develop special counter-forces that act against external loads and are able to compensate for them. The second goal concerns the problem of human adaptation to microgravity. In this framework, using quantitative mathematical modeling we aim to: a) evaluate mechanisms by which blood volume centrally and cranially shifts under microgravity conditions and b) prove there are additional mechanical forces potentially able to modify hemodynamics upon transition from normal Earth gravity to the microgravity environment of space.

Part 1. A Virtual Multicellular Organism VMO

To understand the main research abilities of our technology based on the concept of a VMO, it is necessary to define the main attributes of elements – virtual cells and the rules for these virtual cells must be linked into complex, virtual structural-functional units (VSFU).

A. Virtual Cell

1. A typical virtual cell is an element which has the following main multi-parameter variables $X(x_1, x_2, ..., x_m)$, $Y(y_1, y_2, ..., y_n)$ that imitate chemicals of metabolism, $Z(z_1, z_2, ..., z_l)$, which imitates chemicals or effects (mechanical, electrical), simulator $S(s_1, s_2, ..., s_k)$ and/or inhibitor $I(i_1, i_2, ..., i_p)$ input variables, and special variable of $D(d_1, d_2, ..., d_q)$ (called as disturber of cell homeostasis) that all together provide the relations between X, Y, Z determined as followed (See also fig.1):

$$Y = \begin{cases} Y^s, X > X^{\max} \\ K_1 \cdot X, X^{th} < X \le X^{\max}; \\ 0, X \le X^{th} \end{cases} K_1 = K_1^{St} + K_1^E; \; Z = \begin{cases} Z^s, X > X^{\max} \\ K_2 \cdot X, X^{th} < X \le X^{\max}; \\ 0, X \le X^{th} \end{cases}$$

$$A \cdot X = B \cdot Y + C \cdot Z; \quad B = b_1 S - b_2 \cdot I - b_3 \cdot D; \; C = c_1 \cdot S - c_2 \cdot I. \qquad ...(1)$$

In (1) coefficients K_1^{St}, K_1^E – characterize structural and energetic aspects of assimilation process accordingly, A, B, C, K_2, b_1, b_2, b_3, c_1, c_2 are constants.

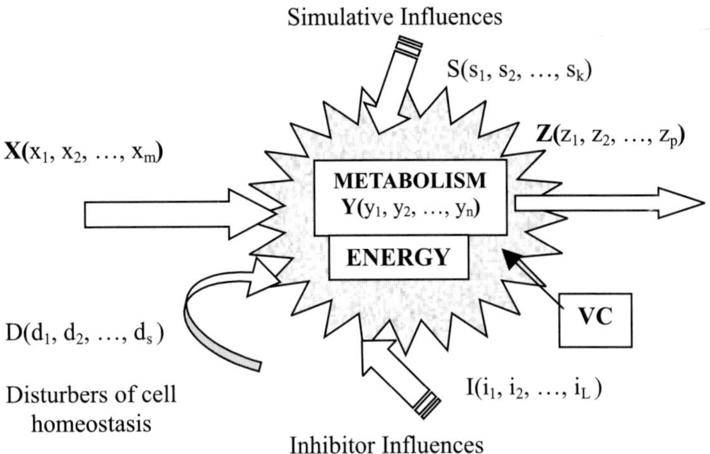

Fig. 1. Schematic illustration of a virtual cell (VC).

According to formula (1), within the variables' threshold (with upper index "th") and saturation (with upper index "S") levels, the typical virtual cell is a linear transformer of input variable X into two variables: internal variable Y - with linear coefficient K_1, and output variable Z - with linear coefficient K_2. If $K_2 > 0$, the VC represents a cell which increases its output function proportional to input incomes. Such a virtual cell is considered as a positive transformer, opposite to the type of virtual cell which has $K_2 < 0$ and is considered as a negative transformer. As shown later, having the both types of virtual cell, one can construct practically all known physiological functional and regulator structures. Scheme on fig. 2 below graphically illustrates the main relations between the variables of VC both under its metabolism and under reactions to influences came from out.

2. As an ideally adapted state of VC to its local intercellular environment is considered the stable internal state (mathematically $dC/dt = 0$) of virtual cell for any combination of A, B, C. Expanding this definition to all cells of a multicellular organism, we come to understanding that to consider multicellular organism as an ideally adapted bio-object to a given environmental state, it is necessary that every cell of the multicellular organism has a steady-state flow of substrates.

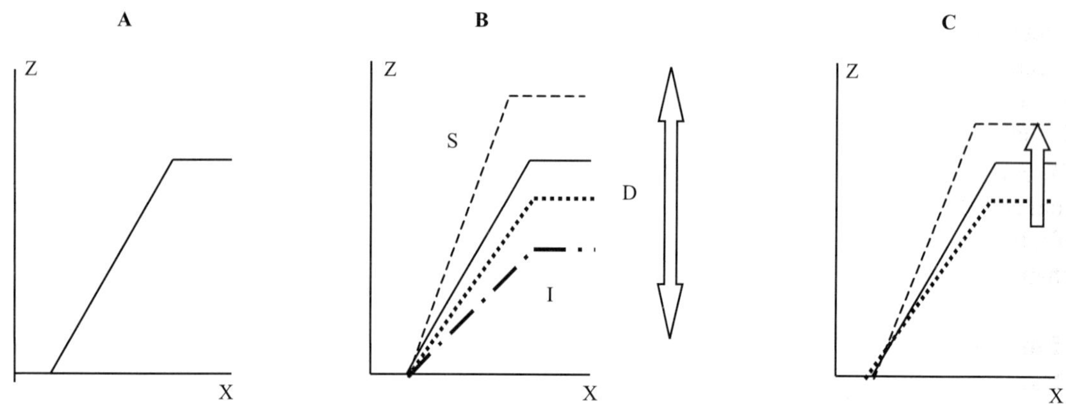

Fig. 2. Schematic illustration of processes in typical Z(X)-transformer: A- current situation; B- changes under simulator (S), inhibitor (I) or disturbing (D) influences. The doubled arrow ⇳ on a middle figure has to show that under external influences the current Z(X)-characteristics may have both up-down and down-up changes while the arrow ⇧ on C underlines that the adaptive trend appeared after D-influences always is directed towards increasing of current Z(X)-function.

3. External influences of type $D(d_1, d_2, ..., d_s)$ are potentially able to induce violations in an ionic un-equality (understood as an important condition for cell homeostasis) between intracellular and extra-cellular environments. To recover its homeostasis, the cell "attacked" by D-influences is compelled to use a part of its current energy, therefore, $D(d_1, d_2, ..., d_s)$ are considered as destructive influences. They are considered as the main direct or indirect acting factors of the changed environment. Being applied to the local "sensitive inputs", these factors are able to origin a cascade of induced changes in links of partial regulator contours finally creating driving forces that provide adaptation trends on a scale of whole organism. So, the reactive process of consequential structural changes in virtual cell (born by its balanced

deficit - dC/dt < 0, and directed towards disappearing of this deficit by increasing of anabolism rate) is considered as the only adaptation process against environmental overloads.

B. Complex Structures of Different Virtual Cell

There may be different specialized versions of virtual cell. Each of them can have its some specific function but they all obligatory will have a common function understood as transformation of their input variable influence X into two other variables: variable Y characterizes the internal state of the virtual cell while Z characterizes the output variable. For further understanding of new technological opportunities, it is useful to imagine why and how the linear characteristics shown on fig. 2 are able to origin well-known nonlinear input-output static transformation characteristics for multi-celled structures such as organs. Here are two basic determinants: (1) even in glandular tissue consisting one specialization of parallel acting relevant cells (let's call this simple form as type 1), the threshold and saturation levels, as well as the recline angles of linear fragment of Z(X)-function of cells of one population practically never have identical characteristics and (2) most of organs include consequential connections of parallel acting populations each consisting of different specialization cells similar to type 1.

Factors that determine the concrete form of such nonlinear behavior of multicellular organs are connected both with the real three-dimensional size of organ (effects of different cells come to the common output with individual time delays) and also by initial asynchronous reaction of different cells that have different reactivity thresholds [4].

To construct more complex functional systems, our constructor will have different specialization virtual cell. Their diversity determines and limits VMO's functions. In frame of this presentation, it is useful to demonstrate the basic functional relations that are possible between two virtual cells, thanks to changes in their connections (see fig. 3).

C. Functional Modeling of Mechanism Providing Reactive Adaptation of Cell to Increased External Loads

Mathematically strong proof of our vision of physiological adaptation process requires modeling of concrete and complex VMO. The limited space doesn't allow us to draw such proof in this presentation. At the same time, the main argument that is used above and concerns substrates' redistribution between cells under their adaptive trends is computationally based using a simple mathematical model representing the interaction of three cells illustrated on fig. 4.

The main modeling limitation here concerns the assumption that inflow of substrates is provided by concentration gradients only. An additional assumption is that the way used by the cell to increase substrates' inflow under cell adaptation against energy deficit is by decreasing cellular membrane resistance.

The model equations connecting flows of substrates between the intercellular (q^{II}), intracellular (q^{Ec}) and mitochondrial (q^{cM}) compartments with the substrates' concentrations in the same compartments (C^E, C^C, C^M) looks like:

$$\frac{dC^E}{dt} q^{II}(t) - \sum_i q_i^{Ec}(t); \quad \frac{dC_i^c}{dt} = q_i^{Ec}(t) - q_i^{cM}(t);$$

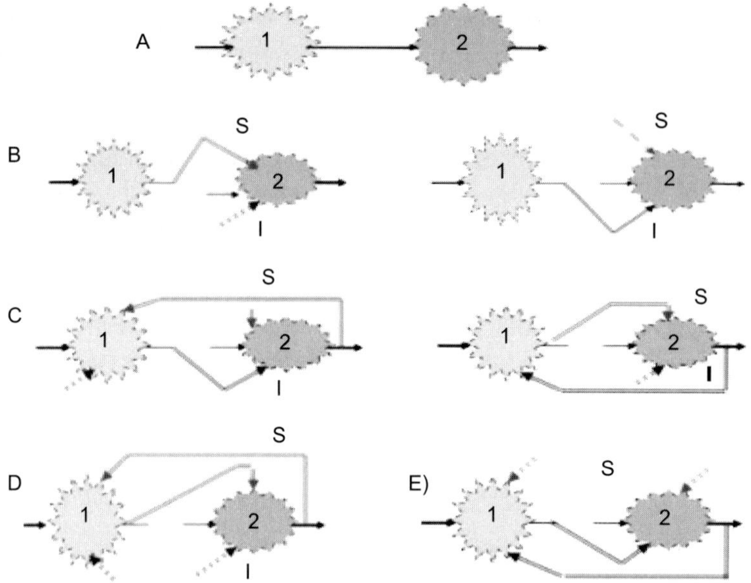

Fig. 3. The main complex functional systems constructed using two VC of different specialization. (A) simple "producer-user" system; (B) systems without homeostatic properties; (C) homeostatic systems; (D) generator system; (E) inhibiting system.

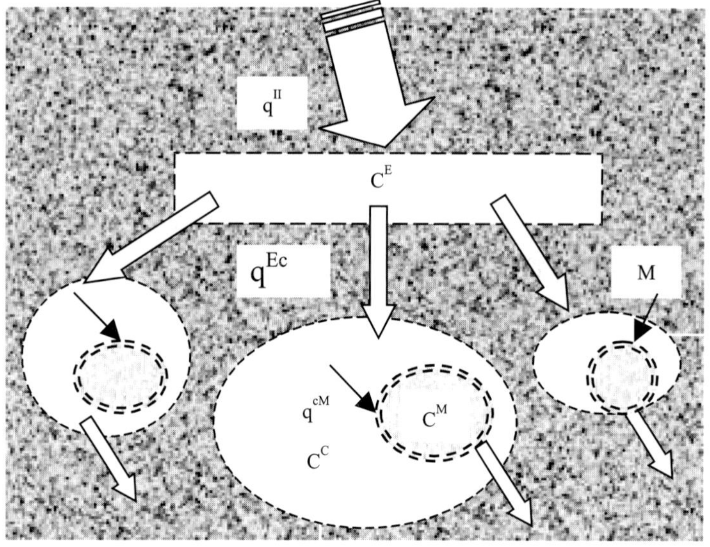

Fig. 4. Schematic illustration of assimilation-dissimilation processes in cells. Different sizes symbolize differences in current state of cells and their mitochondrial productiveness (M). C^E, C^C, C^M – are presenting intercellular, intracellular and mitochondrial substrates' concentrations accordingly; q^{II}, q^{Ec}, q^{cM} – flows of substrates between these compartments.

$$q_i^{Ec}(t) = \frac{C^E(t) - C_i^c(t)}{R_i^c(t)};\ q_i^{cM}(t) = \frac{C_i^c(t) - C_i^M(t)}{R_i^M(t)};\ R_i^M(t) = \frac{\delta}{C_{meti} + \lambda};$$

$$\frac{dC_i^M}{dt} = q_{ij}^{cM}(t) - q_i^{U\,Sum}(t) + k_i^P \cdot P_{mi}(t);$$

$$q_i^{U\,Sum}(t) = \nu_m^{u0}(t) + \nu_{0I}^E + \beta \cdot f_L \cdot M_L$$

where R - the resistance of appropriate membrane, T – temperature, P_m – mitochondrial power, ν_m^{u0} – the mean velocity of mitochondrial destructions, ν_{0I}^E – the mean velocity of mitochondrial energy losses under maintaining of cellular anabolism, f_L – the frequency and M_L –the mean power of external loads, C_{met} concentration of metabolites, β, λ, δ -approximation constants.

Adaptive enlargement of cell energetic power is modeled as:

$$T_{AM}\frac{dP_i^M}{dt} = \begin{cases} -\nu_m^{u0}(T), C_{met} \le Tr \\ a \cdot q_i^{USum}(t) - \nu_m^u(t), C_{met} > Tr \end{cases};$$

$$\frac{dC_{met}}{dt} = k_i^{USum} \cdot q_i^{USum}(t) - k_i^{C_{O_2}} \cdot C_{O_2}(t);$$

where T_{AM} – adaptation time constant, T_r – the threshold value of metabolites' concentration exceeding of which the reactive adaptation mechanism comes active, C_{O2} – oxygen concentration, k^{USum}, k^{CO2}, α – approximation coefficients.

The model described here is planned as a structural unit that will be included into more complex models of VMO. Certainly, the diversity of virtual cell and their schematic connections are the main determinants of resulting VMO's functional properties. At the same time, it was shown [12, 13] that all partial regulator contours (functional, homeostatic) of one multicellular organism theoretically have to modify their reactivity under the slow adaptive trends in an altered environment. To demonstrate the way connecting all cells of one VMO into a system, several semi-quantitative simulation results of our simple model are presented below on fig. 5.

Figure 5 shows that the second cell which had a higher initial mitochondrial power and thus is able to maintain a higher assimilation rate (metabolism) than cell number 1 which remains adaptable under load frequency elevation combined with decreasing of substrates inflow. Under the same conditions, at time moment 800 (the time scale is not absolute but a relative) the less active cell number 1 is finally compelled to decrease its metabolism rate that is still minimal even with elevation of oxygen concentration. So, this simple model visually presents one decipherable aspect of our general hypotheses environment –GHE (it is described below) according to adaptive trends of a group of cells (in minimal case–one cell). By intensively increasing their assimilation rate, the previous configuration of user gradients may be changed on behalf of more active cells. This simulation result allows us to draw an important conclusion concerning the nature of adaptation forces: by means of the re-configuration of user gradients available both under an initial activation of a programmed

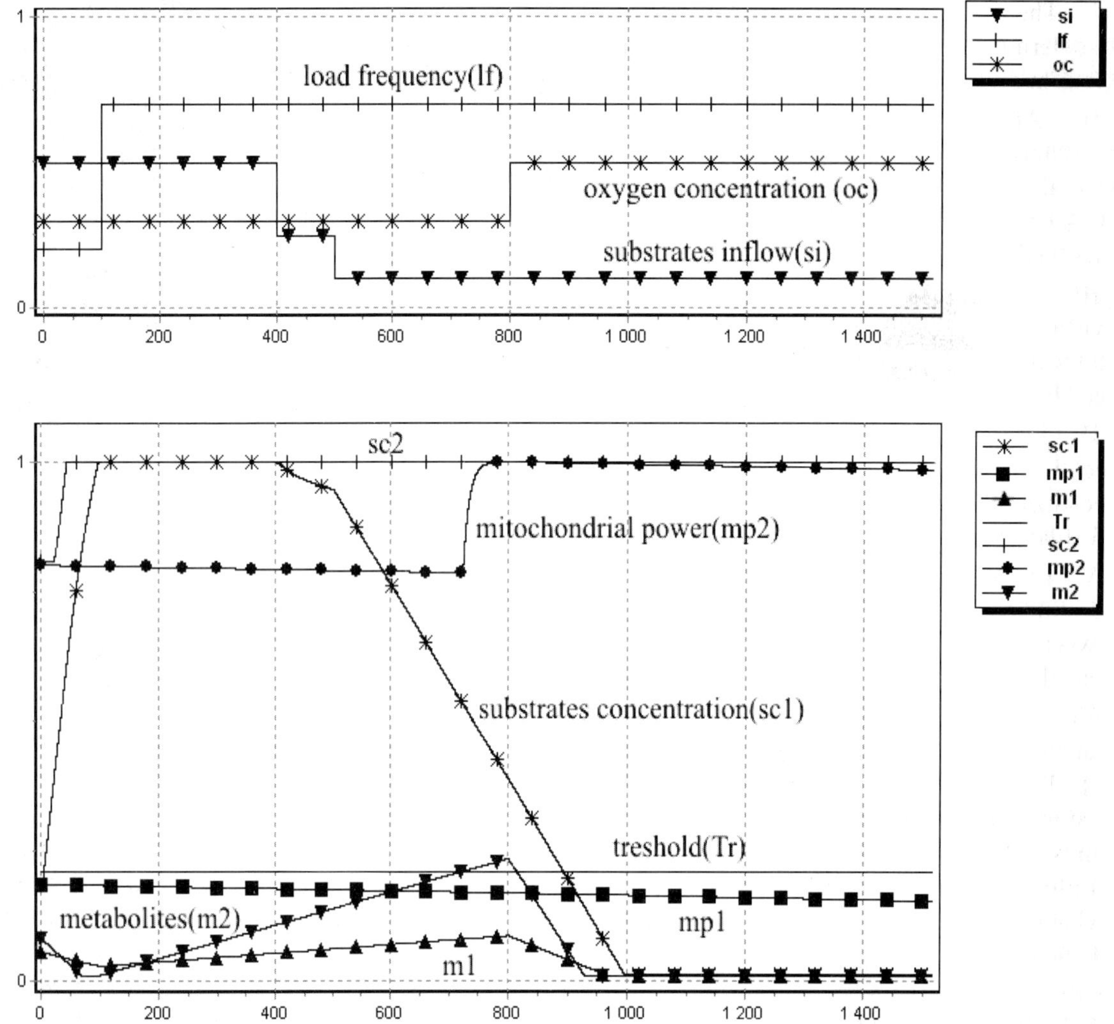

Fig. 5. Computer simulation illustrating different reactions (lower figure) of two cells to combined stroke changes of input variables (upper figure) depending on cells' initial state. Tr- is the threshold level to activate reactive adaptation mechanism.

adaptation mechanism and under its induced activation through the intercellular mechanism of reactive adaptation to increased external loads, the multicellular organism reacts systemically to changes in both external and in internal environments.

D. Conclusion of Part 1

The general vision and basic principles used in constructor of VMO

The scientific platform on which our approach is based is the understanding of all life functions in multicellular organisms as a dynamic process with the following main characteristics:

I – The central chain of multicellular organism is a single "egotistic" cell that has its genetically pre-determined life cycle phases, and is compelled to exist for the welfare of the entire multicellular organism's intercellular environment.

II – Analyzing processes in multicellular organism, we have to differentiate two additional environments– intracellular and extracellular for the whole multicellular organism . Changes in external environment may influence changes in cellular current state using two ways: (a) directly acting on specific external receptor cells and / or (b) through the pathways that are anatomically or functionally connecting receptors with other cells.

III – To maintain its life, the cell uses substrates and oxygen in close proximity to its intercellular environment. Assimilating these input ligands (concentration of that is a time random function), the cell creates its organelles and produces some other chemicals that potentially could be used by other specialization cells. For the last case, the involved combination of cells is considered as a system.

IV –Each cell of the single coupled cells may influence stimulation or inhibition of function of the second cell in the pair. Being anatomically or functionally consequently linked in a chain, different specialization cells are forming the elementary functional unit (EFU). For the simple case the role of EFU plays a single cell (this is characteristic for glandular tissues consisting of parallel acting cells).

V - Every macroscopic organ consists of a huge number of EFUs with some morphological (size and number of organelles) anisotropies. The differences between common specialization cells or between common specialization EFUs are considered as functional heterogeneity (FH). So, the single organ always is a population of cells or a population EFU with their own FH but parallel functioning.

VI – A part of energy synthesized by the cell previously is necessary for the cell to maintain parameters of its intracellular environment within a narrow range (known as cellular homeostasis-CH). The most likely insight of CH is that it is necessary to provide optimal biochemical transformations. The current rate of biochemical assimilative transformations is not a stable but an adaptive variable depending on genetic programs determining anabolism rates for different phases of cell life cycle. This current form of anabolic adaptation is called a cellular adaptation programmed mechanism (CAPM). In case the exogenous (outside of the cell membrane) influences are violating CH, the cell activates a cascade of processes directed to recover the CH. This acute reactivity requires energy. Therefore cells have some limited ability to react using previously accumulated energy.

VII – If under the high frequency reactions, the cell loses a bigger portion of energy than the cell is able to synthesize using its current productiveness (it depends on summary power of mitochondria and ribosomes), a balance deficit (BD) occurs. When an over-threshold level of BD exists for a relatively long term, metabolites are activating a special mechanism called cellular adaptation reactive mechanism (CARM). For every phases of cell life cycle CARM is modifying the current effects of CAPM in one direction only: trying to activate currently passive organelles and or to build new units of mitochondria and ribosomes. By this way, CARM is providing enlargement of cell's current energy productiveness until BD is minimized. Thus, activators of CARM are the primary external factors. It is important to understand that these factors have random time and power distributions.

VIII – The current velocity of ligands' entry into the cell is proportional to their concentration in close environment and to the cell's assimilation potential (figuratively speaking, "appetite") that is proportional to the cell's current productiveness. When the summary concentration of ligands is less than the summary "appetite" of cells in local environment is, the enlargement of some cells' "appetite" decreases the portion of ligands' inflow into the less active cells. As far as all cells of a multicellular organism have a common fluid intercellular environment, every re-configuration of local user gradients has to generate a re-distribution of ligands in frame of the multicellular organism. So CARM is one of the independent functioning mechanisms able to redistribute the currently-existing field of moving forces that determine inflows of substrates to cells. The cells that come compelled to decrease their inflows inhibit their functional activity.

This insight into the role of CARM proposes an explanation of many observations (eg., Fregly and Blatteis;1996) fixed that multicellular organism's adaptation to stable shifts of some concrete parameters such as external temperature and oxygenation that practically always causes reactions in more than one functional system. It also is the basis to understand the hidden mechanism of so called cross effect of adaptation.

IX – For every time moment, within threshold and saturation levels the productivity of the cell is proportional to concentration of chemicals in its nearby intercellular area. However, thanks to FH, each cell has its own threshold and saturation levels, and also its own input-output transformation coefficient. Namely, even in case the input variable has a rapid upstroke, the unevenly-distributed FH is the main originator of nonlinear input-output behavior in a population of independently-functioning cells

The items I-IX above represent our current scientific view that should be considered a general hypothesis environment (GHE) in which our models of partial physiological processes and systems must be "immersed". In the frame of the GHE, every multicellular organism is considered as a semi-hard construction. Being genetically structured as hierarchically functioning object, it is not functioning in an up-down direction and not with given constant characteristics. The GHE shows that the interaction between thermodynamically opened multicellular organism and environment is built on reactions of multicellular organism to the primary, mostly random challenges of environment. The current reactions of multicellular organism are depended on cells' current state only. However, under long-term study we always have to remember that each cell and the entire multicellular organism have some soft - adaptive abilities determined by real energetic limits in environment. This new view emphasizes that the internal interaction between structural macroscopic components of an organism has both up-down and bottom-up directions with a variable intensity. This vision is opposite to widespread ideas concerning the adaptation process to environmental shifts understood as if the adaptation is a special, centrally-organized process. In the framework of GHE, every reactive adaptation is a process that is a sum of local adaptive trends, independently developing under local energy deficits in cells. This view is necessary to understand the nature of internal long-term relations between different organs and systems in dynamic external environment. So, GHE is bundling the acute reactivity of cells, organs and systems with the more slowly acting anabolic-catabolic processes.

It is important to imagine that for a huge number of dynamic cells of an organism, such knowledge can not be constructed using exclusively empirical research results. It is a theoretical insight that is proven using a model approach only. Although a comprehensive examination of the theory requires modeling of the whole multicellular organism, we use the main principles of this vision under modeling of cardiovascular adaptation to microgravity.

The GHE is a scientific platform under modeling and simulation investigations of long-term adaptive interactions both between environment - complex organism interaction and between cell-cell or organ-organ interactions. Its realization supposes that the long-term interaction between every number of organs always is analyzed taking into account both needs of other energy users of organism and the variable rates of energetic-structural synthesis. As to modeling of acute reactions, a stable structural organization is assumed at least on cellular and organic levels. Thus, the most significant aspect of the GHE is connected with nonlinear effects caused by current statistical picture of cellular heterogeneities in organs.

So, there are three aspects of life functions (nonlinear dependencies, adaptation, homeostasis) that have to be involved in one research technology to explain the main rules determining different scale mechanisms in a time dimension.

The following key statements explain the novelty of our understanding:

1. Between threshold and saturation levels, every cell–cell relations are considered as producer-consumer (PC) relations with practically linear dependencies. They have approximately direct or up-down turned S-form explanation of that are described in [10-12]. In physiology, input-output nonlinear dependence within organs is mostly understood as result of spatially-temporally accumulation of elementary and independent cellular acts that are unevenly dispersed statistically. Although one can find many forms of artificially constructed nonlinear dependence between pairs of life variables X and Y, they never have a natural PC-character.
2. Physiological adaptation is an asymmetric process directed against destructive trends. It functions independently in each cell. Adaptation of multicellular organisms to any environmental shift is a simple sum of adaptive acts in its cells. Under adaptation the cell builds new organelles and enlarges those already existing. The current "appetite" of a cell to assimilate chemicals is proportional to its summary mitochondrial and ribosomal power. If a deficit of sources in intercellular environment does not allow all cells to take up substrates according to current needs, these substrates will distribute between cells in proportion to their current "appetite".
3. For the last case, every increase of "appetite" decreases the part of substrates for other cells. Such kind redistribution on organism scale creates violations or stable shifting of variables usually considered as indicators of homeostatic systems' activity.

We know that the measurement techniques today are not able to serve researchers by comprehensive and quantitative information about current FH in each cell from the billions of factors in organs. Nonetheless, we see this theoretical view as a promising framework to understand the nature of the actual forces compelling cells to change their current state and to form a dynamic physiological adaptation trajectory under human adaptation to environmental challenges. Therefore, we have theoretically analyzed the mechanisms responsible for nonlinear (often direct or turned-down S-form) input-output characteristics in cell populations to find methods using that the investigator could operate with organs-populations to qualitative estimate some causal-consequential effects [10-13]. This approach can be applied to the myocardium and arterial mechanoreceptors under cardiovascular function modeling described in part 2.

An equilibrium which existed between assimilation and dissimilation processes on cellular organization level is the ideal state to define it as an adapted state. Under primary stable shifts of environmental state, a imbalanced state able to initiate slow structural adaptation in these cells until a new balanced state is reached will first appear in receptor cells of the multicellular organism. In a given time interval these receptor cells realize their structural adaptation on a way towards their new stable state, all other cells of multicellular organism that have not direct inputs from the external environment, but that have structural-functional connections with the influenced receptor cells, also sense their own imbalance state. So, practically all cells of a multicellular organism are alter their functional regimes and anabolism-catabolism rates forming their own adaptation trajectory.

On an organization level of a multicellular organism, the ideal balanced state practically never exists. At the same time, the regime commonly approached to this ideal condition under stable states of external and internal mental environments is the physiological state in which any cellular anabolism-catabolism imbalance is minimal.

The general rules determining acute reactivity and slow structural adaptation of multicellular organism to sensitive enough external and/or mental loads with stable or even increasing character but obligatorily requiring an elevation of organism's energy losses is schematically illustrated on fig.6.

According to this scheme, the influences come both from external and mental areas are considered as destructive for cell homeostasis if they require mobilization of cellular energetic resources. These differences are not discussed in detail here. We offer a fundamental understanding of how the cellular-reactive adaptation mechanism modifies the current activity of genetically-predetermined mechanism of programmed adaptation. This adaptive increase of cellular function estimated on a scale of general cellular population is illustrated by an increasing cumulative effect of the population (curve 1 -before adaptation, and curve 2 -after adaptation).

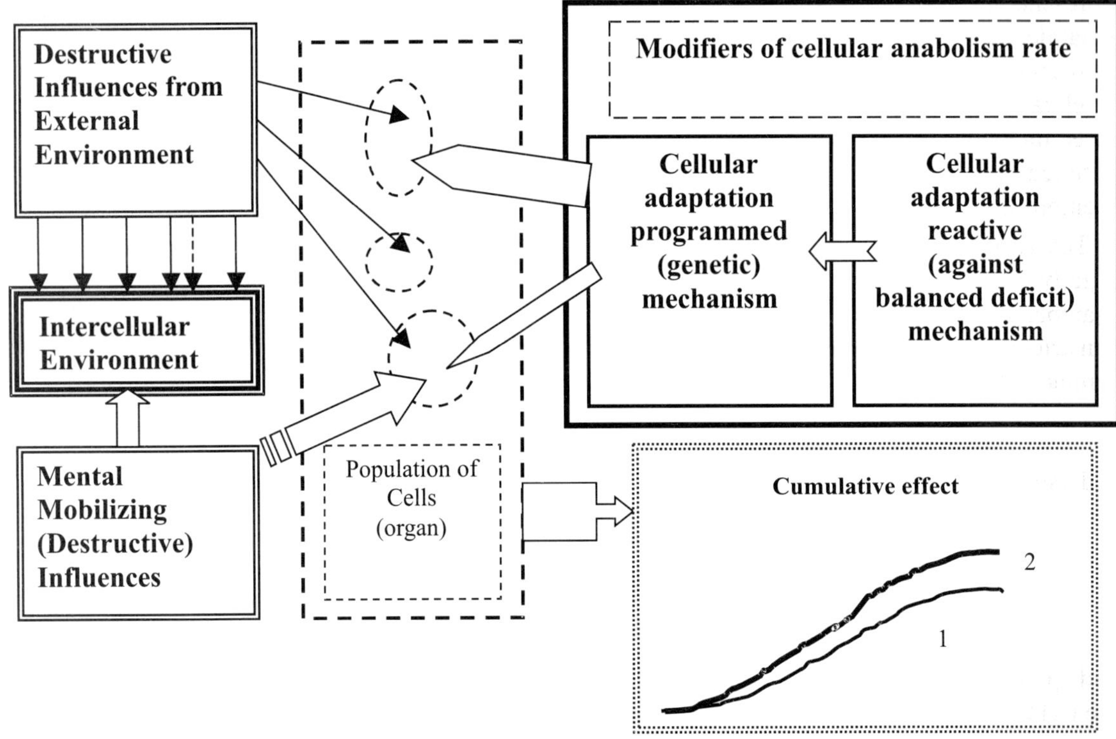

Fig. 6. A general vision of multi-celled organism under its adaptive interaction against external and internal (mental) destructive trends.

Part 2. Modeling of Human Adaptation to Microgravity

The main reasons for this modeling are briefly described below:

1. Data collected during actual space flight are still insufficient to draw conclusions concerning mechanisms responsible for cranial hypervolemia during the first few days of microgravity.
2. Earth-based physical models (anti-orthostasis, water immersion) used to simulate hemodynamic effects of microgravity are not adequate to explain why and how blood is shifted towards the head during space flight.
3. The current paradigm for shifts of body fluid from legs towards central and cranial body segments may require revision using alternative research methods.

Already the first attempt [12, 13] to analyze the problem theoretically indicated in points 1-3 above using three-dimensional cardiovascular model document that the calculated shift of blood volume from the lower body towards thoracic and cranial segments is lower than expected based on Earth-based simulations of microgravity (water immersion, head-down tilt [5, 6, 10, 17]. Having a new general vision of physiological adaptation and a more precise cardiovascular model, we decided to provide a simulation model of this early response to microgravity.

Brief Information about Human Cardiovascular Function Model

The simulation results below and their discussion are based on a quantitative computer model that represents human hemodynamics. The model describes heart pump pulsatile function (HPPF), the dynamics of pressures, volumes and flows in two-dimensional net of vessels (The third dimension representing vessels on left-right line is mostly important under modeling of complex acting accelerations. Therefore, the vascular net was reduced to have maximal computational opportunities based on using of standard personal computers) taking into account rapid reflective control of CVS initiated by different localization vessel mechanoreceptors (in detail described below) and chemoreceptors.

The HPPF is modeled using special presentation of time-depended cardiocyclic changes in the rigidity of ventricles. Our approach is based on the assumption that myocardial cells of one heart chamber are acting parallel. However, the mechanism that describes electrical excitation- mechanical constriction has individual time delays for discrete cells [13,14]. Assuming the durations of the cardiac cycle's (Tc), of systole (τ_s) and of diastole (τ_d) as:

$$\theta_d = t/\tau_d, \quad t \le \tau_d; \quad \theta_s = (Tc - t)/(Tc - \tau_s), \quad \tau_d < t \le Tc,$$

and using the β - function given on interval (0,1) with constants a & b known as:

$$\beta(x) = \begin{cases} 0, x \notin (0;1) \\ C(a,b)x^a(1-x)^b \end{cases}, \text{ where } C(a,b) = \frac{1}{\int_0^1 x^a(1-x)^b\,dx},$$

and approximating the empirical functions $\tau_d = W_d(T_c)$; $\tau_s = W_s(T_c)$, thy dynamics of the rigidities for the right - $D_r(t)$ (i = 1) and left - $D_l(t)$ (i = 2) ventricles can be calculated as (*):

$$D_i(t) = I_i(t)\cdot\begin{cases}\int_0^{\theta_s}\beta_i^s(x)dx, 0 \le t \le \tau_s \\ D_i(\tau_s) - \int_0^{\theta_d}\beta_i^d(x)dx, \tau_s, t \le \tau_d\end{cases} \qquad (*)$$

In formula (*) the inotropic state of concrete heart chamber myocardial population is reflected by a scale coefficient $I_i(t)$. This state depends on the energetic power of cells and allows one to transform the magnitude of normalized D-function into one sufficiently realistic to model pressure-volume dynamic dependencies using the formulae below:

$$P_i(t) = [V_i(t) - U_i(t)]D_i(t);$$

$$Vi(t) = V_i(0) + \int_0^t [Q_{i-1}(t) - Q_i(t)]\,dt;$$

$$Ļ_i dQ_i(t)/dt = [P_i(t) - P_{i+1}(t)]/R_i^V(t) - Q_i(t),\ Ļ_i > 0;$$

$$Q_i(t) = [P_i(t) - P_{i+1}(t)]/R_i^V(t),\ Ļ_i = 0,$$

where P, V and Q – are blood pressure, volume and flow accordingly; $U_i(t)$, – unstressed volume, $Ļ_i$ – flow's inertia. The valves' variable resistance $R_i^V(t)$ is modeled as:

$$R_i^v(t) = \begin{cases} R_i^{vc}, & P_i(t), P_{i+1}(t) \\ R_i^{vo} + R_i^{vc} - R_i^{vo})\cdot e^{\delta\cdot[P_i(t)_i - P_{i+1}(t)}, & P_i(t) < P_{i+1}(t)\end{cases},$$

where δ - an approximation constant, R^{vo}, R^{vc}- resistances of opened and closed valves accordingly.

The vascular net consists of parallel and consequential connected compartments localized along body seven sections (head, neck, thorax, abdomen, thighs, legs and feet and hands). Within the each body section perpendicular to direction of the head-legs axis (in a direction of back-chest axis), there are three (back, axial, chest) levels of arterial and venous compartments localization. The heart is localized in axial level.

Nonlinear static volume-pressure (V-P) dependencies are tuned for each of the 155 vessel compartments. The formula used to calculate pressure gradients and flows ($q_{i,j}$) between every two "I" and "j" vessel compartments connected through resistance $R_{i,j}$ is:

$$q_{i,j}(t) = [(P^T_i + P^E_i + P^G_i) - (P^T_j + P^E_j + P^G_j)]/R_{i,j}(t), \qquad (\#)$$

where upper indexes "T", "E" and "G" indicate the transmural, extravascular and gravitational components of blood pressure, accordingly.

Central nervous control of hemodynamics is modeled by means of additive functioning mechanoreceptor reflexes from reflexogenic zones of aortic arch, carotid sinuses, brain arteries, atriums, lung arteries. Concentrations of adrenalin (C_A), noradrenalin (C_N) and acetylcholine (C_{Ah}) are dynamically connected with mechanoreceptor activity (M_A), activity of heart volume-receptors (Ω) and chemoreceptors (X) as followed:

$$\psi_2 \frac{dC_{Ah}}{dt} = a \cdot M_A + b \cdot \Omega + c \cdot X + \varepsilon^S \cdot H^S - H^I - f_2 \cdot t,$$

$$\psi_3 \frac{dC_N}{dt} = -(a \cdot M_A + b \cdot \Omega + c \cdot X) + \varepsilon^S \cdot H^S - \varepsilon^I \cdot H^I - f_3 \cdot t,$$

$$\psi_1 \frac{dC_A}{dt} = -(a \cdot M_A + b \cdot \Omega + c \cdot X) + \varepsilon^S \cdot H^S - \varepsilon^I \cdot H^I - f_1 \cdot t,$$

where a, b, c, ε^S, ε^I, ψ_1, ψ_2, ψ_3 are reflecting differences between sympathetic and parasympathetic centers of autonomic vegetative nervous system while the last variables in each formula are reflecting appropriate velocities of mediator utilization f_1, f_2, f_3,.

$$T_j^{dY_j(t)} \Big/ dt = Y_j^{\max} \frac{1 - e^{-\gamma_j \cdot Z_j}}{1 + \chi_j \cdot e^{-\gamma_j \cdot Z_j}} - Y_j(t), j = 1, k,$$

$$Tc = Tc^A - \zeta C_A - \rho C_N + \mu C_{Ah} - \omega T\kappa; \quad I_i(t) = I_i^A + \zeta_I C_A + \rho_I C_N + \omega_I T\kappa;$$

$$R_i(t) = R_i(0) \cdot [V(0)/V(t)]^2$$

$$U_i(t) = D_i^b - \sum_{j=1}^{M^\varpi} \eta_{ij}^{\varpi U} \cdot C_i^\varpi + \sum_{l=1}^{M^\upsilon} \eta_{il}^{\upsilon U} \cdot C_i^\upsilon + \hbar^U \cdot (T_k - T_k^n)$$

$$U_i(t) = D_i^b + \sum_{j=1}^{M^\varpi} \eta_{ij}^{\varpi D} \cdot C_i^\varpi + \sum_{l=1}^{M^\upsilon} \eta_{il}^{\upsilon D} \cdot C_i^\upsilon + \hbar^D \cdot (T_k - T_k^n)$$

where D^b, U^b – are the basal values of rigidity and unstressed volume, while the last component reflects vessel tonus dependence from the temperature (T_k) relatively to its normal value ($T^n{}_k$).

To model adaptive changes in cardiovascular mechanoreceptors in order to take into account enlargement of a reception area, their integral static characteristics between the threshold and saturation levels are described by means of standard β-function:

$$N_i(P_i) = N_i^n \cdot \begin{cases} \int_{P_{it}^{\min}}^{P_i} \beta_i^s(x)\,dx, & P_{it}^{\min} \le P_i \le P_{is}^{\max} \\ 0, & Pi \le P_{it}^{\min} \end{cases}$$

Fig.7 presents a typical simulation example reflecting the complex hemodynamic reactions under postural dynamic influences schematically demonstrated on top part of picture.

Shown are pulsatile and mean pressure in aortic arch, the mean pressure in carotid sinus zone, the heart rate (bottom figures) and summary blood volumes in cranial, neck, thoracic (lung blood volume is specially displayed too), abdominal, leg and hand body segments.

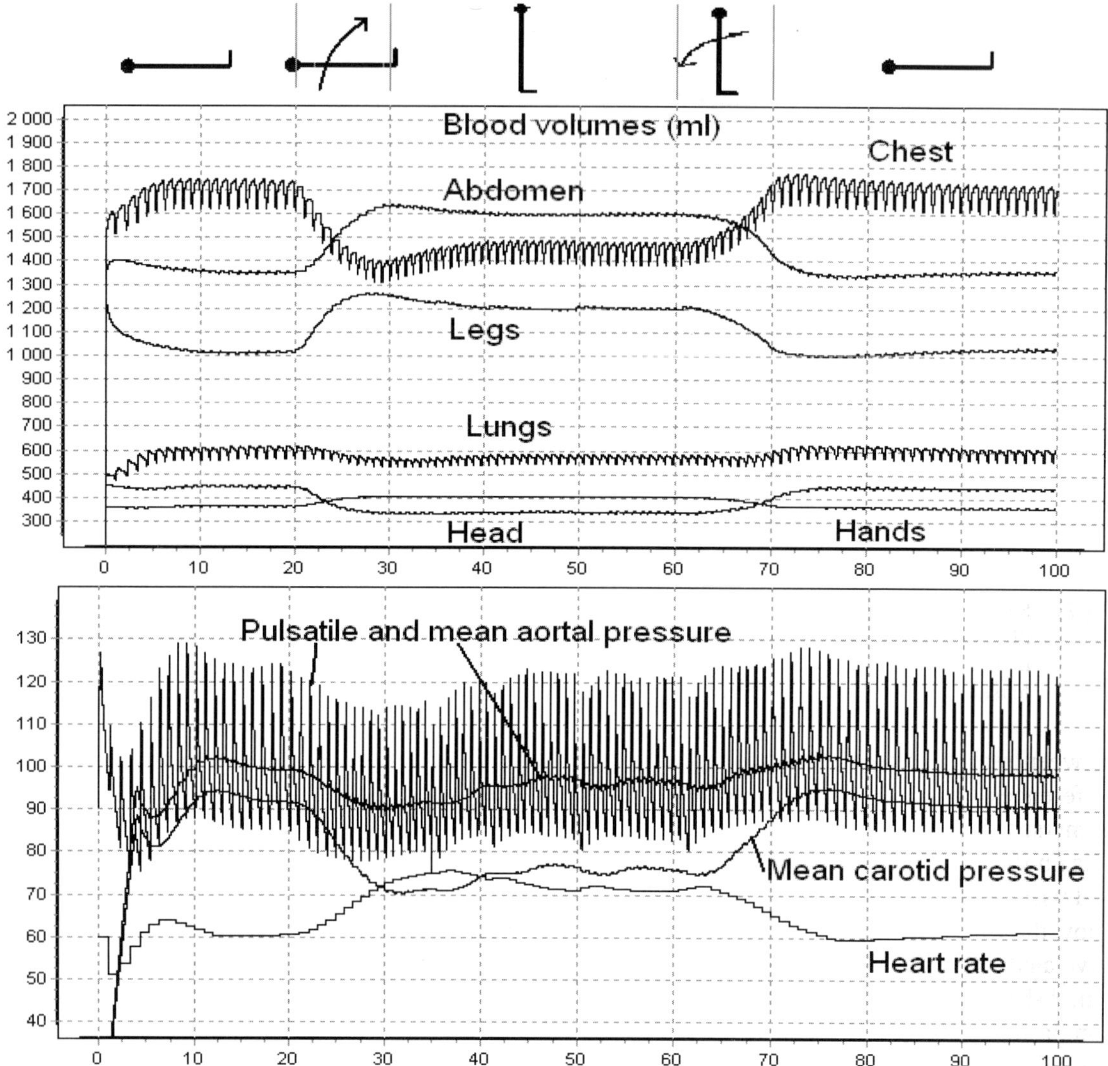

Fig. 7. Heart rate, pulsatile and mead pressures in aortic arch and carotid sinus vascular compartments for human horizontal position (T<20 s), their dynamics under 10 s tilting to a head up (80°) (time interval between 20 and 60 seconds), and under 10 s transition to the horizontal position (simulation data).
*** An initial transitory process between 0 - 20th seconds appears because the steady-state hemodynamics for human horizontal position is calculated from casual initial data.**

Before to analyzing simulation results for microgravity conditions, it is useful to look at fig. 8 that schematically presents a section of thorax-abdomen cavities to show inertial forces' influences on diaphragm line.

To understand the nature of forces that initiate cranial shifts of blood volume, we again review the model formula (#) which determines pressure gradients and flows. It is clear that under posture tests including antiorthostasis, the hydrostatic pressure component P^G determines pressure gradients

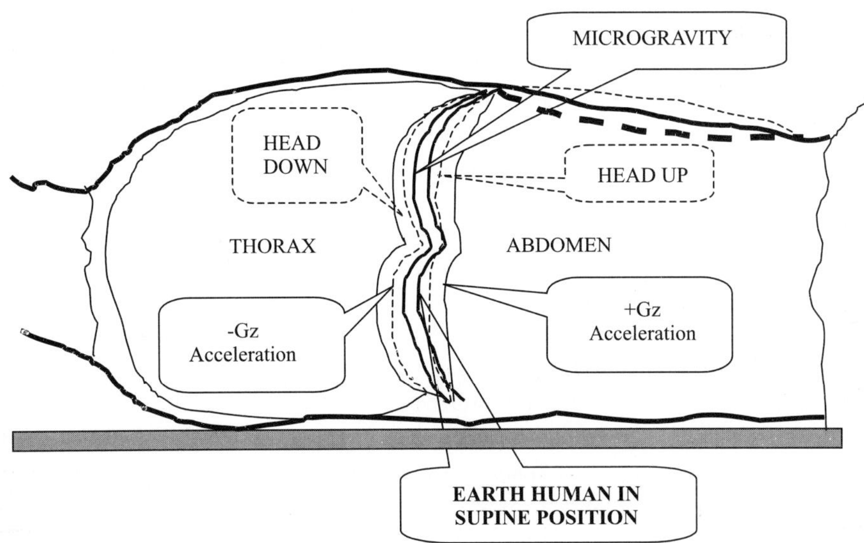

Fig. 8. Schematically are shown diaphragm line shifts between thoracic and abdominal cavities under different directions and amplitudes of inertial forces. Important is to take attention that even under human horizontal on back position the diaphragm is slightly shifted into the abdomen. Weight loss causes a diaphragm muscle tension imbalance and inertial shifting into the thorax.

between reclined axial vessels and shifts blood volumes to cranial zones. As for water immersion, differences between P^T for immersed and above-water compartments creating forces from the immersed tissue compartments towards body dry parts (neck, head). Although a fact is that gravity acts on tissue pressure, the hydrostatic component of pressure in and out of blood vessels is different and tissue specific. To estimate the gravity-depended hemodynamic effects for this case, one need to know the whole picture of tissue pressures both in deep and in skin tissues. Having not such a data, we have analyzed a model situation assuming that these changes in tissue pressure are not sufficient for initial short period of observations. For this case, applying the formula (#) to actual microgravity and comparing it with the human horizontal position, one can see that the only difference in hemodynamics will be initiated by hydrostatic pressure component in back-chest direction. Therefore, reasonable is to expect that under disappearance of this factor in model we will have a hemodynamic regime like that in real space flight. However, as shown in Table 1, this expectance is not true. Although the blood pressures are not indicated in Table 1, the simulation shows that the central venous pressure under microgravity is not elevated. Our simulation mainly maintains the data first described by Kirsch et al [15] and then confirmed by Buckey et al [3]. At the same time, to answer the question why Buckey et al had observed elevation of left ventricle end-diastolic volume under space flight having of central venous pressure decreasing, a discussion is necessary.

Theoretically, there could be two independent factors (or their combination) for increasing of heart end-diastolic volume: (1) increasing of input blood pressure; (2) lengthening of cardiac cycle duration that elevates the period of diastole. The data in publication Buckey et al [3] show that the second factor is presented. However, a more interesting is to interpret the measured values of central venous pressure given as 8.4 cmH_2O in seated before flight, 15.0 cmH_2O –in the supine legs-elevated posture in the shuttle, and 2.5 cmH_2O after 10 min in microgravity. Our interpretation of such data is based on formula # above. The matter is that under measuring of blood pressure, we use the value of surplus pressure relatively to atmospheric pressure. At the same time, local extravascular pressures in several areas of vascular net may differ from the atmospheric pressure. Therefore, as the formula # shows, without knowledge of the local extravascular pressure, we misunderstand the real pressure gradients for blood flows between any pair of vessels located in different cavities. In this case we need to know the value of pleural pressure if it may have independent changes. The reasons for unstability of pleural pressure under changeable gravity were illustrated on fig. 8. So, the second explanation of why Buckey et al [3] have observed an increasing of end-diastolic volume of left ventricle is that the real filling pressure was not the the measured 2.5 cmH_2O but plus to this value an additional - elevated relatively to Earth supine position pleural pressure. Additional data for such interpretation of empirical data can be obtained using our quantitative mathematical model.

To prove that our model is adequate, we have previously tested it for known situations like postural and acceleration influences. The limited space of the article allows not to present all simulation data, but the data collected in left part of table 1 for human four postures (horizontal on back, tilted head-up 90° and head-down 90° and 6° are additional arguments to be sure that the model is able to satisfactorily simulate the regimes of human hemodynamics. The 6° head-down position currently is considered as the hemodynamically best Earth-based model of microgravity, because of cranial hypervolemia, jugular pressure is best approached by using values during microgravity. Our model presents summary blood volumes in body sections along the head to feet direction and in every three layers within each body section. Although space limits do not allow us to compare simulation data with empirical measurements, we have examined model data that are satisfactory to describe real processes. Additionally, to provide more realistic simulated hemodynamic effects under actual space flight, we must consider the launch environment, taking into account both human lodgement position and accompanied emotional stress. The appropriate hemodynamic data at the moment of launching finish just before stopping of space ship engines are presented in Table 1 too.

Subdivided under two additional hypotheses, the hemodynamic effects appearing after weightlessness are shown in four right columns of the Table 1. The first two left columns are presenting data by assuming that the only microgravity effect is the disappearance of blood hydrostatic pressure (the hypothesis H0). As the data show, in comparison with the human horizontal position an increase of head, neck and chest blood volumes is not obtained. In an attempt to approach the hemodynamic picture presumed for actual microgravity, we assume that there also takes place additional changes that are accompanied human transition from the high G environment into a microgravity environment. Taking into account diaphragm shifts illustrated on fig.8 above, we have hypothesized that pleural pressure increases and abdominal pressure decreases simultaneously. This is the basis of our hypothesis H1 that supposes pleural pressure elevation of 12 mm Hg, and an abdominal pressure decrease of 6 mm Hg. Under these parameters, the cranial volume is approximately equal to that for human 88° head-down posture. The jugular venous pressure increases from initial value of 5.4 mm Hg to 7.6 mm Hg. The only problem we see in data for H1 is that the leg volumes still are higher than expected for microgravity conditions. This disagreement is avoided when under weightlessness it is assumed that abdominal pressure decreases 10 mm Hg. As far as real

Table 1. Sic mulated blood volumes.

BLOOD VOLUMES (ml) IN BODY SECTIONS		Earth conditions: Human body positions				Launching Finish *	Microgravity (Hypotheses)**			
		Horizontal on back	Head up 90°	Head down 90°	Head down 6°	Lodgement Position + 3Gz	H0 30 s	H0 3 min	H1 30 s	H1 3 min
Head	Gen	**415**	**333**	**497**	**431**	**350**	**433**	**423**	**438**	**429**
	Chest	135	111	166	141	113	144	141	146	143
	Axis	139	111	166	144	117	144	141	146	143
	Back	141	111	166	146	120	144	141	146	143
Neck	Gen	**143**	**90**	**289**	**152**	**95**	**152**	**147**	**157**	**153**
	Chest	35	21	51	39	21	40	39	42	41
	Axis	68	47	85	71	50	72	69	74	71
	Back	40	21	51	42	24	40	39	42	41
Thoracic	Gen	**1870**	**1575**	**2490**	**1990**	**1620**	**1915**	**1890**	**1620**	**1540**
	Chest	435	388	650	478	340	490	485	390	368
	Axis	920	795	1180	970	810	940	915	845	805
	Back	510	388	650	538	468	490	485	390	368
Abdominal	Gen	**1080**	**1230**	**1050**	**1147**	**1325**	**1005**	**1065**	**1245**	**1347**
	Chest	323	406	357	348	430	332	355	414	446
	Axis	347	415	336	365	445	340	355	419	446
	Back	410	406	357	438	449	332	355	414	446
Hips	Gen	**816**	**983**	**333**	**498**	**895**	**810**	**810**	**835**	**842**
	Chest	245	328	111	183	289	270	270	278	280
	Axis	276	328	111	222	296	270	270	278	280
	Back	295	328	111	394	310	270	270	278	280
Shins + Foots	Gen	**520**	**602**	**310**	**415**	**546**	**509**	**508**	**519**	**521**
	Chest	165	201	103	136	181	170	169	173	174
	Axis	174	201	103	138	182	170	169	173	174
	Back	177	201	103	141	184	170	169	173	174
Hands	Gen	**326**	**351**	**384**	**327**	**337**	**338**	**330**	**345**	**339**

* Data in Launch column are simulated supposing also emotional stress.

** Microgravity Hypotheses:H0- No other accompanied changes but blood hydrostatic pressure; H1- Additional to blood hydrostatic pressures' disappearance are changed the pleural (elevation on 12 mm Hg) and abdominal (decreasing on 6 mmHg) pressures.

measurements are not known both for pleural and for abdominal pressures, we can only fix these simulation results as a basis to think over alternative ways of cardiovascular adaptation to short-term microgravity.

Taking into account the decreased thoracic blood volumes and insufficient increasing of heart volumes under simulated microgravity, we have to conclude that there is not enough basis for essential activation of Henry-Gauer reflex. At the some time, comparing with human head-up erect position, some body fluid loss can take place at initial period of the microgravity. This fluid loss might be mainly explained by increased venous pressure and blood volumes in brain area. Depression of antidiuretic hormone (ADH) secretion under cranial hypervolemia is well-known.

Summarizing theoretical analysis and simulation, provided in frame of the part 2, we can propose an additional view of short-term adaptation to microgravity in a form of scheme shown on fig. 9.

General Conclusion

The theoretical approach used to explain some problems of adaptation physiology documents that mathematical modeling is able to propose alternative scenarios of an organisms' physiological adaptation to altered environments. Starting from the discovery of unknown mechanisms that determine consequential stages of adaptation, we have identified some of the driving forces of

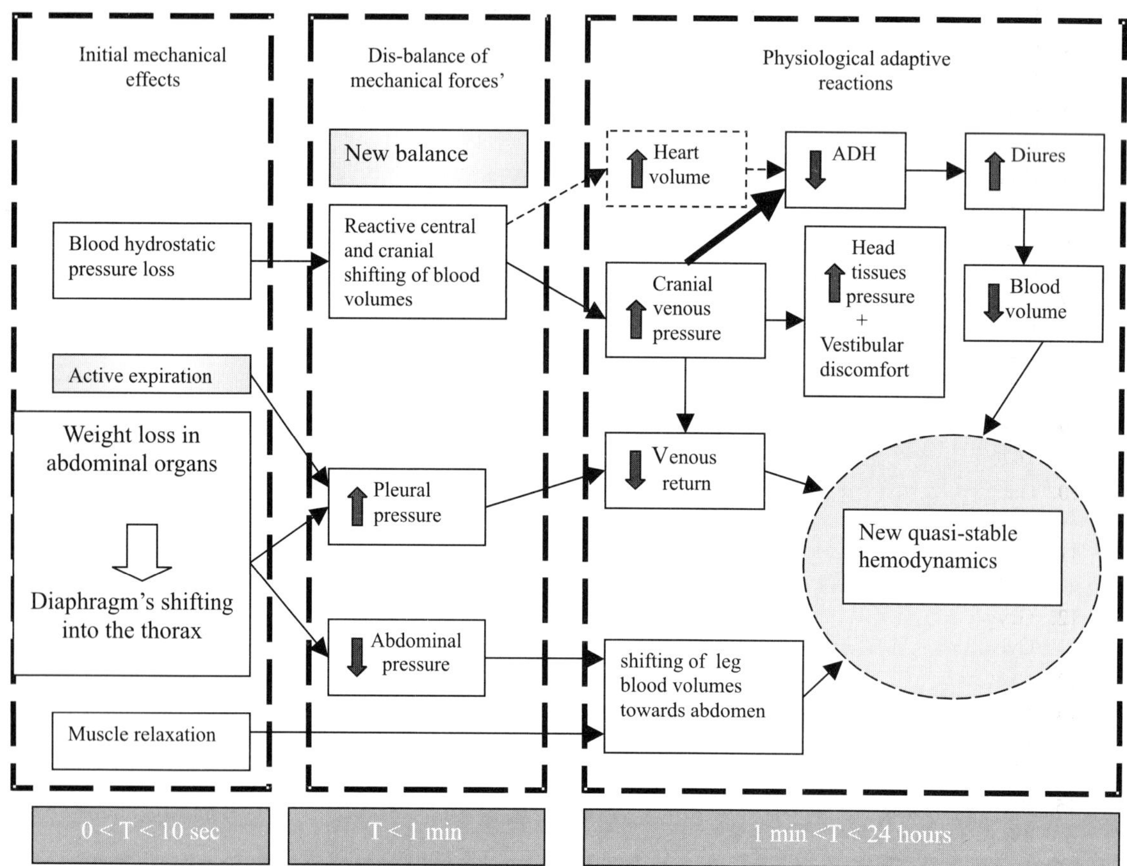

Fig. 9. An alternative vision of short-term adaptation to microgravity.

adaptation. Our approach allows us to propose a conceptually new and promising vision of organism-environment relationships in general. Furthermore, some principles of this general vision are used to model headward fluid shifts of blood during early adaptation to microgravity. Again, concentrating our efforts on understanding adaptation forces, we propose an additional concept of human physiological adaptation to actual microgravity of space flight. This concept could be the basis for further modeling of human long-term adaptation to microgravity and re-adaptation to Earth's gravity.

Acknowledgements

The authors thank programmers Lissov P.M., Aksionova T.V., Tomin O.O. for their participation in software development and simulation research.

References

1. Baldwin KM. Effect of spaceflight on the functional, biochemical, and metabolic properties of skeletal muscle. Med. Sci. Sports Exerc. 1996: 28:983-987.
2. Berry N.M., Rickards C.A., Newman DG. Acute cardiovascular adaptation to repetitive head-up tilting. Aviat Space Environ Med. 2004; 75 (Section II):B114.
3. Buckey J.C. Jr, Gaffney F.A., Lane L.D., Levine B.D., Watenpaugh D.E., Wright S.J., Yancy C.W. Jr, Meyer D.M., Blomqvist C.G. Central venous pressure in space. J Appl. Physiol. 1996 :81(1):19-25.
4. Cavanagh PR, Davis BL, Miller TA. A biomechanical perspective on exercise countermeasures for long-term spaceflight. Aviat. Space Environ. Med. 1992; 63:482-5.
5. Convertino VA. Physiological adaptations to weightlessness: effects on exercise and work performance. Exer. Sport Sci. Rev. 1990; 18:119-66.
6. Charles, J. B., Bungo, M. W. & Fortner, G. W. Cardiopulmonary function. In: Space Physiology and Medicine, ed. Nicogossian A. E., Leach H. C. & Pool S. L., 3rd ed., Lea & Febiger, Philadelphia.1994; 286-304.
7. Cooke W.H., Ames J.E., Crossman A.A., Cox J.F., Kuusel T.A., Tahvanainen K., Levine B.D., Moon B., Blomqvist C.G., Eckberg D.L. Nine months in space: the effects on human autonomic cardiovascular regulation. J. Appl. Physiol. 2000; 89:1039-1045.
8. Cowley, A.W., Jr. Long-Term control of arterial blood pressure. Physiol. Rev. 1992;72: 231–300.
9. Fregly M.J., Blatteis C.M. (Editors) Environmental Physiology . Handbook of Physiology, 1996; Section 4: 1626 p.
10. Gauer O.H., Thron H.S. Postural changes in the circulation. In: Handbook of Physiology.Sec.2.- 1965;2409-2439.
11. Grygoryan R.D. Mathematical principles for analysis of quantum regularities in multi-component objects of biological nature. Reports of Academy of Sciences of USSR, Biophysics. 1990; 314:745-748.
12. Grygoryan R.D. Modeling the interaction of mechanoreflexes in the zones of high and low pressure. Cybernetics and Computing Technology. N. Y., Allerton press. 1990;86:92-96.
13. Grygoryan R.D. Self-organization of adaptation and homeostasis. Academperiodika, Kiev. 2004: 1-501.
14. Grygoryan R.D., Lissov P.N. Internal originators of functions fluctuation in multi-cellular organism. Bioelectromagnetics. Current Concepts. Ed. Ayrapetyan S. & Markov M. Springer. Netherlands-2006; 423-430.
15. Kirsch K.A., Rocker L., Gauer O.H., Krause R. Venous pressure in man during weightlessness. Science; 1984;225:218-219.
16. Kostyuk P.G., Lukyanets E.A. Intercellular calcium signaling – basic mechanisms and possible alterations. Bioelectromagnetics. Current Concepts. Ed. Ayrapetyan S. & Markov M. Springer. Netherlands.2006;87-122.
17. Masoro E.J. (Editor) Aging (1995).Handbook of Physiology, Section 11: .-696 pp.

18. Morey-Holton ER, Whalen RT, Arnaud SB, Van der Meulen MC. The skeleton and its adaptation to gravity. In: Handbook of Physiology: Section 4, Environmental Physiology, edited by Fregly MJ and Blatteis CM. III: The gravitational envionment, volume1: Microgravity. New York: Oxford University Press, Chapt. 31, 1996; 691-719.
19. Newman D.G., Callister R.The effect of baroreflex adaptation on the dynamic cardiovascular response to head-up tilt. Aviat Space Environ Med; 2000;71:255-259.
20. Poyton, R.O., R.P. Dirmeier, K. O'Brien, and E. Spears. A role for the mitochondrion and reactive oxygen species in oxygen sensing and adaptation to hypoxia in yeast. In: Oxygen Sensing: Responses and Adaptation to Hypoxia (ed. S. Schmidt - Nielsen K. (1987) Animal Physiology: Adaptation and Environment. Third Edition. Cambridge UP 2002:619p.
21. Tipton C.M., Hargens A.R. Physiological adaptations and countermeasures associated with long-duration spaceflights. Med.Sci.Sports Exerc.1996; 28:974-976.

Adaptation Biology and Medicine (Volume 5: Health Potentials)
Editors: L. Lukyanova, N. Takeda and P.K. Singal

20. Nitric Oxide (NO) in Modulating Hypoxia Signal Transmission

Bernhard Brüne*, Jie Zhou and Steffen E. Schnitzer

Institute of Biochemistry I, Faculty of Medicine, Johann Wolfgang Goethe-University, 60590 Frankfurt, Germany

Abstract

The ability to sense and respond to changes in oxygen partial pressure represents a fundamental property to assure the cellular oxygen supply to be within a narrow range that balances the risks of oxidative damage vs. oxygen deficiency. The discovery of hypoxia-inducible factor-1 (HIF-1) allowed identifying molecular mechanisms by which changes in oxygenation are transduced to adequate intracellular adaptive responses. The HIF-1 transcriptional system senses decreased oxygen availability and transmits this signal into patho-physiological responses. Among recent advances are the discoveries that nitric oxide (NO) participates in stability regulation of HIF-1α and HIF-1 transactivation during normoxia. Mechanistically, reactive nitrogen intermediates block prolyl-hydroxylase activity, thus attenuating proline hydroxylation of HIF-1α which causes its dissociation from pVHL (von Hippel Lindau protein) with the consequence of protein accumulation due to decreased ubiquitination and proteasomal degradation. Thus, NO mimics a hypoxic response under normoxia. Moreover, we appreciate that hypoxia initiates cell demise by apoptosis/necrosis but also prevents cell death by provoking adaptive responses that, in turn, facilitate cell proliferation or angiogenesis, thereby contributing to tumor progression. We provide evidence that cell adaptive responses to hypoxia attenuate pathways of cell demise, i.e. apoptosis by using HIF-1-dependent but also -independent signaling. HIF-1-independent transducing pathways comprise, among other signals, expression of cyclooxygenase-2 (COX-2). Understanding mechanisms that contribute to HIF-1 stability regulation and pathways how hypoxia affect routs of cell demise will help to uncover the crucial role of oxygen in the decision-making processes of regulating cell death, proliferation and adaptation.

Key words: HIF-1, apoptosis, chemoresistance, tumor, p53, Bcl-2, cyclooxygenase-2, prostanoids, mitochondria

The Oxygen Sensing System

Cellular recognition and appropriate responses towards decreased oxygen availability, i.e. hypoxia is primarily facilitated by the hypoxia-inducible factor (HIF). HIF is a heterodimer, composed of one of the three alpha subunits (HIF-1α, HIF-2α or HIF-3α) and one HIF-1β subunit. Whereas HIF-1β is constitutively expressed and identical to the aryl hydrocarbon receptor (AhR), known as AhR nuclear

*Address of Correspondence: Johann Wolfgang Goethe-University, Faculty of Medicine, Biochemistry I, Theodor-Stern-Kai 7 60590 Frankfurt, Germany. Phone +49-69-6301 7423, Fax +49-69-6301 4203, E-mail: bruene@zbc.kgu.de

translocator (ARNT), the α-subunit (as implicated by its name) is hypoxia responsive. Activation of HIF is achieved by stabilization and/or expression of the alpha subunit, e.g. HIF-1α. Research during the past decade provided insights into HIF-1 stability regulation, expression and transactivation as well as (patho)physiological consequences linked to overexpression of HIF-1α (1-3). Under normoxia the HIF-alpha subunits are usually unstable and virtually undetectable at the protein level. Oxygen-demanding prolyl-4-hydroxylases (designated as PH domain-containing enzymes; PHD) hydroxylate the oxygen-dependent degradation domain (ODD) of HIF at proline residues 402 and 564. As PHDs demonstrate an unusual high K_m for oxygen, which guarantees that small changes in oxygen supply affect enzyme activity, make this system an ideal oxygen sensor. Hydroxylated HIF forms hydrogen bonds with pVHL (von Hippel Lindau protein) side chains, which promotes polyubiquitination of HIF, followed by 26S-proteasomal degradation. Consequently, hypoxia attenuates Pro564/402 hydroxylation, which stabilizes the α-subunit, provokes its association with HIF-1β as well as cofactor recruitment of e.g. p300/CBP. In turn, a functional HIF complex causes expression of those targets that contain HRE (hypoxia-responsive element) sites with the core DNA sequence 5'-ACGTG-3'.

To date more than 70 HIF target genes have been identified (1). Critical adaptive responses affected by HIF target genes are erythropoiesis, angiogenesis and glycolysis, which are examples of systemic, local and cellular responses to hypoxia. In addition, target genes include also some, which are associated with cell proliferation and cell survival as well as cell death via apoptosis. An incomplete list of examples comprise EPO (erythropoietin), VEGF (vascular endothelial growth factors), IGF2 (insulin-like growth factor-2), TGFβ (transforming growth factor-β), IGF-BPs (insulin-like growth factor-binding proteins), MDR1 (multi drug resistance 1), BNIP3 (Bcl-2/adenovirus E1B 19 kDa interacting protein 3), NIX (Nip3-like protein X), NOXA (NADPH oxidase activator) or BNIP3L (Bcl-2 19 kDa interacting protein like) (for references see 4, 5). Included one finds genes affecting pro-survival as well as pro-apoptotic responses to hypoxia, which creates a paradox on initiation of proliferate vs. pro-death pathways and points to the Janus face of oxygen deprivation. Figure 1 provides an overview of the HIF-regulatory system.

Nitric Oxide (NO) in (de)stabilizing HIF-1α

Independent lines of research in a variety of human, pig or bovine cells confirmed that nitric oxide (NO) stabilizes HIF-1α and cause transactivation of HIF-1 under normoxia (for references see 6). This was further proven by activation of the human VEGF promoter by NO under normoxia with the notion that deletion as well as mutation analysis of the VEGF promoter indicated that the NO-responsive cis-elements were the HIF-1 binding site and an adjacent ancillary sequence located immediately downstream within the HRE (7, 8). To address the biological significance of endogenously produced NO in activating HIF, human inducible NO synthase (iNOS) was overexpressed which succeeded in accumulating HIF-1α (9). Supporting evidence came from a transwell co-culture setup of lipopolysaccharide/interferon-γ activated and thus NO producing macrophages and tubular LLC-PK_1 detector cells showing that NO generated by activated but not resting macrophages, provoked a HIF-1α response in LLC-PK_1 cells (10). These experiments suggest that autocrine or paracrine produced NO stabilizes HIF-1α under normoxia and provoke questions on mechanistic explanations considering those concepts, established for hypoxia. In analogy to hypoxia NO decreased ubiquitination of HIF-1α and interrupted the binding of pVHL to HIF-1α (11) which

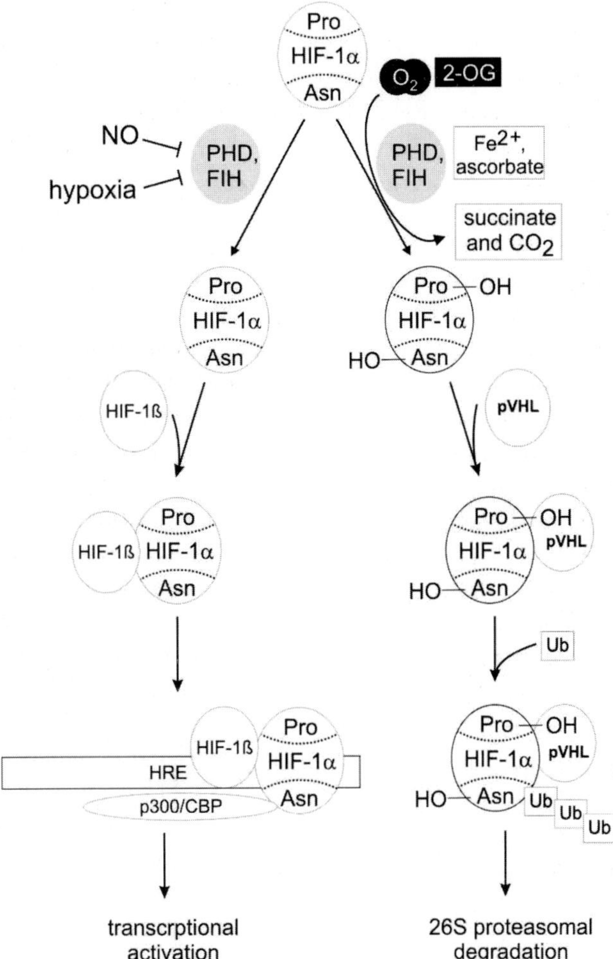

Fig. 1. HIF-1α degradation versus its activation. Prolin hydroxylase (PHD) and asparagin hydroxylase (FIH) are active under normal oxygen supply (pO_2). PHD and FIH, together with the cofactors oxoglutarate (2-OG), O_2, Fe^{2+} and ascorbate hydroxylate HIF-1α. Prolyl-hydroxylated HIF-1α binds to von Hippel Lindau Protein (pVHL), is further ubiquitinated (Ub) and finally degraded by the 26S-proteasome. Low oxygen availability (hypoxia) or NO inhibit PHD to accumulate HIF-1α. Stabilized HIF-1α associates with HIF-1β to recognize the hypoxia response element (HRE) in the promoter of target genes. The co-activator p300/CBP binds only to the HIF-1 complex when Asn803 hydroxylation by FIH is attenuated to support gene activation.

raises the possibility that NO directly attenuates hydroxylation of HIF-1α. An in vitro HIF-1α-pVHL capture assay showed dose-dependent inhibition of PHD activity by the NO donor GSNO which implies that hypoxia and NO use overlapping signaling pathways to attenuate PHD activity.

It is known that NO interacts with iron (II) in heme- or non-heme containing proteins (12), exemplified by spectroscopic studies when NO directly coordinates the ferrous iron in protocatechuate 4,5-dioxygenase, catechol 2,3-dioxygenase (13) or in isopenicillin N synthase (14).

These enzymes coordinate Fe^{2+} in their catalytic site in a 2-histidine-1-carboxylate facial triad which is the defining structural motif of mononuclear non-heme iron(II) enzymes (15). Taking into account that PHDs belong to a non-heme Fe^{2+}-containing family of enzymes, it is rational to propose Fe^{2+}-coordination by NO in the catalytic site of PHD to explain enzyme inhibition. However, pulse labeling studies in the presence of the NO donor NOC-18 implied increased HIF-1α synthesis rather than inhibition of PHD activity as the NO-mechanism (16). Interestingly, when a GST-HIF-1α(429-608) fusion protein was incubated with in vitro translated pVHL in the presence of cell lysates from untreated or NOC-18-exposed cells, NOC-18 did not affect the HIF-1α-pVHL interaction. In contrary, lysate from GSNO-treated cells significantly inhibited the interaction of GST-HIF-1α with pVHL, whereas lysate from (sodium nitroprusside) SNP-treated cells dramatically increased the interaction. This opens the possibility that different redox species derived from chemically distinct NO donors use divergent transmission systems to stabilize/express HIF-1α. Indeed, NOC-18 was reported to use a PI3K, MAPK and cap-dependent translation control system to express HIF-1α (16).

Pathways provoking HIF-1α expression are used by a number of stimuli including growth factors, oncogenes and inflammatory mediators (for references see 2, 17). Considering the proposed role of PI3K in HIF-1α translational regulation it may turn out that in some cells PI3K/Akt is stimulated by NO, which in turn will provoke HIF-1α translation. A similar scenario has also been notice for hypoxia (18). Under conditions where PI3K/Akt stimulation by hypoxia and/or NO occurs, translational control mechanisms may overlap with pathways that attenuate HIF-1α destruction to enhance HIF-1α protein appearance.

Keeping in mind, that the reactivity of NO is affected by the biological milieu it appear logic that HIF-1α stabilization by NO is under the control of the redox environment, i.e. the rate of superoxide formation, especially considering the near diffusion-controlled interaction between NO and O_2^-. Experiments with the redox cycler DMNQ (2,3-dimethoxy-1,4-naphthoquinone) to generate O_2^- and/or H_2O_2 (derived from superoxide dismutase-triggered conversion of O_2^- to H_2O_2) attenuated NO-elicited HIF-1α accumulation (19, 20). In human cerebral vascular smooth muscle cells SNP and SIN-1 inhibited HIF-1α stabilization by GSNO and NOC-18 (21). The conclusion that inhibition were mediated by ROS, particularly $ONOO^-$, was supported by experiments showing that superoxide dismutase (SOD) overcame the inhibitory effects of SNP/SIN-1 and by showing that SOD induced HIF-1a in the absence of hypoxia. These observation predict that the ability of NO to stabilize HIF-1α depends to some extend on the generation of cosignals, i.e. superoxide. However, it remains open whether O_2^- simply reduces effective concentrations of NO and thereby eliminates the stimulus or whether O_2^-, irrespective from its interaction with NO, directly destabilizes HIF-1α.

Moreover, Huang et al. showed the contribution of the ODD of HIF-1α in affecting protein destabilization by NO (22). Lateron studies by Hagen and coworkers revealed that expression of P402A/P564A-HIF-1α, a protein which lacks the hydroxylation and thus destabilization sites, is resistant to destabilization by DETA-NO under hypoxic conditions (23). It was concluded that inhibition of respiration by NO promotes HIF-1α degradation in hypoxia, which implies that increased availability of nonrespiratory oxygen consequently reactivates PHD activity (23). Thus, the ability of NO to suppress HIF-1α under hypoxia depends on inhibition of mitochondrial respiration, since it is absent in p^0-cells but is mimicked by inhibitors of mitochondrial respiration (24) and could be explained by a general decrease of oxygen respiration by respiratory chain inhibitors (25). In addition, the modulation of PHD cofactors such as iron, ascorbate and 2-oxoglutarate may affect the efficacy of hydroxylation and therefore may contribute to altered stability regulation of HIF-1α (26-28). It is known that the addition of ferrous iron antagonized hypoxia- as well as DFX-elicited HIF-1 responses (29) and that a chronic exposure of cells to NO donors was accompanied by an increase in cellular S-nitrosothiol levels and an increase in the labile iron pool (30). This suggests that maneuvers, which increase free iron inside cells might decrease HIF-1α (31, 32).

Hypoxia and Cell Demise

The role of the tumor suppressor p53 as 'guardian of the genome' attracted much interest in coordinating responses towards hypoxia. However, there are conflicting reports whether and to what extend p53 accumulates under hypoxia and if it becomes transcriptionally active. Early observations that p53 induction is HIF-1α dependent (33) were challenged by showing that hypoxic conditions, sufficient for HIF-1α accumulation, were unable to accumulate p53 (34). Moreover, even when p53 is hypoxia inducible it was reported to remain transcriptionally inactive (35). An explanation on these contrasting observations might be that acidosis/nutrient deprivation in combination with hypoxia allowed p53 induction in some cells whereas hypoxia (0.02% oxygen) alone was not sufficient to do so (36). Generally, it can be concluded that p53 accumulation does not overlap with HIF-1α stabilization in response to hypoxia. Instead, p53 stabilization requires conditions of severe hypoxia, close to or equivalent to anoxia, for prolonged periods, probably in association with acidosis (for references see 37). Nevertheless, accumulation of p53 inhibits HIF-1 transcriptional activity and/or lowers the protein content of HIF-1α, although the amount of p53 required to degrade HIF-1α was higher than that needed for p53 to activate its target genes. Mechanistically, low expression of p53 may compete with HIF-1 for the transcriptional coactivator p300 to attenuate HIF-1 transactivation while a high rate of p53 expression may facilitate HIF-1α destruction (for references see 37). This scenario might have implications for hypoxia/anoxia-induced cell death. In the early phase of hypoxia, transactivation of HIF-1 could serve to protect cells from destruction. With time progressing, inhibition of HIF-1 via p53 and further induction of p53 may lead to HIF-1α destruction thereby contributing to cell demise by repressing protective principles and promoting apoptotic ones.

To date a number of p53 target genes have experimentally been proven to execute p53-dependent genotoxic-evoked cell death. However, none of them has been shown to mediate p53-dependent cell death under hypoxia and indeed p53 appears to be incapable of upregulating its established targets during hypoxia, due to a failure to recruit transcriptional coactivators such as p300/CBP (35). Fei and coworkers report that p53 can directly upregulate expression of BNIP3L, a cell death inducer during hypoxia (38). Reduced oxygen (0.1%) provoked BNIP3L upregulation in wild-type p53 cells, in part by recruiting p53 and p300/CBP to the BNIP3L promoter with the notion that hypoxic death is reduced following BNIP3L knockdown. The model implies that BNIP3L functions as a mediator of p53-dependent apoptosis under hypoxia. Further studies need to determine the spatial and temporal interaction between HIF-1, p300/CBP, p53 and its target genes related to hypoxia-evoked cell death and to elucidate how changes in modulators such as pH alterations or nutrient supply contribute to cell demise (36).

BNIP3, NIX and NOXA are pro-apoptotic BH3-only members whose gene and protein expression are hypoxia inducible (for references see 5). BNIP3 regulation under hypoxia might be facilitated via a HRE located in its promoter region. Although BNIP3 expression under hypoxia is noticed in a wide range of cell lines and tumor material, a direct contribution to cell death remains unclear. Often, it fails to induce cell death because pathways of cell demise are antagonized by growth factors which are under the control of HIF. Also, the role of BNIP3 in coordinating a pro-apoptotic cell response towards hypoxia must be approached with caution taking into account that in cardiac myocytes hypoxia induces BNIP3 but acidosis was required to execute cell death (39). Using molecular approaches, the expression of other BH3-only proteins such as the Nip3-like protein (NIX) and NOXA were identified as hypoxia-inducible. It will be the task of future studies to define the relative contribution of the above mentioned proteins in hypoxia-evoked cell death, to determine the precise role of p53 and to critically evaluate the expression of pro-cell death proteins relative to the appearance of HIF-1α and/or HIF-1 signaling. This might clarify whether a direct correlation to HIF-1α expression exists or whether cell death under hypoxia will likely be compounded by other factors such as acidosis.

Pro-survival Actions of Hypoxia

Initial experiments made use of mouse hepatoma cells (Hepa-1) that are wild type or deficient (C4-cells) for the HIF binding partner ARNT. Grafting C4-cells onto athymic mice produced slower growing tumors that showed impaired vascularization compared to wild type cells. Moreover, solid tumor growth in nude mice using HIF-1α-deficient ES cells (embryonic stem cells) is impaired due to diminished adaptation and lack of VEGF upregulation. These results imply that HIF-1 is needed for tumor angiogenesis and/or adaptive responses to sustain tumor growth. HIF activity seems crucial for many aspects of cancer biology (3, 4) and overexpression of HIF-1α has been associated with increased patient mortality. However, the effect of HIF-1α overexpression is dependent on the cancer type with the presence or absence of genetic alterations that affect the balance between pro-apoptotic and pro-survival factors.

A number of studies supported a pro-survival role of hypoxia. When HIF-1α expression had been attenuated by an siRNA approach in HeLa, HCT116 or human umbilical vascular endothelial cells (HUVEC) cells, hypoxia-induced apoptosis was enhanced. In human tongue squamous cell carcinomas (SCC) the treatment with HIF-1α siRNA under normoxia or hypoxia attenuated cell proliferation and induction of apoptosis was associated with increased caspase-activity, decreased expression of Bcl-2 and IAP, but not with p53. Besides the idea that HIF upregulates pro-survival proteins, HIF-1α was also shown to downregulate pro-death protein such as Bid, consistent with a HRE in the *bid* promoter (40). The relative resistance of normoxic cells that showed reduced expression of Bid and the finding that forced expression of Bid in hypoxic cells resulted in increased sensitivity to etoposide demonstrated the contribution of Bid downregulation to drug responsiveness (41).

Attenuated apoptosis under hypoxia correlated with increased expression of Bcl-2, Bcl-X_L and IAP, and/or decreased activity of caspase-3. While the expression of e.g. IAP-2 was HIF independent, other anti-apoptotic regulators such as Mcl-1 responded to hypoxia because of a HRE site. Furthermore, an anti-apoptotic role of HIF-1α was demonstrated when VEGF neutralizing antibodies blocked protective effects of hypoxia on HepG2 cells. It was also shown, that cancer cells secret molecules that inhibit endothelial cell apoptosis and conditioned media obtained from hypoxic adipose stromal cells increased endothelial cell growth and reduced apoptosis. Several HIF targets include growth factors, angiogenic and anti-apoptotic factors such as VEGF, IGF2, TGFβ or PDGF. Their receptor interactions initiate signal transmission that provokes HIF-1α expression and cell survival/proliferation (for references see 4). HIF therefore initiates autocrine signaling that favors pro-survival conditions.

A more indirect pathway to protect from cell death under hypoxia may be associated with activation of nuclear factor κB (NF-κB). Activation of NF-κB has been known for some time as an integral signaling circuit that results in protection from cell demise. In RCC4 cells, NF-kB is constitutively active and its inhibition is required for sensitization towards apoptosis (42). Interestingly, NF-κB has been identified as a novel hypoxia-regulated HIF gene, at least in neutrophils, which suggests that expression of HIF-1α itself can affect NF-κB activity and in turn determine hypoxic survival via NF-κB signaling (43). These considerations imply that under hypoxia multiple pathways operate to attenuate cell death, which provokes the question whether HIF-independent ones synergize with those, which relay on HIF. In A549 cells staurosporine- or etoposide-induced apoptosis was reduced by roughly 70% when cells were preincubated for 24 h under hypoxia (0.5% O_2) (44). Knockdown of HIF-1α by siRNA indicated that besides HIF-1 other protective pathways operate with the further notion that blocking cyclooxygenase-2 (COX-2) with NS398, reduced protection to the same extent as knockdown of HIF-1α. Moreover, a COX-2-dependent but HIF-1α-independent secreted factor transferred chemoresistance to cells even when cultured under normoxia, which suggests that HIF-1- and COX-2-dependent pathways operate in parallel to convey resistance in hypoxic A549 cells.

The interplay between COX-2 and HIF-1 is not finally clear. It is known that IL-1β activates NF-κB signaling which accounts for COX-2 and subsequent HIF-1α expression under normoxia (45). Induction of COX-2 under hypoxia has also been reported although its expression is either unrelated to HIF-1 signaling (46, 47) or HRE-mediated (48). Moreover, up-regulation of COX-2 under hypoxia has been attributed to NF-κB activation (49). High expression of COX-2 is found in many tumors and correlated to chemoresistance. Various human cell types such as macrophages, primary hepatocytes, neutrophils, thymocytes, and/or neurons acquire resistance towards apoptosis upon COX-2 induction (50, 51) Formation of prostaglandin E_2 as a COX-2-derived metabolite under hypoxia has been reported. Moreover, para- and/or autocrine signaling of PGE_2 with activation of adenylyl cyclase and concomitant cAMP generation is an established mechanism to regulate anti-apoptotic proteins such as inhibitors of apoptosis (52).

The use of conditioned medium revealed that a transferable factor(s) is produced under hypoxia, which can protect cells that never experienced hypoxic protection beforehand. Soluble survival factors produced under hypoxia are known and typically comprise HIF-1 target gene products such as VEGF, EPO, insulin-like growth factor and more (4). However, in other systems these molecules can be ruled out, as their production should have been sensitive to knockdown of HIF-1α during the generation of the conditioned medium and their production should not have been antagonized by NS398. Likely, several antiapoptotic mechanisms work in concert to assure efficient control of apoptosis. The expression of intracellular proteins may control receptor- and/or mitochondria-evoked proapoptotic pathways, whereas soluble antiapoptotic factors transmit the protective response to their neighborhoods. This makes sense in order to preserve the integrity of not only single cells but also of tissues organized in organ structures. Figure 2 provides a brief overview on hypoxia in mediating cell protection.

Hypoxia is commonly found in solid tumors of various origins. Selection by hypoxia renders tumor cells resistant to hypoxia induced apoptosis (53). Cells with a reduced apoptotic potential may also

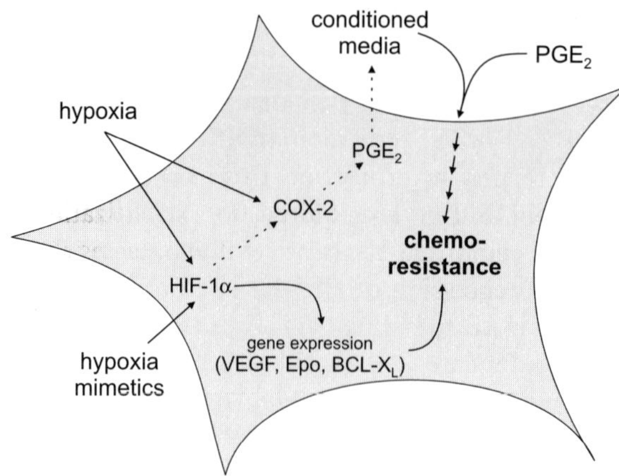

Fig. 2. HIF-1 and COX-2 in chemoresistance. Hypoxia induces HIF-1 and COX-2 chemoprotective pathways operating in parallel. Induction of HIF-1 induces gene expression of protective genes that act intracellular, whereas COX-2 mediates the secretion of 'survival molecules' acting para- and/or autocrine. Hypoxia mimetics, PGE_2 and conditioned media (obtained from hypoxic cells) protect cells form drug-induced apoptosis.

explain the resistance of many solid tumors to chemotherapeutics and the increased potential for invasion, metastasis and patient mortality (54). This view is supported by experimental and clinical data that show the association between increased HIF-1α expression and increased patient mortality. Therefore, anti-cancer drugs that target HIF-1 activity or signaling pathway that provoke HIF-1α stabilization may open new avenues to circumvent chemoresistance.

Concluding Remarks

We have learned that hypoxia activates a wide spectrum of molecular pathways to facilitate adaptation, chemoresistance and apoptosis. A number of variables, supporting as well as antagonizing cell survival decisions at the end determine the fate of cells towards hypoxia.

Some but not all survival mechanisms demand the expression of HIF, the dominant hypoxia responsive system. Therefore, we need to know precisely how HIF-1α is regulated, especially by NO as a bewildering molecule that expands its actions throughout (patho)physiology. It appears that conditions of severe hypoxia/anoxia initiate apoptosis. In contrast, cells often adapt to acute and mild hypoxia and survive. Preventing apoptosis can be explained by the expression of anti-apoptotic regulators either directly attenuating pro-death pathways via Bcl-2 family members or by indirect actions via e.g. the formation and release of growth factors. It appears that HIF-dependent as well as -independent regulatory circuits are interlinked with intracellular and/or autocrine signaling pathways. The dynamic range of cellular responses to decreased oxygen availability may first promote adaptation, which is associated with a pro-survival signal under otherwise unfavourable living conditions. Under severe hypoxia, close or equivalent to anoxia, cell death dominates, in part due to secondary stressors such as acidosis. Chemoresistance and clonal selection may develop as a response of opposing signals delivered by survival and death pathways that allow selection of cells that have a growth advantage either genetically or epigenetically determined. Understanding

Adaptation Biology and Medicine (Volume 5: Health Potentials)
Editors: L. Lukyanova, N. Takeda and P.K. Singal

21. Mitochondrial Proliferation as Adaptation Mechanism in Various Diseases

V.S. Sukhorukov*

Moscow Institute of Pediatrics & Children's Surgery

Summary

The mitochondrion is an important cell organelle known to play a significant role in health and disease. Its mal-adaptation has been highlighted in many clinical conditions. In this report I give evidence that the increase of mitochondrial proliferation is the nonspecific adaptation mechanism during functional tissue deficiency in various diseases.

Keywords: Mitochondrial proliferation, mitochondrial diseases, secondary mitochondrial disorders, diathesis, ragged red fibers, structural myopathies, adaptation

Introduction

The problem of mitochondrial pathology, which includes not only a range of infrequent genetical syndromes but also extensive bunches of the diseases accompanying with different variants of tissue hypoxia, attracts a great attention. Our Institute has been studying mitochondrial disorders for over 15 years. We receive children with mitochondrial diseases from all over Russia and neighboring countries. During this period we have collected considerable and significant data resulting in important conclusions, some of which we would like to share with you now.

Methods

To start with, a few words about the subject of our research and terminology. We believe that mitochondrial disorders should be divided into three groups:

1. Primary mitochondrial disorders (that have become classic within the recent two-three decades, Fig. 1) [1-4];
2. Secondary mitochondrial disorders (the list is actively growing due to recent scientific research, Fig. 2):
   a. Endogenic: accompanying certain diseases as secondary pathogenic links;
   b. Exogenic: mitochondrial disorders connected to external effects (poisons, ecopathogenes, medicines, alimentary insufficiency);
3. Energy deficient diathesis.

*Address for Correspondence: Vladimir Sukhorukov, Moscow Institute of Pediatrics & Children's Surgery, Taldomskaya, 2, Moscow, 125412 Russia. Tel. (495) 483 40 01; fax (495) 483 33 35, E-mail: vsukhorukov@pedklin.ru; vsukhorukov@inbox.ru

Kearns-Seyre syndrome	1972 (mitochondrial genesis assumed)
Complexes I - V deficiency	1974 -1984
MERRF syndrome	1980
MELAS syndrome	1984
Leber disease	1988 (mitochondrial genesis proved)
Pearson syndrome	1988 (mitochondrial genesis proved)
Coenzyme Q_{10} deficiency	1989
NARP syndrome	1990
mtDNA depletion	1991
«Diabetes-deafness» syndrome	1992
Alpers' Disease	1994 (mitochondrial genesis proved)
Inborn myodystrophy with mitochondria disorders	1998

Fig. 1. First descriptions of mitochondrial diseases

Mendelian diseases Hemochromatosis, Wilson-Konovalov, Batten, Hantington, Menkes, Lesch-Nyhan syndromes
Polygenic and multifactor conditions Ageing, diabetes , Parkinson', Alzheimer', Wernicke', Reye' diseases, atherosclerotic cardiosclerosis, migraines, hypercholesterinemia, cardiomyopathias, alcoholic myopathy, Burkitt' lymphoma, irritable bowel syndrome, gastroparesis and others
Autoimmune diseases Multiple sclerosis, lupus erythematosus, rheumatoid arthritis, thyrotoxicosis, primary biliary cirrhosis, Guillain-Barre´ syndrome
Exogenous factors Poisoning of: AZT, Fialuridine, Doxorubicin, Aminoglycosides, Amytal, Amphotericinum, Cyanides, Carbonic oxide, Lead, Mercury, N-Methyl-4-phenyl-1,2,3,6-tetrahydropyridine (MPTP-parkinsonism) Hypoglycemia, avitaminoses

Fig. 2. Secondary mitochondrial disorders

We can see that prevalence of the conditions connected with mitochondrial insufficiency is not limited to the hereditary syndromes caused by mutations of the genes directly responsible for mitochondrial proteins. Moderate disturbances of cell metabolism do not have to manifest themselves as independent diseases, however, they affect the characteristics of the course of other illnesses.

Some time ago Russian pediatricians actively studied the so-called diatheses as individual (constitutive) peculiarities of certain metabolic features that determine individual human physiology and pathological processes.

Two years ago, we had added a new variant to numerous forms of such diatheses, the so-called *energy deficient diathesis*, which represents the latent form of relative individual insufficiency of the cell energy status of an organism. This concept of energy deficient diathesis in humans has substantial theoretic background – successful studies of various forms of individual sensitivity to hypoxia in experimental animals (numerous works by L.D. Lukjanova and co-authors). Another background is our experience in studying the cell energy metabolic parameters during various dysfunctions in children from primary mitochondrial diseases through to weak secondary insufficiency accompanying various diseases (Fig. 3). Our preliminary experience suggests that energy deficient diathesis is spread widely enough (not less than 10-15% of all children) and imposes specific features on the adaptation reactions and duration of various diseases in those children. Thus, for example, children with energy deficient diathesis may suffer more frequently from respiratory diseases as compared to others, they may show specific regeneration processes (e.g., keloid type of wound scar forming), etc. Diagnosed energy deficient diathesis calls for corrected therapeutic approaches to treating such patients which is proved by the efficient use of energotropic medicines within common treatment practices. It also determines the acute importance of developing diagnostic criteria and new methods for treating patients with cell energy metabolism insufficiency as individual background of the disease in various areas of clinical medicine.

As of today, there is a big diagnostic experience in the world science to study mitochondrial disorders. One of the important methodic approaches to such diagnostics is the morphological research. Traditionally from the very first days of "clinical mitochondriology", since 1963, one of the key methods diagnosing mitochondrial diseases was the morphological study of the skeletal muscle tissue in biopsies material. First of all, it was possible due to the description of *ragged red fibers* (RRF) phenomenon. The latter is characterized by abnormal conglomerations of mitochondria in muscle fibers (Figs. 4, 5). Those conglomerations are easily visualized under certain histological coloring methods. Their presence in patients with hereditary mitochondrial diseases allowed many researchers to believe that those conglomerations are manifestations of mitochondrial mutations.

At the same time due to our own data and experience we stick to a different position.

The nature of a "ragged red fibers" (RRF) phenomenon in many respects is determined by a reparation potential of a mitochondria pool. We believe that mitochondria proliferation is, by its nature, an adaptation process. The increase in absolute numbers of those organelles allows the organism to cope with the quality defect of cell energy metabolism.

Let us dwell on the proofs of the statement:

1. Observing families that had one or more patients with primary mitochondrial diseases, we found that their relatives, who did not show any clinical manifestations of mitochondrial

Children with frequent respiratory infections
Children with nondifferentiated development delay
Non-hormonal hereditary syndromes with growth delay (Aarskog, Silver-Russell and others)
Connective tissue disorders (Marfan and Ehlers-Danlos syndromes, connective tissue dysplasia)
Tuberous sclerosis
Children with keloid scarring
Hydronephrosis

Fig. 3. Diseases with mitochondrial insufficiency revealed by author with colleagues (Moscow Institute of Pediatrics and Child Surgery).

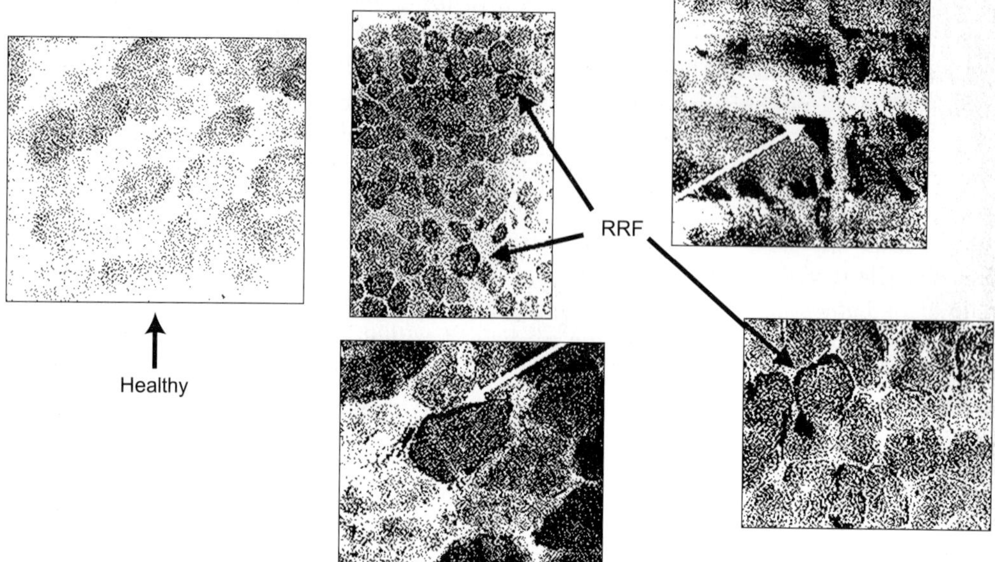

Fig. 4. Ragged red fibers in skeletal muscle of mitochondiral diseases patients (light microscopy histochemical revealing of succinic dehydrogenase activity).

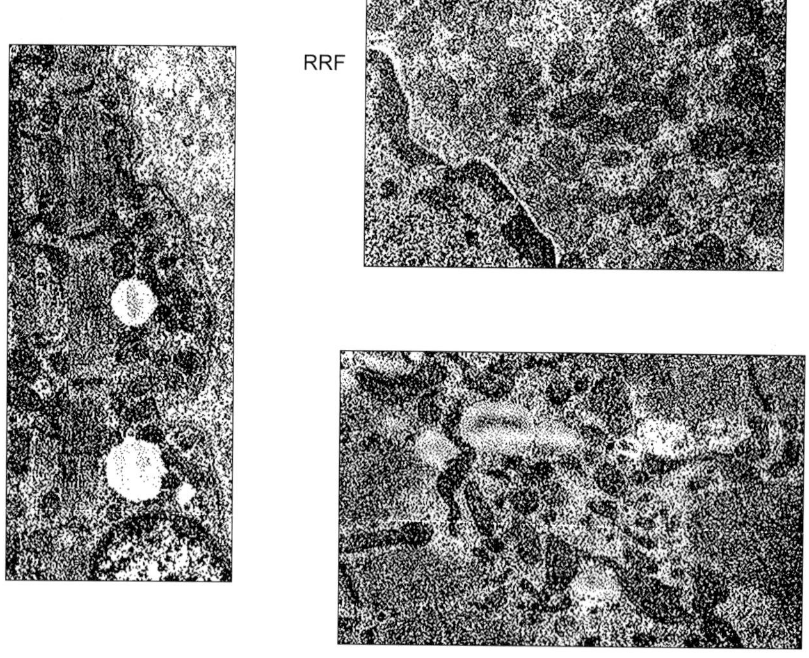

Fig. 5. Ragged red fibers in skeletal muscle of mitochondrial diseases pateints (electron microscopy).

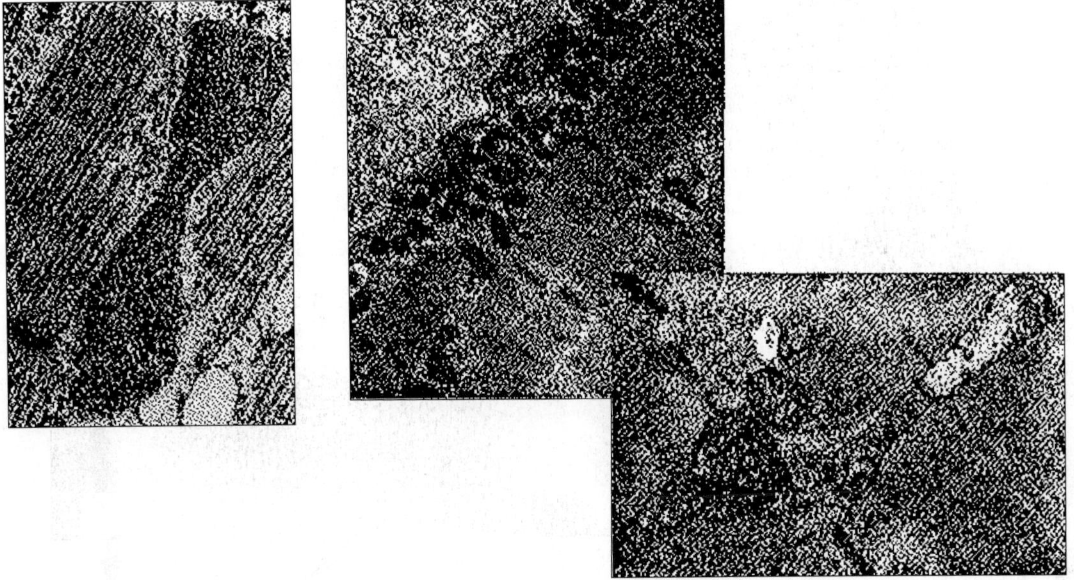

Fig. 6. **Mitochondrial proliferation in "healthy" mothers of mitochondrial patients (electron microscopy).**

deficiency, often might have a considerable amount of RRF, well exceeding the amount in patients themselves (Fig. 6). It was found in many mothers of sick children which is probably natural considering maternal inheritance of the mitochondrial DNA. Those mothers considered themselves absolutely healthy and only under extremely intent clinical observation, it was possible to them increased fatigue, muscle weakness and other signs of energy deficiency.

2. RRF is very often found with non-mitochondrial diseases among which hereditary diseases with well known primary mutation of the nucleus genome (for example, tuberose sclerosis, Marfan and Ehlers-Danlos syndromes) are frequent (Fig. 7). In such cases the accumulation of mitochondria can be hardly related to the mutation of mitochondrial DNA. The availability of abnormal mitochondrial proliferation in diseases, which are obviously not primarily connected with mitochondrial DNA mutation, allows to consider this phenomenon not so much as a marker of such mutations, but as a marker of functional mitochondrial insufficiency.
3. In certain children with structural myopathies (e.g., central core disease) mitochondrial masses are found in muscle bioptate (Fig. 8). According to our preliminary data the presence of such masses correlates with more compensated flow of the myopathy.
4. Another observation that we consider as one more proof, was several patients over forty years old showed for the first time myopathic symptomatic complex. By examining biopsies, we found morphologically straightforward evidence of structural (that is inborn!) myopathy in them. Besides these signs, the muscles also showed considerable abnormal subsarcolemmal masses of mitochondria (Fig. 9). For thirty to forty years, patients had no idea they were suffering with inborn myopathy. We believe that structural and functional defects in those cases

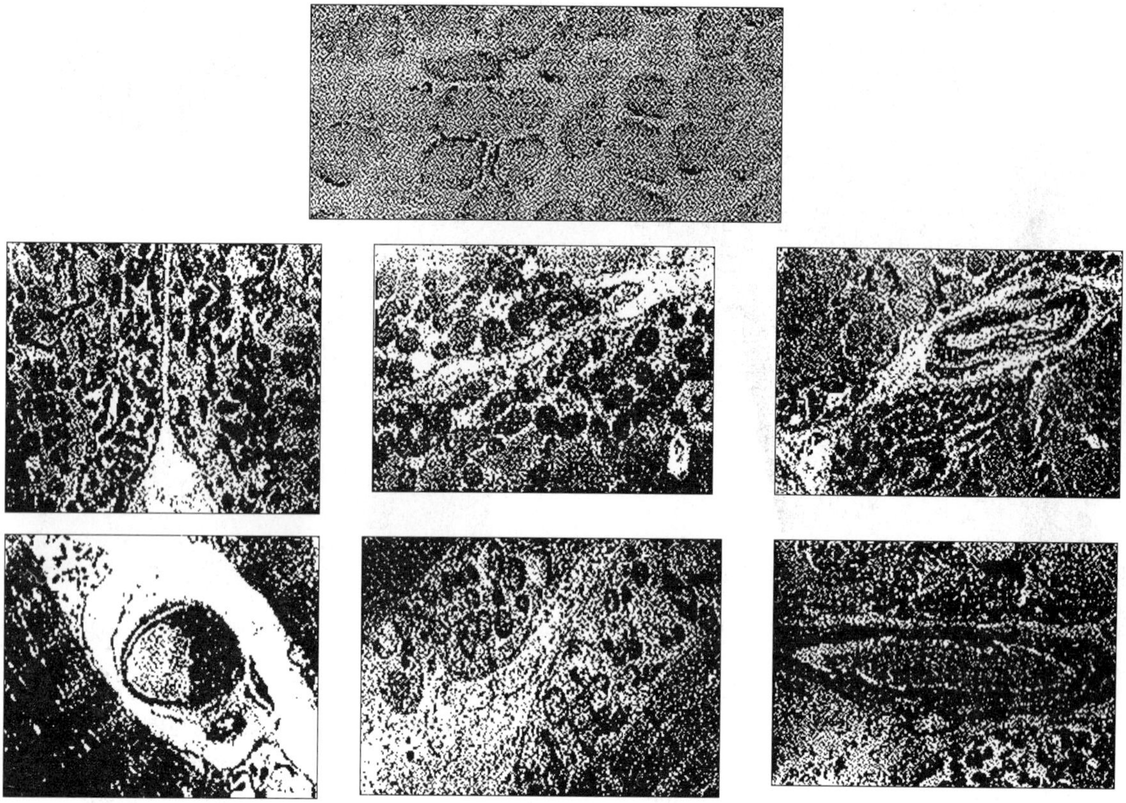

Fig. 7. **Mitochondrial proliferation in "nonmitochondrial" diseases - light microscopy histochemical revealing of succinic dehydrogenase activity (above) and electron microscopy.**

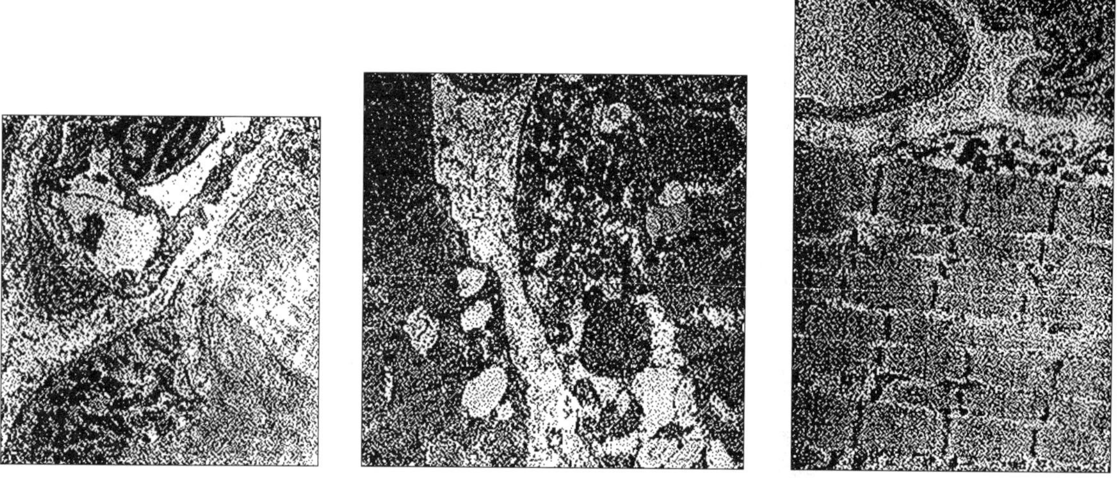

Fig. 8. **Mitochondrial proliferation in "nonmitochondrial" myopathies (electron microscopy).**

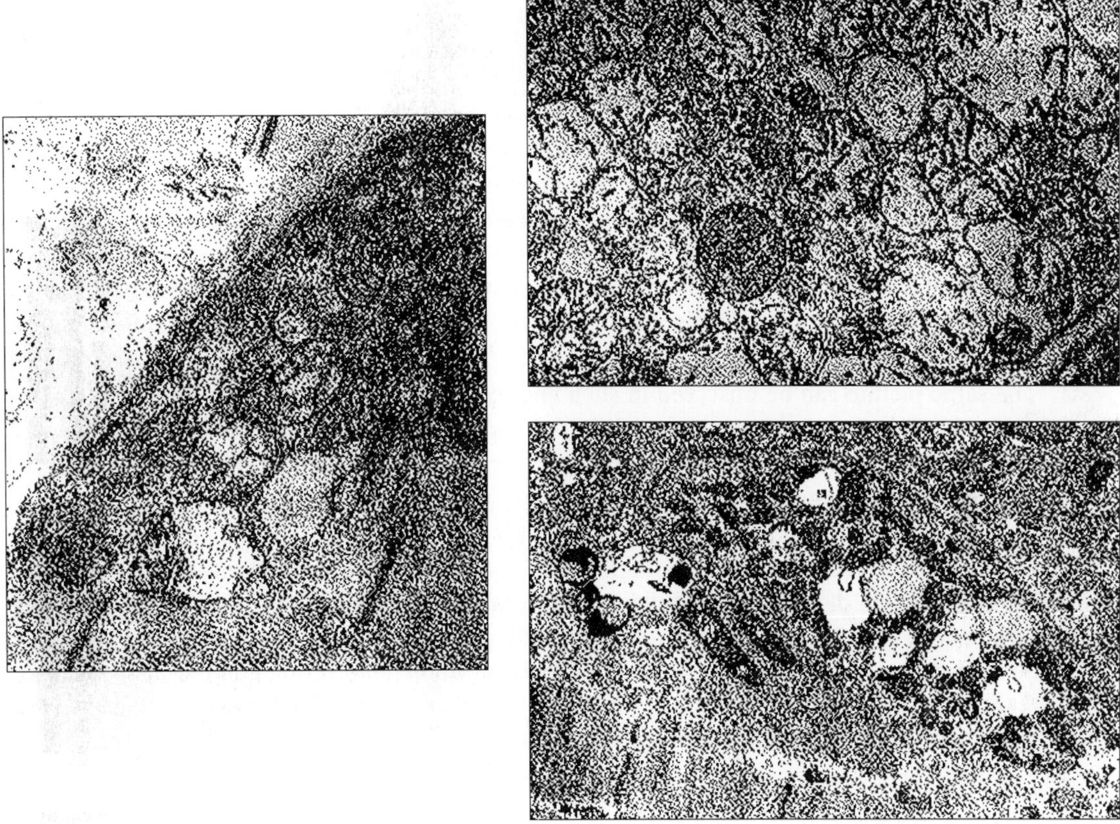

Fig. 9. Mitochondrial proliferation in congenital "nonmitochondrial" myopathies in adult patients (electron microscopy).

were efficiently compensated by mitochondrial proliferation and as a result, the activity of remaining contracting elements of the skeletal muscle was increased.

It should be stressed that according to our data the amount of mitochondria in muscles correlates reliably with age. Although strictly speaking the fact by itself does not prove our theory, we can explain the increase of those organelles by increased pressure on the compensatory cell potential. In this day and age it is a widely recognized fact that mitochondrial dys-adaptation takes part in the process of aging. There are data on the presence of point mutations of mitochondrial DNA specific for elderly age [5]. However, growing dysfunctions call for more strained adaptation processes. In that case, the increase in mitochondrial numbers in tissue elements is quite clear.

All the above proofs deal with what is observed in skeletal muscle tissue. As the spectrum of the tissue elements investigation widens, it shows exactly analogous (from our point of view, to be more precise, homologous) picture in them.

Let's have a look at a few examples:

1. In the smooth muscle tissue of urinary channels (hydronephrosis in children, prostate adenoma in elderly men and some other diseases), it is sometimes possible to observe the following morphological changes: on the background of generally decreased histochemical activity of mitochondrial ferments, one can see separate myocytes with dramatically increased content of the marker (Fig. 10). That is, on the background of the general tissue energy deficiency, there appear cells with drastically greater mitochondrial numbers (same situation as within the skeletal muscle). It is of interest that surgical treatment in such patients is much less efficient: the correction of macroscopic defect (obstruction or widening of the ureter) does not result in considerable restoration of the function as the smooth muscle tissue is functionally defective. And it is also clear that in this connection the adaptation potential of the mitochondrial proliferation must be used to its maximum.
2. The same picture can be observed in such energy dependant structures as kidney epithelium (Fig. 11). The kidney bioptate shown in the slide was taken from a patient suffering with hereditary nephropathy. Histochemical evidence of mitochondrial enzymes activity demonstrates the same picture as in the previous examples.
3. In case of muscle biopsies from a mitochondrial patient, a bit of skin is taken, then the basal layer of epidermis (where physiological cell dystrophy processes do not start yet) shows heterogeneity in distribution of cells with decreased and increased activity of mitochondrial ferments (Fig. 12). As in the presented examples, we talk in particular about succinic dehydrogenase (SDH) strictly connected to inner mitochondrial membrane, that means in reality the histo-chemical exposure of mitochondria themselves, which makes sure that we deal again in all probability with mitochondria proliferation.
4. The peripheral blood cells mitochondria behave more or less the same. The invasivity of muscle biopsy has been forcing us for a long time to look for an alternative method of mitochondria morphological research. As an alternative material, we tried to analyze various types of leukocytes. In Russia the cyto-chemical analysis of lymphocyte, certain ferments activity (including mitochondrial) has been used for a long time for various diagnostic

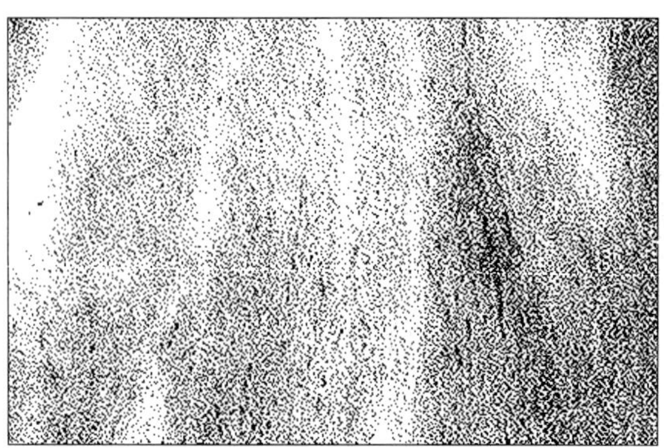

Normal distribution

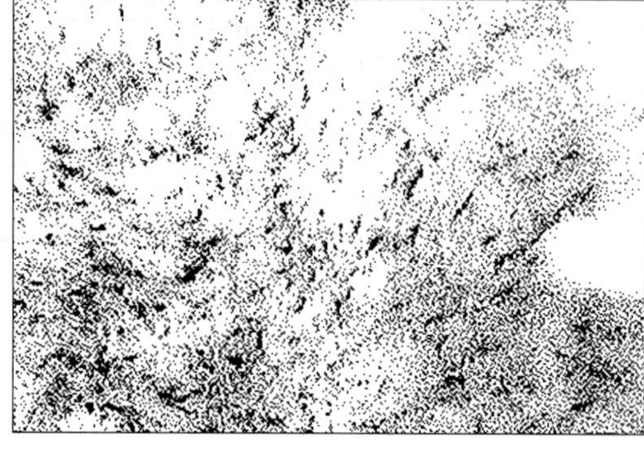

Pathological distribution

Fig. 10. Infringement of SDH activity distribution in smooth muscles (ureter in hydronephrosis) (light microscopy).

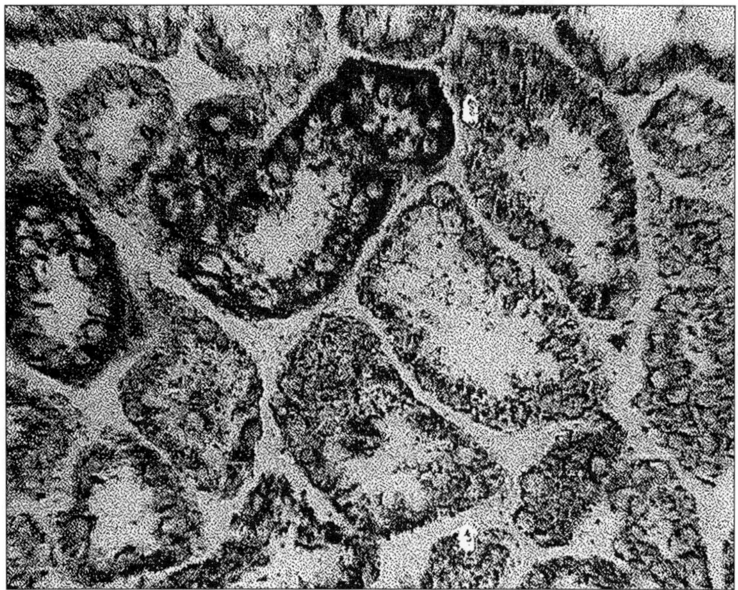

Fig. 11. SDH activity in kidney of a child with heredity nephropathy (light microscopy).

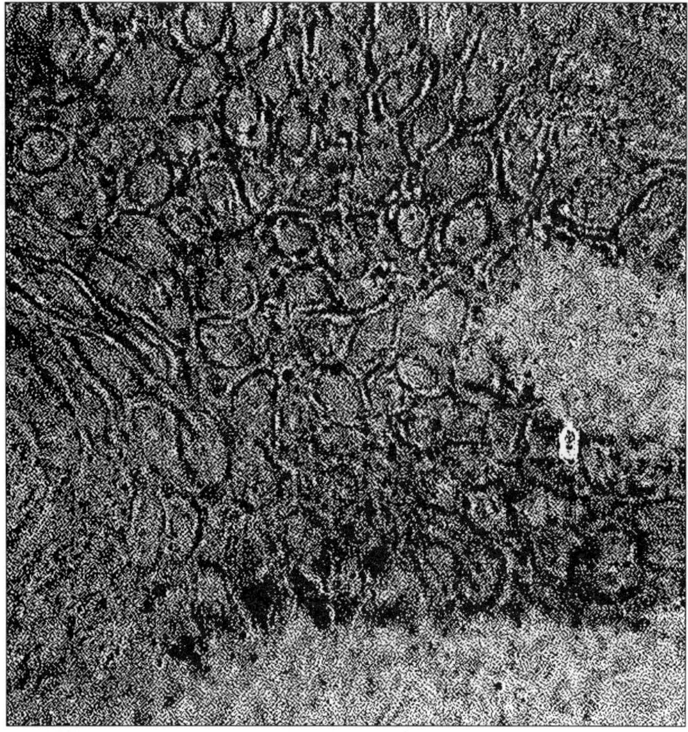

Fig. 12. SDH activity in epidermis of a child with MELAS syndrome (light microscopy).

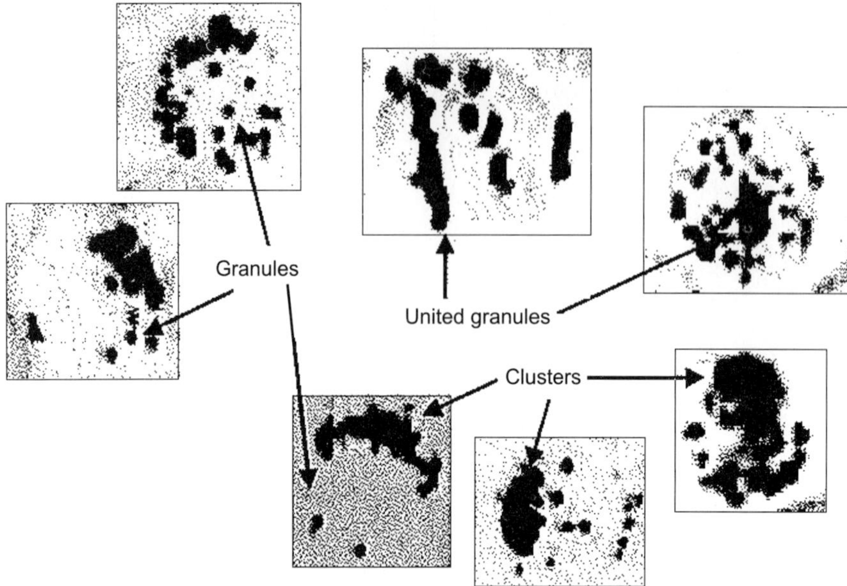

Fig. 13. Cytochemical analysis of mitochondrial enzymes activity in blood lymphocytes.

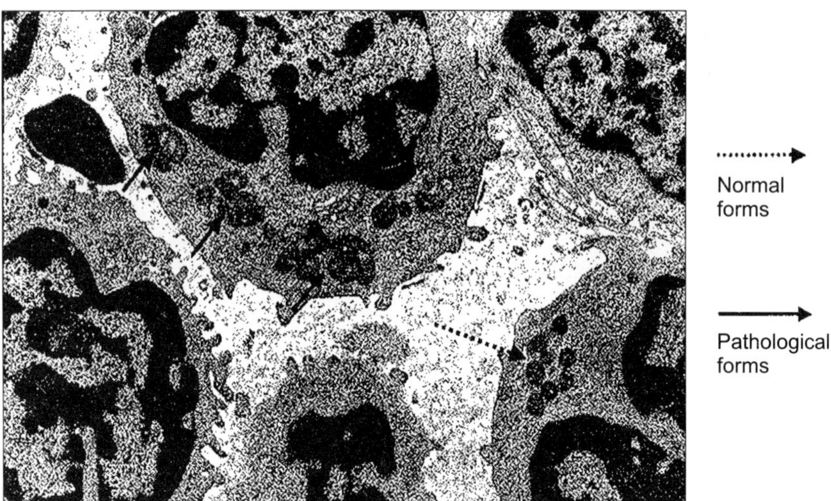

Fig. 14. Ultrastructural changes of lymphocytes mitochondria in energy deficient children.

purposes in a number of hospitals (Fig. 13). The credit goes to Prof. Nartsyssov and his works at the times no one heard about mitochondrial diseases. Back in 1999-2000 Prof. Nartsyssov and ourselves analyzed the cytochemical activity of mitochondrial ferments in neutrofiles, monocytes, lymphocytes (as well as trombocytes) in muscle biopsies taken from a large group of patients. The results turned out similar in various blood tissue elements, and only for the reason of technical convenience of working with lymphocytes, I will from now on refer only to them. Thus, the results of lymphocyte cyto-chemical analysis correlated with the data on the amount and quality of RRF skeletal muscle for over eighty per cent. The information value of this method increased dramatically when the visual analysis was backed by computer morphometry. The increased number of granules corresponding to the activity of certain mitochondria ferments is probably a result of the same cell reaction phenomenon at general cell energy metabolism deficiency in patients with polysystemic mitochondrial diseases. Further use of this method allowed to dramatically cut the number of diagnostic biopsies of muscle tissues and proved the diagnostic value of cyto-chemical analysis for exposing mitochondrial deficiency in various adult and children diseases. Last year to verify the data received by that sort of analysis, we conducted a series of electron microscopic lymphocyte research in children with polysystemic mitochondrial disorders. The results showed that under general energy deficiency disorders, mitochondria in those cells behave the same as in all other tissue elements described above. You can see corresponding results in the slide (Fig. 14).

We believe the above facts are a convincing proof of adaptation potential of proliferating mitochondria under functional tissue deficiency, including the cases not connected etiologically with primary mitochondrial defect. All this defines the acute importance of working out the methods for stimulating mitochondrial proliferation to solve the problem of increasing adaptation abilities of the organism.

References

1. Luft R., Ikkos D., Palmieri G., Ernster L., Afzelius B. A. A case of severe hypermetabolism of non-thyroid origin with a defect in the maintenance of mitochondrial respiratory control: a correlated clinical, biochemical, and morphological study. Journal of Clinical Investigation, 1962, 41:1776-1804.
2. Wallace D.C., Singh G., Lott M.T., et al. Mitochondrial DNA mutation associated with Leber's hereditary optic neuropathy. Science 1988, 242, 1427±1431.
3. Wallace D.C., Zheng X., Lott M.T., et al. Familial mitochondrial encephalomyopathy (MERRF): genetic, pathophysiological, and biochemical characterization of a mitochondrial DNA disease. Cell 1988, 55, 601±610.
4. Holt I.J., Harding A.E., Morgan Hughes J.A. Deletions of muscle mitochondrial DNA in patients with mitochondrial myopathies. Nature 1988 331, 717±719.
5. Attardi G. Role of mitochondrial DNA in human aging. Mitochondrion 2002; 2: 27-37.

Adaptation Biology and Medicine (Volume 5: Health Potentials)
Editors: L. Lukyanova, N. Takeda and P.K. Singal

22. Increase in Resistance of Membrane Structures in Adaptation to Changing Oxygen Level

Tatyana G. Sazontova[1], Natalya A. Anchishkina[2], Anna G. Zhukova[2], Elena N. Tkachuk[3] and Yuriy V. Arkhipenko[1]*

[1]M.V.Lomonosov Moscow State University; [2]Institute of General Pathology and Pathophysiology, Russian Academy of Medical Sciences; [3]Hypoxia Medical Academy; Moscow, Russia. [2]**N.A.Anchishkina,** Ph.D. student **and A.G.Zhukova,** Ph.D., Sci.D.,Institute of General Pathology and Pathophysiology, Baltiyskaya 8, Moscow 125315, Russia, tel. +7(495)932-881. [3]**E.N.Tkachuk,** Ph.D., Hypoxia Medical Academy, Ivankovskoe shosse 3, Moscow 123367, Russia, tel. +7(495)913-2999; E-mail: user@hypoxia.msk.ru. [1]**Yu.V.Arkhipenko,** Ph.D., D.Sc.,Faculty of Fundamental Medicine, M.V.Lomonosov Moscow State University, Lomonosovsky pr., 31, korp. 5, Moscow 119192, Russia, tel. +7(495)932-8814; E-mail: arkhipenko@fbm.msu.ru.

Abstract

The principle of intermittent adaptation effects of changing oxygen level (short-term hypoxia and hyperoxia) has been substantiated and confirmed experimentally for the first time. Adaptation of male Wistar rats was carried out in different modes: 1) hypoxia-normoxia (HP+N) (5-min inhalation of a gas mixture with 10% O_2, 3-min intervals of normoxia; 1 h daily for 15 days); 2) hyperoxia-normoxia (HO+N) (the same protocol, but the O_2 content in the inhaled air was 30%) and 3) a new training regime (HP+HO): 5-min hypoxia (10% O_2) + 5-min hyperoxia (30% O_2), 1 h daily for 15 days. Short-term (15 days) adaptation to HP+N had a protective effect on brain cortex membranes. However, in contrast to long-term (23 days) HP+N, short-term adaptation was concomitant with increased sensitivity of liver cells to free radical oxidation with simultaneous reduction of antioxidant protection components. The use of HO+N caused similar, although less pronounced changes, whereas combined application of HP+HO had a membrane-stabilizing action on the both organs, this effect being more apparent in comparison with other adaptation regimes. In the case of HP+HO, the level of the inducible isoform of hemoxygenase was decreased in comparison with control and HO+N. Moreover, contrary to other adaptation regimes, adaptive protective responses in HP+HO were observed already after 15 days. As a result the resistance of liver and brain tissues to enhanced production of reactive oxygen species increased. The use of a new adaptation regime, which combines intermittent hypoxia and hyperoxia, ensures a strong membrane-protective effect.

Keywords: Adaptation, hypoxia, hyperoxia, reactive oxygen species, antioxidant enzymes, brain, liver.

*Address for Correspondence: T.G. Sazontova, Ph.D., D.Sc. Faculty of Fundamental Medicine, M.V. Lomonosov Moscow State University Lomonosovsky pr., 31, korp. 5, Moscow 119192, Russia Tel/fax. +7(495)932-8814; E-mail: arkhipenko@fbm.msu.ru

Introduction

Therapeutic correction of various diseases associated with excessive activation of free radical oxidation and increased level of reactive oxygen species (ROS) is one of the topical problems of modern-day medicine. There exist two similar, at the first glance but actually radically different approaches to the solution of this problem, viz., a decrease in the intensity of free radical oxidation and an increase in the level of antioxidant protection - both endogenous and exogenous.

Correction of ROS-mediated pathological processes by decreasing the intensity of free radical oxidation is a basic principle underlying the therapeutic application of long-term high-altitude adaptation to hypoxia. Indeed, low pressure in the highland associated with a low level of ROS production in the organisms has a beneficial effect on an unhealthy organism, since it eliminates or corrects the pathological processes induced by excess ROS. For example, the beneficial effect of graded high-altitude hypoxia on patients with non-specific lung illnesses such as pneumonia, bronchial asthma, bronchitis, was demonstrated more than 30 years ago by Nikolai Sirotinin and his associates (Sirotinin N.N. 1973; Kolchinskaya *et al.*, 1999; Ivashkevich & Serebrovskaya, 2000). But the free radical origin of this disorders was established only 10-15 years ago (Meltzer, 1989; Doelman & Bast, 1990; Zenkov *et al.*, 2001).

The second approach to the correction of pathological states associated with high ROS levels is the use of exogenous antioxidants for patients with acute hypoxic, ischemic and reperfusion injuries where intrinsic protective systems of the organism are insufficient to produce complete protection. However, among a vast array of experimental data related to antioxidant therapy, the number of publications, which cast some doubt on the previous statement that natural and synthetic antioxidants do not always produce a beneficial effect, is steadily increasing. This is particularly true of remote, frequently neglected consequences of long-term application of antioxidants, since the majority of drugs including artificial ones are capable to provoke inhibition of endogenous antiradical protection systems (Sazontova *et al.*, 1987; Belykh *et al.*, 1992; Mosconi *et al.*, 1988; Arkhipenko & Sazontova, 1995; Jaeschke, 1995; Fehrenbach, Northoff, 2001). Thus, the crucial role in effective protection from ROS should be sought in the formation of intrinsic endogenous protective systems including, in particular, a multicomponent antioxidant protective system rather than in the use of exogenous antioxidants. In this context, the use of methods allowing restriction of excessive activation of free radical oxidation due to stimulation of intrinsic antiradical systems acquires special importance. These methods employ the intermittent normobaric hypoxic training (INHT) based on the gas mixtures with reduced oxygen content (Meerson, 1993).

The protective effect of INHT is based on the activation of mechanisms of adaptation to hypoxia and reoxygenation during the return of air respiration from hypoxic to normal oxygen level. As it is known, reoxygenation and reperfusion provoke the generation of ROS which exert a damaging action on the cell and/or trigger a cascade of intracellular redox signaling reactions (Das, 2001). Activation of transcription factors and synthesis of protective intracellular components including those possessing antioxidant activities represent a specific response of the cell to free radical signals and initiation of the redox signal cascade. The free radical signals generated by INHT as a result of alternating hypoxic-normoxic influences increases the resistance of cells to damaging factors including ROS-mediated ones. It is of note that such periodic weak generation of ROS provides a mechanism for increasing the resistance of the organism to various factors, such as stress (Sazontova *et al.*, 1987), physical training (Powers *et al.*, 1994; Sazontova *et al.*, 1996), cold adaptation (Spasic *et al.*,

1993), dietary adaptogens (Cai *et al*., 1995; Sanz *et al*.,1994; Singh *et al*., 1994), n-3 PUFA enriched diets (Hu *et al*., 1989; Sazontova *et al*., 1995), intermittent hyperoxygenation (Kiselev, 2002) or interval hypoxic training which produce pronounced membrane-stabilizing effects (Arkhipenko *et al*., 1997a).

Periodic activation of free radical oxidation upon interval hypoxia is still one of the main challenges in this area. Whereas under stress, physical training or hyperoxia activation of free radical oxidation presents no problem owing to the direct formation of excess ROS, in case of hypoxia the situation is different. Although hypoxia does not lead to direct accumulation of ROS, it seldom occurs by itself, being mostly coupled with reoxygenation, i.e., and restoration of the initial oxygen level in the cell. This suggests that any hypoxic influence followed by normoxia or reoxygenation is an integral part of and a triggering pulse for ROS activation. Reoxygenation leads to the elevation of ROS, while interval hypoxic training represents a manifold repetition of this process. In their turn, alternating interval hypoxia-reoxygenation sessions cause periodic activation of free radical oxidation similar to that observed under the periodic influence of direct ROS inducers.

It was found that in contrast to continuous high altitude hypoxia, intermittent adaptation not only had a wider therapeutic effect, but also afforded potent protection of the organism from subsequent injuries (Arkhipenko *et al*, 1997; Meerson *et al*., 1992; Costa *et al*., 1993; Meerson, 1993; Sazontova *et al*., 1994, 1995, 1996; Radjpurohit *et al*., 1996; Arkhipenko *et al*, 1997; Serebrovskaya, Safronova, 1999; Sazontova, 2002). This implies that the intermittent adaptation to hypoxia has advantage over high altitude continuous adaptation as a method of the different type of the prophylaxis. Indeed, INHT is used now as precautions against various diseases (e.g., before periodical acute conditions of the chronic desease as bronchial asthma) or training of athletes against excessive hypoxic loads, and an effective tool for creating a protective preconditioning effect also, e.g., during preoperative preparation, etc (Tkachouk *et al*, 1994). Now the range of applications of interval hypoxic training is gradually expanding by virtue of its capacity for increase of the resistance of the organism to damaging hypoxic factors.

An analysis of effects of various adaptation factors showed that the formation of stable adaptive protective responses is a long-lasting (3–5 weeks) process, which creates certain problems for both clinical (conservative therapy, surgical procedures, prophylaxis in remission, etc.) and sports (training of athletes) medicine. This, in turn, generates a problem of reduction of the time for achieving long-term adaptation with concomitant increase of adaptation efficiency. This problem can be solved through enhancement of signal intensity which, in terms of the redox signaling theory, is related to more intensive production of ROS. An increase in signal intensity, e.g., augmentation of the degree or duration of hypoxia, is fraught with generation of deleterious side effects. We investigated this problem in series of experiments where increased intensity of hypoxic training (Arkhipenko *et al*., 1997; Sazontova *et al*., 1994) was used. The resistance of membrane structures of the heart and, especially, the liver was decreased.

Here we actually deal with a vicious circle where the efficiency of long-term adaptation can be increased only through an increase in signal intensity. This, in turn, generates tissue-specific side effects. One possible way to the solution of this problem is the development of approaches able to *increase the intensities of adaptation signals without affecting the hypoxic component*. Stipulating that under intermittent hypoxic-normoxic conditions adaptation signals are mediated by periodic activation of weak free radical signals inducing the synthesis of protective systems of the cell, we set ourselves the task to increase the intensity of free radical-induced signaling through an addition of the hyperoxic component rather than through an increase in the intensity of the hypoxic component. To this end, we developed an original interval normobaric training procedure combined alternating hypoxic and hyperoxic phases. Such a training mode entails cyclic fluctuations of oxygen levels on the both sides of normoxia, i.e., the use of an intermittent cycle combines moderate hypoxia with hyperoxia. Interval normobaric hypoxic and hyperoxic training (INH-HÒ) ensures variations in oxygen levels which cannot be achieved through the use of hypoxia alone (Scheme).

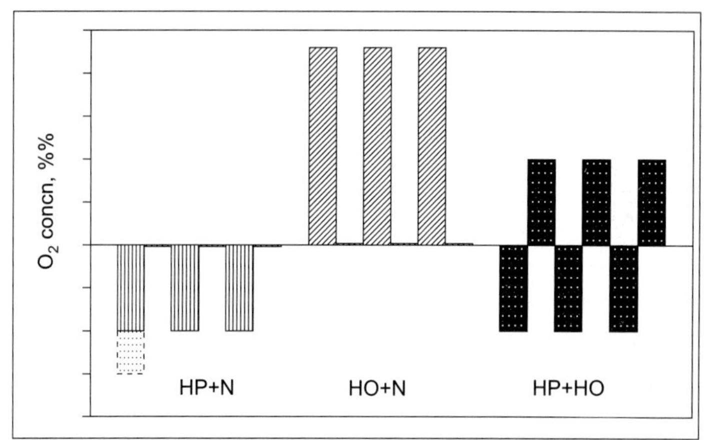

Scheme. Combination of low and high oxygen levels in the different modes of intermittent adaptation. HP+N - hypoxia + normoxia, HO+N -hyperoxia + normoxia and HP+HO - hypoxia + hyperoxia.

The present work is the first to corroborate the possibility to achieve stable adaptation effects of intermittent training through increasing the interval of variation of oxygen level.

Methods

Animals and Adaptation

The experiments were performed on 36 male Wistar rats weighing 250 g. INHT is a variant of soft adaptation based on periodic low-intensity induction of ROS, therefore after its long-term (23 days) application the resistance of membrane structures in all tested organs was either increased or unchanged (Sazontova *et al.*, 1994). Stipulating that the primary goal of these studies was to establish differences in the specificities of various tissues in different adaptation regimes and manifestations of miscellaneous side effects under conditions of changing oxygen levels, we used a more short-lasting training regime. The experimental animals were divided into control and three groups under different short-term adaptation regimes. (1) Hypoxia – normoxia (HP+N; Scheme), 5-min inhalation of a gas mixture containing 10% O_2 with 3-min intervals of normoxia, 60 min daily for 15 days. (2) Hyperoxia – normoxia (HO+N; Scheme), 5-min inhalation of a gas mixture containing 30% O_2 with 3-min intervals of normoxia, 60 min daily for 15 days. (3) Hypoxia-hyperoxia (HP+HO, Scheme), 5-min inhalation of a gas mixture containing 10% O_2 with 5-min intervals of hyperoxia, 30% O_2; 60 min daily for 15 days). The air mixtures with low and high content of O_2 were obtained with the help of a device (Hypoxia Medical Academy) operating on the membrane gas partition principle with on-line computer control of O_2 levels in isolated animal cage.

Homogenate Preparation

The animals were decapitated on the next day after cessation of the training course. The brains and livers were extracted, frozen in liquid nitrogen and homogenized in a Teflon-glass homogenizer in 8 (brain tissue) and 12 (liver tissue) volumes of a medium contained 30 mM Trizma® and 100 mM NaCl pH 7.4. Protein concentration was determined according to method (Lowry *et al.*, 1951).

Antioxidant Enzymes Assays

Catalase activity was determined by the method of Luck (1963) at 240 nm. SOD activity was measured by the rate of inhibition of superoxide radical formation in the xanthine–xanthine oxidase system (Beauchamp, Fridovich, 1971) at 560 nm by formazan formation from tetranitrotetrazolium blue. Glutathione reductase activity was determined by NADPH consumption in the presence of GSSG (Beutler, 1984).

Lipid Peroxidation

The resistance of brain and liver tissues to free radical oxidation was studied *in vitro* by following oxidation in ROS-inducing systems. Freshly prepared diluted (1:5) homogenates (exactly 3 mg protein) were incubated at 37°C in 20 mM Trizma® (pH 7,0), 100 mM NaCl and 0,3 mM sodium ascorbate. The concentrations of ROS-induced oxidation products were determined by following characteristic absorption spectra using the method of Ohkawa *et al.* (1979), in the modification of Kikugawa *et al.* (1992).

Western-blot analysis

HSP70 and HSP32 or hemoxygenase – 1 (HO-1) were determined in the cytoplasmic fractions of the brain and liver. The tissues were homogenated in a Teflon-glass homogenizer (4°C) in a lyzing buffer which contained 50 mM Tris-HCl, 5 mM EDTA, 1 mM DTT and 1% Triton X-100 pH 7.5 (buffer:tissue ratio = 6:1, by weight). The homogenate was filtered through six layers of gauze and centrifuged for 30 min at 12000 g (4°C). The proteins were separated in 10% polyacrylamide gel and transferred to a PVDF membrane by electroelution (overnight). Preincubation of Western blots was carried out in TBST supplemented with 1% Tween-20 containing 5% (w/v) defatted milk (1 h). The samples prepared from the thymuses of rats and from H35 hepatoma cells subject to heat shock (41,5°C, 30 min) were used as a positive control. The Western blots were incubated in the presence of first polyclonal antibodies against HSP32 (Stressgen) (1:1000) and first monoclonal antibodies (Stressgen) against HSP70 in the same dilution for 1 h, washed and incubated in the presence of second antibodies (Jackson Immuno Research) conjugated with horseradish peroxidase (1:2000) for 1 h. Detection of HSP32 and HSP70 was performed in a reaction with ECL reagents (Kodak film).

Statistical Analysis

All data are expressed as mean ± S.E. Treatment of experimental results was carried out on the basis of parametric and nonparametric statistic criteria (Instat-2, Sigma).

Results

HP+N Regime

The activities of antioxidant protection enzymes in liver and brain are shown in Fig 1. Under HP+N conditions, SOD activities were increased in both organs - by 37% (liver) and 22% (brain), respectively. The changes in the catalase activity were far less pronounced and manifested in a decrease (by 14%) in the liver. After adaptation to HP+N, a significant (by 33%) decrease in glutathione reductase activity was found observed both in the liver and the brain.

Accumulation of free radical products during induction of *in vitro* oxidation in liver and brain homogenates is shown in Fig. 2. As can be seen, the sensitivity of liver tissues to ROS-mediated processes increased after adaptation to HP+N. The level of ROS products increased by 41% after 20-min, by 43% after 40-min and by 25% after 1 h 15 min of oxidation. In contrast to the liver, short-term adaptation to HP+N did not cause any appreciable increase in the intensity of ROS-induced free-radical oxidation in the brain.

HO+N Regime

In the next step we studied the effect of the hyperoxic component (HO+N) on the resistance of membrane structures of the liver as an organ manifesting the highest sensitivity to HP+N. Fig.1 and 2 demonstrates that in contrast to the HP+N regime the use of the hyperoxic component alone under moderate HO+N did not induce any significant changes either in the activities of antioxidant enzymes or in their sensitivities to free radical oxidation. Apparently, this protective effect is due to the synthesis of other urgent response proteins. The levels of two proteins whose synthesis is mediated by ROS, namely: HSP70, which are determined by the ROS-mediated effects of various stress-inducing factors (Nanji *et al.*, 1995), and hemoxygenase-1, a highly inducible form whose synthesis is initiated by oxidative factors (Ryter, Tyrrell, 2000), were studied by Western blotting.

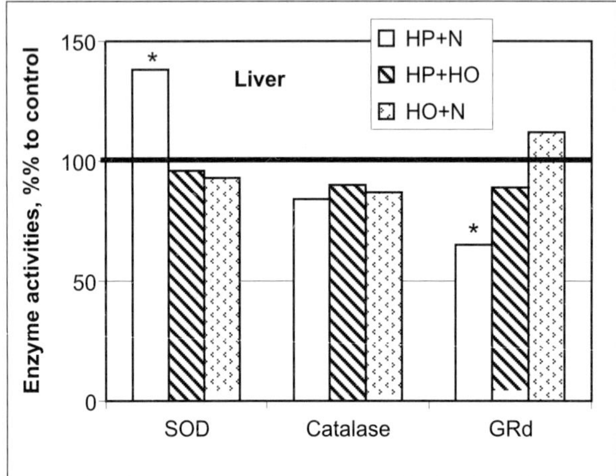

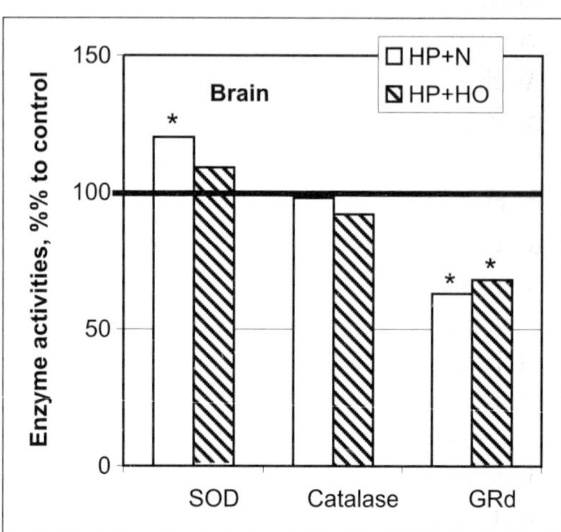

Fig. 1. Activity of the antioxidant enzymes - catalase, SOD and glutathione reductase in liver and brain after adaptation to HP+N – hypoxia+normoxia; HP+HO – hypoxia+hyperoxia, HO+N – hyperoxia+normoxia, with compare to control (100%). (P<0,05): * - significance of differences in comparison with control.

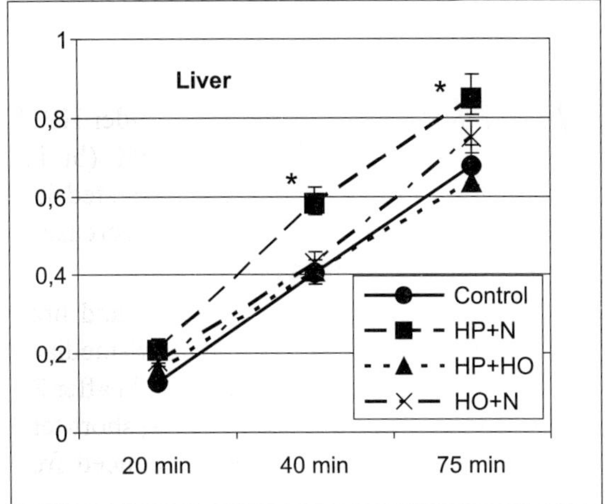

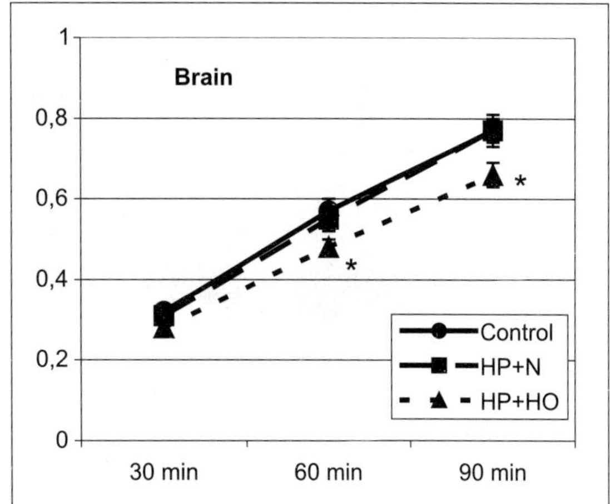

Fig. 2. Effects of three different normobaric training regimes (C – control, HP+N – hypoxia+normoxia; HP+HO – hypoxia+hyperoxia, HO+N – hyperoxia+normoxia) on the levels of 2-thiobarbiturate-reactive products in liver and in brain cortical tissues during induction of free radical oxidation in vitro. Ordinate - O.D (532 nm)/mg protein; abscissa - time of incubation. (P<0,05): * - significance of differences in comparison with control.

Under HO+N conditions, we observed the synthesis of inducible stress proteins (e.g., HSP70) and enhanced synthesis of the inducible form HO-1 (Fig. 3). Taking into account the dual role of HO-1 as an oxidative stress marker and a suppressor of activity of free heme in prooxidant reactions, which induces its transformation into biliverdin (Ryter & Tyrrell, 2000), it becomes evident why free radical processes in this experimental group are compensated and unaccompanied by any significant synthesis of antioxidant protection enzymes. These data suggest that low-intensity hyperoxia performed in an interval regime neither increases, nor decreases the resistance of liver tissues to ROS in comparison with the control level.

With this in mind, in the final step of our experiments we used a special kind of adaptation which combined hypoxic and hyperoxic episodes

HP+HO Regime

It was found that in contrast to HP+N where the sensitivity of liver tissues to free radical oxidation increased, adaptation in the HO+HP regime did not decrease the resistance of these tissues to ROS, but even increased it. It should be noted that the decrease in the intensity of free radical oxidation in the HP+HO group (Fig. 2) was paralleled by preservation of SOD and glutathione reductase activities, whereas under HP+N the activities of these enzymes varied widely (Fig. 1). Moreover, the levels of urgent response proteins testify to a complete lack of both stress component of adaptation to HP+HO and excessive free radical signals because the synthesis of inducible forms of HSP70 and HO-1 was absent (Fig. 3). On the whole, these data point to an increase of ROS-sensitivitive liver to ROS-dependent oxidation.

In the brain, the changes in the activities of antioxidant protection enzymes under the influence of HP+N and HP+HO were essentially identical and manifested in a lowered activity of glutathione

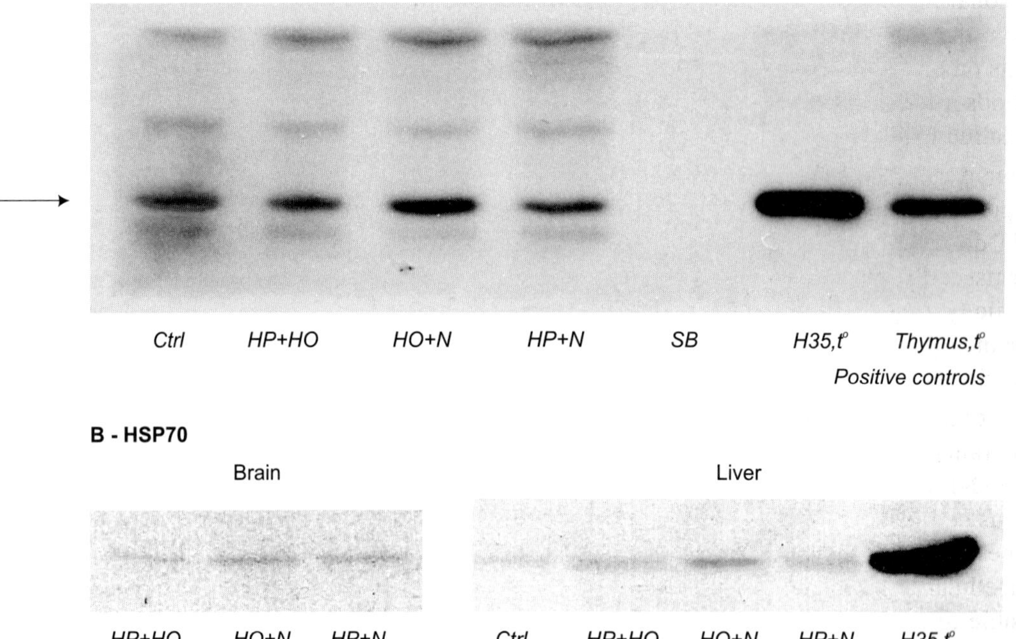

Fig. 3. Western blot analysis of HSP32 (A) in liver and HSP70 (B) in liver and brain in control, after adaptation to HP+N, HO+N and HP+HO.
The samples prepared from the thymuses of rats and from H35 hepatoma cells subject to heat shock (41,5°C, 30 min) were used as a positive control. The Western blots were incubated in the presence of first polyclonal antibodies against HSP32 (Stressgen) (1:1000) and first monoclonal antibodies (Stressgen) against HSP70 in the same dilution for 1 h, washed and incubated in the presence of second antibodies (Jackson Immuno Research) conjugated with horseradish peroxidase (1:2000) for 1 h.

reductase – by 32% in comparison with control (Fig. 1). The synthesis of the inducible form HO-1 was not observed either for HP+N, or for HP+HO, whereas accumulation of HSP70 was equally small in the both experimental series (Fig. 3). However, the levels of the oxidation products formed during LPO induction *in vitro* varied. Whereas under HP+N the resistance of brain tissues to ROS did not differ from control, under HP+HO conditions this parameter was increased from control. Thus, the addition of the hyperoxic component (HP+HO) increased the resistance of brain tissues to free radical oxidation.

DISCUSSION

Our previous studies established pronounced tissue specificity of adaptation to stress, physical exercise and intermittent hypoxia (Sazontova, 2002). It was shown in particular that factors of similar origin and intensity may produce quite different effects on tissues which vary from protection to damage. This can be attributed to the difference in the sensitivities of concrete tissues, cells or intracellular structures to ROS, different ratios of prooxidant and antioxidant systems of the cell at the moment of its exposure to exogenous factors. Activation of free radical oxidation may give rise to miscellaneous intracellular events from compensation of oxidative processes and formation of additional antiradical protective systems to apoptosis or necrosis (Semenza, 1999). For this very reason, the conclusion on the beneficial effect of adaptation cannot be based on the analysis of its effects on one organ only. In this context, detection of the most sensitive organs and development of methods protecting against possible side effects become the main challenge for modern-day adaptation medicine.

Our experiments suggest that the differently-directed changes in the activities of individual components of antioxidant protection systems observed in brain tissue after completion of short-term INHT do not influence the resistance of membrane structures to ROS-induced processes, whereas in liver tissues the sensitivity of intracellular structures to free radical oxidation increases. Such tissue specificity was observed due to reduction of adaptation time under classical INHT. This and the use of resistance tests enabled us to establish the potential abilities of tissues to be protected against the deleterious effects of damaging factors.

In contrast to HP+N regime adaptation in the HO+HP regime increased the resistance to ROS of both brain and liver tissues. It is important that the level of the synthesis of inducible forms of HSP70 and HO-1 was absent. Both these proteins are known to play a dual role by acting as markers of various cell injuries and by protecting the cell from the damaging effects of ROS. Despite this, their synthesis is indicative of the entry of the redox signal into the cell and initiation of a cascade of intracellular signaling reactions. Simultaneous measurements of HSP70 and HO-1 content made it possible to estimate both the level of stress responses and the intensity of free radical signals. The results suggest that addition of the hyperoxic component to the classical INHT and use even short-term training in HP+HO regime lead to a complete lack of both stress component and excessive free radical signals. These data point to an increased resistance after HP+HO of the brain and ROS-sensitivitive liver tissues to free radical oxidation.

The main findings of this study can be summarized as follows. First, stable protective effects of adaptation can be achieved through a combination of hypoxic and hyperoxic training episodes. Second, addition of the hyperoxic component to a classical INHT regime increases the resistance of liver tissues, which are especially sensitive to hypoxia. And, finally, combined therapy including hypoxic and moderately hyperoxic episodes provides stable adaptive protection with simultaneous reduction of the training time. This suggests that an addition of the hyperoxic component makes it possible to attain the protective effect of classical INHT at earlier times and under milder conditions for the whole organism.

This research was supported by the Russian Foundation for Basic Researches (grant No 05-04-08102) and the Federal Agency for Education (grant No RNP2.1.1.9493).

References

1. Arkhipenko Yu.V., and Sazontova T.G. (1995). Mechanisms of the cardioprotective effect of a diet enriched with n-3 polyunsaturated fatty acids. Pathophysiology. 2:131-140.
2. Arkhipenko Yu.V., Sazontova T.G., and Rice-Evans C. (1997). Hypertrophy and regression of rat heart: free radical related metabolic systems. Pathophysiology. 4:241-248.
3. Arkhipenko Yu.V., Sazontova T.G., Tkachouk E.N., and Meerson F.Z. (1997a). Adaptation to continuous and intermittent hypoxia: role of the active oxygen-dependent system. In: Adaptation Biology and Medicine (Vol.1. Subcellular Basis), edited by B.K.Sharma *et al.*), Narosa Publishing House, New Delhi *et al.*, pp. 251-259.

4. Beauchamp C. and Fridovich I. (1971). Superoxide dismutase: Improved assay and an assay applicable to acrylamide gels. Analyt.Biochem. 44:276-287.
5. Belykh A.G., Gukasov V.M. and Chukaev S.A. (1992). Free padical oxidation during normobaric hypoxia. Physiol.J. 38:73-76 (in Russian).
6. Beutler E. (1984). Red cell metabolism. A manual the biochemical methods, Gune and Stratton, Orlando, Fl., USA.
7. Cai Y.N., Appelkvist E.L., and Deplerre J.W. (1995). Hepatic oxidative stress and related defenses during treatment of mice with acetylsalicylic acid and other peroxysome proliferators. J.Biochem.Toxicol. 10:87-94.
8. Costa L.E., Llesuy S., and Boveris A. (1993). Active oxygen species in the liver of rats submitted to chronic hypobaric hypoxia. Am. J. Physiol. 264:C1395-C1400.
9. Das D.K. (2001). Redox regulation of cardiomiocyte survival and death. Antiox. Redox. Signal. 3:23-37
10. Doelman C.J.A. and Bast A. (1990). Oxygen radicals in lung pathology. Free Radical Biol. Med. 9:384-400.
11. Fehrenbach E., Northoff H. (2001). Free radicals, exercise, apoptosis, and heat shock proteins. Exerc. Immunol. Rev. 7:66-89
12. Hu M.L., Frankel E.N., Leibowitz B.E., and Tappel A.L. (1989). Effect of dietary lipids and vitamin E on in vitro lipid peroxidation in rat liver and kidney homogenates. J.Nutr. 119:1574-1582.
13. Ivashkevich A.A. and Serebrovskaya T.V. (2000). Scientific avtivities of academician Nikolai Sirotinin. High Altitude Med. Biol., 1:75-80.
14. Jaeschke H. (1995). Mechanisms of oxidant stress-induced acute tissue injury. Proc.Soc.Exp.Biol.Med. 209:104-111.
15. Kikugawa K., Kojima T., Yamaki S., and Kosugi H. (1992). Interpretation of the thiobarbituric acid reactivity of rat liver and brain homogenates in the presence of ferric ion and ethylenediaminetetraacetic acid. Analyt.Biochem. 202:249-255.
16. Kiselev S.O. (2002). The mechanism of action of HBO on the organism. Hyperbaric Physiol. Med. 2:3-7 (in Russian).
17. Kolchinskaya A.Z., Khatsukov B.Kh., and Zakusilo M.P. (1999). Oxygen shortage, destructive and constructive effects. Russian Academic Press, Nalchik, Russia.
18. Lowry O.M., Rosebrough N.J., Farr A.L. and Randall R.J. (1951). Protein measurement with the folin phenol reagent. J. Biol. Chem. 193:265-275.
19. Luck H. (1963). Catalase. In: Methods of enzymatic analysis, edited by Bergmeyer H.U., Verlag-Chemie Academic Press, New York, pp. 885-888.
20. Meerson F.Z. (1993). Adaptation to intermittent hypoxia: mechanisms of protective effects. Hypoxia Med. J. 3:2-8.
21. Meerson F.Z., Arkhipenko Yu.V., Rozhitskaya I.I., Didenko V.V., and Sazontova T.G. (1992). Opposite effect on antioxidant enzymes caused by adaptation to continuous and intermittent hypoxia. Bull. Exp. Biol. Med. 64:14-15 (in Russian).
22. Meltzer S., Goldberg B., Lad P., and Easton J. (1989). Superoxide generation and its modulation by adenosine in neutrofils og subjects with asthma. J. Allergy Clin. Immunol. 83:960-966.
23. Mosconi C., Colli S., Medini L., Stragliotto E., Maderta P., Tremoli E., and Galli C. (1988). Vitamin E influences the effects of fish oil on fatty acids and eicosanoid production in plasma and circulating cells in the rat. Biochem.Pharmacol. 37:3415-3421.
24. Nanji A.A., Griniuviene B., Yacoub L.K., Sadrzadeh S.M., Levitsky S and McCully J.D. (1995) Heat-shock gene expression in alcoholic liver disease in the rat is related to the severity of liver injury and lipid peroxidation. Proc. Soc. Exp. Biol. Med.210:12-19.
25. Ohkawa H., Ohishi N., and Yagi K. (1979). Assay for lipid peroxides in animal tissues by thiobarbituric acid reaction. Analyt.Biochem. 95:351-358.

26. Padros E., Dunach M., Morros A., Sabes M., and Manosa J. (1984). 4-th-derivative spectrophotometry of proteins. Trends in Biochem. Sci. 9:508-510.

27. Powers S.K., Criswell D., Lawler J., Ji L.L., Martin D., Herb R.A., and Dudley G. (1994). Influence of exercise and fiber type on antioxidant enzyme activity in rat skeletal muscle. Am.J.Physiol. 266: No 2, Pt 2:R375-R380.

28. Rajpurohit R., Koch C.J., Tao Z., Teixeira C.V., and Shapiro I.M. (1996). Adaptation of chondrocytes to low oxygen tension: relationship between hypoxia and cellular metabolism. J.Cell.Physiol. 168:424-432.

29. Ryter S.W., Tyrrell R. (2000). The heme synthesis and degradation pathways: role in oxidant sensitivity. Free Rad. Biol. Med. 28:298-309.

30. Sanz M.J., Ferrandiz M.L., Cejudo M., Terencio M.C., Gil B., Bustos G., Ubeda A., Gunasegaran R., and Alcaraz M.J. (1994). Influence of a series of natural flavonoids on free radical generating systems and oxidative stress. Xenobiotica. 24:689-699.

31. Sazontova T.G. (1996). The opposite effect of adaptation to short-term stress exposures and adaptation to intermittent hypoxia on the activity of liver plasma membrane Na,K-ATPase. Bull. Exper. Biol. Med. 121:383-386 (in Russian).

32. Sazontova T.G., Arkhipenko Yu.V., and Meerson F.Z. (1987). Increase in activities of heart antioxidant protection enzymes during adaptation of rats to short-term stress exposures. Bull. Exp. Biol. Med. 103:411-413 (in Russian).

33. Sazontova T.G., Arkhipenko Yu.V., and Meerson F.Z. (1995). Adaptation to intermitten hypoxia and a diet (PUFA class p-3) have a cardioprotective effect and increase the resistance of myocardial sarcoplasmic reticulum Ca2+ transport to free radical oxidation. Bull. Exp. Biol. Med. 120:42-45 (in Russian).

34. Sazontova T.G., Tkatchouk E.N., Kolmykova S.N., Ehrenbourg I.V., Meerson F.Z., and Arkhipenko Yu.V. (1994). Comparative analysis of peroxidation and antioxidant enzyme activities in rats adapted to different regimes of normobaric hypoxia. Hypoxia Med. J. 1:4-7.

35. Sazontova T.G. (2002). Regularities in the modulation of antioxidant status of a cell in response to activation of free radical oxidation. Hypoxia Med. J. 1:2-10.

36. Semenza G.L. (1999). Perspectives on oxygen sensing. Cell. 98:281-284

37. Serebrovskaya T.V., Safronova O.S., and Gordiy S.K. (1999). Free radical processes under differenr oxygen supply conditions. Physiol. J. 45:92-104 (in Ukrainian).

38. Singh B., Sharma S.P., and Goyal R. (1994). Evaluation of Geriforte, an herbal geriatric tonic, on antioxidant defense system in Wistar rats. Ann. N.Y. Acad.Sci. 717:170-173.

39. Sirotinin N.N. (1973). Pathophisiology of extreme states. Meditsina, Moscow, Russia.

40. Spasic M.B., Saicic Z.S., Buzadzic B., Korac B., Blagojevic D. and Petrovic V.M. (1993). Effect of long term exposure to cold on the antioxidant defense system in the rat. Free Radic. Biol. Med. 15:291-299.

41. Tkachouk E.N., Gorbachenkov A.A., Kolchinskaya A.Z., Ehrenburg I.V., and Kondrykinskaya I.I. (1994). Adaptation to interval hypoxia for profilaxis and therapy. Hypoxia Medical, Ltd, Moscow, Russia.

42. Zenkov N.K., Lankin V.Z., and Menschikova E.B. (2001). Oxidative Stress. MAIK “Science/Interperiodica”, Moscow, Russia.

Adaptation Biology and Medicine (Volume 5: Health Potentials)
Editors: L. Lukyanova, N. Takeda and P.K. Singal

23. Plant Adaptogens Activate Cellular Adaptive Mechanisms by Causing Mild Damage

F.A.C. Wiegant[1]*, G. Limandjaja[2], S.A.H. de Poot[2], L.A. Bayda[3], O.N. Vorontsova[3], T.A. Zenina[3], M. Langelaar-Makkinje[1], J.A. Post[1] and G. Wikman[4]

[1]Faculty of Science, Department of Biology, Utrecht University, The Netherlands, [2]University College Utrecht, The Netherlands, [3]Institute of Pathology and Pathophysiology, Moscow, Russian Federation, [4]Swedish Herbal Institute, Göteborg, Sweden

Abstract

The molecular mechanism in which plant adaptogens induce a state of adaptive resistance in cells is unclear. In this paper we used different mammalian cell lines to study the effect of *Eleutherococcus senticosus* and of *Rhodiola rosea* on the expression of seven different heat shock proteins (hsp27, hsp32, hsp47, hsp60, hsp70i, hsc70 and hsp90). We could demonstrate that plant adaptogens only induce hsp32 (also know as heme oxygenase or HO-1), which performs a crucial function in cellular antioxidative defense. Moreover, it is demonstrated that the enhanced synthesis of heme oxygenase is induced by a pro-oxidant action of the plant adaptogens in cells. Plant adaptogens are shown to initially consume glutathione (GSH) and to cause oxidation of lipids as well as of cytoplasmic components. As a result of this damage endogenous cellular defense and repair mechanisms are activated leading to an enhanced synthesis of heme oxygenase and to an increased synthesis of GSH. As a consequence of these induced protective mechanisms an enhanced cellular (oxidative) stress resistance is observed in different mammalian cell cultures 24 hours after exposure to plant adaptogens. It is proposed that part of the action of plant adaptogens can be explained by an hormetic induction of defense mechanisms.

In summary, we have demonstrated that plant adaptogens enhance cellular resistance by causing mild damage first. The detection of damage leads to an activation of endogenous cellular repair- and resistance programs, such as an enhanced presence of heme oxygenase and increased levels of GSH, which explains the enhanced adaptive state induced by plant adaptogens.

Keywords: Adaptogens, eleutherococcus, rhodiola, adaptive response, hormesis

Introduction and Historical Background

A state of enhanced cellular adaptive resistance is the current focus of much research effort. Different ways to achieve such a state have been documented, such as the use of extracts from plant adaptogens. The terminology 'adaptogen' was originally used for various pharmacologically active compounds

*Address for Correspondence: Dr. F.A.C. Wiegant, Cellular Architecture & Dynamics, Institute of Biomembranes, Utrecht University, P.O.Box 80056, 3508 TB Utrecht, The Netherlands. E-mail: f.a.c.wiegant@uu.nl

that could elicit a state of non-specifically raised resistance in mammals allowing them to counteract stressors and to adapt to exceptional strain (see Brekhman 1969, 1980; Meerson 1984). Later this term was also applied to a group of plants that exert adaptogenic properties (Brekhman 1969, 1980; Wagner *et al.* 1994). A review article by Brekhman (1969), one of the pioneers in plant adaptogen research, summarized almost 20 years of Russian research and made Western scientists aware of the research and the concepts. Indeed a large number of experiments indicated that physical performance and adaptive ability were increased when extracts from plants such as *Eleutherococcus senticosus* and *Rhodiola rosea* were applied to a variety of model systems (Wagner *et al.* 1994; Wagner 1995; Panossian *et al.* 1999; Davydov and Krikorian 2000; Kelley 2001; Panossian 2003, 2005 for reviews). In the definition of Brekhman (1980) adaptogens were described as having the following three characteristics: (1) it must show a non-specific activity, i.e. an increase in power of resistance against physical, chemical or biological noxious agents; (2) an adaptogen must have a normalizing influence independent of the nature of the pathological state; and (3) it must be harmless and must not influence normal body functions more than required.

In this way adaptogens are thought to reinforce the non-specific adaptive capacity to withstand situations of stress, to encourage a faster response towards stress condition and to facilitate a more rapid return to normality (Wagner *et al.* 1994).

Although many different aspects of a possible mechanism of action have been reported, an overall and comprehensive picture remains elusive. In previous research, our lab has clearly shown that plant adaptogens (*Eleutherococcus senticosus* and *Rhodiola rosea*) were also able to induce beneficial effects in 'simple' model organisms such as snails (*Lymnaea stagnalis*) and worms (*Caenorhabditis elegans*). In these models plant adaptogens were shown to be protective in acute (and chronic) stress conditions (Boon-Niermeyer *et al.* 2000), to be supportive in repair processes (unpublished) and in *C.elegans* they were even able to enhance the lifespan as well as to enhance stress resistance (Wiegant *et al.* 2006).

These data are in agreement with observations in mice and rats, where plant adaptogens are also able to enhance stress resistance. Apparently plant adaptogens are able to induce basic evolutionary conserved mechanisms leading to enhanced resistance and adaptation.

In order to focus on the molecular background of the beneficial effect exerted by adaptogens, we used mammalian cells in culture. At the cellular level the synthesis of individual heat shock proteins (hsps) is considered to be specifically linked to processes of defense, recovery and adaptation. Therefore, the hypothesis has been proposed that adaptogens may exert their beneficial effect through a modulation of hsp synthesis (eg Afanasiev 1996). However, no systematic research has been performed yet on the involvement of the different stress proteins in the action of plant adaptogens.

In this paper we initially focus on the question whether plant adaptogens are able to induce the synthesis of the different hsp families. These stress proteins act as chaperones that are involved in repair of damaged/denatured proteins. When these stress proteins are available at higher levels in the cell, they exert a protective action against stress-induced damages. Therefore, the possible induction of a variety of stress proteins was studied when different cell types in culture were exposed to plant adaptogens. We will elaborate on the background of how the synthesis of hsp32 is enhanced, which appears to be uniquely induced by adaptogens. Finally, we will discuss one of the general properties of plant adaptogens, which states that it should be harmless, in the light of our findings.

Materials and Methods

Effect of adaptogen extracts on cells in culture: The effect was studied of standardized extracts from two different plant adaptogens (*Eleutherococcus senticosus* (SHE/3) and *Rhodiola rosea* (SHR/5)) on the induction of six different heat shock proteins (hsp27, hsp32, hsp47, hsp60, hsp70i and hsp90) in four different mammalian cell lines (Rat H9c2 heart cells, Rat-1 fibroblasts, Reuber H35 hepatoma cells and Mouse HT22 neuronal hippocampus cells). The different cell lines were maintained under tissue culture conditions in the laboratory. The extracts were provided by the Swedish Herbal Institute.

In order to correlate the synthesis of hsps with survival characteristics, the effect of increasing concentrations of the plant extracts on heat shock protein levels was studied in relation to the possible toxicity that these extracts might exert. Toxicity was measured using the MTT-assay, whereas protein levels were determined with Western blotting.

Gel electrophoresis and Western Blotting: Protein samples were obtained for each set of conditions. The same amount of protein was loaded on gels. Gel electrophoresis, blotting and detection of specific proteins were performed with standard procedures. Antibodies against the different heat shock proteins were obtained from Stressgen.

Glutathione assay: Cells were either or not incubated with BSO (Buthionine Sulfoximine) and subsequently a modified Tietze-recycling assay (Baker *et al.* 1990, Tietze 1969) was used as described before (de Wit *et al.* 2001).

Detection of lipid peroxidation and oxidation of cytoplasmic components: The Bodipy assay was used to detect lipid peroxidation according to Drummen *et al.* (2004), whereas the DCF assay was used according to Wang and Joseph (1999) to evaluate the oxidative damage of cytoplasmic components.

Results

Effect on Cell Survival by Increasing Concentrations of Plant Adaptogens

In order to establish a possible toxicity of plant adaptogens, different cell types were exposed for 24 hours to increasing concentrations of *Eleutherococcus senticosus* or *Rhodiola rosea*. The effect was studied on a number of mammalian cell lines in culture which were chosen as a model for heart cells, connective tissue, liver and brain cells (H9c2 heart cells, Rat-1 fibroblasts, Reuber H35 rat hepatoma cells and Mouse HT22 neuronal hippocampus cells, respectively). Cell viability was assayed using the MTT test. A dose-dependent increase in toxicity of both *Eleutherococcus* as well as of *Rhodiola* was observed in all cell lines tested. Moreover, it was observed that *Rhodiola* already exerted a toxic effect at lower concentrations in comparison with *Eleutherococcus*. As an example, the results obtained in H35 rat hepatoma cells are shown in Figure 1.

In general, taking the results of all mammalian cell types into consideration, *Rhodiola* starts to affect the viability of about 20% of the cell population in concentrations higher than 100 μg/ml, whereas *Eleutherococcus* affects viability in concentrations higher than 1000 μg/ml in case of a 24-hour exposure. Upon exposures of shorter time duration, these concentrations had to be much higher in order to affect the viability of 20% of the cell population.

Concentrations in excess of 2.5 mg/ml *Eleutherococcus* or 250 μg/ml *Rhodiola* were shown to represent the LD50 value in most cell lines studied when exposed for 24 hours.

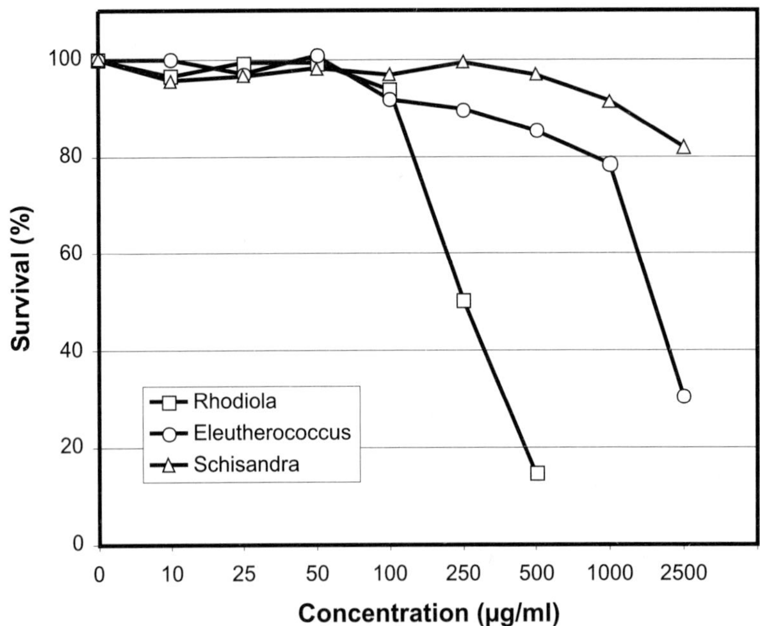

Fig. 1. Toxicity of plant adaptogen extracts in H35 cells. H35 cells were exposed for 24 hours to increasing concentrations of *Eleutherococcus senticosus* (SHE/3), *Rhodiola rosea* (SHR/5) or *Schisandra chinensis*. After exposure cells were rinsed, survival was established using the MTT assay and expressed as percentage of control.

Effect of Increasing Concentrations of Eleutherococcus and of Rhodiola on the Levels of Heat Shock Proteins in Different Cell Types

A range of concentration was selected to evaluate the effect of plant adaptogens on the induction of hsps. Concentrations were included that caused a small reduction in survival as well as concentrations that did not (yet) affect survival characteristics. It is of special interest whether these latter concentrations are able to enhance the level of hsps in cells. Cell cultures were exposed to the selected concentrations of *Eleutherococcus* or *Rhodiola* during a period of either 24 hours or 6 hours, after which cells were processed to determine the levels of the various hsps, using Western blot analysis.

Most of the well-known stress proteins (hsp27, hsp47, hsp60, hsp70i, hsc70 and hsp90) were not induced by plant adaptogens. Not even when high concentrations were applied that started to become toxic to the cells. There was only one stress protein which was clearly, consistently and significantly induced in all cell types studied in a concentration-depended way. This is hsp32, also known as heme oxygenase (HO-1). This hsp was already induced at concentrations of plant adaptogens which were not yet toxic (as established using the MTT assay).

As an example, the results are shown on the synthesis of hsc70, hsp70i and of hsp32 when H35 cell cultures were incubated for 24 hours in the presence of increasing concentrations of *Eleutherococcus* or of *Rhodiola* (Figure 2).

In order to determine whether the induction of hsps already occurred on a shorter time scale, we incubated cells with different concentrations of plant adaptogens during a period of 6 hours. As can be

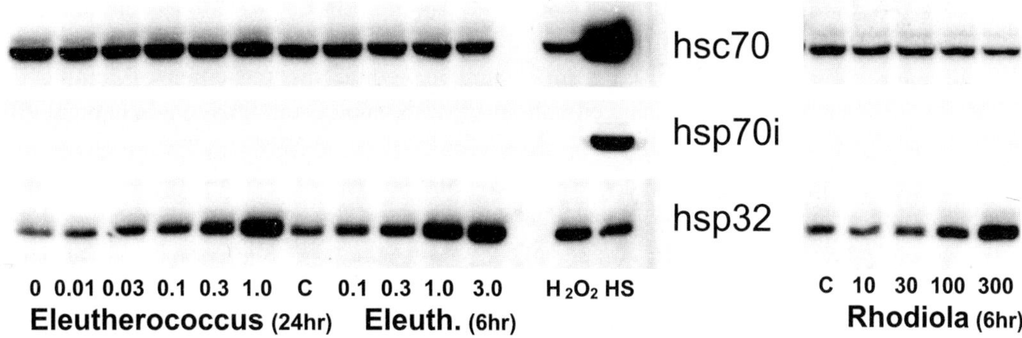

Fig. 2. Effect of plant adaptogens on the level of different heat shock proteins (heat shock cognate protein with a molecular weight of 70 kDa (hsc70), inducible heat shock protein 70 (hsp70i) and hsp32 (or heme oxygenase)). H35 rat hepatoma cells were exposed to increasing concentrations of *Eleutherococcus* or *Rhodiola* for either 24 hours or 6 hours. After the indicated time, cells were rinsed, solubilized and prepared for gel electrophoresis. As positive control, cells were exposed to either hydrogen peroxide (100 µM H_2O_2) for 1 hour or to a heat treatment of 30 minutes at 42°C after which the cells were rinsed and incubated in stress-free conditions for 6 hours before being processed for gel electrophoresis. The same amount of proteins was put in each lane. The presence and level of the indicated proteins was determined using specific antibodies according to standard Western blotting procedures.

observed in Figure 2, hsp32 is already induced during an exposure of 6 hours to *Eleutherococcus* or to *Rhodiola*.

What do We Know About hsp32 and why Would It be Induced by Adaptogens?

Hsp32 or heme oxygenase (HO-1) is the rate-limiting enzyme that catalyzes the metabolic conversion of heme to the bile pigments (e.g. biliverdin and bilirubin which has strong anti-oxidative properties), carbon monoxide (CO) and free iron. The beneficial effect for the cell of these breakdown products has been demonstrated (Maines 1988, 1997 for reviews). In short, cytoprotective properties of the degradation products of heme include antioxidant, anti-inflammatory, antiproliferative and anti-apoptotic effects (Prawan *et al.* 2005, Ryter *et al.* 2006). In addition, its induction has usually been related to an enhanced resistance against further oxidative stress. There are many different (oxidative) stress conditions that lead to the induction of heme oxygenase, such as heat, hydrogen peroxide, superoxide stress (menadione), nitric oxide (NO), heavy metal stress, etc. (Maines 1988, 1997, Prawan *et al.* 2005). Currently a number of review papers have emphasized the importance as well as the clinical relevance of heme oxygenase since it leads to an enhanced resistance against oxidative stress conditions and alleviates a number of pathological conditions (Kirkby and Adin 2006, Ryter *et al.* 2006). All this definitely supports the view that it is worthwhile to study the effect of plant adaptogens on the enhanced expression of heme oxygenase and its possible clinical applications.

With respect to the question why and how the synthesis of HO-1 is induced by plant adaptogens, two possible explanations may be given:

1) A pro-oxidant action of adaptogens. The extract may cause mild oxidative damage in cells, which subsequently elicits the normal cellular reaction, i.e. an initiation of a defence- and repair response aimed at counteracting the disturbed intracellular redox balance of which the

enhanced synthesis of HO-1 (hsp32), but also of glutathione (GSH) are important elements in mammalian cells. In this first option, plant adaptogens might be viewed as 'hormetic' compounds that, by causing some mild damage, induce the activation of cellular defense mechanisms leading to an enhanced non-specific resistance and thus to a beneficial effect.

2) An alternative explanation might be that, in the absence of any oxidative damage, (some unknown compound in) the extract is able to stimulate the signaling pathway that leads to an activation of the gene transcription process. The subsequent enhanced synthesis of HO-1 may thus lead to an enhanced overall resistance, without the prior involvement of an increase in oxidative stress damage.

In a large set of experiments we were clearly able to support the first possibility and to exclude the second one: Plant adaptogens do cause mild damage to macromolecules in the cells (at the level of proteins, lipids, and DNA). It is postulated that this mild damage is detected by the cell and in response to this a variety of cellular defense mechanisms is initiated of which the synthesis of HO-1 (hsp32) is an obvious element. The cellular response induced by plant adaptogens leads to an enhanced ability to cope with (oxidative) stress, which at least partly explains their beneficial effect.

Do adaptogens Cause Oxidative Stress?

In our experiments the following effects were observed upon exposure of cells to plant adaptogens:

- Fall and rise in GSH levels. The glutathione (GSH) concentration is a crucial element in the anti-oxidative defense of the cytoplasm. Upon incubation with adaptogens there is an initial reduction in the GSH concentration followed by an increased synthesis and an enhanced level 24 hours later (figures 3 and 4).
- Oxidative damage to cytoplasmic components. Using DCF measurements it was shown that plant adaptogens cause oxidative damage at the cytoplasmic level (figures 5 and 6).
- Oxidative damage to the membrane. Lipid peroxidation was identified using the Bodipy probe in response to incubation with adaptogens.
- Induction of hsp32 due to a pro-oxidant action of adaptogens (figure 7).

A. Plant Adaptogens Induce a Fall and a Subsequent Rise in the Level of GSH

Chiu *et al.* (2002) recently reported that the plant adaptogen *Schisandra chinensis* enhances the level of GSH, which can be defined or interpreted as a positive effect, since it enhances the overall cellular antioxidative defense capacity. But how do adaptogens cause this enhancement? In our experiments we were able to demonstrate that plant adaptogens enhance the levels of GSH in cells by first consuming it. Upon exposure to adaptogens, the cellular level of GSH initially falls during the first hours of its presence. As an example, the change in GSH content is shown in HT22 cells as a result of exposure to 500 μg/ml *Eleutherococcus* (Figure 3). This fall subsequently triggers the activation of the GSH producing enzyme (GSH synthetase). Once the level of GSH is measured the next day, the concentration is higher as in control cell cultures (Figure 3). Only this latter, beneficial, effect has been published by Chiu et al. (2002), whereas the direct, 'negative', effect in the first hours after addition was not checked/reported by these researchers.

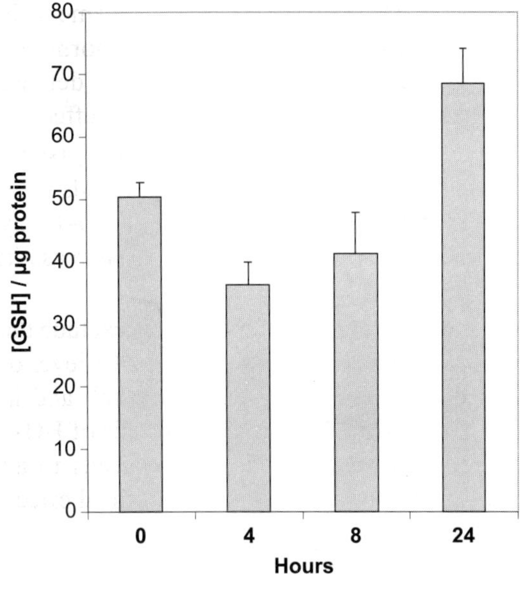

Fig. 3. **Effect of *Eleutherococcus* (500 µg/ml) on the GSH content in HT22 cells. Cell cultures were exposed to this adaptogen during different time periods (0, 4, 8 or 24 hours) after which cells were rinsed and processed to determine the GSH content per µg protein.**

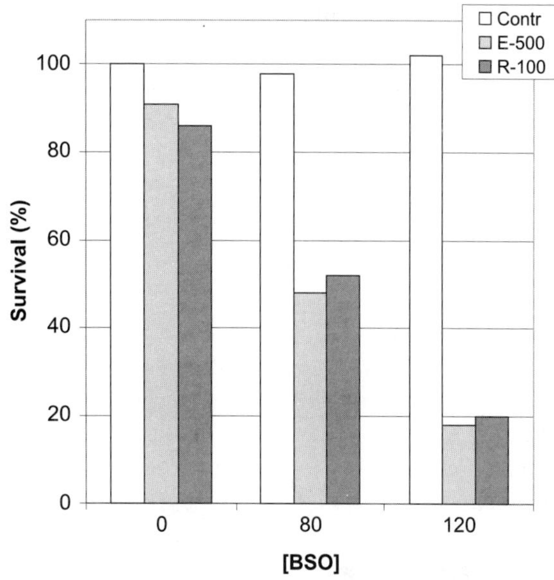

Fig. 4. **Effect of adaptogens on survival in GSH-depleted H9C2 heart cells. Cell cultures were depleted of GSH by exposing them to BSO (either 80 or 120 µM) for 3 days. During the last day, *Eleutherococcus* (500 µg/ml) or *Rhodiola* (100 µg/ml) was added. In order to establish survival, cell cultures were rinsed and processed for MTT analysis.**

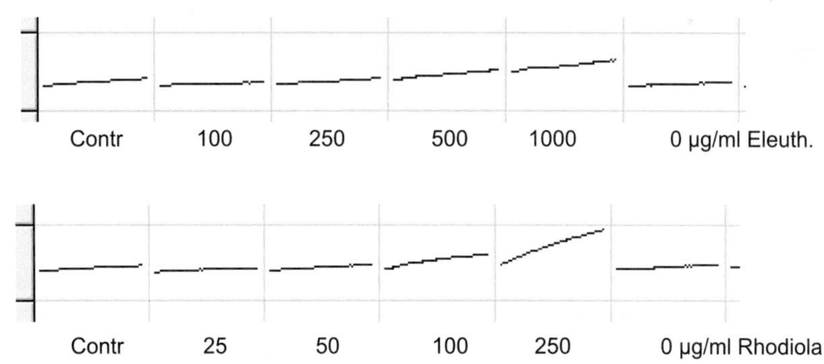

Fig. 5. **Effect of adaptogens on oxidation of cytoplasmic components using the DCF (Dichlorofluorescein) assay. Cell cultures, cultivated in 96-well ELISA plates, were exposed to increasing concentrations of *Eleutherococcus* or *Rhodiola*. A possible change in fluorescence (as a sign of oxidation) was evaluated in a fluorescence reader during a period of 1 hour. The pro-oxidant action of plant adaptogens can be compared with the background oxidation that occurs in the cytoplasm of control cultures. An increase in slope represents a higher rate of oxidation of cytoplasmic components.**

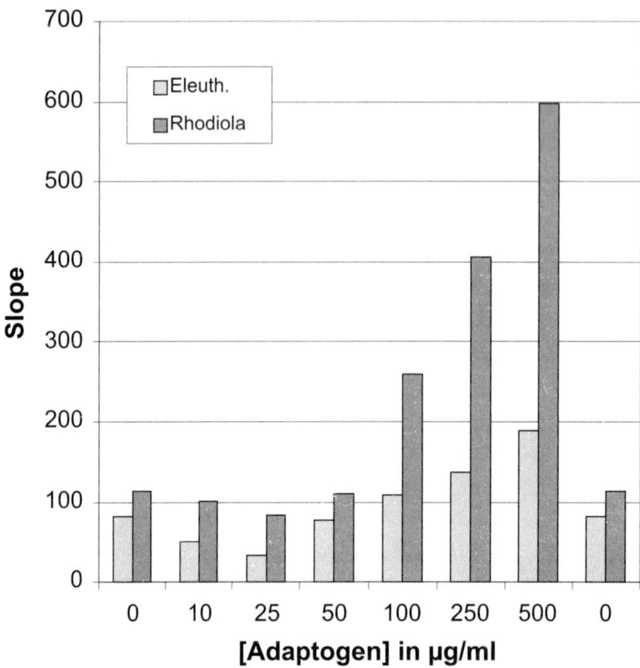

Fig. 6. Effect of adaptogens on oxidation of cytoplasmic components in Rat-1 fibroblasts. Cell cultures, grown in 96-well ELISA plates, were loaded with the DCF probe, rinsed and exposed to increasing concentrations of *Eleutherococcus* or *Rhodiola* (0-500 µg/ml) after which a change in fluorescence of the probe was monitored in a fluorescence reader. The slope represents the rate of oxidation of the DCF probe. At a concentration of 25 µg/ml both plant adaptogens show antioxidant properties followed by a pro-oxidant action at a concentration of 100 µg/ml and higher.

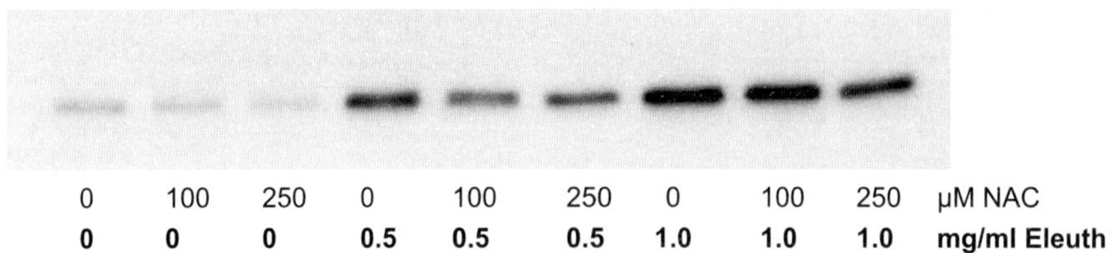

Fig. 7. Effect of *Eleutherococcus* (0.5 or 1.0 mg/ml) on the induction of hsp32 (HO-1) in the presence or absence of the antioxidant N-Acetyl Cystein (NAC)(100 or 250 µM). H35 cells were exposed to the indicated concentration of adaptogen and/or antioxidant for 6 hours, rinsed and processed for gel electrophoresis and Western blotting. The same amount of protein was loaded in each lane. HO-1 (hsp32) was visualized using specific antibodies obtained from StressGen. It can be observed that increasing concentrations of NAC reduce the enhanced synthesis of HO-1 in adaptogen treated cultures as well as the basic level in control cell cultures.

How to confirm that adaptogens consume GSH and cause initial (mild) oxidative damage? We chose a number of techniques to demonstrate the pro-oxidant action of plant adaptogens: the BSO-technique, DCF- and Bodipy measurements.

B. Plant Adaptogens are Toxic in GSH-depleted Cells

In our lab, assays are used to determine the anti-oxidative potential of specific compounds. A nice way is to use BSO (Buthionine Sulfoximine), a compound that inhibits the enzyme involved in the production of GSH (GSH synthetase). If BSO is added to cells, the level of GSH slowly and gradually decreases, simply because no new GSH can be synthesized. Thus, slowly and gradually the cells will be depleted from their GSH and hence will get more and more vulnerable to (oxidative) stress. Usually, when cell cultures are incubated for two or three days with 100 µM of BSO, the cells can already be killed by mild oxidative stress (which in control non-BSO treated cultures is harmless). Under these conditions the tests are performed, since the cells can still be saved by adding compounds that act as antioxidants. The more effective these anti-oxidative compounds are, the better they can 'replace' or 'mimic' GSH, the better the BSO-treated cells can withstand (oxidative) stress such as hydrogen peroxide.

The question was asked whether in these circumstances of GSH-depletion, plant adaptogens are able to save the GSH-depleted cells from a (oxidative) stress condition.

However, in stead of observing any antioxidative potential of plant adaptogens in the above mentioned test, a strong pro-oxidant action was measured. GSH-depleted cells became very sensitive to plant adaptogens. The remaining GSH pool was depleted even faster and since there was no way to activate the GSH synthesis (being blocked by BSO) the cells were killed in stead of being saved. This observation confirms that plant adaptogens consume GSH.

When plant adaptogens are added overnight to cell cultures that were treated with BSO for 2 or 3 days, it can be observed that the mere presence of plant adaptogens kills part of the cell population (Figure 4). Whereas BSO in itself does not affect cell survival, an overnight incubation with *Eleutherococcus* (500 µg/ml) or *Rhodiola* (100 µg/ml) kills about 50% of the cell culture (in the presence of 80 µM BSO) or even up to 80% (in the presence of 120 µM BSO). The chosen concentration of plant adaptogens did not affect more than 10% of control cell cultures (Figure 4).

In conclusion, at a concentration of BSO that lowers GSH, but which hardly affects survival, concomittant incubation with *Eleutherococcus* or *Rhodiola* strongly affects survival, which confirms the suggested pro-oxidant action of plant adaptogens.

C. Adaptogens Exert a Pro-oxidant Action on Lipids and on Cytoplasmic Components

In order to determine whether adaptogens exert a pro-oxidant action on membranes and in the cytoplasm, measurements were performed using the Bodipy-membrane probe (Drummen *et al.* 2004) to determine lipid peroxidation, whereas the DCF probe was used to establish the occurrence of oxidation of cytoplasmic components when cells are exposed to increasing concentrations of plant adaptogens. Using these measurements we could conclude that there are compounds in the extract that have a pro-oxidant action.

Do Plant Adaptogens Cause Any Lipid Peroxidation?

Membrane lipids are oxidatively damaged when adaptogens are added. An immediate increase in lipid peroxidation could be detected using the Bodipy-probe (Drummen *et al.* 2004). This probe is built into the membrane and, upon oxidation, changes its fluorescent characteristics which occurred upon exposure of cells to plant adaptogens.

But how to be sure that this is a direct effect caused by adaptogens? And not indirectly by way of triggering other mechanisms or by an interaction with cellular compounds that will lead to lipid peroxidation in an indirect way? A direct action could be confirmed in an experiment using Bodipy loaded lipid vesicles in a reaction tube to which adaptogens were added directly. In these vesicles only the lipid membrane components are incorporated and there are thus no interfering factors present, no cytoplasm, no enzymes, no proteins. And then exactly the same happened. Both *Eleutherococcus* and *Rhodiola* caused lipid peroxidation. This confirmed the observation that adaptogens exert a direct oxidative damage to the lipid molecules, the building blocks of the membranes.

Do Plant Adaptogens Cause any Oxidation of Cytoplasmic Components?

To answer this question we used the DCF (Dichlorofluorescein) assay to quantify cellular oxidative stress. The theory behind this technique is that nonfluorescent fluorescein derivatives will emit fluorescence after being oxidized. The probe is trapped inside the cell and the emitted fluorescence is proportional to the amount of oxidative stress (Wang and Joseph 1999). Cells were cultured in 96 well plates, loaded with DCF and subsequently exposed to increasing concentrations of plant adaptogens while a possible change in signal was detected in a fluorescence reader. As is shown in Figure 5, an increasing level of fluorescence is measured in separate wells upon exposure to higher concentrations of *Eleutherococcus* or *Rhodiola*. Measurements took place during 60 minutes. The slope of the signal was calculated and is shown in Figure 6. A bigger slope represents a higher level of pro-oxidant action. As an example the pro-oxidant effect of the plant adaptogens is shown in Rat1 fibroblasts. *Rhodiola* clearly exerts a higher level of pro-oxidant action in comparison with the same concentrations of *Eleutherococcus*.

D. Hsp32 Induction by Adaptogens is Reduced in the Presence of an Antioxidant

Finally, we determined whether the induction of heme oxygenase (HO-1 or hsp32) by plant adaptogens indeed occurred through a pro-oxidant action of the adaptogens. To this end, cells were exposed to increasing concentrations of the adaptogens but now in the presence of increasing concentrations of the antioxidant NAC (N-Acetyl-Cystein). What can be expected? The antioxidant may prevent some of the possible oxidative damage by the adaptogens, which may imply that a lower level of induction of hsp32 will be observed. However, when adaptogens exert a direct effect on the induction of heme oxygenase, even in the absence of any oxidative damage, then the antioxidant will not interfere with this direct induction and logically will not affect the level to which heme oxygenase is induced.

As can be observed in Figure 7, the presence of NAC reduces the hsp32-inducing effect of adaptogens. Thus, hsp32 appears to be induced through a pro-oxidant action of the plant adaptogens and not through influencing signaling pathways leading to change in gene expression directly. This confirms the idea that adaptogens cause (mild) oxidative intracellular damage to which cells react with an activation of their defense systems.

An important question is whether cells become resistant against oxidative stress when incubated with plant adatogens for 24 hours? Yes, they do. In order to prove this, cells were exposed to adaptogens for 24 hours, then they were loaded with DCF and subsequently exposed to different concentrations of hydrogen peroxide. As shown in Figure 8, it can clearly be observed that increasing

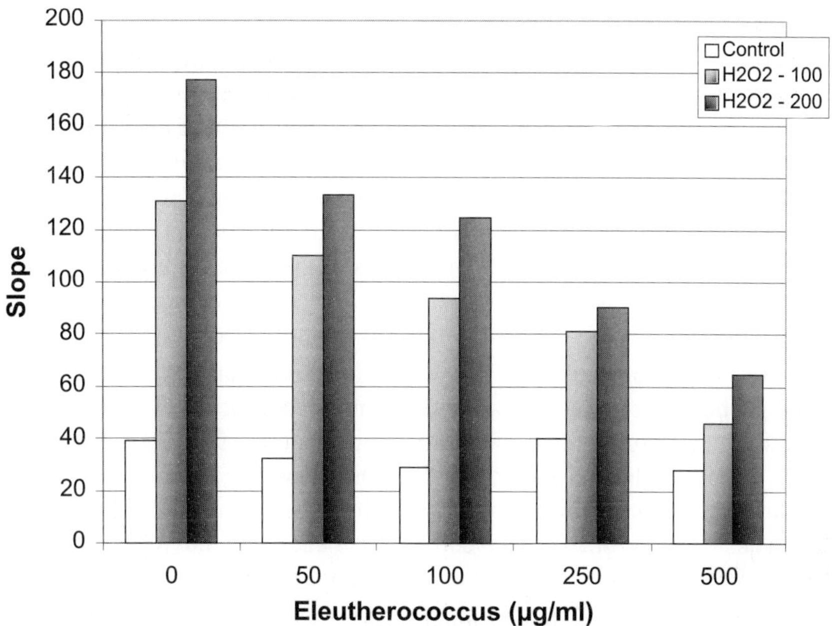

Fig. 8. **Protective effect of *Eleutheroccus* on oxidative stress induced damage in Rat-1 fibroblasts. Cell cultures were grown in 96-well plates and incubated with increasing concentration of adaptogen (0-500 μg/ml) for 24 hours. After this time period, cells were rinsed, loaded with DCF (see legends to Figures 5 and 6) and subsequently exposed to hydrogen peroxide (100 or 200 μM H_2O_2). The rate of oxidation is expressed as the slope. A decrease in slope reflects an enhanced resistance against oxidative stress. Higher concentrations of the adaptogen thus induce a higher capacity to resist oxidative damage.**

concentrations of *Eleutherococcus* (up to 500 μg/ml) reduce the slope, which means that there is less oxidative damage in comparison with control and the adaptogen thus did enhance the antioxidative cellular capacity. Based on our results it can be suggested that the enhanced level of GSH as well as the enhanced activity of heme oxygenase are crucial elements underlying this enhanced cellular resistance.

Discussion

The aim of our studies was to analyse whether extracts from plant adaptogens are able to enhance the level of heat shock proteins in cell cultures. In this paper we demonstrated that *Eleutherococcus senticosus* as well as *Rhodiola rosea* were able to specifically induce the synthesis of heme oxygenase (HO-1), which is a 32 kDa heat shock protein, whereas the other members of the heat shock protein family were not induced by plant adaptogens. Heme oxygenase is the rate-limiting enzyme required for the degradation of heme. Ryter *et al.* (2006) reviewed evidence on the antioxidative, antiapoptotic, antiinflammatory and antiproliferative mechanisms that were triggered by the three degradation products of heme: the bile pigments biliverdin and bilirubin, the gas carbon monoxide (CO) and free iron. In this respect, the enhanced presence or activity of HO-1 can be interpreted as a positive sign.

The enhanced presence of heme oxygenase in biological systems is frequently used as a marker for experienced oxidative stress. It is part of an endogenous cellular defence mechanism to increase protection against oxidative stress and to reach an adapted state.

With respect to the molecular background of HO-1 induction by plant adaptogens we have shown in this paper that adaptogens exert a pro-oxidant effect: they consume glutathione, they oxidize cytoplasmic components and they induce lipid peroxidation. This 'mild' damage triggers endogenous cellular defense- and repair mechanisms, the enhanced activity of which leads to a state in which the cells are more resistant to stress.

An important question is which compounds in the extracts, in addition to free radical quenching compounds, are the ones that have a pro-oxidant effect? And under what circumstances do the pro-oxidant compounds overrule the antioxidant compounds in the extracts leading to a net pro-oxidant action? And under which conditions an optimal beneficial effect can be obtained using the pro-oxidant action?

Another aspect relates to the definition of plant adaptogens, and especially the property that adaptogens should be harmless, which is mentioned in the definition by Brekhman (1980). Also Dardymov and Krikorian (2000) mention the harmlessness as one of the four general properties of a plant adaptogen: (1) it is harmless to the host; (2) it has a general, rather nonspecific, effect; (3) it increases the resistance of the recipient to a variety of physical, chemical, or biological stressors; and (4) for the user, it acts as a general stabilizer/normalizer. Panossian and Wagner (2005) changed the original criterion of 'being without harm', into 'an adaptogen must have a beneficial effect on an organism without inflicting more than 'marginal harm''. Based on our results, it could even be suggested that plant adaptogens should cause cellular harm in order to trigger defense- and repair mechanisms which brings cells into the adaptive state of enhanced resistance.

Without harm, cells are not alerted, no activation of defense mechanisms takes place and no adaptation will occur. When this scenario at the cellular level is proven to be true, then plant adaptogens may be described as hormetic agents. Hormesis is a term describing the beneficial effects of mild and repeated stimulation or stress, which ultimately bolsters defences against deleterious processes (Calabrese and Blain 2005). There has been an increase in interest with respect to research on the phenomenon of hormesis over the past two decades in different fields of science. In the field of anti-aging and rejuvenation research for instance, its use has even led to the proposition of clinical strategies (Rattan 2004, Kyriazis 2005).

An understanding of mild damage induced by adaptogens leading to enhanced resistance can also be compared with the effect of training on muscular activity, -resistance and endurance. Only by training (and repeated induction of mild damage and/or oxidative stress), muscles can become prepared for intensive use in which they are able to resist a high level of oxidative stress. The molecular mechanism of enhanced oxidative resistance due to exercise has recently been described in terms of an hormetic action (Li *et al.* 2006).

Another way to put it is that plant adaptogens can be viewed as 'stress vaccines'. A mild form of stress induces resistance or 'immunity' to subsequent severe forms of stress. A main difference with vaccination being that adaptogens do not trigger memory formation against stress in cells. Repeated application of adaptogens is required to maintain an adaptive state.

Acknowledgement

The Swedish Herbal Institute is gratefully acknowledged for support throughout the years.

References

1. Afanasiev SA, krylatov AV, Lasukova TV, Lishmanov YB. The role of inducible stress proteins in cardioprotective effects of Rhodiola rosea. J. Biokhimiya 1996; 61: 1779-1784.
2. Baker MA, Cerniglia GJ, and Zaman A. Microtiter plate assay for the measurement of glutathione and glutathione disulfide in large numbers of biological samples. Anal. Biochem. 1990; 190: 360-365.
3. Boon-Niermeijer EK, van den Berg A, Wikman G and Wiegant FAC. Phyto-adaptogens protect against environmental stress-induced death of embryos from the freshwater snail Lymnaea stagnalis. Phytomedicine 2000; 7(5): 389-399.
4. Brekhman II and Dardymov IV. New substances of plant origin which increase non-specific resistance. Ann. Rev. Pharmacol. 1969: 9: 419-430.
5. Brekhman II. Man and biologically active substance: the effect of drugs, diet and pollution on health. Pergamon Press, Oxford, 1980.
6. Calabrese EJ and Blain R. The occurrence of hormetic dose responses in the toxicological literature, the hormesis database: an overview. Toxicol. Appl. Pharmacol. 2005; 202(3): 289-301.
7. Chiu PY, Mak DH, Poon MK and Ko KM. In vivo antioxidant action of a lignan-enriched extract of Schisandra fruit and an anthraquinone-containing extract of Polygonum root in comparison with schisandrin B and emodin. Planta Med. 2002; 68(11): 951-6.
8. Drummen GPC, Makkinje M, Verkleij AJ, Op den Kamp JAF, and Post JA. Attenuation of lipid peroxidation by antioxidants in rat-1 fibroblasts: comparison of the lipid peroxidation reporter molecules *cis*-parinaric acid and C11-BODIPY$^{581/591}$ in a biological setting. Biochimica et Biophysica Acta 2004; 1636: 136-150.
9. Davydov M and Krikorian AD. Eleutherococcus senticosus (Rupr. & Maxim.) Maxim. (Araliacea) as an adaptogen; a closer look. J. Ethnopharmacology 2000; 72(3): 345-393.
10. Kelly GS. Rhodiola rosea: a possible plant adaptogen. Altern. Med. Rev. 2001; 6(3): 293-302.
11. Kirkby KA and Adin CA. Products of heme oxygenase and their potential therapeutic applications. Am. J. Physiol. Renal Physiol. 2006; 290: F563-F571.
12. Kyriazis M. Clinical anti-aging hormetic strategies. Rejuvenation Research 2005; 8(2): 96-100.
13. Li JJ, Gomez-Cabrera MC and Vina J. Exercise and hormesis: activation of cellular antioxidant signaling pathways. Ann. NY. Acad. Sci. 2006; 1067: 425-435.
14. Maines MD. Heme oxygenae: function, multiplicity, regulatory mechanisms and clinical applications. FASEB J. 1988; 2(10): 2557-2568.
15. Maines MD. The heme oxygenase system: a regulator of second messenger gases. Annu. Rev. Pharmacol. Toxicol. 1997; 37: 517-554.
16. Meerson F. Adaptation, stress and prophylaxis. Springer Verlag, 1984.
17. Panossian A, Wikman G and Wagner H. Plant adaptogens III. Earlier and more recent aspects and concepts on their mode of action. Phytomedicine 1999; 6: 287-300.
18. Panossian AG. Adaptogens: tonic herbs for fatigue and stress. Alternative & Complentary Therapies 2003; 9(6): 327-331.
19. Panossian A and Wagner H. Stimulating effect of adaptogens: an overview with particular reference to their efficacy following single dose administration. Phytotherapy Research 2005; 19(10): 819-838.
20. Prawan A, Kundu JB and Surh Y-J. Molecular basis of heme oxygenase-1 induction: implications for chemoprevention and chemoprotection. Antioxidants & Redox Signaling 2005; 7(11-12): 1688-1703.
21. Rattan SI. Hormetic mechanisms of anti-aging and rejuvenating effects of repeated mild heat stress on human fibroblasts in vitro. Rejuvenation Research 2004; 7(1): 40-48.
22. Ryter SW, Alam J and Choi AMK. Heme oxygenase-1/carbon monoxide: from basic science to therapeutic applications. Physiol. Rev. 2006: 86: 583-650.

23. Tietze F. Enzymatic method for quantitative determination of nanogram amounts of total and oxidized glutathione: applications to mammalian blood and other tissues. Anal. Biochem. 1969; 27: 502-522.
24. Wagner H, Nörr H, and Winterhoff H. Plant adaptogens. Phytomedicine 1994;1:63-76.
25. Wagner H. Immunstimulants and adaptogens from plants. Phytochemistry of medicinal plants. In: Arnason J. ed. Plenum Press, New York, 1995. pp 1-18.
26. Wang H and Joseph JA. Quantifying cellular oxidative stress by dichlorofluorescein assay using microplate reader. Free Radical in Biology & Medicine 1999; 27: 612-616.
27. Wiegant FAC, Šurinová S, Ytsma E, Langelaar-Makkinje M, Post JA and Wikman G. Life-span enhancing effects of plant adaptogens in *C.elegans*. Proceedings of 8th World Congress of the International Society of Adaptive Medicine, Moscow, 2006, pp 61.
28. Wit de R, Makkinje M, Boonstra J, Verkleij AJ and Post JA. Hydrogen peroxide reversibly inhibits epidermal growth factor (EGF) receptor internalization and coincident ubiquitination of the EGF receptor and Eps15. FASEB J. 2001; 15: 306-308.

Adaptation Biology and Medicine (Volume 5: Health Potentials)
Editors: L. Lukyanova, N. Takeda and P.K. Singal

24. Interactions between Cytokines and Protease Inhibitors in the Mechanisms Responsible for Thermoregulatory Responses During Systemic Inflammation

Valery N. Gourine* and Alexander V. Gourine

Institute of Physiology, National Academy of Sciences of Belarus, Academicheskaya Str. 28, Minsk 220725, Belarus

ABSTRACT

Recent data suggest that protease inhibitors such as α_1-antitrypsin (α_1A) and α_2-macroglobulin (α_2M) may play an important role in the development of the thermoregulatory responses during systemic inflammation. We have showed that: (i) protease inhibitors α_1A and α_2M induce febrile response in rats and mice when injected in amounts similar to these observed during the development of the acute phase response; (ii) α_1A and α_2M induce rise in plasma concentrations of pyrogenic cytokines interleukin-6 (IL-6) and tumor necrosis factor-α (TNFα) in mice; (iii) α_1A and α_2M induce production of IL-6 and TNFα by rat blood mononuclear cells *in vitro*; (iv) α_2M gene knockout mice develop a short-lasting and markedly attenuated febrile responses to endotoxin compared with their wild-type counterparts; (v) during febrile response, plasma concentration of TNF-α is significantly lower in the α_2M knockout mice compared with their wild-type controls. These data support the idea that protease inhibitors are essential for the normal development of the febrile response and cytokine responses during systemic inflammation. The results also suggest that putative mechanism of protease inhibitors involvement in febrile response is through the regulation of cytokine production and/or clearance.

Keywords: α_1-antitrypsin, α_2-macroglobulin, interleukin, tumor necrosis factor, endotoxin, lipopolysaccharide, systemic inflammation, thermoregulation

Introduction

It is generally accepted that fever is an adaptive response of the organism to infection, injury or trauma aimed at facilitating host resistance and slowing the growth of the pathogen (Kluger, 1991). Considerable evidence accumulated by now indicate that cytokines interleukin(IL)-1β, tumor necrosis factor-α(TNFα), IL-6 and others are key mediators of the regulated rise in body temperature,

*Address for Correspondence: Professor Valery N. Gourine, Institute of Physiology National Academy of Sciences of Belarus, Academicheskaya Str. 28 Minsk 220725, Belarus. E-mail: srnalex@yahoo.com

i.e. fever, essential for the development of the febrile response by altering the "set point" for T_b regulation (Kluger, 1991; Kluger *et al.*, 1995).

There is evidence that protease inhibitors such as α_1-antitrypsin (α_1A) and α_2-macroglobulin (α_2M) regulate cytokine metabolism (Borth and Luger, 1989; Crookston *et al.*, 1994; Gonias *et al.*, 1994; LaMarre *et al.*, 1991; Matsuda *et al.*, 1989). Serine protease inhibitors constitute the third largest group (after albumin and immunoglobulins) of functional blood proteins. α_1A and α_2M are the major plasma protease inhibitors in mammals. α_1M is also specific cytokine carrier which binds major pro- and anti-inflammatory cytokines implicated in the febrile response, including IL-1β, IL-6, and TNF-α (Borth and Luger, 1989; Crookston *et al.*, 1994; LaMarre *et al.*, 1991; Matsuda *et al.*, 1989). In a vast amount of literature (impossible to list here) protease inhibitors have been implicated in regulation of the activity of the immune system and cytokine metabolism. However, the precise role of protease inhibitor-cytokine interactions *in vivo* is far from clear.

Protease inhibitors are produced in various tissues, and production of these factors markedly increases during infection and inflammation (Geiger *et al.*, 1992; Gourine and Gourine, 2004; Koj, 1985, 1996; Sharma *et al.*, 1992). The pattern of production of different protease inhibitors during the acute phase response differs significantly among species (Koj, 1996). However, as a rule, in different species at least one of the protease inhibitors (either α_1A, α_2M or others) is among the acute phase proteins and during inflammation its concentration rises markedly in the blood, in a variety of peripheral tissues and also under certain conditions in the brain (Geiger *et al.*, 1992; Koj, 1985, 1996; Sharma *et al.*, 1992).

Protease inhibitors are often considered to be nonspecific defense molecules. One of the main functions of protease inhibitors is to protect tissues from unwanted proteases released by pathogenic microorganisms or from the dying cells of the host. In this brief review we present and discuss data obtained recently in our laboratory, suggesting that protease inhibitors, and α_2M in particular, play an important role in the febrile response by regulating levels of pyrogenic cytokines. These findings indicate a novel role for plasma protease inhibitors *in vivo.*

Experimental Evidence

Protease inhibitors α_1A and α_2M induce febrile response and sickness syndrome when injected (Gourine et al., 1999a, 1999b; Gurin et al., 1999a, 1999b; Gourine and Gourine, 2004). In experiments in mice we showed that intravenous administration of α_1A and α_2M in amounts similar or even smaller than those observed during the development of the inflammatory response results in a significant elevation in body temperature. Injection of α_2M also induces a significant decrease in food intake and decrease in body weight in mice. In experiments in rats we showed that α_1A after intracerebroventricular administration induces long-lasting (more than 7h) fever.

Protease inhibitors α_1A and α_2M induce rise in plasma concentrations of pyrogenic cytokines interleukin-6 (IL-6) and tumor necrosis factor-α (TNFα) (Gourine et al., 1999a). The rise in body temperature after injection of α_1A or α_2M is also accompanied by a significant increase in plasma concentrations of IL-6 and TNFα. It was also found that the IL-6 activity was significantly higher in blood plasma of IL-6 knockout mice injected with IL-6 together with α_2M as compared to mice injected with IL-6 alone.

Protease inhibitors $\alpha_1 A$ and $\alpha_2 M$ induce production of IL-6 and TNFα by rat blood mononuclear cells (BMC) in vitro (Sandakov et al., 1999, 2000). As described above, administration of α_1A and α_2M results in a profound increase in plasma levels of IL-6 and TNF in mice. It has been also shown that α_1A and α_2M are able to induce IL-6 and TNF production by cultured rat BMC, suggesting that these protease inhibitors are potent inducers of cytokine production *in vitro*.

$\alpha_2 M$ is essential for the normal development of LPS-induced febrile response (Gourine et al., 2000). The role of α_2M in regulation of the febrile and cytokine responses during the development of the systemic inflammatory response induced by endotoxin was studied using α_2M gene knockout mice developed by Umans *et al*., (1995). These transgenic animals provide us with an excellent model to investigate the role of the major plasma protease inhibitor in fever and acute phase response. α_2M knockout mice develop a short-lasting and markedly attenuated febrile response in response to endotoxin compared with their wild-type counterparts. During febrile response plasma concentration of TNF-α, but not IL-1β or IL-6, is significantly lower in the α_2M knockout mice compared with their wild-type controls. Interestingly, endotoxin-induced increase in hepatic TNF-α, IL-1β and IL-6 mRNA levels are identical in α_2M knockout and wild-type mice at 1.5 and 4 h after injection.

Discussion

Results described above indicate that plasma protease inhibitors play an important role in the development of febrile response during systemic inflammation. For a substance to be legitimately categorized as an endogenous mediator of fever, it should fulfill several criteria (Kluger, 1991): (1) Criterion 1 is that application of the putative mediator of the febrile response at the hypothetical site of action results in a rise in body temperature; (2) Criterion 2 is that during the febrile response, the released putative mediator of fever has some quantitative relationship to the magnitude of the rise in body temperature, and the putative mediator of the febrile response is released in amounts similar to those that cause an equivalent rise in body temperature when the putative mediator of fever is given; (3) Criterion 3 is that febrile response, induced by the putative mediator, should be associated with appropriate thermoregulatory effector responses, that are indicative of an increase in thermoregulatory set point (e.g., induction of peripheral vasoconstriction, rise in shivering metabolism); (4) Criterion 4 is that substances that prevent the production of the putative mediator of febrile response should cause smaller rises in body temperature than those that do not; (5) Criterion 5 states that substances that prevent the action of the putative mediator of the febrile response should cause smaller rises in body temperature than those that do not.

The results of our studies taken together with data from the literature indicate that most of the above criteria are completely or partially fulfilled for protease inhibitors and for α_2M, in particular. Two protease inhibitors α_1A and α_2M are able to induce febrile response after intravenous injection. α_1A and α_2M induce febrile response when injected in amounts similar (and even smaller) to those that observed during the development of the acute phase response. Thus, criterion 1 is satisfied.

There is ample evidence that during development of the inflammatory response the concentration of α_1A and/or α_2M rises significantly in the blood and in a variety of peripheral tissues and also under certain conditions in the brain (Geiger *et al*., 1992; Koj, 1985, 1996; Sharma *et al*., 1992). However, a quantitative relationship of α_1A and/or α_2M release to the magnitude of the rise in body temperature during febrile response has not been specifically investigated. Thus, criterion 2 is only partially fulfilled for protease inhibitors.

Sandakov (2000) showed that in rabbits febrile response induced by α_1A is accompanied by marked peripheral vasoconstriction. Because the rabbit relies to a large extent on modulations in ear blood flow to regulate body temperature, these results support criterion 3.

We have also shown that α_2M gene knockout mice develop short-lasting and markedly attenuated febrile response in response to endotoxin, indicating that α_2M is essential for the normal development of the febrile response. Thus, criteria 4 and 5 are also fulfilled for α_2M.

Our data also indicate that protease inhibitors modulate febrile response by regulating levels of pyrogenic cytokines that are responsible for initiation and maintenance of the febrile response. Evidence of close interrelationships between protease inhibitors and cytokines (Borth and Luger, 1989; Crookston *et al.*, 1994; Gonias *et al.*, 1994; LaMarre *et al.*, 1991; Matsuda *et al.*, 1989) strongly support this hypothesis.

In experiments in mice we have observed a profound increase in plasma levels of IL-6 and TNF-α shortly after α_1A or α_2M treatment. The time-course of changes in IL-6 and TNF-α plasma levels following α_1A or α_2M treatment was similar to that observed during endotoxin-induced febrile response. These data suggest that pyrogenic cytokines may be responsible for the development of fever and acute phase response induced by α_1A and α_2M. It is, however, unclear whether this increase in plasma IL-6 and TNF-α levels is due to stimulation of the production of these cytokines *de novo* or is due to inhibition of rapid clearance of low-levels of IL-6 and TNF-α, which are produced even under basal, "non-pathological" conditions. The later mechanism is possible since α_1A or α_2M are potent inhibitors of serine proteases involved in cytokine degradation. Future studies of changes in hepatic IL-6 and TNF-α mRNA levels following α_1A or α_2M injection will ultimately answer this question.

In experiments *in vitro* we have found that treatment of cultured rat blood mononuclear cells with α_1A and α_2M results in a marked increase in IL-6 and TNF concentration in the incubation media. Assuming that proteolytic degradation of cytokines in the *in vitro* system is low, we speculate that α_1A and α_2M are directly stimulating BMC to produce and release IL-6 and TNF. IL-6 activity was also significantly higher in blood plasma of IL-6 knockout mice injected with IL-6 together with α_2M as compared to mice injected with IL-6 alone. Taken together the data indicate that α_1A and α_2M may be involved in regulation of both cytokine production as well as cytokine clearance.

Experiments in α_2M gene knockout mice provided convincing evidence that protease inhibitors, and α_2M in particular, play important roles in regulation of plasma cytokine concentrations *in vivo*. It was found that smaller febrile response in α_2M gene knockout mice was accompanied by lower plasma concentration of TNF-α, which is considered as one of the major proinflammatory cytokine and its lower levels in plasma of knockout mice may indicate attenuated inflammatory response. There were no difference in endotoxin-induced increase in hepatic TNF-α mRNA levels between α_2M gene knockout and wild-type mice. Since endotoxin-induced TNF-α gene expression in the liver is not affected by α_2M elimination we hypothesize that a putative mechanism of α_2M involvement in febrile response is through the inhibition of TNF-α clearance.

Our studies identified protease inhibitors as novel endogenous mediators essential for the normal development of the febrile response to bacterial infection. The putative mechanism of protease inhibitors involvement in the febrile response is through the regulation of cytokine production and/or clearance. α_2M, as one of the major plasma protease inhibitors, plays an important role in the febrile response, presumably by inhibiting clearance of TNF-α - one of the major proinflammatory cytokines. This information could ultimately lead to the development of novel approaches of modulating inflammatory response, which may be important when excessive production of proinflammatory cytokines is harmful for the host.

Acknowledgments

We thank Prof. M. Kluger, Dr. D. Sandakov, Mrs. K. Rudolph, Dr. Y. Tesfaighi and Dr. W. Kozak for their valuable contribution to this study and Dr. F. Van Lueven from Experimental Genetics Group, Department of Human Genetics (Leuven, Belgium) for providing α_2M gene knockout mice. This study was supported by National Academy of Sciences of Belarus and by the Fogarty International Center, NIH Grant # R03 TW00992-01.

References

1. Borth W, Luger TA. Identification of α_2-macroglobulin as a cytokine binding plasma protein. Binding of interleukin-1β to "F" α_2-macroglobulin. *J Biol Chem.* 1989; 264: 5818-5825.
2. Crookston KP, Webb DJ, Wolf BB *et al.* Classification of α_2-macroglobulin-cytokine interactions based on affinity of noncovalent association in solution under apparent equilibrium conditions. *J Biol Chem.* 1994; 269: 1533-1540.
3. Geiger T, Fischer M, Jagher B *et al.* Changes in hepatic mRNA levels of acute phase proteins during rat adjuvant arthritis. *Clin Exp Rheum.* 1992; 10: 467-473.
4. Gonias SL, LaMarre J, Crookston KP *et al.* α_2-Macroglobulin and the α_2-macroglobulin receptor/LRP: A growth regulatory axis. *Ann N Y Acad Sci.* 1994; 737: 273-290.
5. Gourine AV, Gourine VN, Sandakov DB, Rudolph K, Kluger MJ. Febrile and cytokine responses induced by acute phase proteins α_2-macroglobulin and α_1-antitrypsin (serine protease inhibitors) in mice. *Neurosci Abstr.* 1999; 1695.
6. Gourine AV, Rudolph K, Umans L, Van Leuven F, Gourine VN, Kluger MJ. Febrile response in mice lacking α_2-macroglobulin gene. In: Gourine VN. ed., *Basic and Applied Thermophysiology,* Polibig, Minsk, 2000; pp. 189-191.
7. Gourine VN, Gourine AV. *Biologically active substances of the blood and thermoregulation.* Polibig, Minsk, 2004; 215 p. (in Russian).
8. Gourine VN, Sandakov DB, Gourine AV, Rudolph K, Kluger MJ. Acute phase proteins, α_1-antitrypsin and α_2-macroglobulin (serine protease inhibitors) induce acute phase response in mice and rats. In: *Abstracts of the XI International Symposium Pharmacology of Thermoregulation.* Seville, 1999.
9. Gurin VN, Sandakov DB, Gurin AV. Brain proteinase activity and the mobilization of endogenous proteolytic inhibitors as factors in the development of the functional state of the body. *Fiziol Cheloveka.* 1999; 25: 71-77 (in Russian).
10. Gurin VN, Sandakov DB, Gurin AV. Role of proteinase inhibitors in the process that provide coupling between body's functional systems. *Vestn Ross Akad Med Nauk.* 1999; 6: 25-29 (in Russian).
11. Kluger MJ. Fever: role of pyrogens and cryogens. *Physiol Rev.* 1991; 71: 93-127.
12. Kluger MJ, Kozak W, Leon LR, Soszynski D, Conn CA. Cytokines and fever. *Neuroimmunomodulation.* 1995; 2: 216-223.
13. Koj A. Biological functions of acute phase proteins. In: Gordon AH, Koj A, eds., *The acute phase response to injury and infection*, Elsevier, Amsterdam, 1985: p. 145-160.
14. Koj A. Initiation of the acute phase response and synthesis of cytokines. *Biochim Biophys Acta.* 1996; 1317: 84-94.
15. LaMarre J, Wollenberg GK, Gonias SL, Hayes MA. Cytokine binding and clearance properties of proteinase activated α_2-macroglobulin. *Lab Invest.* 1991; 65: 3-14.

16. Matsuda T, Hirano T, Nagasawa S, Kishimoto T. Identification of α_2-macroglobulin as a carrier protein for IL-6. *J Immunol.* 1989; 142: 148-152.
17. Sandakov DB. *Functional state of the thermoregulatory system upon action of the serine proteinase inhibitors α_1-antitrypsin and* a_2*-macroglobulin. Thesis cand. biol. sciences.* Inst.Physiol., Natl. Acad. Sci., Minsk. 2000.
18. Sandakov DB, Rudolph K, Gourine AV, Gourine VN, Kozak W, Kluger MJ. The acute phase proteins, α_1-antitrypsin and α_2-macroglobulin, induce production of cytokines by rat leukocytes in vitro. In: Gourine V.N, ed., *Recent Advances in Thermal Biology,* Polibig, Minsk, 1999, pp. 89-92.
19. Sandakov DB, Rudolph K, Gourine AV, Gourine VN, Kluger MJ. The influence of protease inhibitors α_1-antitrypsin and α_2-macroglobulin on interleukin-6 and tumor necrosis factor production by rat mononuclear cells *in vitro*. *Arch Clin Exp Med.* 2000; 9: 93-94 (in Russian).
20. Sharma RJ, MacAllan DC, Sedgwick P, *et al.* Kinetics of endotoxin-induced acute-phase protein gene expression and its modulation by TNF-α monoclonal antibody. *Am J Physiol.* 1992; 31: R786-R793.
21. Umans L, Serneels L, Overbergh L, *et al.* Targeted inactivation of the mouse a_2-macroglobulin gene. *J Biol Chem.* 1995; 270: 19778-19785.

Adaptation Biology and Medicine (Volume 5: Health Potentials)
Editors: L. Lukyanova, N. Takeda and P.K. Singal

25. Impairment of Adaptive Responses to Heat Exposure: Aging and Bacterial Translocation

Toshihiko Katafuchi*

Department of Integrative Physiology, Graduate School of Medical Sciences, Kyushu University, Fukuoka 812-8582, Japan

Summary

To investigate the age-related changes in the expression of interleukin-1β (IL-1β) and its related substances in the brain during heat stress, amounts of mRNAs for IL-1β, cyclooxegenase-2 (COX-2), and an inhibitor of nuclear factor (NF)-κB-β (IκB-β) that is known to reflect an activation of NF-κB, in the cortex, cerebellum and hippocampus were measured using a quantitative real-time capillary PCR method. The basal level of IL-1β mRNA in aged rats (108-110 w.) was significantly higher than that in young animals (10-11 w.) in these brain regions. Heat exposure (33°C) for 1 hr enhanced the expression of IL-1β and COX-2 mRNAs in aged rats but not in young ones. The amount of lipopolysaccharide (LPS) assessed by its bioactivity in the cortex increased by heat exposure only in aged rats. However, an intraperitoneal pretreatment with endotoxin inhibitor (EI), a synthetic peptide that detoxifies LPS by binding to a toxic component of LPS, lipid A, significantly attenuated the heat exposure-induced increases in mRNAs for IL-1β, COX-2, IκB-β, and the LPS activity. Administration of EI also debilitated the heat exposure-induced hyperthermia and responses of plasma ACTH and catecholamines. These findings suggest that the heat exposure-induced bacterial translocation is one of the mechanisms of the increased expression of mRNAs for IL-1β and its related substances in aged rats. It is considered that heat exposure induces greater redistribution of blood flow away from the gut to the superficial regions of the body for heat loss in aged rats than in young animals, thereby resulting in gut ischemia-induced bacterial translocation. This may explain at least in part the mechanisms of the age-related impairment in the adaptive responses to environmental stress.

Keywords: Bacterial translocation, lipopolysaccharide, heat exposure, endotoxin, inhibitor, inter lenkin-1β

Introduction

Cytokines in the brain have various influences on the autonomic, neuroendocrine, and behavioral systems, including the sympathetic nerve activity, the hypothalamic-pituitary-adrenal (HPA) axis, feeding, sleep, nociception, and peripheral immunity (Rothwell & Hopkins, 1995; Hori *et al.*, 2001). It is also known that these homeostatic functions are affected by aging, suggesting that the age-related changes in production and/or action of cytokines may be involved at least in part in the central

*Address for Correspondence: Toshihiko Katafuchi, M.D., Ph.D., Department of Integrative Physiology, Graduate School of Medical Sciences, Kyushu University, Fukuoka 812-8582, Japan. Tel: +81-92-642-6087 Fax: +81-92-642-6093, E-mail: kataf@physiol.med.kyushu-u.ac.jp

mechanisms of aging. For example, it has been reported that neuronal expression of a proinflammatory cytokine, interleukin-1β (IL-1β), increases with aging so as to contribute as a trigger for the age-induced impairment in long-term potentiation (Murray & Lynch, 1998), and that the increased neuronal expression of IL-1β is suggested to be a feature of the age-related neurodegeneration (Campbell *et al.*, 1998).

One of the characteristic aspects of aging is a progressive impairment in the ability to adapt to environmental challenges. In this article, to explore an impairment of adaptive responses to the environmental stress in the aged animals, expression of mRNAs in the brain for IL-1β and its related substances such as cyclooxygenase-2 (COX-2), an enzyme induced by IL-1β and activation of nuclear factor (NF)-κB to produce prostaglandins, and the inhibitor of NF-κB-β (IκB-β) that is known to contribute to persistent NF-κB activation (Stasiolek *et al.*, 2000) were examined in young and aged rats. Furthermore, an involvement of enteric flora-derived lipopolysaccharide (LPS) in the induction of IL-1β mRNA during heating was investigated by using an endotoxin inhibitor (EI) (Rustici *et al.*, 1993; Katafuchi *et al.*, 2003b).

Enhanced Expression of IL-1β, COX-2, and IκB-β mRNAs in the Brain after Heat Exposure in Aged Rats

Heat exposure (33°C, 1 hour) resulted in an increase in rectal temperature (2-3°C) in both young (10-11 weeks old) and aged (108-110 weeks old) Wistar rats, which was not statistically different between them. When amounts of Il-1β mRNA in the cortex were measured quantitatively using a real-time capillary PCR method (Morrison *et al.*, 1998; Katafuchi *et al.*, 2003a), only aged but not young animals showed an increase the mRNA following heat stress (Fig. 1A), suggesting an elevated sensitivity of cytokine production (Katafuchi *et al.*, 2003b). It has been shown that the LPS-stimulated production of IL-1β in gingival fibroblast cells is enhanced in aged rats compared with that in the young animals (Okamura *et al.*, 1999). This finding is compatible with our result showing an elevated sensitivity of IL-1β mRNA expression in aged rats. Furthermore, the basal level of IL-1β mRNA was significantly higher in aged rats than in young ones (Fig. 1A). This may be consistent with the previous reports demonstrating an increase in neuronal expression of IL-1β in aged rats (Murray & Lynch, 1998), and the increased intracellular levels of IL-1β in monocytes of older subjects (O'Mahony *et al.*, 1998).

The levels of COX-2 and IκB-β mRNAs in the cortex also markedly increased after heat exposure in aged rats but not in young ones (Katafuchi *et al.*, 2003b). It has been suggested that IκB-β is a stress protein inducible by hyperthermia and that an increased IκB-β expression in a non-phosphorylated form contributes to persistent activation of NF-κB (Stasiolek *et al.*, 2000). Since transcription of COX-2 gene is enhanced by activation of NF-κB, the increase in COX-2 mRNA after heat exposure in aged rats might result from an elevation of IL-1β in an IL-1β-dependent manner (Crofford *et al.*, 1997; Sorli *et al.*, 1998).

Increase in LPS Activity Following Heat Exposure in Aged Rats

It has been reported that IL-1β in various brain regions increases in rats, mice and rabbits after systemic administration of LPS (Ban *et al.*, 1992; Hillhouse & Mosley, 1993; Laye *et al.*, 1994; Nakamori *et al.*, 1994; Quan *et al.*, 1994). IL-1 synthesizing cells in the central nervous system have

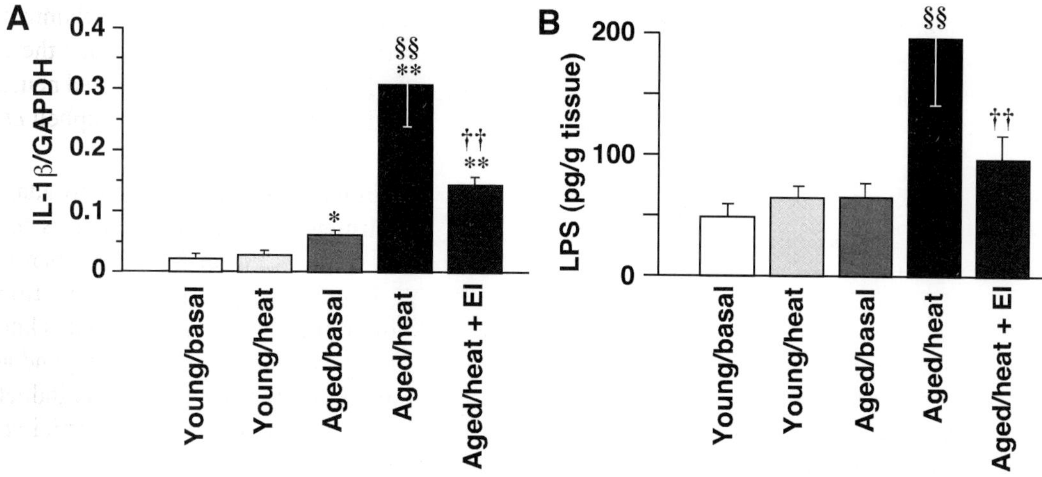

Fig. 1. Effects of heat exposure on IL-1β mRNA and LPS activity in the cortex. Two-way ANOVA showed an overall difference between groups in IL-1β mRNA (A) and LPS activity (B). Amounts of IL-1β mRNA in each sample were normalized to the corresponding concentration of a house keeping gene, glycealdehyde 3-phosphate dehydrogenase (GAPDH). Number of each group was 6. *$P<0.05$ and **$P<0.01$, vs. Young/basal; §§$P<0.01$, vs. Aged/basal; ††$P<0.01$, vs. Aged/heat (Newman-Keuls' post hoc test).

been reported to be glial cells in the parenchyma including perivascular regions following peripheral administration of LPS (Buttini & Boddeke, 1995; Quan *et al.*, 1998b). In addition, it has been also shown that systemic injection of LPS induces COX-2 mRNA in cerebral endothelial cells (Matsumura *et al.*, 1998; Quan *et al.*, 1998a) as well as perivascular microglia (Elmquist *et al.*, 1997). Although all of these experiments were performed in the young adult animals, we found that an expression of brain IL-1β mRNA during heat stress was different in young and aged rats (Katafuchi *et al.*, 2003b). These findings suggest that the heat exposure-induced enhancement of brain IL-1β expression is due to an increase in LPS activity in the blood or brain, which may occur only in aged animals. In fact, we found that LPS in the cortex significantly increased after heat exposure in aged rats but not in the young, while the basal level of LPS was not different between them (Fig. 1B). The method for LPS measurement we used was sensitive to lipid A that is the active component of LPS endotoxins responsible for the stimulation of immune cells (Katafuchi *et al.*, 2003b).

Attenuation of IL-1β Expression and LPS Activity by Endotoxin Inhibitor

To examine whether the increased LPS activity in the cortex of aged rats was responsible for the expression of IL-1β, COX-2 and IκB-β mRNAs during heat exposure, an EI, a synthetic peptide that binds to lipid A with high affinity, detoxifies LPS, and prevents the LPS-induced cytokine release *in vivo* (Rustici *et al.*, 1993) was administered before heat exposure. This peptide has been also shown to suppress the febrile response to LPS, raise the survival rate after LPS injection, but have very low toxicity and lethality compared with a cyclic peptide antibiotic, polymyxin B (Fig. 2). As shown in Fig. 1A and B, pretreatment with EI significantly attenuated the heat exposure-induced increase in I IL-1β and LPS activity in the cortex. Increases in COX-2 and IκB-β mRNAs induced by heat exposure were also suppressed by pretreatment with EI (Katafuchi *et al.*, 2003b). These findings, taken

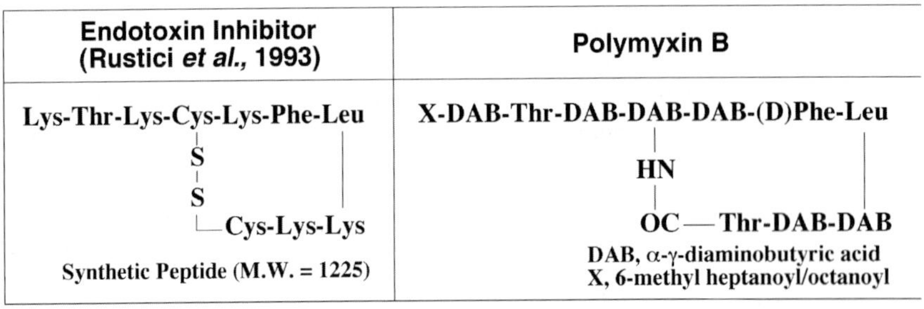

Endotoxin Inhibitor (Rustici *et al.*, 1993)	Polymyxin B
Lys-Thr-Lys-Cys-Lys-Phe-Leu S S Cys-Lys-Lys Synthetic Peptide (M.W. = 1225)	X-DAB-Thr-DAB-DAB-DAB-(D)Phe-Leu HN OC — Thr-DAB-DAB DAB, α-γ-diaminobutyric acid X, 6-methyl heptanoyl/octanoyl

	Endotoxin Inhibitor	Polymyxin B
Antibiotic Activity	<<	<<
Pyrogenicity of Peptide-LPS Complex	(-)	(-)
Lethality (Toxicity)	0% (60 mg/kg iv)	100% (10 mg/kg iv)
LPS-induce TNF-α and IL-6 Release in vivo (% Inhibition)	TNF-α 93% IL-6 84%	TNF-α 92% IL-6 90%
Sruvival 72 hr after LPS (ip) (control group = 5%)	40%	30%

Fig. 2. Structure of endotoxin inhibitor (EI) and polymyxin B. EI suppresses the LPS-induced fever and cytokine release, and raises the survival rate after LPS like polymyxin B, but has much smaller antibiotic activity and lethality.

together, suggest that the enhancement of mRNA expression is attributable to an increase in LPS activity in the brain. Furthermore, pretreatment with EI attenuated the hyperthermia and ACTH response induced by heat exposure (Katafuchi *et al.*, 2003b), indicating that the responses of body temperature and the hypothalamo-pituitary-adrenal axis to heat exposure are also mediated, at least in part, by the increased LPS activity in the brain.

Possible Mechanisms of Heat Exposure-induced Increase in IL-1β Expression and LPS Activity in Aged Rats

It is quite possible that the increased LPS in the brain may be derived from the indigenous bacteria in the gastrointestinal tract (bacterial translocation). Generally, three primary mechanisms that promote the bacterial translocation have been proposed: (1) intestinal bacterial overgrowth due to excessive oral antibiotics, bowel obstruction etc., (2) deficiencies in host immune defenses due to immunosuppressive agents, lymphoma, etc., and (3) increased permeability or damage to the intestinal mucosal barrier due to thermal injury, hemorrhagic shock, etc. (Berg, 1995).

Endotoxemia has been reported in adult patients during heatstroke (Bouchama *et al.*, 1991). During heat exposure or heatstroke, the activation of heat loss mechanisms such as an increase in skin blood flow and sweating (salivation in rats) evokes an increase in the systemic blood volume and a decrease in the central blood volume. This redistribution of the blood results in a decrease in the splanchnic blood flow, i.e., gut ischemia, leading to the increase in plasma endotoxin. It has shown that the decrease in blood flow in the gastrointestinal tract is much larger in aged rabbits than in young ones, while the increase in the cardiac output is smaller in aged animals (Hales & Sakurada, 1998). Therefore, it is possible that the greater redistribution of blood flow away from the gut of the aged rabbit is a compensatory response for the relatively poor capacity to increase cardiac output to increase or maintain blood flow in heat loss tissues such as skin, tongue and nasal mucosa. This hypothesis may well account for the heat-exposure-induced increase in LPS activity in the brain in aged but not in the young rats.

LPS has been reported to activate immune cells either after binding to the LPS binding protein, then to CD14 molecule and one of the toll-like receptor superfamily, (TLR-4) (Triantafilou & Triantafilou, 2002), or direct binding to LPS receptors such as CD11c/CD18 (Ingalls & Golenbock, 1995) and L-selectin (Malhotra *et al.*, 1996), thereby inducing an activation of NF-κB (Wyckoff *et al.*, 1998), which leads to the enhanced expression of NF-κB-inducible cytokine genes including IL-1β (May & Ghosh, 1998). It is thus suggested that expression of mRNAs for IL-1β, and its related substances such as COX-2 and IκB-β is enhanced in the brain during heat stress probably due to the bacterial translocation in aged but not in the young rats. These findings may explain at least in part the mechanisms of the age-related impairment in the adaptive responses to environmental stress (Fig. 3).

Acknowledgments

This work was supported by Grant-in Aid for Scientific Research (17590207 (C) to T.K.) from the Ministry of Education, Culture, Sports, Science and Technology, the Japanese Government.

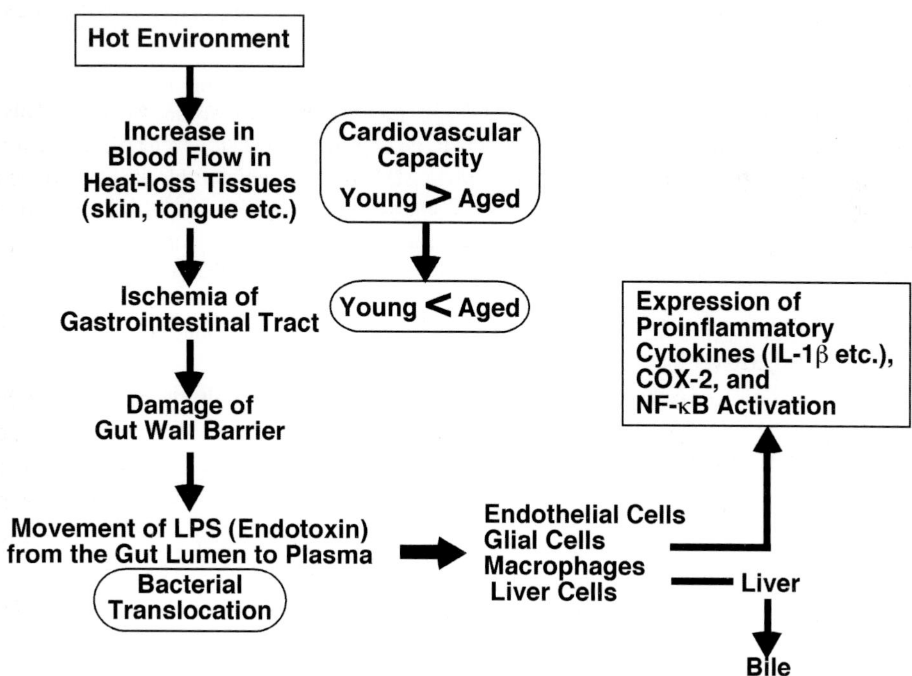

Fig. 3. Possible mechanisms of the heat exposure-induced expression of IL-1β, COX-2, and IκB-β mRNAs. For details, see the text.

References

1. Ban E, Haour F, Lenstra R. Brain interleukin 1 gene expression induced by peripheral lipopolysaccharide administration. *Cytokine.* 1992; 4: 48-54.
2. Berg RD. Bacterial translocation from the gastrointestinal tract. *Trends Microbiol.* 1995; 3: 149-154.
3. Bouchama A, Parhar RS, el-Yazigi A, Sheth K, al-Sedairy S. Endotoxemia and release of tumor necrosis factor and interleukin 1α in acute heatstroke. *J Appl Physiol.* 1991; 70: 2640-2644.
4. Buttini M, Boddeke H. Peripheral lipopolysaccharide stimulation induces interleukin-1β messenger RNA in rat brain microglial cells. *Neuroscience.* 1995; 65: 523-530.
5. Campbell VA, Segurado R, Lynch MA. Regulation of intracellular Ca^{2+} concentration by interleukin-1β in rat cortical synaptosomes: an age-related study. *Neurobiol Aging.* 1998; 19: 575-579.
6. Crofford LJ, Tan B, McCarthy CJ, Hla T. Involvement of nuclear factor κB in the regulation of cyclooxygenase-2 expression by interleukin-1 in rheumatoid synoviocytes. *Arthritis Rheum.* 1997; 40: 226-236.
7. Elmquist JK, Breder CD, Sherin JE, Scammell TE, Hickey WF, Dewitt D, Saper CB. Intravenous lipopolysaccharide induces cyclooxygenase 2-like immunoreactivity in rat brain perivascular microglia and meningeal macrophages. *J Comp Neurol.* 1997; 381: 119-129.
8. Hales JR, Sakurada S. Heat tolerance. A role for fever? *Ann N Y Acad Sci.* 1998; 856: 188-205.
9. Hillhouse EW, Mosley K. Peripheral endotoxin induces hypothalamic immunoreactive interleukin-1β in the rat. *Br J Pharmacol.* 1993; 109: 289-290.
10. Hori T, Katafuchi T, Oka T. Central cytokines: effects on peripheral immunity, inflammation and nociception. In: Ader R, Felten DL, Cohen N, eds., *Psychoneuroimmunology*, Academic Press, San Diego, 2001: Vol. 1, 517-545.
11. Ingalls RR, Golenbock DT. CD11c/CD18, a transmembrane signaling receptor for lipopolysaccharide. *J Exp Med.* 1995; 181: 1473-1479.
12. Katafuchi T, Kondo T, Yasaka T, Kubo K, Take S, Yoshimura M. Prolonged effects of polyriboinosinic:polyribocytidylic acid on spontaneous running wheel activity and brain interferon-α mRNA in rats: a model for immunologically induced fatigue. *Neuroscience.* 2003a; 120: 837-845.
13. Katafuchi T, Takaki A, Take S, Kondo T, Yoshimura M. Endotoxin inhibitor blocks heat exposure-induced expression of brain cytokine mRNA in aged rats. *Brain Res Mol Brain Res.* 2003b; 118: 24-32.
14. Laye S, Parnet P, Goujon E, Dantzer R. Peripheral administration of lipopolysaccharide induces the expression of cytokine transcripts in the brain and pituitary of mice. *Brain Res Mol Brain Res.* 1994; 27: 157-162.
15. Malhotra R, Priest R, Bird MI. Role for L-selectin in lipopolysaccharide-induced activation of neutrophils. *Biochem J.* 1996; 320 (Pt 2): 589-593.
16. Matsumura K, Cao C, Ozaki M, Morii H, Nakadate K, Watanabe Y. Brain endothelial cells express cyclooxygenase-2 during lipopolysaccharide-induced fever: light and electron microscopic immunocytochemical studies. *J Neurosci.* 1998; 18: 6279-6289.
17. May MJ, Ghosh S. Signal transduction through NF-kB. *Immunol Today.* 1998; 19: 80-88.
18. Morrison TB, Weis JJ, Wittwer CT. Quantification of low-copy transcripts by continuous SYBR Green I monitoring during amplification. *Biotechniques.* 1998; 24: 954-958, 960, 962.
19. Murray CA, Lynch MA. Evidence that increased hippocampal expression of the cytokine interleukin-1β is a common trigger for age- and stress-induced impairments in long-term potentiation. *J Neurosci.* 1998; 18: 2974-2981.
20. Nakamori T, Morimoto A, Yamaguchi K, Watanabe T, Murakami N. Interleukin-1β production in the rabbit brain during endotoxin-induced fever. *J Physiol.* 1994; 476: 177-186.
21. O'Mahony L, Holland J, Jackson J, Feighery C, Hennessy TP, Mealy K. Quantitative intracellular cytokine measurement: age-related changes in proinflammatory cytokine production. *Clin Exp Immunol.* 1998; 113: 213-219.

22. Okamura H, Yamaguchi M, Abiko Y. Enhancement of lipopolysaccharide-stimulated PGE_2 and IL-1β production in gingival fibroblast cells from old rats. *Exp Gerontol.* 1999; 34: 379-392.
23. Quan N, Sundar SK, Weiss JM. Induction of interleukin-1 in various brain regions after peripheral and central injections of lipopolysaccharide. *J Neuroimmunol.* 1994; 49: 125-134.
24. Quan N, Whiteside M, Herkenham M. Cyclooxygenase 2 mRNA expression in rat brain after peripheral injection of lipopolysaccharide. *Brain Res.* 1998a; 802: 189-197.
25. Quan N, Whiteside M, Herkenham M. Time course and localization patterns of interleukin-1b messenger RNA expression in brain and pituitary after peripheral administration of lipopolysaccharide. *Neuroscience.* 1998b; 83: 281-293.
26. Rothwell NJ, Hopkins SJ. Cytokines and the nervous system II: Actions and mechanisms of action. *Trends Neurosci.* 1995; 18: 130-136.
27. Rustici A, Velucchi M, Faggioni R, Sironi M, Ghezzi P, Quataert S, Green B, Porro M. Molecular mapping and detoxification of the lipid A binding site by synthetic peptides. *Science.* 1993; 259: 361-365.
28. Sorli CH, Zhang HJ, Armstrong MB, Rajotte RV, Maclouf J, Robertson RP. Basal expression of cyclooxygenase-2 and nuclear factor-interleukin 6 are dominant and coordinately regulated by interleukin 1 in the pancreatic islet. *Proc Natl Acad Sci U S A.* 1998; 95: 1788-1793.
29. Stasiolek M, Gavrilyuk V, Sharp A, Horvath P, Selmaj K, Feinstein DL. Inhibitory and stimulatory effects of lactacystin on expression of nitric oxide synthase type 2 in brain glial cells. The role of IkB-β. *J Biol Chem.* 2000; 275: 24847-24856.
30. Triantafilou M, Triantafilou K. Lipopolysaccharide recognition: CD14, TLRs and the LPS-activation cluster. *Trends Immunol.* 2002; 23: 301-304.
31. Wyckoff TJ, Raetz CR, Jackman JE. Antibacterial and anti-inflammatory agents that target endotoxin. *Trends Microbiol.* 1998; 6: 154-159.

Adaptation Biology and Medicine (Volume 5: Health Potentials)
Editors: L. Lukyanova, N. Takeda and P.K. Singal

26. Facts that Prove that Adaptation to Life at Extreme Altitude (8848 m) is Possible

G. Zubieta-Castillo*, G.R. Zubieta-Calleja, L. Zubieta-Calleja and Nancy Zubieta-Calleja

Abstract

With adequate and gradual adaptation, life is possible even at the hypoxic levels of the summit of Mt. Everest. We describe some further examples that provide proof of such a statement: (1) Life at high altitude in the city of La Paz between 3100 and 4100 m where people sustain variable degrees of tissue hypoxia having a very low arterial oxygen tension (PaO2 = 37 mmHg) due to respiratory and/or cardiac disease. (2) It is possible to perform maximal work at extreme altitudes, as evidenced by a soccer match played at 6542 meters, on the summit of Mount Sajama. (3) Severe high altitude pulmonary edema (HAPE) occurred in a rugby player within 72 hours at 3600 m (PaO2 of 27 mmHg and a SaO2 of 45 %) upon arrival from Portugal. (4) In the Triple Hypoxia Syndrome (THS) where Chronic Mountain Sickness (CMS) polyerythrocythemic patients with gradual adaptation to hypoxia, with a high hematocrit can occasionally tolerate a PaO2 of 30 mmHg, for a week or longer, a severe hypoxic condition similar to that on the summit of Mount Everest. (5) Human fetus under normal conditions develop at oxygen tension values equal to the altitude of Mount Everest until delivery (PaO2 = 28 mmHg). They are naturally capable of living in hypoxic enviromental conditions present on our planet.

Consequently, normal subjects with full capacity for adaptation will show that life is possible at any existing altitude on planet Earth, provided that the following conditions are met: adequate environmental temperatures, lodging, food, slow and progressive adaptation. This seems possible in only one generation, as the human organism is provided with the adequate compensation mechanisms thereby acquiring the capacity for reproduction on site.

Keywords: Pulmonary edema, polyerythrocythemia, hematocrit, habitat and acclimatization

Introduction

Based on the observation of man's highest habitats, 5000 m above sea level is considered the highest altitude man can adapt to. Likewise, it is well known that acute exposure of humans to this altitude and above, without any kind of adaptation for tissue oxygenation is not possible.

Short permanence of humans, under acute conditions, with partial and incomplete adaptation with eminent risk of death demonstrate that even at the highest places of the planet Earth, enough mechanisms of tissue oxygenation are still present. The only objective of the ascent is the conquest of

*Address for Correspondence: G. Zubieta-Castillo, High Altitude Pathology Institute (IPPA), La Paz, Bolivia. E-mail: zubieta@altitudeclinic.com, www.altitudeclinic.com

the summit, even for just a few minutes, demonstrating the extraordinary physical capacity of the human being. This is done even under adverse climatic conditions like extreme cold and storms.

As knowledge and technology improve, man is able to accomplish the most difficult challenges previously considered impossible. The conquest of the summit of Mt. Everest (8848 m) the highest point of the planet, for example, was originally thought to be impossible. Careful preparation of the expeditions, experience, persistence and extraordinary stamina allowed Sir Edmund Hillary and Tensing to stand on top of the mountain. Up to the year 2006, hundreds have reached such extreme altitude, and after Messner and Habeler, some of them without supplementary oxygen. But the question remains: what is the highest possible point of permanent residence?

In the study entitled: "A consideration of the Possibility of Ascending Mt. Everest" Kellas proposes that the "limits of permanent acclimatization to high altitudes", based on his own experience was 6096 m (1). Several decades later most physiologists believed it was much lower but in the north of Chile, the Aucanquilcha mine is at 5950 m.

West, as a distinguished medical historian and scientist who participated actively in high altitude research, is the author of the fascinating book entitled: "High Life" (2). In it, methodic references can be found. He and Houston (3) have also worked on the tolerable limits of hypoxia [West, 1988 #8863]. Roads, transportation facilities, food supplies and improved housing facilities permit life in places never imagined before.

The hypothesis: "man can adapt to the altitude of the summit of Mt. Everest", was born based on experience gained over 36 years. Ever since its creation, in 1970, the High Altitude Pathology Institute, Clinica IPPA (3510 m), has studied and treated patients (permanent residents and newcomers) in the cities of La Paz and El Alto (3100 to 4100 m) and the surrounding areas in Bolivia (4). It is estimated that in Bolivia more than 5,000,000 people (2/3 of the total population) live above 2000 m. Additionally, the proposed hypothesis is also based on a study of the history of high life and exponentially growing scientific literature on high altitude around the world.

In this paper, we also address the issue of the hypothesis: is man able to adapt to life on the summit of Mt Everest in only one generation? The later affirmation pretends to show that man is already genetically prepared to handle life at the highest point of our planet.

Studies and Arguments in Support of the Hypothesis

1 Life at High Altitude

The city of La Paz and El Alto (3100 to 4100 m) with over 1 million inhabitants are two metropolises full of activity in all areas of human behavior. In fact, most people born there don't even realize that they are high altitude dwellers. Through our observation of patients with ventilatory or respiratory impairment developed during many years at high altitude (4), we arrived to the conclusion that even if oxygen supply to the tissues is scarce, life is possible. Polyerythrocythemic patients* are present in around 10% of the population living at high altitude in Bolivia. Depending on the nature of the diseases, from which they were treated leaving sequelae, they are able to live as well as healthy people at high altitude. Hence, adaptation to life even with disease is normal at this altitude.

---

* new terminology explained further down, under Triple Hypoxia Syndrome.

After the acute phase of exposure to hypoxia where the cardiopulmonary system plays the fundamental role, the increase in oxygen content through the increase in the number of red blood cells is one of the most important progressive mechanisms of adaptation to high altitude along with the cellular and molecular changes as the increase in the number of mitochondria demonstrated by many authors (25).

2 Exercise Studies and Oxygen Consumption of Soccer Players who Carried out a Match at 6542 M.

Aymara residents of La Paz in active duties in the Army (n = 17) were compared to mountain guides and Aymara native porters (n = 7), residents of the Sajama town (4300 m) at IPPA (3510 m). The later group achieved the same exercise through lower oxygen consumption. (5) These subjects then went on to play a soccer match at 6542 m on the summit of Sajama. Bolivian Aymara natives performed this on August 2nd 2001, within 24 hours of arrival including ascent and preparation of the field. They climbed from their town at 4300 to 6542 m. in 9 hours, set up a pitch and played 20 minutes per side and returned in 7 hours. All this was carried out in only 16 hours (6). They interrupted their stay at such altitude only due to stormy weather, which is one of the unfavorable environmental factors that mountain climbers have to face, and that frequently frustrate their goals.

Fig. 1. Bolivian Aymara Porters and guides carrying out soccer match on the summit of Mt. Sajama 6542 m.

This remarkable feat and hypoxic endurance demonstration shows that hypoxia is not necessarily a hindrance to sport. Furthermore, it is possible to perform maximal work at extreme altitudes and under acute conditions of adaptation. There are, naturally, differences in ethnic groups and in individuals, but if sea level residents are previously adapted to an intermediate altitude, they can perform as well as the Aymara players, without the need of generations for adaptation.

3 Severe High Altitude Pulmonary Edema and Tolerance to Extreme Hypoxia.

A Portuguese rugby player arrived on a tour to the city of La Paz (3510 m). He felt slightly short of breath and dyspneic, without evident symptomatology of disease. His mental state was lucid and able to carry out a normal conversation at all times. Upon initial consultation a full-blown high altitude pulmonary edema (HAPE) compromising both lungs was observed. His blood gases reported a PaO2 of 27 mmHg, PaCo2 = 36 mmHg, pH = 7.41 and a SaO2 of 45%. He sustained these hypoxic levels during three days, and fully recovered to later continue to play normally at sea level, upon returning home.

4 The Triple Hypoxia Syndrome in Chronic Mountain Sickness

Chronic Mountain Sickness during many years of studies by different authors has been referred to as Polycythemia, Erythremia, Excessive Erythrocytosis, Increased Polycythemia, terms that etiologically do not adequately relate to the multiple causes of the disease. That is why the most precise term seems to be Polyerythrocythemia (Poly = many, Erythros = red cell, Hemia = blood). Up to now, no one was able to show any deformity or change in the molecular structure of red blood cells that would explain the use of the suffix “osis”, a term more suitable for hematological disease. The only objective sign of this disease is the increase in the number of red blood cells.

CMS is fundamentally caused by pulmonary disease of different ethiopathogenesis (7) although it can disturb different organs and systems (9, 10). These patients have lower than normal PaO2 for that altitude. Consequently it is inferred that the tissues of these patients are functioning as if they were normal people at higher altitudes. Figure 1 gives the calculated altitude of residence of these CMS patients that are actually living at around 3500 m.

During the “Triple Hypoxia Syndrome” (THS) in CMS with gradual adaptation to hypoxia, severe hypoxic conditions can be tolerated for a week or longer (12, 13). All these patients have very low arterial oxygen tensions which turn out to be similar to the values (PaO2 = 35 mmHg) calculated from subjects breathing at the altitude of the summit of Mt Everest 8848 m (8), but the later under severe acute conditions (Fig. 2). (14, 15) They evolve over many years and their high hematocrit is regularly controlled at their altitude of residence.

This syndrome, an acute complication in CMS, is actually the superposition of 3 hypoxias. The first hypoxia is that of high altitude, the second hypoxia is that of CMS due to multiple ethiopathogenesis but primarily of pulmonary origin. The third hypoxia is an acute state also due to multiple over imposed causes such as lung disease, influenza, and others. This third stage is reversible with adequate treatment of the underlying cause at the same altitude of residence.

As time passed and we gained more experience, we confirmed that tolerance to chronic hypoxia at extreme altitude is possible, as previously expressed (12). Few days before we wrote this paper, we had a long-standing CMS 84-year-old patient, who came in walking for a consultation. He carried out a

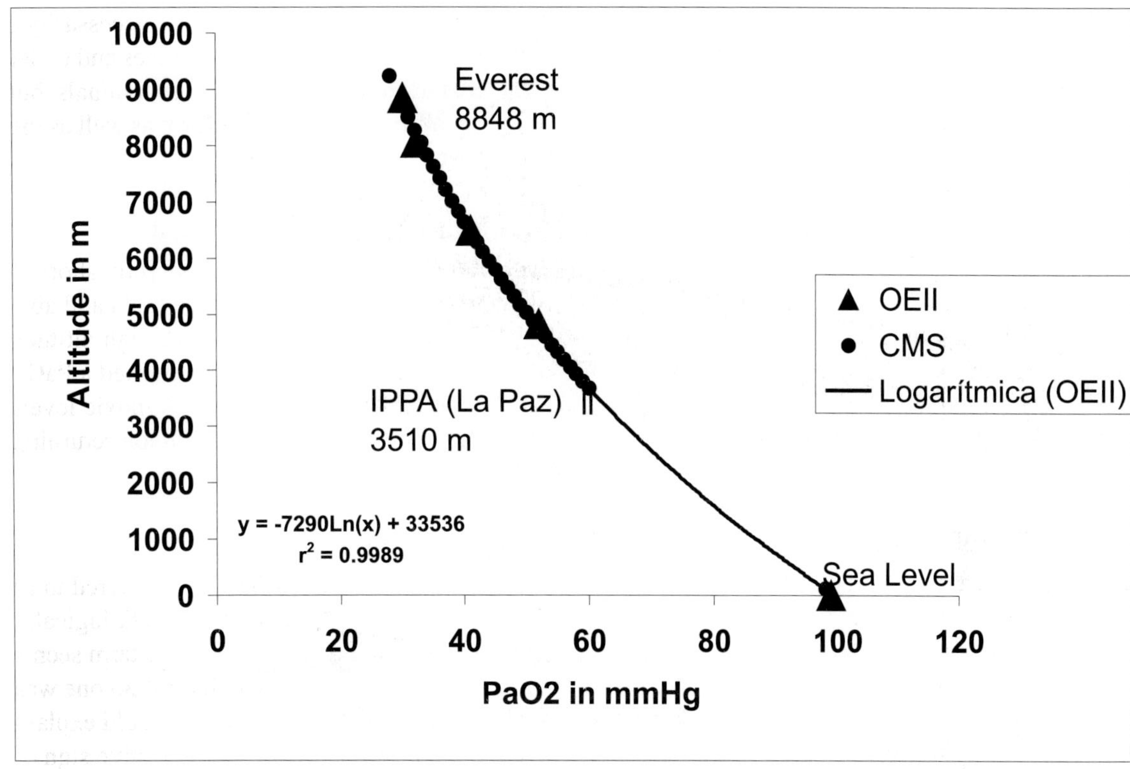

Fig. 2. Curve calculated from data by Houston of Operation Everest II (OEII) and applied to plot the relative altitude of CMS patients living in the city of La Paz and El Alto (3100-4100 m) whose PaO2 were below the normal present at 3500 m.

normal conversation with unquestionable memory. During several weeks his resting blood gases were: PaO2 was 21 mmHg, PaCO2 = 45 mmHg and pH = 7.43, along with a hematocrit of 65%, an extreme case of THS. His FVC was 57% of predicted. Cyanosis and dyspnea were evident along with a very limited exercise capacity. His pulmonary artery hypertension was at 47 mmHg as measured by Doppler Echocardiography. In this patient, the pulse oximeter BCI SensePro was unable to read the SaO2 values. He refused intensive care therapy. Yet he was still attending social activities such as card playing with his friends at home. He had been observed by us 5 years before when his hematocrit was 66%, his PaO2 = 46 mmHg, PaCO2 = 34 mmHg, pH = 7.39 and FVC 66% of predicted. At that time he carried on a "completely normal life".

After a periodical oxygen therapy that lasted several weeks, he died at home, from cardiac insufficiency. This condition showed that his mental capacity was not deteriorated nor impaired at this extreme level of hypoxia. In fact, these values question the current concepts of tolerance to hypoxia. They also contribute an explanation of the existence of oxygen in the lower saturation levels of oxyhemoglobin, that up to now are considered of no use. We are certain that a sea level physician would naturally consider this as impossible.

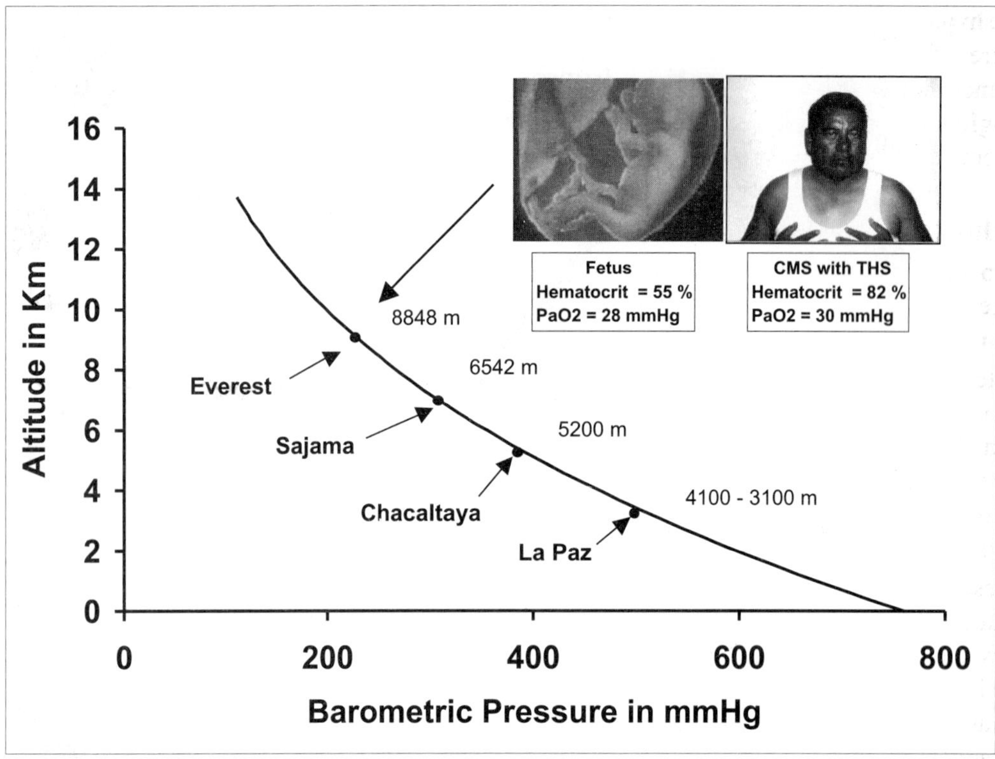

Fig. 3. **Active Chronic Mountain Sickness (CMS) with Triple Hypoxia Syndrome (THS) patient living in the city of La Paz between 3100 and 4100 m above sea level, whose tissue hypoxia is comparable to the values calculated for the summit of Mt. Everest showing similar values of PaO2 in a fetus.**

5 The Oxygen Levels in the Human Fetus

The fetus lives in optimal conditions in the uterus. He has adequate body temperature, optimal metabolic conditions, and nutrition. His blood gases as reported by as low as 25 mmHg in the ascending aorta (28). In the city of La Paz (3510 m), a normal delivery has been found to give the following umbilical chord values: PaO2 = 28 mmHg, PaCo2 = 29 mmHg and pH = 7.28; PvO2 = 19 mmHg, PvCO2 = 47 mmHg and pH = 7.12.

6 Oxygen Content and Oxygen Affinity

Polyerytrocythemia, that maintains oxygen content, is one of the most important mechanisms of adaptation to chronic hypoxia avoiding a high cost in energy (7).

It is another one of the important mechanisms facilitating the capture of oxygen by hemoglobin. Changes in oxygen affinity are significant mechanisms of adaptation, but are cyclical and constantly fluctuating throughout the circulation. Increased oxygen affinity is not completely defined if it pertains only to change in the molecular aspects of hemoglobin or other factors, which participate in this supposedly beneficial change as described in high altitude animals and man (21, 22, 23). The combination of hemoglobin with oxygen is directly and fundamentally related to the difference in oxygen tension.

The hypothesis is based mainly in the fact that cells at tissue level, can catch oxygen with a very low-pressure difference. It is evident that cells have the same affinity for oxygen as hemoglobin with the difference that it can be used immediately in the metabolic process to produce energy in the ribosome. Hemoglobin captures oxygen in order to gradually release it according to cellular metabolic requirements. Mitochondria are increased in number at high altitude (25).

Conclusions

Short permanence of humans on the highest places of the planet Earth, under acute conditions, without complete adaptation and with eminent risk of death, demonstrates that even at such altitudes, enough mechanisms of tissue oxygenation are still present.

Note, that in severe CMS and in the example of the 84-year-old man with CMS and triple hypoxia syndrome, these walk-in patients can arrive to consultation with a PaO2 of around 21 mmHg. This extreme hypoxic state can be sustained over several weeks if not treated promptly. Such low PaO2 is below that estimated on the summit of Mt. Everest (24). Polyerythrocythemia is not only related to the severity of the respiratory insufficiency, showing that other factors also play a role as many authors point out (7).

These are the facts that gave rise to the hypothesis that animals and human beings can adapt to the altitude and hypoxic conditions similar to that of Mt. Everest (26). If and only if, they avoid the adverse factors such as cold, inadequate housing, inadequate nutrition, associated with altitude and facing exclusively the hypoxia but gradually and after long periods of adaptation to incremental altitudes. Cells and other microorganisms utilize oxygen while there is a small tension gradient between the environment and the cell structure. Yeast cells, for example, consume oxygen until the level of oxygen tension is 0 mmHg. Only 1 mmHg of pressure gradient is enough to sustain life at the tissue level at 37 degrees centigrade. Under these conditions life is possible at such extreme high altitude. Fetal circulation is a physiological remarkable adaptation in order to tolerate low levels of oxygen. A PaO2 of 25 mmHg has been found tolerable in a low cardiac output due to its relatively high fetal hemoglobin concentration (27).

If we accept that sick people with advanced pathological lesions are able to adapt to such hypoxic conditions as shown above, normal people should be able to do that more easily, accepting that there are different capabilities, in different races and groups of people. Furthermore, when a patient with CMS has a low PaO2, it is as if he would actually be living at a higher altitude. In order to show this, a graph plotting the PaO2 at different altitudes was adapted to depict at what relative altitude CMS patients seem to be living (fig. 1).

We are not pretending or suggesting to send people to live on the summit of Mt Everest, of course. But the biological conditions of man and animals allow for the possibility of adaptation to extreme hypoxia. So life is possible anywhere on this planet, provided that adequate nutrition and housing are available, and enough time is allowed for slow adaptation. This knowledge permits us to understand the mechanisms of adaptation to tissue hypoxia. Maybe some day human beings will confront such adverse environmental conditions where this knowledge will be beneficial.

In this paper we explain the foundations of the hypothesis: man is able to adapt to life on the summit of Mt Everest in one generation. Reproduction will be feasible once adaptation is acquired. The first generation will continue increasing the population. The later affirmation pretends to show that man is already genetically prepared for life on the highest point of planet Earth. But not every single human being, as some will be genetically incompatible or incapable. For example those suffering pulmonary fibrosis or Down's Syndrome or thousands of genetic variations that can be present in people living at sea level and that are incompatible with hypoxic environments

References

1. West, J.B., *Alexander M. Kellas and the physiological challenge of Mt. Everest.* J Appl Physiol, 1987. 63 (1): 3-11.
2. West, J.B., *High Life: A History of High-Altitude Physiology and Medicine*. 1998: American Physiological Society Oxford University Press.
3. Houston, C.S. and A. Cymerman, *Hypoxia: the tolerable limits: Operation Everest II*, in *Hypoxia. The Tolerable Limits*, J.R. Sutton, C.S. Houston, and G. Coates, Editors. 1988, Benchmark Press: Indianapolis, IN. 3-8.
4. Zubieta-Castillo, G. and G.R. Zubieta-Calleja, *www.altitudeclinic.com.* 1996.
5. Zubieta-Calleja, G.R., Zubieta-Castillo, G., Zubieta-Calleja, L., Zubieta, N., *Exercise performance of Bolivian Aymara in 3 conditions: at La Paz 3510 m, breathing a hypoxic mixture simulating Chacaltaya and at Chacaltaya 5200 m. (Abstract).* HAMB, 2002. 3 (1): 114.
6. Zubieta-Castillo, G.R., Zubieta-Calleja, G.R., Zubieta-Calleja L., Zubieta, N., *Bolivian Aymara that played soccer at 6542 m maintain higher oxygen saturation and lower oxygen uptake during maximal exercise (Abstract).* HAMB, 2002. 3 (1): 114.
7. Zubieta-Castillo, G. and G. Zubieta-Calleja, *New Concepts on chronic mountain sickness.* Acta Andina, 1996. 5 : 3-8.
8. West, J.B. and P.D. Wagner, *Predicted gas exchange on the summit of Mt Everest.* Respir Physiol, 1980. 42 : 1-16.
9. Zubieta-Castillo, G. and G.R. Zubieta-Calleja, *Chronic mountain sickness and miners (Spanish).* Revista de la Academia Nacional de Ciencias de Bolivia, 1985. 4 : 109-116.
10. Zubieta-Castillo, G. and G. Zubieta-Calleja, *Pulmonary diseases and chronic mountain sickness (Spanish).* Revista de la Academia Nacional de Ciencias de Bolivia, 1986. 5 : 47-54.
11. Zubieta-Calleja, G.R., Zubieta-Castillo, G., Zubieta-Calleja, L., Zubieta, N., *Exercise performance in chronic mountain sickness (CMS) patients at 3510 m. (Abstract).* HAMB, 2002. 3 (1): 115.
12. Zubieta-Castillo, G. and G. Zubieta-Calleja, *Triple hypoxia syndrome.* Acta Andina, 1996. 5 (1): 15-18.
13. Zubieta-Castillo, G., Zubieta-Calleja, G., *The triple hypoxia syndrome at altitude (Abstract).* Amer Rev Respir Dis, 1988. 137 (4): 509.
14. Zubieta-Castillo, G., Zubieta-Calleja G.R. , Zubieta-Calleja, L., *Exercise performance in a woman with CMS, following triple hypoxia syndrome treatment (Abstract).* HAMB, 2002. 3 (1): 114.
15. Zubieta-Calleja G.R., Z.-C., G., *CMS exercise (Abstract).* 1992.
16. Cruz, J.C., *et al.*, *Phlebotomy improves pulmonary gas exchange in chronic mountain polycythemia.* Respiration, 1979. 38 : 305-313.
17. Dayton, L.M., *et al. Symptomatic and pulmonary response to acute phlebotomy in secondary polycythemia.* Chest, 1975. 68 (6): 785-790.
18. Winslow, R.M., C.C. Monge, and N.J. Statham, *In vivo blood oxygen affinity in high altitude natives (Abstract)*, in *Hypoxia: Man at Altitude*, J.R. Sutton, N.L. Jones, and C.S. Houston, Editors. 1982, Thieme-Stratton: New York, NY. 202.
19. Aste-Salazar, H. and A. Hurtado, *The affinity of hemoglobin for oxygen at sea level and at high altitudes.* Am J Physiol, 1944. 142 : 733-743.

20. Samaja, M., A. Veicsteinas, and P. Cerretelli, *Oxygen affinity of blood in altitude Sherpas.* J Appl Physiol, 1979. 47 : 337-341.
21. Hebbel, R.P., *et al.*, *Hemoglobin oxygen affinity and adaptation to altitude: Evidence for pre-adaptation to altitude in humans with left-shifted oxyhemoglobin dissociation curves.* J Clin Invest, 1977. 60 : 213-228.
22. Eaton, J.W., T.D. Skelton, and E.M. Berger, *Survival at extreme altitude: Protective effect of increased hemoglobin-oxygen affinity.* Science, 1974. 183 : 43-44.
23. Hebbel, R.P., *et al.*, *Human llamas: Adaptation to altitude in subjects with high hemoglobin oxygen affinity.* J Clin Invest, 1978. 62 : 593-600.
24. West, J.B., *Man on the summit of Mount Everest*, in *High Altitude and Man*, J.B. West and S. Lahiri, Editors. 1984, American Physiological Society: Bethesda, MD. 5-18.
25. Lukianova LD. Molecular Mechanisms of tissue hypoxia and organism adaptation. J Fiziol 2003 49: 3; 17-35.
26. Zubieta-Castillo G, Zubieta-Calleja GR, Zubieta-Calleja L, Zubieta-Calleja, N. Adaptation to life at the altitude of the summit of Everest. Fiziol Zh. 2003; 49(3): 110-7.
27. Huikeshoven FJ, Hope ID, Power GG, Gilbert RD, Longo LD. A comparison of sheep and human fetal oxygen delivery systems with use of a mathematical model. Am J Obstet Gynecol. 1985 Feb 5; 151(4): 449-55.
28. Huikeshoven FJ, Hope ID, Power GG, Gilbert RD, Longo LD. Mathematical model of fetal circulation and oxygen delivery. Am J Physiol. 1985 Aug; 249(2 Pt 2): 192-202.